TIME SERIES METHODS IN HYDROSCIENCES

PROCEEDINGS OF AN INTERNATIONAL CONFERENCE HELD AT CANADA CENTRE FOR INLAND WATERS, BURLINGTON, ONTARIO, CANADA, OCTOBER 6–8, 1981

DEVELOPMENTS IN WATER SCIENCE, 17

OTHER TITLES IN THIS SERIES

1 G. BUGLIARELLO AND F. GUNTER
COMPUTER SYSTEMS AND WATER RESOURCES

2 H.L. GOLTERMAN
PHYSIOLOGICAL LIMNOLOGY

3 Y.Y. HAIMES, W.A. HALL AND H.T. FREEDMAN
MULTIOBJECTIVE OPTIMIZATION IN WATER RESOURCES SYSTEMS:
THE SURROGATE WORTH TRADE-OFF-METHOD

4 J.J. FRIED
GROUNDWATER POLLUTION

5 N. RAJARATNAM
TURBULENT JETS

6 D. STEPHENSON
PIPELINE DESIGN FOR WATER ENGINEERS

7 V. HÁLEK AND J. ŠVEC
GROUNDWATER HYDRAULICS

8 J. BALEK
HYDROLOGY AND WATER RESOURCES IN TROPICAL AFRICA

9 T.A. McMAHON AND R.G. MEIN
RESERVOIR CAPACITY AND YIELD

10 G. KOVÁCS
SEEPAGE HYDRAULICS

11 W.H. GRAF AND C.H. MORTIMER (EDITORS)
HYDRODYNAMICS OF LAKES: PROCEEDINGS OF A SYMPOSIUM
12–13 OCTOBER, 1978, LAUSANNE, SWITZERLAND

12 W. BACK AND D.A. STEPHENSON (EDITORS)
CONTEMPORARY HYDROGEOLOGY: THE GEORGE BURKE MAXEY
MEMORIAL VOLUME

13 M.A. MARIÑO AND J.N. LUTHIN
SEEPAGE AND GROUNDWATER

14 D. STEPHENSON
STORMWATER HYDROLOGY AND DRAINAGE

15 D. STEPHENSON
PIPELINE DESIGN FOR WATER ENGINEERS
(completely revised edition of Vol. 6 in the series)

16 W. BACK AND R. LÉTOLLE (EDITORS)
SYMPOSIUM ON GEOCHEMISTRY OF GROUNDWATER

TIME SERIES METHODS IN HYDROSCIENCES

PROCEEDINGS OF AN INTERNATIONAL CONFERENCE HELD AT
CANADA CENTRE FOR INLAND WATERS, BURLINGTON, ONTARIO, CANADA,
OCTOBER 6–8, 1981

EDITED BY

A.H. EL-SHAARAWI

National Water Research Institute, Burlington, Ont. L7R 4A6 (Canada)

IN COLLABORATION WITH

S.R. ESTERBY

National Water Research Institute, Burlington, Ont. L7R 4A6 (Canada)

ELSEVIER SCIENTIFIC PUBLISHING COMPANY
Amsterdam–Oxford–New York 1982

ELSEVIER SCIENTIFIC PUBLISHING COMPANY
Molenwerf 1
P.O. Box 211, 1000 AE Amsterdam, The Netherlands

Distributors for the United States and Canada:

ELSEVIER SCIENCE PUBLISHING COMPANY INC.
52, Vanderbilt Avenue
New York, N.Y. 10017

Library of Congress Cataloging in Publication Data
Main entry under title:

Time series methods in hydrosciences.

(Developments in water science ; 17)
Proceedings of the International Conference on Time Series Methods in Hydrosciences, Canada Centre for Inland Waters, Burlington, Ont., Oct. 6-8, 1981.

1. Hydrology--Mathematical models--Congresses.
2. Time-series analysis--Congresses. I. El-Shaarawi, A. H. II. Esterby, S. R. III. International Conference on Time Series Methods in Hydrosciences (1981 : Canada Centre for Inland Waters) IV. Series.
GB656.2.M33T55 1982 551.48'0724 82-11378
ISBN 0-444-42102-5 (U.S.)

ISBN 0-444-42102-5 (Vol. 17)
ISBN 0-444-41669-2 (Series)

Printed in The Netherlands

FOREWORD

This volume constitutes the proceedings of the International Conference on Time Series Methods in Hydrosciences which was held at the Canada Centre for Inland Waters, Burlington, Ontario, Canada, October 6, 7 and 8, 1981. The participants of the Conference came from Canada, USA, UK, Holland, West Germany, Italy, Saudi Arabia, Australia, Japan, Turkey, Norway and Greece. Invited papers were presented by:

Dr. D.E. Cartwright	-	Institute of Oceanographic Sciences, UK
Dr. E.H. Lloyd	-	University of Lancaster, UK
Dr. D.R. Brillinger	-	University of California, Berkeley, USA
Dr. R.L. Bras	-	Massachusettes Institute of Technology, USA
Dr. A.M. Mathai	-	McGill University, Canada
Dr. T.E. Unny	-	University of Waterloo, Canada
Dr. I.B. MacNeill	-	University of Western Ontario, Canada

Unfortunately, Dr. Bras' paper is not included in this proceedings. All papers included in the proceedings have been refereed.

I express my sincere thanks to all those who helped me in organizing the Conference, especially -

Dr. B. Scott	-	National Water Research Institute
Dr. P. Budgell	-	Bayfield Laboratory for Marine Science and Surveys
Dr. R.J. Kulperger	-	McMaster University
Mr. R.E. Kwiatkowski	-	Water Quality Branch, Inland Waters Directorates
Mr. J. KinKead	-	Water Resources Branch, Ontario Ministry of Environment
Dr. V. Klemes	-	National Hydrology Research Institute

Financial, logistic and moral support were provided by (1) the National Water Research Institute, Environment Canada, (2) Bayfield Laboratory for Marine Science and Surveys, Department of Fisheries and Oceans, (3) Water Quality Branch, Environment Canada, and (4) Water Resources Branch, Ontario Ministry of Environment. For all this, I am indebted to Dr. Keith Rodgers, Director of the National Water Research Institute, Mr. F. Elder, Chief of Aquatic Physics and Systems Division (NWRI) and Mr. T.D.W. McCulloch, Director General, Bayfield Laboratory for Marine Science and Surveys.

Last, but not least, I express my thanks to Mrs. Joanne Mann, Mrs. Ann Harvey and Mrs. B. Muir, for their secretarial assistance.

A.H. El-Shaarawi

CONTENTS

Table of Contents

Table of Contents

Table of Contents

SOME CONTRASTING EXAMPLES OF THE TIME AND FREQUENCY DOMAIN APPROACHES TO TIME SERIES ANALYSIS

DAVID R. BRILLINGER
University of California, Berkeley

ABSTRACT

Two distinct approaches to the analysis of time series data are in common use - the time side and the frequency side. The frequency approach involves essential use of sinusoids and bands of (angular) frequency, with Fourier transforms playing an important role. The time approach makes little use of these. Certain useful techniques are hybrids of these two approaches. This work proceeding via examples, compares and contrasts the two approaches with respect to modelling, statistical inference and researchers' aims.

1 INTRODUCTION

Many, many time series analyses have been carried out at this point in time. Some of these analyses have been carried out totally in the time domain, some have proceeded essentially in the frequency domain, and some have made substantial use of both domains. There are numerous examples in hydrology of each type of analysis. It seems useful to examine some time series analyses to attempt to recognize the strengths and weaknesses of each approach and to try to discern just what lead the researchers involved to adopt the particular approach that they did.

This work presents descriptions of a number of time series analyses that the author has been involved with. Some of these have been frequency side, some have been time side and some have been hybrids. Some have been parametric, some have been nonparamet-

Reprinted from *Time Series Methods in Hydrosciences*, by A.H. El-Shaarawi and S.R. Esterby (Editors)

ric. Some have involved linear systems, some have been concerned with nonlinear systems. None of the studies are hydrological, however it is clear that analagous situations do arise in hydrology. It seemed best to present examples that the author knew all details concerning.

2 TIME SERIES ANALYSIS

Tukey (1978) defines our field of study as follows:

> "Time series analysis consists of all the techniques that, when applied to time series data, yield, at least sometimes, either insight or knowledge, AND everything that helps us choose or understand these procedures."

In that paper he further lists some of the aims of time series analysis. These are: 1. discovery of phenomena, 2. modelling, 3. preparation for further inquiry, 4. reaching conclusions, 5. assessment of predictability and 6. description of variability. As one attempts to understand the relative merits of the various approaches and techniques of time series analysis, it is worthwhile to keep the above definition and aims in mind.

Most researchers would seem agreed on what is a time side analysis. There is uncertainty over just what constitutes the frequency side. The following variant of a statement in Bloomfield *et al* (1981) is helpful: frequency side analysis is thinking of systems, their inputs, outputs and behavior in sinusoidal terms.

It is easier to list techniques that are time side, frequency side or hybrids. On the time side one may list: state space, autoregressive moving average (ARMA) and econometric modelling, trend analysis, regression, pulse probing of systems and empirical orthogonal functions among other things. On the frequency side one may list: spectral and cepstral analysis, seasonal adjustment, harmonic decomposition and sinusoidal probing of systems. Hybrid techniques include: complex demodulation, moving spectrum analysis and the probing of systems by chirps. In practice it seems that there is usually a frequency version of a time side procedure, and vice versa. It further seems that these techniques are generally allies, rather than competitors.

A number of practical time series analyses will now be described and their type of analysis commented on.

3 THE CHANDLER WOBBLE

The point of intersection of the Earth's axis of rotation with the polar cap does not remain fixed, rather it wanders about within a region of the approximate size of a tennis court. Let $(X(t),Y(t))$ denote the coordinates of the point at time t, relative to its long run average position. Set $Z(t) = X(t) + iY(t)$, then (from Munk and MacDonald (1960)) the equations of motion are

$$\frac{dZ(t)}{dt} = \alpha Z(t) + \frac{d\Phi(t)}{dt}$$

with $\Phi(t)$ the excitation function whose increments $d\Phi(t)$ describe the change in the Earth's inertia tensor in the time interval $(t,t+dt)$. Supposing the process Φ to have stationary increments, the power spectrum of the series Z is given by

$$f_{ZZ}(\lambda) = |i\lambda - \alpha|^{-2} f_{\Phi\Phi}(\lambda) \quad .$$

What is of interest here is to derive an estimate of α and to derive characteristics of the excitation process Φ. It is known that the excitation process contains an annual component, due to the alternation of seasons in the southern and northern hemispheres. To build a specific model, suppose that the increments of seasonally adjusted Φ are white noise with variance σ^2. The spectrum of the seasonally adjusted Z is then $|i\lambda - \alpha|^{-2}\sigma^2/2\pi$. The data available for analysis is $Z(t)$ perturbed by measurement error for $t = 0,\ldots, T-1$. (In Brillinger (1973) it is monthly data from 1902 to 1969.) Supposing the variance of the measurement error series to be ρ^2, the power spectrum of the series of first differences of the seasonally adjusted discrete data is given by

$$\frac{\sigma^2}{2\pi}\,\frac{1-e^{-2\beta}}{2\beta}\,\frac{1}{1-2e^{-\beta}\cos(\lambda-\gamma)+e^{-2\beta}} + \frac{\rho^2|1-e^{-i\lambda}|^2}{2\pi} = f(\lambda)$$

where $\alpha = -\beta + i\gamma$. Given the data one would like estimates of the parameters $\alpha, \beta, \gamma, \rho, \sigma$ and to examine the validity of the model. These things are possible on the frequency side.

Let $d^T(\lambda)$ denote the finite Fourier transform of the series of first differences of the seasonaly corrected data. The periodogram of this data is then $I^T(\lambda)$. The periodogram ordinates $I^T(2\pi s/T)$, $s = 1,2,\ldots$ being approximately independent exponential variates with means $f(2\pi s/T)$, $s = 1,2,\ldots$ respectively, estimating the parameters by maximizing the "Gaussian" likelihood

$$\prod_s f\left(\frac{2\pi s}{T}\right)^{-1} \exp\left(-I^T\left(\frac{2\pi s}{T}\right)/f\left(\frac{2\pi s}{T}\right)\right)$$

is one way to proceed. (In essence this procedure is suggested in Whittle (1954).) Estimates derived in this fashion, and estimates of their standard errors are presented in Brillinger (1973). Figure 4 of that paper is an estimate of the power spectrum derived by smoothing the periodogram together with the estimated above functional form. The fit is quite good.

However the nonparametric estimate does show a minor peak at frequency .154 cycles/month that is suspiciously large. This frequency was further investigated by the method of complex demodulation. Complex demodulation is a hybrid frequential-temporal technique. If $X(t)$ denotes the series of concern, then the steps involved are: i. form $U(t) = X(t)\exp(-i\lambda t)$, for λ the frequency of interest, ii. smooth the series $U(t)$ to obtain the series $V(t)$, this is the complex demodulate at frequency λ , iii. graph $|V(t)|^2$ and $\arg V(t)$. One of the important uses of complex demodulation is the detection of changes with time in a frequency band of interest. For the frequency .154 special activity seems to be present only for the period 1905 to 1914 .

The above analysis took place principally in the frequency domain, but partially in the hybrid domain as well. The advantages of the frequency domain included: a. operations on the series (sampling, seasonal adjustment, differencing) could be handled directly, b. measurement noise was easily dealt with, c. estimation became a problem of maximizing an elementary function, d. standard errors were a byproduct of the estimation procedure. It is further evident that a frequency component present for only a restricted time period could only be discovered by a hybrid procedure. This was why complex demodulation was so useful.

4 FREE OSCILLATIONS OF THE EARTH

For a time interval after a major earthquake the Earth rings at certain fundamental frequencies. This motion is called its free oscillations. The frequencies are called its eigenfrequencies. The estimation of the values of the eigenfrequencies and their associated decay rates is a problem of fundamental importance to geophycists building models of the Earth. The problem is that of how to estimate these parameters given the seismogram of a major earthquake. The frequency domain provides an effective means of doing this. Complex demodulation provides an effective means of checking the mechanical model.

Dynamical considerations suggest the following model for the seismogram,

$$X(t) = \sum_{k=1}^{K} \alpha_k \exp(-\beta_k t) \cos(\gamma_k t + \delta_k) + \varepsilon(t)$$

for $t > 0$, with the γ_k the eigenfrequencies of interest, the β_k their decay rates, α_k and δ_k constants and ε a noise series. Crude estimates of the γ_k may be derived by graphing the periodogram of a data stretch. The model may be examined by complex demodulating at estimated γ_k. If the smoothing filter has a bandwidth small enough to exclude other eigenfrequencies, and if the above model holds with the noise not too substantial, then a graph of log $|V(t)|$ will fall off in a linear fashion (slope approximately $-\beta_k$) and arg $V(t)$ will be approximately constant (if the estimated frequency is close enough to the true one.) Bolt and Brillinger (1979) present such graphs for the record made at Trieste of the great Chilean earthquake of 1960. The model seems confirmed. What is needed now are precise estimates of the unknown parameters and estimates of their standard errors. These may be constructed as follows.

Let $a = \gamma + i\beta$, $b = \alpha\exp(i\delta)$,

$$d_X^T(\lambda) = \sum_{t=0}^{T-1} X(t)\exp(-i\lambda t) \quad \text{and} \quad \Delta^T(\lambda) = \sum_{t=0}^{T-1} \exp(-i\lambda t) .$$

For λ in an interval I_k near γ_k, one has $d_X^T(\lambda) \doteq b_k \Delta^T(\lambda - a_k) + d_\varepsilon^T(\lambda)$. Now if the noise series, ε, is stationary and such that well-separated values are only weakly dependent, then the finite Fourier transform values $d_\varepsilon^T(2\pi s/T)$, for s an integer with $2\pi s/T$ near λ, will be approximately independent complex normal variates with mean 0 and variance $2\pi T f_{\varepsilon\varepsilon}(\lambda)$. (See Brillinger (1981) for example.) The maximum likelihood estimates of the unknown parameters are thus the least squares estimates found by minimizing

$$\sum \left| d_X^T\left(\frac{2\pi s}{T}\right) - b_k \Delta^T\left(\frac{2\pi s}{T} - a_k\right)\right|^2$$

where the summation is over frequencies $2\pi s/T$ in I_k. Further the asymptotic distribution of these estimates may be found directly and so standard errors estimated and confidence intervals constructed. Details are given in Bolt and Brillinger (1979).

Once again, by going over to the frequency domain a direct estimation procedure has been found. Because estimates of standard errors are part of the procedure, estimated eigenfrequencies from different seismograms may now be combined efficiently. Further the approximate sampling properties of the estimates are clear, being based on normal variates. A hybrid procedure allowed confirmation of the model.

5 THE HUMAN PUPILLARY SYSTEM

The pupil of the eye exhibits a number of nonlinear characteristics. When it is probed with narrow bandwidth sinusoidal light, the motions of its diameter display second and possibly third order harmonics of the fundamental frequency. Further the shape of the transfer function estimated by such sinusoidal probing changes as the amplitude of the stimulus is varied and finally a dynamic asymmetry is exhibited between responses to on and off stimuli. It is apparent that a nonlinear model needs to be developed in order to describe the pupillary system.

A useful model for nonlinear systems is the following one discussed by Tick (1961),

$$Y(t) = \alpha + \int a(t-u)X(u)du + \iint b(t-u,t-v)X(u)X(v)dudv + \varepsilon(t)$$

with X, the system input stationary and Gaussian, with Y the system output and with ε a stationary noise series. Let A and B denote the linear and quadratic transfer functions of this system,

$$A(\lambda) = \int a(u)\exp(-i\lambda u)du$$

$$B(\lambda,\mu) = \int b(u,v)\exp(-i\lambda u - i\mu v)dudv \quad ,$$

then, in this case of Gaussian stimulation, one has the relationships

$$f_{YX}(\lambda) = A(\lambda)f_{XX}(\lambda)$$

$$f_{XXY}(-\lambda,-\mu) = 2B(\lambda,\mu)f_{XX}(\lambda)f_{XX}(\mu) \quad .$$

Here f_{XX} is the power spectrum of the input, f_{YX} the cross-spectrum of the input and the output and f_{XXY} the cross-bispectrum of the input and the output. (This last is the Fourier transform of the third order cross-moment function.)

These last relationships allow the computation of estimates of

A and B once estimates of the spectra involved have been computed. The spectral estimates may be based directly on the Fourier transforms of the data stretches available. As a final step a and b may be estimated by back Fourier transforming the estimates of A and B, taking care to insert convergence factors in the process. Hung et al (1979) present the specific computational formulas involved and present an example of this system identification procedure for the human pupillary system. The estimated a and b are found to make sense physiologically and to be consistent with characteristics noted in other types of experiment with the system.

The extent of linearity of the system may be measured by the (linear) coherence

$$|R_{YX}(\lambda)|^2 = |f_{YX}(\lambda)|^2/f_{XX}(\lambda)f_{YY}(\lambda) \quad ,$$

with $|R|^2 \leq 1$ and the nearer it is to 1, the more strongly linear the system. Setting $W(t) = \iint b(t-u,t-v)X(u)X(v)dudv$, the quadratic coherence is defined as

$$|R_{YW}(\lambda)|^2 = \frac{1}{2f_{YY}(\lambda)} \int \frac{|f_{XXY}(\lambda-\mu,\mu)|^2}{f_{XX}(\lambda-\mu)f_{XX}(\mu)} d\mu \quad .$$

This too is bounded by 1, with its nearness to 1 indicating how purely quadratic the system is. The strength of linear plus pure quadratic relationship is measured by $|R_{YX}|^2 + |R_{YW}|^2$. Estimates of the linear and quadratic coherence for the human pupil are presented in Hung et al (1979). The linear coherence is larger, but the quadratic is important as well.

The above analysis is a frequency domain one. Had the input series been Gaussian white noise, a and b could have been estimated directly by cross-correlation, however in the experiments of Hung et al X could not be taken to be white noise. (A side remark is that even in the white noise case, the cross-correlations might be better computed via a (fast) Fourier transform.) In the non-white case a form of deconvolution needs to be carried out and this is done effectively via frequency domain procedures.

Proceeding via the frequency domain lead to the definition of the linear and quadratic coherences. These are frequency side parameters that prove exceedingly useful in practice. There seem to be no useful time side analogs.

6 A LINEAR DESCRIPTION OF NEURON FIRING

In an important class of neurophysiological experiments a sequence of constant amplitude electrical impulses is taken as input to a neuron. The neuron in turn emits a train of near constant amplitude electrical impulses. The neurophysiologist is interested in describing and understanding the process by which an input train is converted to an output train.

To develop a formal description of such a process it is convenient to assimilate the input and output pulse trains to point processes M and N with $M(t)$ the number of input pulses in the time interval $(0,t]$ and $N(t)$ the corresponding number of output pulses. A linear model relating two point processes is described by

$$\text{Prob}(N \text{ point in } (t,t+h) \mid M) \sim [\mu + \sum_j a(t-\sigma_j)]h$$

for small h, where the σ_j are the times of input pulses. It is of interest to estimate the function a and to construct a measure of how appropriate this model is in practical situations. These things may be done by means of a frequency side approach.

The basic statistic is once again a finite Fourier transform,

$$d_M^T(\lambda) = \sum_{0<\sigma_j<T} \exp(-i\lambda\sigma_j) = \int_0^T \exp(-i\lambda t)dM(t) \quad .$$

The periodogram of the data is defined as $I_{MM}^T(\lambda) = (2\pi T)^{-1}|d_M^T(\lambda)|^2$. The power spectrum, $f_{MM}(\lambda)$, may be defined for $\lambda \neq 0$ as the limit, T going to ∞, of $E\, I_{MM}^T(\lambda)$. At $\lambda = 0$ it may be defined by continuity. As in the case of ordinary time series, the power spectrum may be cross-spectrum may be defined and estimated in a similar fashion.

The model leads directly to the relationship $f_{NM}(\lambda) = A(\lambda)f_{MM}(\lambda)$, with A the Fourier transform of a . This relationship provides estimates of A and a in turn. Quite a number of such estimates are given in Brillinger et al (1976) for neurons of the sea hare. One factor causing the forms of A and a to vary substantially is whether the synapse is excitatory (input tends to increase the output rate) or inhibitory (input decreases the output rate). The time lag from input to output shows up in the estimates as well, as does the refractory period (output pulses may not be spaced arbitrarily closely together).

The degree to which the output train may be determined from the input via the model presented is conveniently measured by the coherence function, $|R_{NM}(\lambda)|^2 = |f_{NM}(\lambda)|^2/f_{NN}(\lambda)f_{MM}(\lambda)$, once again. In the examples of Brillinger et al (1976) this function is found to vary substantially with frequency. Generally it is much larger at the lower frequencies. It is surprisingly large in many cases given the essential nonlinearity of the system under study.

The frequency side approach is naturally effective in detecting periodicities that are present and in one of the Brillinger et al (1976) examples the estimated power spectrum displays a minor peak corresponding to a periodicity that really could not be seen on the time side. However, as the above development makes clear, the frequency approach further allows the deconvolution of input from system characteristics and leads to the definition of a useful measure of linear time invariant association.

The cited reference presents a frequency side solution to an important problem for which no other solution is presently known. It concerned the physiological connections of three neurons, L2, L3 and L10, of the sea hare. The three neurons were clearly related (there was substantial coherence between all pairs of covarying pulse trains). It was known that L10 was the driving neuron; however it was not known if the neurons were in series L10 →L2→L3 or L10 →L2 → L3 or if L3 and L2 had no direct connection, but L10 →L2 and

L10 $\longrightarrow$ L3 only.

Partial coherence analysis is a useful tool for examining such questions. Denote the spike trains by A, B, C respectively. The partial coherence between trains A and B is defined to be the coherence between the trains A and B with the linear time invariant effects of C removed. It is given by the modulus-squared of

$$(f_{CC}f_{AB} - f_{AC}f_{CB})/\sqrt{(f_{CC}f_{BB} - f_{BC}f_{CB})(f_{CC}f_{AA} - f_{CA}f_{AC})} \quad .$$

In this expression dependence on λ has been suppressed for convenience. In the case referred to, the partial coherence of L3 and L2 with the effects of L10 removed was not significant and the presence of a direct L2 to L3 connector could be ruled out essentially.

7 THE THRESHOLD MODEL OF NEURON FIRING

Suppose that a neuron receives as input the fluctuating electrical signal $X(t)$. Physiological considerations suggest the following description of its firing. A membrane potential

$$U(t) = \int_0^{B(t)} a(u)X(t-u)du$$

is formed internally, where $a(\cdot)$ describes a summation process and $B(t)$ denotes the time, at t, since the neuron last fired. The neuron then fires when $U(t)$ crosses a threshold $\theta + \varepsilon(t)$, ε being a noise process. Given experimental data it is of interest to verify and fit this model.

Frequency analyses may be carried out in the manner of the previous section given stretches of input and corresponding output data. However given the essential nonlinearity of the system and the feedback from output to input (due to the presence of $B(t)$) these may not be expected to be effective. (As will be mentioned later, in the case that X can be taken to be Gaussian stationary they are of some use.) In Brillinger and Segundo (1979) a time side solution is provided.

Let X_t, U_t, Y_t, $t = 0, \pm 1, \cdots$ denote the sampled versions of the series involved. One has $Y_t = 1$ if the neuron fired at time t and $Y_t = 0$ otherwise. Suppose that the noise is Gaussian white, then

$$\text{Prob}(Y_t = 1 \mid U_t) = \Phi(U_t - \theta)$$

with Φ the normal cumulative. Further, conditional on the given input the likelihood function of the data is

$$\prod_{t=0}^{T-1} \Phi(U_t - \theta)^{Y_t} \left(1 - \Phi(U_t - \theta)\right)^{1-Y_t} \quad .$$

The parameters a_u and θ may now be estimated by maximizing this likelihood. Brillinger and Segundo (1979) present a number of estimates found in this fashion for the neurons R2 and L5 of the sea hare. Once these estimates have been obtained, the function $\text{Prob}(Y = 1 \mid u)$ may be estimated. This was done. It was found to have the sigmoidal shape of Φ.

In the case that the input X is Gaussian and the feedback effect is not large, it may be shown that the estimated a_u derived via cross-spectral analysis are, up to sampling fluctuations, proportional to the desired a_u. (See Brillinger (1977).) Such estimates are given in Brillinger and Segundo (1979) and good agreement found.

For this problem, a frequency analysis could not suffice. The system had a nonlinearity and a feedback was present. By choice of special input, (Gaussian), and if the feedback was not strong, the frequency analysis gave approximate answers; however it is better to address the system directly.

8 NICHOLSON'S DATA ON SHEEP BLOWFLIES

During the 1950's the Australian entomologist A.J. Nicholson carried out an extensive series of experiments concerning the population variation of Lucilia cuprina (the sheep blowfly) under various conditions. Nicholson maintained populations of the flies on various diets (some constant, some fluctuating), experiencing different forms of competition (between larvae and adults, for egg laying space, etc.), and other many other conditions. The paper Brillinger et al (1980) reports the analysis of population data for a cage maintained under constant conditions. The basic data were the numbers of flies emerging and flies dieing in successive two day intervals. From these series, and the initial conditions, the number of adults alive at time t could be computed. The amount of food provided the flies was constant and limited. This caused the population size to oscillate dramatically, for when many flies were present the females did not receive enough protein to realize their maximum fecundity. In consequence many fewer eggs were laid and the next generation smaller. Nicholson ran the experiment for approximately 700 days.

The life cycle of a blowfly lasts 35 to 40 days. The aggregate numbers displayed an oscillation with this period throughout much of the experiment. Oster (1977) presents the Fourier spectrum of the data and a peak does stand out. However, while the data does have substantial stationary features, it also has a chaotic appearance in one stretch. Spectrum analysis does not take notice of alternate behavior in separate stretches. Complex demodulation was not especially informative either. A cross-spectrum analysis of the number of deaths, D_t, on the number of emergences, E_t, led to a plausible shape for the impulse response, however the coherence was not high. It seemed that a much better description must be obtainable for such fine experimental data.

Considerations of the biology involved suggested that the probability of a blowfly dieing, in a two day period, would depend on its age, i, on the number, N, it was competing with, and the number, N−, it had competed with last time period. An expression that worked well was

$$q_{i,N,N-}(\theta) = (1 - \alpha_i)(1 - \beta N)(1 - \gamma N-)$$

with θ denoting the unknown parameter values α_i, β, γ. A state space approach, (Gupta and Mehra (1974), Lipster and Shiryayev (1978)), was then taken for the description of the data. A state vector $\underset{\sim}{N}_t$ was defined whose entries gave the (unobservable except for age 0) members of each age group. The Kalman-Bucy filter was set up for

$$\underset{\sim}{m}_t = E(\underset{\sim}{N}_t \mid N_u, u \leq t)$$

and maximum likelihood estimation came down to choosing θ to minimize

$$\sum_{t=2}^{T} (D_t - \sum_{i=1}^{I} q_{i-1,N_{t-1},N_{t-2}}(\theta)\, m_{i-1,t-1})^2 / N_{t-1}^2 \ .$$

Specific details may be found in Brillinger et al (1980). The model was found to provide an effective description.

For this data nonlinearities were present. Further different subgroups of the population were behaving differently. Despite the presence of understood oscillations, a frequency approach was not very revealing.

9 DISCUSSION

This paper has described a number of time series analyses proceeding from a frequency side analysis to a time side analysis with some hybrid analyses in between. In each case no initial commitment was made to one side or the other, rather at some stage one approach became much more revealing than the other.

Because of space limitations some of the bases for deciding on the final approach will simply be listed. These are: goals and circumstances, ease of (physical) interpretation, simplicity and parsimony sampling fluctuations, computational difficulty, sensitivity, physical theory (versus black box), data quality, data quantity, ease in dealing with complications, expertness available, real time versus dead time, efficiency, dangers (eg. overtight parameterization), bandwidth of phenomenon, presence and type of nonlinearities, type of nonstationarity.

10 REFERENCES

Bloomfield, P., Brillinger, D.R., Cleveland, W.S. and Tukey, J.W., 1981. The Practice of Spectrum Analysis. In preparation, 233 pp.

Bolt, B.A. and Brillinger, D.R., 1979. Estimation of uncertainties in eigenspectral estimates. Geophys. J. R. astr. Soc.,59:593-603.

Brillinger, D.R., 1973. An empirical investigation of the Chandler wobble. Bull. Internat. Statist. Inst., 45: 413-433.

Brillinger, D.R., 1977. The identification of a particular nonlinear time series system. Biometrika, 64: 509-515.

Brillinger, D.R., 1981. Time Series: Data Analysis and Theory. Holden-Day, San Francisco,540 pp.

Brillinger, D.R., Bryant, H.L. and Segundo, J.P., 1976. Identification of synaptic interactions. Biol. Cybernetics, 22: 213-228.

Brillinger, D.R., Guckenheimer, J., Guttorp, P. and Oster, G., 1980. Lectures on Math. in Life Sci. 13. Amer. Math. Soc., Providence.

Brillinger, D.R. and Segundo, J.P., 1979 . Empirical examination of the threshold model of neuron. Biol. Cybernetics, 35: 213-220.

Gupta, N.K. and Mehra, R.K., 1974 . Computational aspects of maximum likelihood estimation. IEEE Trans. Aut. Control, AC-19: 771-783.

Hung, G., Brillinger, D.R. and Stark, L., 1979 . Interpretation of kernels. II. Math. Biosciences, 46: 159-187.

Lipster, R.S. and Shiryayev, A.N., 1978. Statistics of Random Processes. Springer, New York.

Munk, W.H. and MacDonald, G.J.F., 1960 . The Rotation of the Earth. Cambridge Press, Cambridge.

Oster, G., 1977. Modern Modelling of Continuum Phenomena, R. DiPrima (Editor). Amer. Math. Soc., Providence.

Tick, L.J., 1961. The estimation of transfer functions of quadratic systems. Technometrics, 3: 563-567.

Tukey, J.W., 1978. Can we predict where "time series". In: D.R. Brillinger and G.C. Tiao (Editors) Directions in Time Series. IMS.

Whittle, P., 1954. Some recent contributions to the theory of stationary processes. In: H. Wold, A Study in the Analysis of Stationary Time-Series. Almqvist and Wiksell, Uppsala.

DETECTION OF INTERVENTIONS AT UNKNOWN TIMES

IAN B. MACNEILL

Statistical Laboratory, The University of Western Ontario,
London, Ontario, Canada

1 INTRODUCTION

Models for hydrological time series are characterized by parameters which may stay constant or which may change over the course of time. While the detection of changes in parameters when the time of change is specified is a relatively standard statistical problem, the detection of changes when the time of change is unknown is a non-standard problem that is currently receiving considerable attention.

A method of detecting change of regression parameters at unknown times is presented. A derivation is presented of a likelihood ratio type statistic for detecting changes in regression parameters at unknown times. Distributional properties of the statistic are discussed. The statistic is then applied to several periodic series.

Models are then considered for improving the short and intermediate-term forecasting capacity of periodic models and yet preserving both long-term predictive capacity and the clear meaning of the model parameters. The models are constructed so as to have certain of the properties of autoregressive schemes but yet to retain the basic properties of periodicity. For autoregressions, one uses the entire observed data to estimate the parameters. Then one uses these estimates to estimate the next observation from several of those immediately preceding it. The residuals formed by the differences between the actual and estimated observations may be used to test goodness-of-fit. This regime is applied to the amplitude of periodic components. The basic amplitude is first estimated and then is modified by adaptive means by the addition of components defined by the immediately preceding

Reprinted from *Time Series Methods in Hydrosciences*, by A.H. El-Shaarawi and S.R. Esterby (Editors)

observations. Thus, for example, the amplitude may shrink if present and previous observations indicate the presence of a low amplitude cycle. The effect of present observations on the amplitude disappears in the long-term; hence the basic cycle defines the long-term prediction.

2 DOUBLY STOCHASTIC MODELS

A doubly stochastic regression model may be defined as follows. Let $\{\varepsilon(j),\ j \geq 1\}$ be a sequence of independent and identically distributed error terms each normally distributed with zero mean and variance $\sigma^2 > 0$, and let $\{f_i(t),\ t > 0,\ i = 0, 1, \ldots, p\}$ denote a set of regressor functions. Also, let $\{\beta(j),\ j \geq 1\}$ denote a sequence of regression coefficients. If $\beta'(j) = (\beta_0(j), \ldots, \beta_p(j))$ is stochastic, then the dependent variables denoted by $\{Y(j),\ j \geq 1\}$ in a doubly stochastic regression model may be defined as follows:

$$Y(j) = \sum_{i=0}^{p} \beta_i(j)\, f_i(j) + \varepsilon(j),\ j \geq 1.$$

An approach to the estimation of variable regression parameters involving recursive least squares regression analysis is given by Young [1974] who discusses several stochastic models which possess specified structure for regression parameters. Harrison and Stevens [1976] define a dynamic linear model which is characterized by stochastic parameters whose estimation utilizes the recursive procedures originally formulated by Kalman [1963]; again, the structure for the stochastic models for the regression parameters is assumed known. Brown, Durbin and Evans [1975] use recursive residuals to attack the problem of detecting changes over time in regression parameters. MacNeill [1978a, 1978b] discusses properties of raw regression residuals that can be used for detection of changes at unspecified times in regression parameters. For the same problem, MacNeill [1980] also discusses an alternative statistic. Priestley [1965], Priestley and Subba Rao [1969], and Subba Rao and Tong [1974] treat similar problems using a spectral approach. A number of authors, including Haggan and Ozaki [1979], have discussed problems involving non-linear phenomena and related non-linear models.

3 A TEST FOR CHANGE OF REGRESSION AT UNSPECIFIED TIME

In the first instance, it is probably appropriate to consider the regression problem to be stationary in the sense that

$\beta(1) = \beta(2) = \ldots = \beta$

and then test the null hypothesis implicit in this assumption against that of change at unspecified time. A formulation of this problem discussed by MacNeill [1980] is as follows. If we let $Y_n' = (Y(1), \ldots, Y(n))$, $\varepsilon_n' = (\varepsilon(1), \ldots, \varepsilon(n))$, and X_n be the design matrix whose ij^{th} component is $f_j(t_i)$, then, in standard matrix form, we may write the regression equation as

$Y_n = X_n\beta + \varepsilon_n$,

and the Gauss-Markov estimator for β, denoted by $\hat{\beta}$, then is

$\hat{\beta} = (X_n' X_n)^{-1} X_n'Y_n$.

The subscripts on the vectors and matrices are omitted where no confusion results. The vector of regression residuals is defined to be $Y - \hat{Y}$ where the ith component of $\hat{Y}$ is:

$\hat{Y}_i = \hat{\beta}' f(t_i)$.

The alternative hypothesis requires changes in β at unknown times. To specify alternatives we let $\delta'(i) = \{\delta_0(i), \delta_1(i), \ldots, \delta_p(i)\}$ represent the changes in the vector of regression coefficients effected between i^{th} and the $(i+1)^{th}$ observations. That is, if $\beta(i)$ is the vector of regression coefficients for the ith observation, then $\beta(i+1) = \beta(i) + \delta(i)$. So that the Bayes-type argument introduced by Chernoff and Sacks [1964] may be used to eliminate nuisance parameters, we assume that β has a multivariate normal distribution with zero mean and covariance matrix $\tau^2 I$ where $\tau^2 > 0$. We then let a particular change sequence be defined by:

$\omega' = \{\omega_1, \omega_2, \ldots\}$

where ω_i is 1 if a change in β occurs between the ith and $(i+1)^{th}$ observations and is zero otherwise. Thus, a single change through the series of observations would require one component of $\underset{\sim}{\omega}$ to be 1 and the rest zero. The assignation of a prior distribution to the collection of all possible change sequences, $\underset{\sim}{\omega}$, then makes it possible to formulate the problem in a way introduced by Gardner [1969]. The nuisance parameters $\underset{\sim}{\delta}(i)$ can then be integrated out and, with τ^2 small, the likelihood ratio statistic for testing the null hypothesis against change sequences $\underset{\sim}{\omega}$ with a uniform prior can be shown to be approximately proportional to:

$$Q_n = \frac{1}{\sigma^2} \sum_{k=1}^{n-1} \underset{\sim}{Y}'(\underset{\sim}{I}-\underset{\sim}{X}(\underset{\sim}{X}'\underset{\sim}{X})^{-1}\underset{\sim}{X}')\, \underset{\sim}{X}^k \underset{\sim}{X}^{k'}\, (\underset{\sim}{I}-\underset{\sim}{X}(\underset{\sim}{X}'\underset{\sim}{X})^{-1}\underset{\sim}{X}')\underset{\sim}{Y},$$

where $\underset{\sim}{X}^k$ is $\underset{\sim}{X}$ with the first k rows identically equal to zero. The approximation becomes exact as τ^2 vanishes. Note that:

$$Q_n = \frac{1}{\sigma^2} \sum_{k=1}^{n-1} ||\underset{\sim}{\varepsilon}'(\underset{\sim}{I}-\underset{\sim}{X}(\underset{\sim}{X}'\underset{\sim}{X})^{-1}\underset{\sim}{X}')\underset{\sim}{X}^k||^2 = \frac{1}{\sigma^2} \sum_{k=1}^{n-1} ||(\underset{\sim}{Y}'-\hat{\underset{\sim}{Y}}')\underset{\sim}{X}^k||^2$$

where, if $Z' = \{Z_1, Z_2, \ldots, Z_\ell\}$, $||Z||^2 = Z_1^2 + Z_2^2 + \ldots + Z_\ell^2$.

Associated with the sequence of partial sums of regression residuals is a generalized Brownian Bridge (see MacNeill 1978b) which we shall denote by $\{\beta_f(t),\ t \in [0,1]\}$. The stochastic integral

$$\int_0^1 \beta_f^2(t)dt$$

is then related to a Crámer-von Mises type statistic defined upon the sequence of partial sums of regression residuals; some examples are considered by MacNeill [1978a]. Let μ_f and σ_f^2 denote the mean and variance of the stochastic integral. Then it may be shown that:

$$E(Q) \simeq \sigma^2 \mu_f \sum_{i=2}^{n} (i-1)(\underset{\sim}{X}_i \cdot \underset{\sim}{X}_i)$$

and

$$Var(Q) \simeq 2\sigma^4 \sigma^2_f \sum_{i=2}^{n} \sum_{j=2}^{n} [\min\{(i-1), (j-1)\}]^2 (\underset{\sim}{X}_i \cdot \underset{\sim}{X}_j)^2$$

where $\underset{\sim}{X}_i$ is the ith row of the design matrix $\underset{\sim}{X}$, and

$$(\underset{\sim}{X}_i \cdot \underset{\sim}{X}_j) = \sum_{\ell=0}^{p} X_{i\ell} X_{j\ell} = \sum_{\ell=0}^{p} f_\ell(t_i) f_\ell(t_j).$$

We next discuss distribution theory for Q_n which is a quadratic form in independent normal variates. The matrix of the quadratic

TABLE 1

Selected quantiles for $P[Q_n \leq \alpha]$ when $Y(j) = \beta_0/2+\beta_1\cos(2\pi\ j/n) + \beta_2\ \sin(2\ \pi j/n) + \varepsilon(j)$

$P[Q_n \leq X]$ \ n	10	12	14	16	18	20
.01	1.062	1.796	2.700	3.765	4.990	6.369
.025	1.451	2.366	3.476	4.773	6.255	7.922
.05	1.884	2.992	4.323	5.870	7.632	9.608
.10	2.537	3.930	5.589	7.512	9.699	12.145
.50	8.023	11.887	16.480	21.804	27.860	34.648
.90	30.409	44.764	61.827	81.599	104.081	129.280
.95	41.803	61.490	84.890	112.006	142.838	177.391
.975	53.669	78.909	108.909	143.672	183.199	227.486
.99	69.841	102.648	141.641	186.824	238.198	295.766

form is of the form $\underset{\sim}{P}\ \underset{\sim}{M}\ \underset{\sim}{P}$ where $\underset{\sim}{P}$ is the usual regression projection matrix. Assume that $\{\lambda_j\}_{j=1}^{n}$ is the set of eigenvalues of the matrix $\underset{\sim}{P}\ \underset{\sim}{M}\ \underset{\sim}{P}$ ordered from largest to smallest. Then the characteristic function for Q_n is:

$$\phi(s) = \prod_{k=1}^{n} (1 - 2i\, s\, \lambda_k)^{-1/2}, \quad i = \sqrt{-1} \ .$$

If we let $D(2is) = \phi^{-2}(s)$, $\lambda = 2is$, and assume that there are 2n observations, then, provided the eigenvalues are distinct, it can be shown that:

$$P[Q_{2n} \leq \alpha] = 1 - \frac{1}{\pi} \sum_{j=1}^{n} (-1)^{n-1} \int_{\frac{1}{\lambda_{2j-1}}}^{\frac{1}{\lambda_{2j}}} \frac{e^{\frac{-\lambda}{2}\alpha}}{\lambda(-D(\lambda))^{1/2}}\, d\lambda. \tag{1}$$

If some of the eigenvalues are not distinct then one may: compute (1) with these eigenvalues removed; compute the χ^2 distribution associated with the repeated eigenvalues; and convolve the resulting distributions to obtain the distribution of Q_{2n}. We let M be the matrix whose (i,j)th component is $(\underset{\sim}{X}_i \cdot \underset{\sim}{X}_j)\min[(i-1),(j-1)]$, let $\underset{\sim}{P} = (\underset{\sim}{I} - \underset{\sim}{X}(\underset{\sim}{X}'\underset{\sim}{X})^{-1}\underset{\sim}{X}')$, and consider

$$Q_n = \frac{1}{\sigma^2}\, \underset{\sim}{\varepsilon}_n'\ \underset{\sim}{P}\ \underset{\sim}{M}\ \underset{\sim}{P}\ \underset{\sim}{\varepsilon}_n .$$

It is then a straightforward numerical problem to compute using a package such as EISPACK the eigenvalues of $\underset{\sim}{P}\ \underset{\sim}{M}\ \underset{\sim}{P}$ and to obtain the characteristic function which we may then invert. Some results for a single sinusoid appear in Table 1.

4 ESTIMATION OF REGRESSION PARAMETER PROCESSES

When the test for change of regression parameters at unknown time rejects the null hypothesis, one focuses attention on the parameter process. The method of estimation of $\{\underset{\sim}{\beta}(t),\ t > 1\}$ used below is that of recursive regression whereby $\underset{\sim}{\beta}(t)$ is estimated by least squares using a segment of length k of the observations centred around t. This is the method of "moving regression" referred to by Brown *et al* [1975]. The least squares estimator for $\underset{\sim}{\beta}(t)$, is given by the following equation:

$$\hat{\underset{\sim}{\beta}}(t,k) = (\underset{\sim}{X}'(t,k)\ \underset{\sim}{X}(t,k))^{-1}\underset{\sim}{X}'(t,k)\ \underset{\sim}{Y}(t,k)$$

where the design matrix and vector of dependent variables utilize only the k observations centred around t. If $P_{k,t}$ and $C_{k,t}$ denote $(\underset{\sim}{X}'(t,k)\ \underset{\sim}{X}(t,k))^{-1}$ and $\underset{\sim}{X}'(t,k)\underset{\sim}{Y}(t,k)$, respectively, and $\underset{\sim}{X}'(t) = (f_0(t), f_1(t), \ldots, f_p(t))$, then one may use the following recursive relations to compute estimates of the regression process:

$$\underset{\sim}{P}_{k+1,t} = \underset{\sim}{P}_{k,t} - \underset{\sim}{P}_{k,t}\ \underset{\sim}{X}(t+k+1)(1+\underset{\sim}{X}'(t+k+1)\underset{\sim}{P}_{k,t}\ \underset{\sim}{X}(t+k+1))^{-1}$$
$$\underset{\sim}{X}'(t+k+1)\ \underset{\sim}{P}_{k,t},$$

$$\underset{\sim}{P}_{k,t+1} = \underset{\sim}{P}_{k+1,t} + \underset{\sim}{P}_{k+1,t}\underset{\sim}{X}(t)(1-\underset{\sim}{X}'(t)\underset{\sim}{P}_{k+1,t}\underset{\sim}{X}(t))^{-1}\underset{\sim}{X}'(t)\underset{\sim}{P}_{k+1,t},\ \text{and}$$

$$\underset{\sim}{C}_{k,t+1} = \underset{\sim}{C}_{k,t} + Y(t+k+1)\underset{\sim}{X}(t+k+1) - Y(t)\underset{\sim}{X}(t).$$

This is the algorithm of Plackett [1950] the use of which in a time series context is discussed by Young [1974], in a regression context by Brown *et al* [1975], and in a short-term forecasting context by Harrison and Stevens [1976].

Each component of the estimator, $\hat{\underset{\sim}{\beta}}(t,k)$, is a linear combination of all components of $\underset{\sim}{\beta}(\cdot)$ for all k values of the time parameter centred about t. More precisely, if $\underset{\sim}{B}(t,k)$ is a $(p+1)\times 1$ vector whose ℓ^{th} component is

$$\sum_{y=0}^{p}\ \sum_{i=0}^{k}\ f_\ell(t+i)f_j(t+i)\beta_j(t+i),$$

then

$$\hat{\underset{\sim}{\beta}}(t+k/2,k) = (\underset{\sim}{X}'(t,k)\underset{\sim}{X}(t,k))^{-1}\underset{\sim}{X}(t,k)\underset{\sim}{\varepsilon}(t,k) + (\underset{\sim}{X}(t,k)\underset{\sim}{X}(t,k))^{-1}\underset{\sim}{B}(t,k).$$

Thus the relation between $\hat{\underset{\sim}{\beta}}(t)$ and $\hat{\underset{\sim}{\beta}}(t,k)$ is complicated by the presence of moving averages in the noise process and by the presence of correlation between the components of the estimation vector induced by the estimation procedure.

We proceed to explore the empirical properties of the estimators of the stochastic process of regression coefficients by fitting ARIMA models (see Box and Jenkins [1970]) to the various components of this process. We first consider the Wolfer sunspot series for

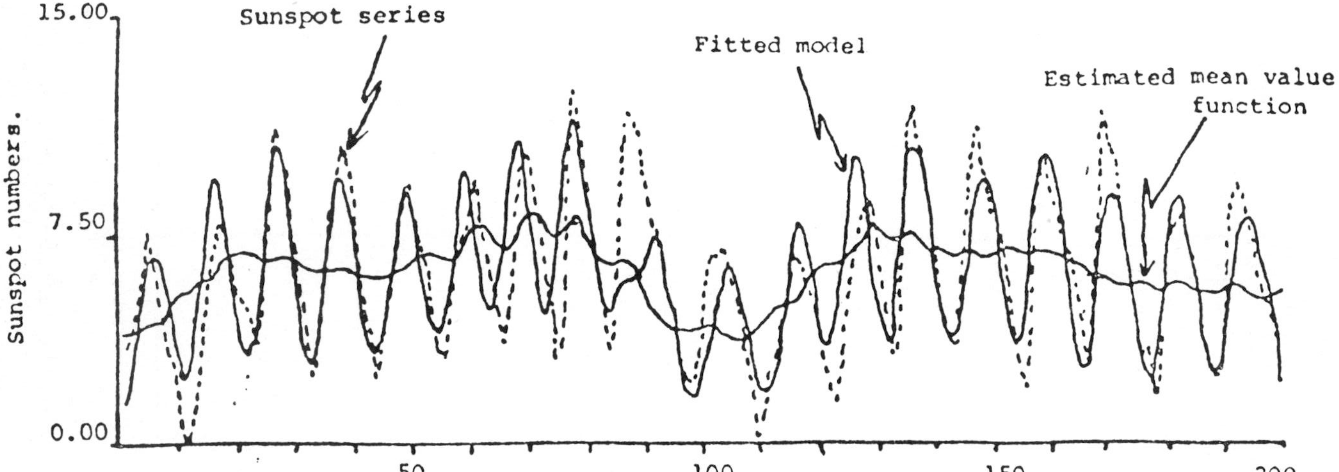

Figure 1. Square root of yearly sunspot numbers: 1701 - 1900

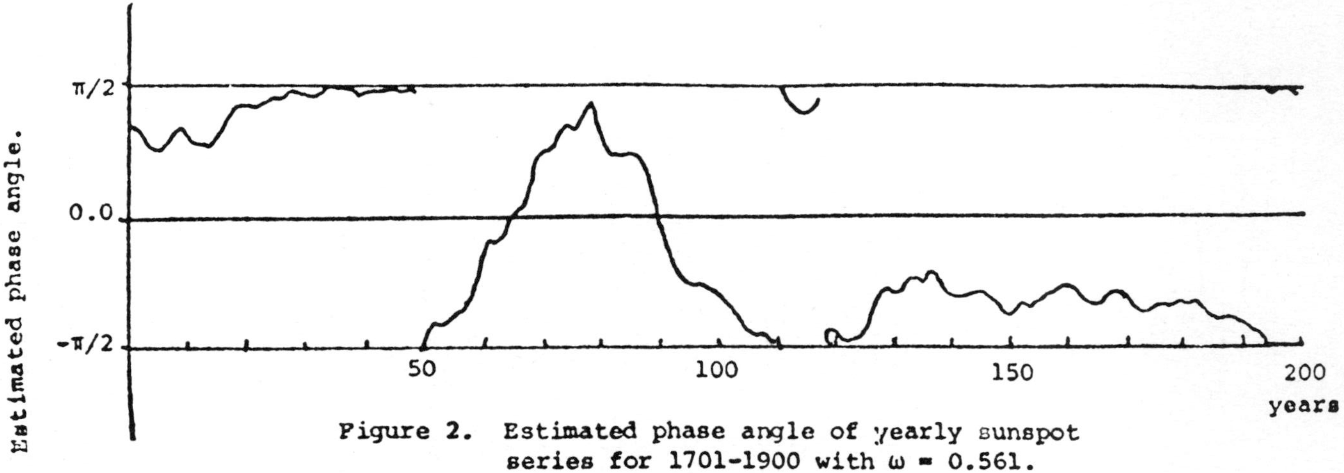

Figure 2. Estimated phase angle of yearly sunspot series for 1701-1900 with $\omega = 0.561$.

the period 1700-1960. We fit the following periodic model; $\omega = 0.561$:

$$Y(t) = \beta_0 + \beta_1 \cos \omega t + \beta_2 \sin \omega t + \varepsilon(t)$$
$$= \beta_0 + \gamma \sin(\omega t + \psi) + \varepsilon(t).$$

The relevant Q statistics are as follows: $Q = 27.09 \times 10^5$, $E(Q) \simeq 2.05 \times 10^5$, and $\sqrt{Var(Q)} \simeq 0.75 \times 10^5$. Hence the hypothesis of no change of regression parameters is rejected.

The recursive regression procedure is applied for various values of k. As one might expect, the process $\{\hat{\beta}(t,k),\ t > 0\}$ is ragged for small values of k and smooth for large values. Figure 1 contains plots of the yearly sunspot series for the period 1701-1900. Figure 2 contains a plot of the estimated phase angle for the series. We make the assumption that each component process has a mean value and fit ARIMA models to the deviations from this mean. The forecast functions regress to these means so one does not lose the long-term predictive properties of a simple sunusoid. In the short-term, the model is adaptive.

5 CONCLUSION

An adaptive harmonic regression model of a doubly stochastic nature fitted to data indicates that such models are capable of improving both fits to the data and forecasts of future observations.

REFERENCES

Box, G.E.P. and Jenkins, G.M., 1970. Time Series Analysis: Forecasting and Control. Holden-Day, San Francisco.

Brown, R.L., Durbin, J. and Evans, J.M., 1975. Techniques for testing the constancy of regression relationships over time. J. Roy. Statist. Soc. Ser. B 37: 149-192.

Chernoff, H. and Zacks, S., 1964. Estimating the current mean of a normal distribution which is subject to changes in time. Ann. Math. Statist. 35: 999-1018.

Gardner, L.A., 1969. On detecting changes in the mean of normal variates. Ann. Math. Statist. 40: 116-126.

Haggan, V. and Ozaki, T., 1969. Modelling non-linear random vibrations using an amplitude-dependent autoregressive time series model. UMIST Technical Report #115.

Harrison, P.J. and Stevens, C.F., 1976. Bayesian forecasting. J. Roy. Statist. Soc. B 38: 205-247.

Kalman, R.E., 1963. New methods in Wiener filtering theory. In Proceedings of the First Symposium on Engineering Applications of Random Function Theory and Probability. J.L. Bagdanoff and F. Kozin (Editors). Wiley, New York.

MacNeill, I.B., 1974. Tests for change of parameter at unknown time and distributions of some related functionals on Brownian motion. Ann. Statist. 2: 950-962.

MacNeill, I.B., 1978a. Properties of sequence of partial sum of polynomial regression residuals with applications to tests for change of regression at unknown time. Ann. Statist. 6: 422-433.

MacNeill, I.B., 1978b. Limit processes for sequences of partial sums of regression residuals. Ann. Prob. 6: 696-698.

MacNeill, I.B., 1980. Detection of changes in the parameters of periodic or pseudo-periodic systems when the change times are unknown. In: S. Ikeda (Editor), Statistical Climatology. Elsevier, Amsterdam.

Plackett, R.L., 1950. Some theorems in least squares. Biometrika 37: 149-157.

Priestley, M.B., 1965. Evolutionary spectra and non-stationary processes. J. Roy. Statist. Soc. B 27: 204-237.

Priestley, M.B. and Subba Rao, T., 1969. A test for non-stationarity of time series. J. Roy. Statist. Soc. B 31: 140-149.

Subba Rao, T. and Tong, H., 1974. Linear time dependent systems. I.E.E.E. Trans. on Aut. Cont., AC-19: 735-737.

Young, P.C., 1974. Recursive approaches to time series analysis. Bull. Inst. Maths. & Applic. 10: 209-224.

DISTRIBUTION OF PARTIAL SUMS WITH APPLICATIONS TO DAM CAPACITY AND ACID RAIN

A.M. MATHAI

McGill University, Montreal, Canada

SUMMARY

Distribution of the total random input over a period of time into a dam or storage is considered in this article. The inputs could be the amounts of sediments carried into a dam on different occasions over a period of time or the amounts of water in excess of the normal flow due to rains on different occasions or the amounts of acid deposited in a lake by clouds carrying acid vapours etc. Exact distribution of the total input is obtained when the individual inputs are of a general gamma type. Possible applications of these results to various practical problems are also pointed out.

1. INTRODUCTION

Consider independent random inputs $X_1,\ldots,X_n$ into a storage tank. The total input over a period of n occasions is then

$S_n = X_1+\ldots+X_n$

These random inputs may be the excess amounts of water flowing into a dam due to rains of different durations at the waterhead, they could be the amounts of pollutants discharged into a lake from different sources or from one source on different occasions, they could be the amounts of sediments carried to a dam on different occasions etc. If $X_1,X_2,\ldots$ represent excess flow due to rains then an appropriate practical model would be exponential distributions with different mean values because the durations

Reprinted from *Time Series Methods in Hydrosciences*, by A.H. El-Shaarawi and S.R. Esterby (Editors)

of the rainfalls are usually different for different occasions. In such a case S_n is a sum of independent but not identically distributed exponential random variables. If water is collected at different waterheads on different occasions before flowing into a dam then an appropriate model for S_n is that it is a sum of n independent gamma distributed random variables. In order to answer the problems of overflow one needs to study the probability that S_n exceeds a preassigned number. For this and other purposes one needs the exact distribution of S_n.

Consider another problem of a refinery emitting sulphur dioxide into the air. The concentration of the pollutants may be maximum at a certain height from the chimney and then it may thin out in all directions. Under some mild conditions it is reasonable to assume that the distribution of the pollutants in the air is a three dimensional Gaussian type with the origin being at an optimal height from the chimney. Suppose that a patch of cloud is passing through that place. How much of the pollutants will be carried away by that cloud? In other words how much acid rain one can expect from that patch of cloud? Assuming that the patch of cloud can be looked upon as an ellipsoid, with respect to the water molecule content, with a centre of its own then the amount of acid rain is proportional to the probability content of a disoriented ellipsoid in a 3-dimensional Gaussian distribution.

In the general case of this type of problem one has the following situation. Consider a p-variate vector random variable distributed according to a p-variate normal. That is

$$X \sim N_p(\mu,\Sigma)$$

where $\mu = E(X)$ and Σ is the covariance matrix and E denotes the expected value. A disoriented ellipsoid will then be of the form

$$A = \{(X-\alpha)'C(X-\alpha) \leq \delta\}$$

where C is a symmetric positive definite matrix, α is a p-vector of known constants, a prime denotes a transpose and δ is a pre-assigned number. If $\mu=\alpha$ then the ellipsoid is centred at the expected value of X. Then the probability content of A, denoted by P(A) is available from the multinormal density as follows.

$$P(A) = P\{(X-\alpha)'C(X-\alpha) \leq \delta\}$$

where X is a p-variate random vector normally distributed and α, C, δ are all known. For convenience we shall reorient and relocate the ellipsoid by making the transformation

$$Y = \Sigma^{-1/2}(X-\mu) \text{ or } X = \Sigma^{1/2}Y + \mu = \Sigma^{1/2}(Y+\Sigma^{-1/2}\mu)$$

where $\Sigma^{1/2}$ is the symmetric square root of Σ. Then

$$Y \sim N_p(0, \underset{\sim}{I}) \text{ and } X-\alpha = \Sigma^{1/2}(Y+\Sigma^{-1/2}(\mu-\alpha))$$

$$(X-\alpha)'C(X-\alpha) = (Y+\beta)'V(Y+\beta)$$

where

$$V = \Sigma^{1/2}C\Sigma^{1/2}, \quad \beta = \Sigma^{-1/2}(\mu-\alpha)$$

Since V is symmetric and positive definite there exists an orthogonal matrix Q such that $Q'VQ = D = \text{diag}(\lambda_1, \ldots, \lambda_p)$. Make the transformation Z=Q'Y or QZ=Y and then $Z \sim N_p(0, \underset{\sim}{I})$ and

$$(X-\alpha)'C(X-\alpha) = (Y+\beta)'V(Y+\beta) = (Z+\gamma)'D(Z+\gamma) = \sum_{j=1}^{p} \lambda_j(z_j+\gamma_j)^2 = U, \text{ say} \tag{1.1}$$

where $\gamma = Q'\beta$, γ_j is the jth element in $Q'\beta$ and z_j is the jth element in Z. Since $z_j \sim N(0,1)$, $j=1,\ldots,p$ and mutually independent $(z_j+\gamma_j)^2$ is a noncentral chisquare with one degree of freedom. What is the probability content of the ellipsoid $(X-\alpha)'C(X-\alpha) \leq \delta$? It is nothing but the probability $P\{U \leq \delta\}$ where U is a linear combination of independent noncentral chisquare variables with one degree of freedom each. If we are considering the amounts of acid rain that can be expected from several cloud patches passing through the chimney area then the problem reduces to the evaluation of a probability of the type $P\{S_n \leq \delta\}$ with $S_n = X_1 + \ldots + X_n$ where $X_1, \ldots, X_n$ are independent weighted noncentral chisquare variables with different degrees of freedom and δ is a preassigned number.

In this paper we will consider the distribution of $Y = X_1 + \ldots + X_p$ when $X_1, \ldots, X_p$ are mutually independent and (1) gamma distributed with different parameters (2) noncentral chisquare distributed with different degrees of freedom.

2. GAMMA INPUT

Consider p inputs into a dam which are mutually independently gamma distributed with the ith input denoted by X_i and having the density

$$f_i(x_i) = (\Gamma(\alpha_i)\lambda_i^{\alpha_i})^{-1} x_i^{\alpha_i-1} e^{-x_i/\lambda_i} \quad ,\alpha_i>0,\ \lambda_i>0,\ x_i>0, \qquad (2.1)$$

and zero elsewhere, so that the total input is $Y=X_1+\ldots+X_p$. Since the moment generating functions exist and invertible in the following problems we will work with the moment generating functions for convenience. The moment generating function of X_i is

$$M_{x_i}(t) = (1-\lambda_i t)^{-\alpha_i}$$

and that of Y is

$$M_y(t) = \prod_{i=1}^{p}(1-\lambda_i t)^{-\alpha_i} \qquad (2.2)$$

The density of Y is available by inverting $M_y(t)$. There are many ways of doing this but we will proceed as follows.

$$(1-\lambda_2 t) = (1-\lambda_1 t)(\lambda_2/\lambda_1)(1-(1-\lambda_1/\lambda_2)/(1-\lambda_1 t))$$

Hence

$$(1-\lambda_1 t)^{-\alpha_1}(1-\lambda_2 t)^{-\alpha_2} = (1-\lambda_1 t)^{-(\alpha_1+\alpha_2)}(\lambda_2/\lambda_1)^{-\alpha_2}$$
$$(1-(1-\lambda_1/\lambda_2)/(1-\lambda_1 t))^{-\alpha_2}$$

But

$$(1-(1-\lambda_1/\lambda_2)/(1-\lambda_1 t))^{-\alpha_2} = \sum_{r=0}^{\infty}(\alpha_2)_r\,(1-\lambda_1/\lambda_2)^r(1-\lambda_1 t)^{-r}/r!$$

for $t < 1/\lambda_2,\ 1/\lambda_1$, where for example $(\alpha)_r = (\alpha+1)\ldots(\alpha+r-1)$. Hence the density of $Y=X_1+X_2$, denoted by $g_2(y)$, is

$$g_2(y) = (\lambda_2/\lambda_1)^{-\alpha_2}\, y^{\alpha_1+\alpha_2-1}\, e^{-y/\lambda_1} \sum_{r=0}^{\infty}(\alpha_2)_r(1-\lambda_1/\lambda_2)^r\, y^r/$$
$$(\lambda_1^{\alpha_1+\alpha_2}\, r!\, \lambda_1^r\, \Gamma(\alpha_1+\alpha_2+r))$$
$$= (\lambda_1^{\alpha_1}\lambda_2^{\alpha_2}\Gamma(\alpha_1+\alpha_2))^{-1}\, y^{\alpha_1+\alpha_2-1}\, e^{-y/\lambda_1}$$
$${}_1F_1(\alpha_2;\alpha_1+\alpha_2;(\lambda_1^{-1}-\lambda_2^{-1})y)\ ,\quad y>0 \qquad (2.3)$$

where ${}_1F_1$ is a confluent hypergeometric function. Proceeding exactly the same way one has the following result in the general case.

$$g_p(y) = (\lambda_1^{\alpha_1} \ldots \lambda_p^{\alpha_p}\, \Gamma(\alpha_1+\ldots+\alpha_p))^{-1}\, y^{\alpha_1+\ldots+\alpha_p-1}\, e^{-y/\lambda_1}$$

$$\sum_{r_2=0}^{\infty} \ldots \sum_{r_p=0}^{\infty} (\alpha_2)_{r_2} \ldots (\alpha_p)_{r_p} ((\lambda_1^{-1} - \lambda_2^{-1})y)^{r_2} \ldots$$

$$((\lambda_1^{-1} - \lambda_p^{-1})y)^{r_p} / (r_2! \ldots r_p! (\alpha_1+\ldots+\alpha_p)_{r_2+\ldots+r_p}). \quad (2.4)$$

The multiple sum appearing here is a confluent hypergeometric function of p-1 variables, namely $\emptyset_2$ (see Mathai & Saxena, 1978, p.163). The density is therefore,

$$g_p(y) = (\lambda_1^{\alpha_1} \ldots \lambda_p^{\alpha_p}\, \Gamma(\alpha_1+\ldots+\alpha_p))^{-1}\, y^{\alpha_1+\ldots+\alpha_p-1}\, e^{-y/\lambda_1}$$

$$\emptyset_2(\alpha_2,\ldots,\alpha_p;\alpha_1+\ldots+\alpha_p;(\lambda_1^{-1}-\lambda_2^{-1})y,\ldots,(\lambda_1^{-1}-\lambda_p^{-1})y), \quad (2.5)$$

$$y > 0$$

Evidently this case includes independent exponential inputs with different mean values in which case $\alpha_1=\ldots=\alpha_p=1$ and $\lambda_1, \ldots, \lambda_p$ are not all the same. If some of the λ_j's are equal then evidently the number of arguments in $\emptyset_2$ of (2.5) is reduced. In order to make the convergence of the multiple series in (2.4) faster one can use the following technique. Replace y by y/β $\beta>0$, and choose β so that $\emptyset_2$ converges fast. It is well-known that for small values of the arguments $\emptyset_2$ approximates to unity. This replacement is equivalent to considering the density of βY.

Instead of simplifying the various factors of $\Pi(1-\lambda_j t)$ in terms of $(1-\lambda_1 t)$ one can also proceed as follows. Consider the identity

$$(1-\lambda_i t) = (1-\gamma t)(\lambda_i/\gamma)(1-(1-\gamma/\lambda_i)/(1-\gamma t))$$

for some $\gamma>0$. Then proceeding as before one can invert the moment generating function for $t < 1/\gamma,\ 1/\lambda_i,\ i=1,\ldots,p$ to get the density in the following form.

$$g_p(y) = \lambda_1^{\alpha_1}\ldots\lambda_p^{\alpha_p}\ \Gamma(\alpha_1+\ldots+\alpha_p))^{-1}\ y^{\alpha_1+\ldots+\alpha_p-1}\ e^{-y/\gamma}$$

$$\emptyset_2(\alpha_1,\ldots,\alpha_p;\alpha_1+\ldots+\alpha_p;(\gamma^{-1}-\lambda_1^{-1})y,\ldots,(\gamma^{-1}-\lambda_p^{-1})y), \quad (2.6)$$

$y>0$, in which case the $\emptyset_2$ is of p variables $(\gamma^{-1}-\lambda_i^{-1})y$, $i=1,\ldots,p$. If γ is chosen such that

$$\gamma^{\alpha_1+\ldots+\alpha_p} = \lambda_1^{\alpha_1}\ldots\lambda_p^{\alpha_p}$$

then the coefficient part to $\emptyset_2$ is a gamma density with the parameters $\alpha_1+\ldots+\alpha_p$ and γ. Now if y is replaced by y/β, $\beta>0$ and choose β so that $\emptyset_2$ is approximated to unity then the density of βY is approximated to a gamma density with the parameters $\alpha_1+\ldots+\alpha_p$ and $\gamma\beta$.

3. ANOTHER REPRESENTATION FOR THE DENSITY WITH GAMMA INPUTS

Again consider the case where $Y=X_1+\ldots+X_p$ with $X_1,\ldots,X_p$ being independent gamma variables with the densities given in (2.1). Consider the case when $\alpha_1,\ldots,\alpha_p$ are all equal to α. In this case

$$\prod_{j=1}^{p}(1-\lambda_j t)^{-\alpha_j} = |\underset{\sim}{I}-tB|^{-\alpha} \quad (3.1)$$

where $\underset{\sim}{I}$ is an identity matrix, B is a symmetric positive definite matrix with the eigen values λ_j, $j=1,\ldots,p$ and $|\underset{\sim}{I}-tB|$ denotes the determinant of $\underset{\sim}{I}-tB$. But

$$|\underset{\sim}{I}-tB|^{-\alpha} = |B|^{-\alpha}\gamma^{\alpha p}(1-\gamma t)^{-\alpha p}\ |\underset{\sim}{I}-(\underset{\sim}{I}-\gamma B^{-1})/(1-\gamma t)|^{-\alpha}$$

$$= |B|^{-\alpha}\ \gamma^{\alpha p}\ \sum_{k=0}^{\infty}\sum_{K}(\alpha)_K\ C_K(\underset{\sim}{I}-\gamma B^{-1})(1-\gamma t)^{-(k+\alpha p)}/k!, \quad (3.2)$$

for $t < 1/\gamma$, $1/\lambda_i$, $i=1,\ldots,p$ where $K=(k_1,\ldots,k_p)$, $k_1 \geq k_2 \geq \ldots \geq k_p \geq 0$, $k=k_1+\ldots+k_p$, $k_1,\ldots,k_p$ are nonnegative integers and C_K denotes a zonal polynomial of order k and $(\alpha)_K = \prod_{i=1}^{p}(\alpha-(i-1)/2)_{k_i}$. For details regarding zonal polynomials see Mathai & Saxena (1978) and the references therein. By inverting the above moment generating function one gets the density as follows.

$$g_p(y) = |B|^{-\alpha} \gamma^{\alpha p} \sum_{k=0}^{\infty} \Sigma_K (\alpha)_K C_K(\underset{\sim}{I}-\gamma B^{-1}) y^{\alpha p+k-1} e^{-y/\gamma} /$$

$$(k! \gamma^{\alpha p+k} \Gamma(\alpha p+k)) \quad , y > 0.$$

But $|B|^{-\alpha} = (\lambda_1 \dots \lambda_p)^{-\alpha}$ and $\Gamma(\alpha p+k) = (\alpha p)_k \Gamma(\alpha p)$. Hence

$$g_p(y) = ((\lambda_1 \dots \lambda_p)^{\alpha} \Gamma(\alpha p))^{-1} y^{\alpha p-1} e^{-y/\gamma} \sum_{k=0}^{\infty} \Sigma_K (\alpha)_K (y/\gamma)^k$$

$$C_K(\underset{\sim}{I}-\gamma B^{-1})/(k! (\alpha p)_k) , \quad y > 0 \tag{3.3}$$

This is another representation for the density of Y. Since the density of Y is unique, by comparing (3.3) and (2.6) we get the following

Theorem 1. $\sum_{k=0}^{\infty} \Sigma_K (\alpha)_K (y/\gamma)^k C_K(\underset{\sim}{I}-\gamma B^{-1})/(k!(\alpha p)_k)$

$$= \emptyset_2(\alpha,\dots,\alpha; \alpha p; (\gamma^{-1}-\lambda_1^{-1})y,\dots,(\gamma^{-1}-\lambda_p^{-1})y)$$

for $y > 0$, $\gamma > 0$, $\lambda_i > 0$, $i=1,\dots,p$.

Similar results can be obtained when $\alpha_1,\dots,\alpha_p$ are all integers in which case one can look upon λ_i being repeated α_i times, $i=1,\dots,p$.

If the inputs $X_1,\dots,X_p$ represent the inputs of sediments into a dam over a period of p occasions then one can compute the probability that the total sedimentation is less than a pre-assigned number δ, that is, $P\{Y \le \delta\}$ by using any one of the explicit computable representations for $g_p(y)$ given above. That is,

$$P\{Y \le \delta\} = \int_0^{\delta} g_p(y)dy$$

where term by term integration is valid.

4. AMOUNT OF ACID RAIN

Consider the situation of sulphur dioxide or other such pollutants distributed in a certain region in the atmosphere according to a 3-variate normal distribution with the centre at the point $\mu' = (\mu_1,\mu_2,\mu_3)$ where $\mu_i=E(X_i)$, $i=1,2,3$, with X_i denoting the ith co-ordinate or the ith variable with the concentration of the pollutants proportional to the probability content. Consider a cloud patch in the region in the shape of an ellipsoid with a centre and axes of symmetry of its own. The amount of acid vapour carried by the cloud is proportional to the probability content of this ellipsoid. The amount of acid rain that can be expected out of this cloud is proportional to the probability of this ellipsoid. Thus if the ellipsoid is described by $(X-\alpha)'C(X-\alpha) \leq \delta$ then the question to be answered is what is the probability that $(X-\alpha)'C(X-\alpha) \leq \delta$ for a preassigned δ where X is a 3-variate normal with mean vector μ and covariance matrix Σ where μ, α are known vectors and Σ is a known matrix? In section I we have transformed this problem in the general case of p co-ordinates to the evaluation of the probability $P\{U \leq \delta\}$ where $U = \sum_{j=1}^{p} \lambda_j (z_j + \gamma_j)^2 = \sum_{j=1}^{p} \lambda_j u_j$ where $\lambda_1, \ldots, \lambda_p$ are known constants and u_j is a noncentral chisquare with one degree of freedom and noncentrality parameter $\gamma_j^2/2$. The moment generating function of u_j is therefore

$$M_{u_j}(t) = e^{-\gamma_j^2/2} (1-2t)^{-1/2} e^{(\gamma_j^2/2)/(1-2t)}$$

The moment generating function of U is therefore

$$M_u(t) = e^{-\Phi} \prod_{j=1}^{p} (1-2\lambda_j t)^{-1/2} e^{\beta_j/(1-2\lambda_j t)} \qquad (4.1)$$

where $\beta_j = \gamma_j^2/2$, $\Phi = \sum_{j=1}^{p} \beta_j$. The inversion of (4.1) gives the density of U. This moment generating function in (4.1) or the corresponding Laplace transform of the density of U comes in a wide variety of problems connected with geometric probabilities, distributions of quadratic forms, communication problems, radar performance etc., see for example Helstrom (1978), Rice (1981), Gilliland and Hansen (1974), Ruben (1962) and the many references therein. It also comes in certain time series problems, see for example, MacNeill (1974).

If there are n independent and identical occasions of such cloud formations then the density for the total amount of acid rain will have the moment generating function

$$M_u(t) = e^{-\Phi_1} \cdot \prod_{j=1}^{p} (1-2\lambda_j t)^{-n/2} e^{\beta_j/(1-2\lambda_j t)} \qquad (4.2)$$

This expression also comes in a variety of problems involving traces of noncentral Wishart matrices, see for example Mathai (1980). In this case it is also shown in Mathai(1980) that the quantity $(U-n\ trB)/(2n\ trB^2)^{1/2}$ goes to a standard normal when n goes to infinity where B is a symmetric positive definite matrix with $\lambda_1,\ldots,\lambda_p$ being its eigen values and trB denotes the trace of B.

If there are a number of independent occasions but not all identically distributed then we will have the general situation where $M_u(t)$ will have the following form

$$M_u(t) = e^{-\Phi_2} \prod_{j=1}^{p} (1-2\lambda_j t)^{-n_j/2}\, e^{\beta_j/(1-2\lambda_j t)} \qquad (4.3)$$

where $n_1,\ldots,n_p$ are positive integers.

We will consider the inversion of (4.1) here. Other cases of (4.2) and (4.3) can be handled in the same fashion.

$$\prod_{j=1}^{p} e^{\beta_j(1-2t\lambda_j)^{-1}} = \sum_{r=0}^{\infty} \sum_{r_1+\ldots+r_p=r} \beta_1^{r_1}\ldots\beta_p^{r_p}(1-2t\lambda_1)^{-r_1}$$

$$\ldots(1-2t\lambda_p)^{-r_p}/(r_1!\ldots r_p!)$$

Hence

$$M_u(t) = e^{-\Phi} \sum_{r=0}^{\infty} \sum_{r_1+\ldots+r_p=r} \beta_1^{r_1}\ldots\beta_p^{r_p} \prod_{j=1}^{p}(1-2t\lambda_j)^{-(r_j+1/2)}$$

$$(r_1!\ldots r_p!)^{-1} \qquad (4.4)$$

Now compare the product containing t with the expression in (2.2). Hence the density of U, denoted by f(u), is given as follows.

$$f(u) = e^{-\Phi} \sum_{r=0}^{\infty} \sum_{r_1+\ldots+r_p=r} \beta_1^{r_1}\ldots\beta_p^{r_p}\, u^{r_1+\ldots+r_p+p/2-1}\, e^{-u/\gamma}$$

$$\emptyset_2(r_1+1/2,\ldots,r_p+1/2;r_1+\ldots+r_p+p/2;(\gamma^{-1}-(2\lambda_1)^{-1}u,\ldots,$$

$$(\gamma^{-1}-(2\lambda_p)^{-1}u))((r_1!\ldots r_p!)(2\lambda_1)^{r_1+1/2}\ldots(2\lambda_p)^{r_p+1/2}$$

$$\Gamma(r_1+\ldots+r_p+p/2))^{-1}, \quad u > 0$$

where $\emptyset_2$ and γ are the same quantities appearing in (2.6).

REFERENCES

Gilliland, D.C., and Hansen,E.R. 1974. A note on some series representations of the integral of a bivariate normal distribution over an offset circle. Naval Research Logistics Quarterly, 21,No.1: 207-211.

Helstrom,C.1978. Approximate evaluation of detection probabilities in radar and optical communications.IEEE Trans. Aerospace Electro.Systems.,AES 14: 630-640.

MacNeill,I.B.1974. Tests for change of parameter at unknown times and distributions of some related functionals on Brownian motion. Ann.Statist., 2: 950-962.

Mathai,A.M.1980. Moments of the trace of a noncentral Wishart matrix.,Commun.Statist.(Theor.Meth.)A9(8): 795-801.

Mathai,A.M. and Saxena,R.K.1978. The H-function with Applications in Statistics and Other Disciplines.Wiley Halsted, New York.

Rice,S.O.1981. Distribution of quadratic forms in normal random variables - Evaluation by numerical integration. SIAM J.Sci. Stat. Comput.1,No.4: 438-448.

Ruben,H.1962. Probability content of regions under spherical normal distribution IV: The distribution of homogeneous and nonhomogeneous quadratic functions of normal variables, Ann. Math.Statist., 33: 542-570.

TESTING FOR NON-LINEAR SHIFTS IN STATIONARY ϕ-MIXING PROCESSES

R. J. KULPERGER

Department of Mathematical Sciences, McMaster University, Hamilton, Ontario, Canada, L8S 4K1

INTRODUCTION

Let $\underset{\sim}{X} = \{X_n; n \geq 1\}$ and $\underset{\sim}{Y} = \{Y_n; n \geq 1\}$ be two independent stationary processes, and suppose we observe data $\{X_1, \ldots, X_m\}$ and $\{Y_1, \ldots, Y_n\}$. It is of interest to know if the processes $\underset{\sim}{X}$ and $\underset{\sim}{Y}$ differ in a specified manner. Suppose X_1 and Y_1 have continuous distribution functions (df's) F and G respectively. One type of change is a generalized shift, so that

$$F(x) = G(x + \Delta(x)). \tag{1.1}$$

$\Delta(x)$ is a shift function, and (1.1) says Y_1 and $X_1 + \Delta(X_1)$ have the same distribution. In the iid case, Doksum (1974) considered this and obtained an estimate $\Delta_N(x)$ of $\Delta(x)$, under some identifiability conditions, where $N = n + m$. Doksum showed $\sqrt{N}(\Delta_N(x) - \Delta(x))$ converged weakly to a Gaussian process. Weak convergence is essentially convergence in distribution for processes. See Billingsley (1968) for more details.

Now suppose $\underset{\sim}{X}$ and $\underset{\sim}{Y}$ are ϕ-mixing processes (see Billingsley (1968) for definitions). Mixing in general is a type of asymptotic independence. Here the $\underset{\sim}{X}$ process is mixing means that the random variables $f_1(X_1, \ldots, X_k)$ and $f_2(X_s; s \geq k + n)$ are approximately statistically independent for large n, where f_1 and f_2 are measureable functions, but otherwise arbitrary. In other words, the random variables far apart in time behave approximately independently, but random variables near by in time may be dependent. Many time series models should have

Reprinted from *Time Series Methods in Hydrosciences*, by A.H. El-Shaarawi and S.R. Esterby (Editors)

this type of property. A special type of mixing is ϕ-mixing. Kulperger (1981) showed $\sqrt{N}(\Delta_N(x) - \Delta(x))$ converges weakly to a Gaussian process $A(x)$. This result is given in section 2. In the iid case, $A(x)$ is a scaled version of a Brownian bridge; but not so in the ϕ-mixing case. In section 3, we carry out the estimation procedure on some simulated data. This requires estimating the covariance structure of $A(x)$. The covariance structure of a Brownian bridge $B(t)$ is given by $\text{cov}(B(t), B(s)) = t(1 - s)$, $0 \le t \le s \le 1$, which simplifies the calculations in the independent data case.

A possible type of application is in a time series two sample problem. If a pollutant, or some other type of intervention, is made or added into a system, by taking segments far apart in time before and after this incident, one can estimate $\Delta(x)$. This happens if, for example, a factory is built and dumps sewage or some other pollutant into the system, for example a river system. Suppose the underlying series is somehow connected to temperature. Then $\Delta(x)$ has the physical interpretation of being the reaction to this intervention. It may be that $\Delta(x)$ is near zero for low temperatures, but has a large effect for higher temperatures. Looking only for constant shifts or average changes may not show any change, when the above case may be happening. The two sample procedure presented here also avoids many parametric assumptions.

THE ESTIMATES

The identifiability condition on $\Delta(x)$ is that $\Delta(x) + x$ is non-decreasing. From (1.1) we obtain $\Delta(x) = G^{-1}(F(x)) - x$, where for a df H, $H^{-1}(x) = \inf(y: H(y) \ge x)$.

From the data, we can obtain empirical distribution functions (edf's)

$$F_m(x) = \frac{1}{m}\sum_1^m I_{(-\infty,x)}(X_j), \quad \text{and} \quad G_n(x) = \frac{1}{n}\sum_1^n I_{(-\infty,x)}(Y_j),$$

where I_A is the indicator function of the set A. These estimate F and G respectively. Thus a natural estimate of $\Delta(x)$ is

$$\Delta_N(x) = G_n^{-1}(F_m(x)) - x. \tag{2.1}$$

From Billingsley (1968), under certain conditions, as $m \to \infty$,

$$\sqrt{m}(F_m(x) - F(x)) \to W(F(x)) \quad \text{weakly,} \tag{2.2}$$

where W is a continuous Gaussian process on $[0,1]$ with $E(W(t)) = 0$ and

$$\operatorname{cov}(W(t), W(s)) = E(h_t(U_1)h_s(U_1)) + \sum_{h=2}^{\infty} E(h_t(U_1)h_s(U_k) + h_t(U_k)h_s(U_1)) \tag{2.3}$$

where $U_j = F(X_j)$ and $h_t(x) = I_{[0,t]}(x) - t$.

Similarly as $n \to \infty$

$$\sqrt{n}\,(G_n(x) - G(x)) \to V(G(x)) \quad \text{weakly,} \tag{2.4}$$

where V is a continuous Gaussian process on $[0,1]$, independent of W, and $E(V(t)) = 0$ and

$$\operatorname{cov}(V(t), V(s)) = E(h_t(Z_1)h_s(Z_1)) + \sum_{k=2}^{\infty} E(h_t(Z_1)h_s(Z_k) + h_t(Z_k)h_s(Z_1)) \tag{2.5}$$

where $Z_j = G(Y_j)$.

Theorem 2.1 (Kulperger (1981))

Suppose G has a positive density g. Then under conditions for which (2.2) and (2.4) hold, and if $\frac{m}{n} \to \lambda \in (o,1)$, then $\sqrt{N}(\Delta_N(x) - \Delta(x)) \to A(x)$ weakly, where

$$A(x) = \frac{1}{g(G^{-1}(F(x)))}\left(\frac{1}{\sqrt{1-\lambda}}\, V(F(x)) + \frac{W(F(x))}{\sqrt{\lambda}}\right).$$

In the iid case, V and W are independent Brownian bridges. In the ϕ-mixing case they are not Brownian bridges, since the covariances (2.3) and (2.5) depend on F and G respectively. In general, V and W are not even deterministically time changed Brownian bridges.

The covariance functions of V and W must be estimated, as well as g, the density function of G. Kulperger (1981 , Theorem 4.1) obtained consistent estimates of the limiting covariance functions of the V and W processes.

To illustrate these computations, as well as to give some indication of how well they perform in some sense in a specific case, we simulate one realization of such an $\underset{\sim}{X}$ and $\underset{\sim}{Y}$ process. This is described in the next section.

A SIMULATION EXAMPLE AND NUMERICAL COMPUTATIONS

For this simulation, we consider a 10-dependent process. Let $X_1^*, \ldots, X_{m+9}^*$ be iid N(0,1) r.v.'s and $X_i = \sum_{j=0}^{9} X_{j+i}^*$, $i = 1, \ldots, m$. Marginally X_i is a N(0,10) r.v.. Let $Y_1^*, \ldots, Y_{m+9}^*$ be iid N(0,1) r.v.'s $\tilde{Y}_i = \sum_{j=0}^{9} Y_{i+j}^*$ and $Y_i = \tilde{Y}_i + \Delta(\tilde{Y}_i)$, $i=1, \ldots, m$. We take $\Delta(x) = \frac{10e^x}{1+e^x}$. In particular, notice that the $\underset{\sim}{Y}$ process is not Gaussian. The iid N(0,1)'s are generated by the Box-Muller (1958) method. We take $n = m = 200$, so that $\lambda = \frac{1}{2}$.

Using Theorem 2.1, we obtain approximate or asymptotic marginal confidence intervals for $\Delta(x)$, for $x = -5, -4, \ldots, 5$. To do some we must compute consistent estimates of the corresponding Var(A(x)) terms. Suppose we wish to construct a confidence interval at x_0. Estimate $g(G^{-1}(F(x_0))$ by $g_n(G_n^{-1}(F_m(x_0)))$, where

$$g_n(x) = \frac{1}{2h_n}(G_n(x+h_n) - G_n(x-h_n)) \tag{3.1}$$

and $h_n = \beta n^{-\alpha}$. We take $\beta = 1$, $\alpha = \frac{1}{5}$, so that (3.1) is a consistent estimate of $g(x)$.

From (2.3), for each t, $\text{var}(W(t)) = 2\pi f_{t,t}^{(x)}(0)$, where $f_{t,t}^{(x)}(\gamma) = \frac{1}{2\pi} \sum_k e^{i\gamma} \text{cov}(h_t(U_1), h_t(U_k))$ is the spectral density of the time series $\{h_t(U_k)\}$ evaluated at frequency γ, and $i = \sqrt{-1}$. We now estimate this spectral density at frequency 0 from Kulperger (1981).

Let $\Gamma_m = \{\lambda_j: \;_j = \frac{2\pi j}{m}, j = 1, \ldots, [\beta\sqrt{m}]\}$, where $[\cdot]$ is the greatest integer function. Let $h_{t,m,j,X} = h_t(F_m(X_j))$. Define the estimated finite Fourier transforms for $\gamma \in \Gamma_m$ by

$$d_{t,m,X}(\gamma) = \sum_{j=1}^{m} e^{-i\gamma j} h_{t,m,j,X}.$$

Notice that $\sum_{1}^{m} e^{-i\gamma j} = 0$ for $\gamma \in \Gamma_m$. Let

$$2\pi \hat{f}^{(x)}_{t,t}(0) = \frac{1}{M_m} \sum_{j=1}^{M_m} \frac{1}{m} |d_{t,m,X}(\gamma_j)|^2 \qquad (3.2)$$

where $M_m = [\beta\sqrt{m}]$ and the sum is over $\gamma_j \in \Gamma_m$. We take $\beta = .45$. Under the assumption that $\mathrm{Law}(\{U_j\}j\geq 1)$ is absolutely continuous with respect to Lebesque measure on $[0,1]^\infty$, $2\pi\hat{f}^{(x)}_{t,t}(0) \to 2\pi f^{(x)}_{t,t}(0)$ in probability as $m \to \infty$. We then estimate $\mathrm{var}(W(F(x_0)))$ by $2\pi\hat{f}^{(x)}_{t,t}(0)$, with $t = F_n(x_0)$.

Similarly observe that $\mathrm{var}(V(t)) = 2\pi f^{(y)}_{t,t}(0)$ where $f^{(y)}_{t,t}(\gamma)$ is the spectral density of $\{h_t(G(Y_1))\}$ evaluated at frequency γ. Define the estimated finite Fourier transform

$$d_{t,n,Y}(\gamma) = \sum_{j=1}^{n} e^{-i\gamma j} h_t(G_n(Y_j))$$

where $\gamma \in \Gamma_n = \{\gamma_j : \gamma_j = \frac{2\pi j}{n},\ j = 1, \ldots, [\beta\sqrt{n}]\}$. Estimate $\mathrm{Var}(V(t))$ by $2\pi\hat{f}^{(y)}_{t,t}(0)$, the analogue of (3.2), with m replaced by n and $t = G_n(x)$. Estimate $\mathrm{Var}(A(x))$ by

$$\hat{\sigma}^2(x) = \frac{2\pi\hat{f}^{(x)}_{t,t}(0)}{\sqrt{1 - 1/2}} + \frac{2\pi\hat{f}^{(y)}_{s,s}(0)}{\sqrt{1/2}} \cdot \frac{}{g_n(G_n^{-1}(F_m(x)))},$$

and $t = F_m(x)$, $s = G_n(x)$.

Figure 1 contains the true and estimated $\Delta(\cdot)$ and $\Delta_N(\cdot)$, the solid line being $\Delta(\cdot)$. Figure 2 contains $\Delta_N(\cdot)$ with the estimated marginal 95% confidence intervals. The lengths of these confidence intervals are of the right size for these processes with these sample sizes, in the sense that they are close to the intervals we would hav

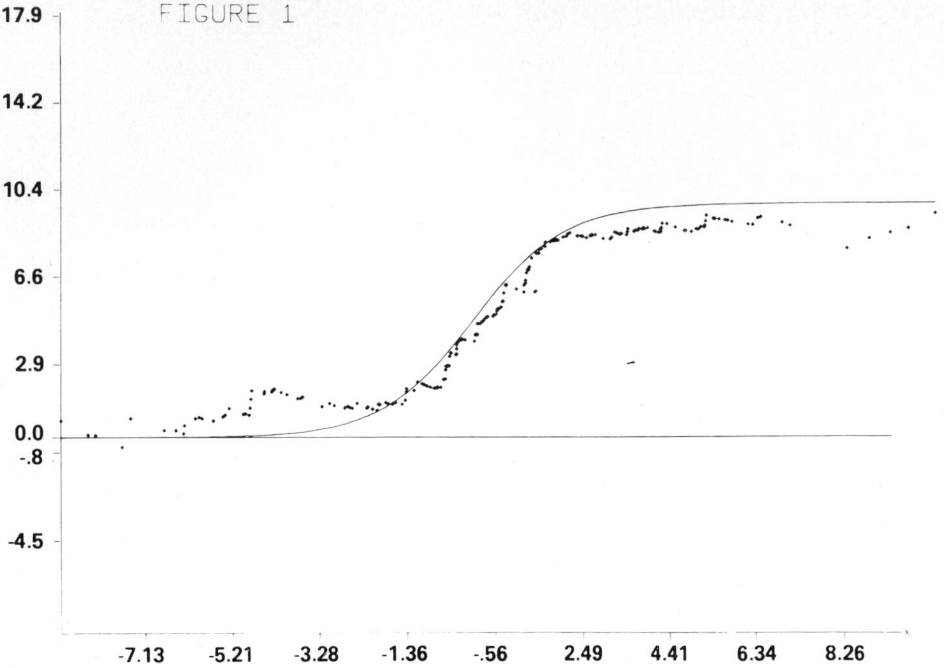
FIGURE 1
17.9
14.2
10.4
6.6
2.9
0.0
-.8
-4.5
-7.13
-5.21
-3.28
-1.36
-.56
2.49
4.41
6.34
8.26

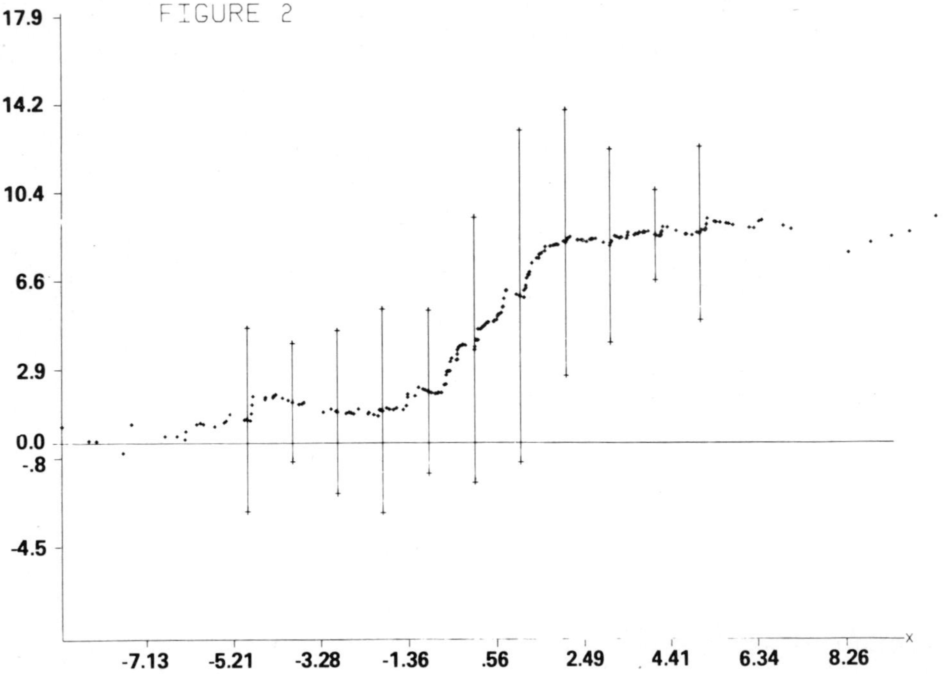
FIGURE 2
17.9
14.2
10.4
6.6
2.9
0.0
-.8
-4.5
-7.13
-5.21
-3.28
-1.36
.56
2.49
4.41
6.34
8.26

obtained if var(A(x)) were really known. For our simulation $G(x + \Delta(x)) = F(x) = \Phi(x/\sqrt{10})$, where Φ is the standard normal d.f., with density ϕ. Therefore $g(x + \Delta(x))(1 + \Delta'(x)) = \phi(\frac{x}{\sqrt{10}})/\sqrt{10}$.

Notice that $g(G^{-1}(F(x))) = g(x + \Delta(x))$.

The sum in (2.3) cannot easily be obtained in closed form, since it involved terms of the form $\int(ax)\,\phi\,(x)dx$. However observations of the X process, $X_1, \ldots, X_{11}$ allows us to obtain $U_1 = F(X_1), \ldots, U_{11} = F(X_{11})$ and hence compute the random variables

$$D(t_i) = h_{t_i}(U_1)^2 + 2\sum_{j=2}^{11} h_{t_i}(U_1)h_{t_i}(U_j).$$

where $t_i = \Phi(x_i/\sqrt{10})$, $x_i = i - 6$, $i = 1, \ldots, 11$.

Simulating many replications of this allows us to estimate (2.4) by the Law of Large Numbers.

The ratios of the lengths of the estimated to the tru asymptotic confidence intervals are thus obtained, and recorded in Table 1. These ratios are correct to one decimal place, since the Monte-Carlo estimates of Var(W(t)) and Var(V(t)) are based on sample sizes 200. The ratios seem to be quite reasonable in this example.

ACKNOWLEDGEMENTS

I wish to thank N̈. Sorokowsky for programming the simulation. I also wish to thank Jackie Collin for typing this paper. This work was partially supported by grant number A5176 from the Natural Science and Engineering Research Council of Canada.

TABLE 1

x	$\Delta(x)$	$\Delta_N(x)$	$\hat{\sigma}^2(x)/\sigma^2(x)$	Estimated 95% marginal confidence interval	
-5	.0669	1.008	1.53	-2.88,	4.90
-4	.1799	1.7914	0.72	-2.10,	5.68
-3	.4743	1.4446	1.09	-2.01,	4.90
-2	1.1920	1.5561	.88	-2.77,	5.88
-1	2.6894	2.1666	.32	-1.29,	5.62
0	5.000	4.0825	.63	-1.54,	9.71
1	7.3106	6.2555	1.35	- .79	13.30
2	8.8080	8.6029	1.77	2.97,	14.24
3	9.5257	8.4722	1.57	4.37,	12.57
4	9.8201	8.8200	.37	6.91,	10.73
5	9.9331	9.0295	1.09	5.35,	12.70

$\sigma^2(x)$ is estimated from 200 Monte-Carlo runs. The ratios $\hat{\sigma}^2(x)$ to $\sigma^2(x)$ are accurate to 1 decimal place.

REFERENCES

Billingsley, P. (1968). Convergence of Probability Measures. Wiley, New York.

Box, G. and M.E. Muller (1958). A note on the generation of random normal deviates. Ann. Math. Statist., 29, 610.

Doksum, K. (1974). Empirical probability plots and statistical inference for nonlinear models in the two sample case. Ann. Statist., 2, 267-277.

Kulperger, R. (1981). Estimating non-linear shifts in stationary ϕ-mixing processes in the two sample case. Preprint.

A ROBUST STATISTIC FOR TESTING THAT TWO AUTOCORRELATED SAMPLES COME FROM IDENTICAL POPULATIONS

M.L. TIKU

Department of Mathematical Sciences, McMaster University, Hamilton, Canada

ABSTRACT

Testing that two independent samples come from identical populations is a common statistical problem. If the random variables within the samples are iid (independently and identically distributed), Tiku (1980) gives a robust statistic (that is, a statistic whose null distribution is fairly insensitive to underlying populations) which is also remarkably powerful. In this paper, we give an analogous statistic which can be used if the random variables are moderately autocorrelated; this statistic is shown to be robust and powerful.

1. INTRODUCTION

Testing that two independent samples come from identical populations is a common statistical problem. If the underlying populations are normal and if one has simple random samples, one would naturally employ the Student's t statistic $t = (\bar{x}_1 - \bar{x}_2)/[s\sqrt{\{(1/n_1)+(1/n_2)\}}]$. This statistic, however, is not robust to most nonnormal populations prevalent in practice; see Tiku (1971) and Subrahmaniam et al. (1975), for example. A statistic (based on Tiku's modified maximum likelihood estimators), analogous to t, which is robust and is also remarkably powerful was given by Tiku (1980); see also Tiku and Singh (1981). In this paper, we propose a statistic which can be used if the random variables are moderately autocorrelated through a first-order stationary stochastic process. The proposed statistic

Reprinted from *Time Series Methods in Hydrosciences*, by A.H. El-Shaarawi and S.R. Esterby (Editors)

is shown to be robust to both symmetric and skew populations and remarkably powerful against location shifts.

2. THE TEST STATISTIC

Let

$$y_{1,i} = \mu_1 + u_{1,i}, \quad i = 1,2,\ldots,n_1+1$$

$$\text{and } y_{2,i} = \mu_2 + u_{2,i}, \quad i = 1,2,\ldots,n_2+1, \tag{1}$$

$$\text{where } u_{1,t} = \emptyset_1 u_{1,t-1} + e_{1t},$$

$$\text{and } u_{2,t} = \emptyset_2 u_{2,t-1} + e_{2t}. \tag{2}$$

We assume that e_{1t} and e_{2t} are iid with mean $E(e) = 0$ and variance $V(e)=\sigma^2$. Note that $V(y_{1,i}) = \sigma^2/(1-\emptyset_1^2)$ and $V(y_{2,i}) = \sigma^2/(1-\emptyset_2^2)$; $0 \leqslant |\emptyset_i| < 1$, $i = 1, 2$. One wants to test the null hypothesis,

$$H_0: \mu_1 = \mu_2 \text{ and } \emptyset_1 = \emptyset_2, \tag{3}$$

that is, the samples $y_{1,i}$, $i = 1, 2,\ldots,n_1+1$, and $y_{2,i}$, $i = 1,2,\ldots, n_2+1$, come from identical populations. Write $n = \min(n_1,n_2)$ and define

$$z_i = y_{1,i} - y_{2,i}, \quad i = 1,2,\ldots,n+1. \tag{4}$$

Under H_0, $z_i = \emptyset(u_{1,i-1} - u_{2,i-1}) + e_{1i} - e_{2i}$

$$= \emptyset z_{i-1} + e_i, \quad i = 2,3,\ldots,n+1; \tag{5}$$

Let $\hat{\emptyset}$ be the least squares estimator of $\emptyset$; then

$$\hat{\emptyset} = \sum_{i=2}^{n+1} z_{i-1} z_i / \sum_{i=2}^{n+1} z_{i-1}^2 \tag{6}$$

which is consistent and asymptotically unbiased (Chatfield, 1975; Christopeit and Helmes, 1980). Define

$$x_{1i} = y_{1,i+1} - \hat{\emptyset} y_{1,i}, \quad i = 1,2,\ldots,n_1$$

$$\text{and } x_{2i} = y_{2,i+1} - \hat{\emptyset} y_{2,i}, \quad i = 1,2,\ldots,n_2. \tag{7}$$

To test H_0, the proposed statistic is ($A_1 = n_1-2r_1$, $A_2 = n_2-2r_2$)

$$T = (\hat{\mu}_1-\hat{\mu}_2)/\hat{\sigma}\sqrt{\{(1/m_1) + (1/m_2)\}}, \quad (8)$$

where $m_1 = n_1-2r_1+2r_1\beta_1$ and $m_2 = n_2-2r_2+2r_2\beta_2$ and $\hat{\sigma}^2 = \{(A_1-1)\hat{\sigma}_1^2 + (A_2-1)\hat{\sigma}_2^2\}/(A_1+A_2-2)$; $\hat{\mu}_1, \hat{\mu}_2, \hat{\sigma}_1$ and $\hat{\sigma}_2$ are Tiku's (1967,1978,1980) MML (modified maximum likelihood estimators; these estimators are calculated from equations (A.1) to (A.4) given in the appendix with x_i's replaced by x_{1i}'s and x_{2i}'s and r replaced by $r_1 = [0.5 + 0.1n_1]$ and $r_2 = [0.5 + 0.1n_2]$, respectively.

Theorem: If the underlying populations are normal, the asymptotic (A_1 and A_2 both tend to infinity) null distribution of T is normal N(0,1).

Proof: Since n is larger than min (A_1, A_2), therefore, as A_1 and A_2 tend to infinity, n also tends to infinity, in which case $\hat{\emptyset}$ converges to $\emptyset$; consequently, x_{1i}, i = 1,2,...,n_1, and x_{2i}, i = 1,2,...,n_2, are iid normal $N(\mu,\sigma)$ under H_0 and the theorem follows immediately from the fact that $E(\hat{\sigma}_1) = E(\hat{\sigma}_2) = \sigma$ and $\sqrt{m_1}\ (\hat{\mu}_1-\mu)/\sigma$ and $\sqrt{m_2}\ (\hat{\mu}_2-\mu_2)/\sigma$ are independently distributed as normal N(0,1); see Tiku (1978, Lemmas 1 and 2; 1981).

Even if the underlying populations are non-normal but with existent means and variances, $\hat{\emptyset}$ converges to $\emptyset$ as n tends to infinity and, consequently, x_{1i} and x_{2i} are iid. In such situations, the estimators $\hat{\sigma}_1$ and $\hat{\sigma}_2$ converge (Tiku, 1980, p.134) to their expected values $E(\hat{\sigma}_1) = E(\hat{\sigma}_2) = k\sigma$ (under H_0), and since $\hat{\mu}_1$ and $\hat{\mu}_2$ are linear functions of order statistics they are asymptotically normally distributed under some very general regularity conditions; see Stigler (1974). For most symmetric populations (the family of Student's t distributions with degree of freedom greater than 2, for example), Tiku (1980, p.134) showed that $V(\sqrt{m}\ \hat{\mu}/k\sigma) \approx 1$ for large samples; the asymptotic null distribution of T for such symmetric non-normal populations is, therefore, also approximately Normal N(0,1).

TABLE 1

Simulated Values of the Probability $P(T>h|H_0)$:

	$n_1=n_2=10$					$n_1=20,\ n_2=10$					$n_1=n_2=20$				
	ϕ					ϕ					ϕ				
δ	-0.5	0.0	0.1	0.5	0.7	-0.5	0.0	0.1	0.5	0.7	-0.5	0.0	0.1	0.5	0.7
								Normal							
5	.054	.045	.042	.035	.031	.049	.042	.052	.052	.050	.050	.047	.043	.044	.029
1	.012	.009	.007	.005	.004	.009	.008	.010	.013	.010	.008	.008	.008	.004	.002
								Double-Exponential							
5	.051	.046	.048	.038	.040	.050	.054	.050	.058	.063	.057	.056	.053	.050	.039
1	.011	.007	.009	.006	.008	.012	.012	.011	.013	.018	.010	.014	.011	.009	.004
								Student's t, df 4							
5	.053	.044	.047	.037	.034	.046	.046	.049	.046	.055	.047	.051	.052	.040	.039
1	.014	.007	.008	.005	.005	.010	.009	.012	.012	.016	.010	.006	.008	.005	.005
								Student's t, df 3							
5	.050	.047	.045	.040	.036	.050	.053	.053	.053	.062	.050	.052	.058	.041	.043
1	.012	.008	.008	.006	.006	.010	.011	.011	.012	.014	.012	.009	.009	.007	.005
								Student's* t, df 2							
5	.053	.047	.044	.040	.042	.057	.053	.055	.060	.059	.059	.057	.054	.053	.046
1	.012	.008	.008	.009	.010	.010	.011	.010	.015	.019	.012	.009	.011	.009	.007
								0.90N(0,1) + 0.10N(0,3)							
5	.053	.045	.042	.035	.036	.050	.049	.051	.052	.050	.048	.045	.047	.046	.032
1	.012	.008	.008	[illegible]	.007	.011	.010	.013	.013	.013	.012	.009	.006	.006	.005

TABLE 1 (CONTINUED)

δ	-0.5	0.0	0.1	0.5	0.7	-0.5	0.0	0.1	0.5	0.7	-0.5	0.0	0.1	0.5	0.7
								(n-1)N(0,1) & 1N(0,3)**							
5	.050	.047	.044	.034	.033						.051	.046	.047	.046	.032
1	.013	.010	.009	.006	.005	—	—	—	—	—	.009	.008	.007	.006	.002
								χ_6^2-6							
5	.049	.041	.041	.034	.026	.038	.041	.042	.043	.048	.051	.046	.047	.036	.032
1	.011	.007	.008	.005	.003	.009	.007	.005	.009	.012	.010	.010	.007	.003	.003
								χ_4^2-4							
5	.046	.042	.038	.033	.031	.038	.036	.039	.037	.046	.046	.049	.053	.034	.034
1	.010	.007	.007	.005	.004	.007	.008	.007	.006	.009	.009	.008	.006	.003	.003
								χ_2^2-2							
5	.047	.048	.041	.034	.030	.035	.037	.036	.038	.050	.050	.051	.049	.044	.033
1	.011	.008	.006	.006	.004	.005	.006	.005	.005	.010	.009	.009	.009	.005	.005

* This distribution has infinite variance but finite mean

** One observation chosen at random, say x_i, is replaced by $3x_i$.

Expecting the small sample null distribution of T to be approximately Student's t with $A_1+A_2 - 3$ degrees of freedom, we simulated the mean, variance, skewness $\beta_1^* = \mu_3^2/\mu_2^3$ and kurtosis $\beta_2^* = \mu_4/\mu_2^2$ of T for numerous underlying populations (both symmetric and skew) and, to our surprise, found them very close to 0, 1, 0 and 3, respectively, rather than close to the corresponding values of the Student's t distribution, within the range $-1.0 < \emptyset \leq 0.5$, indicating approximate normality of T; for $\emptyset > 0.5$, the null distribution of T turned out to be intractable with values of the variance and β_2^* much different than 1 and 3, respectively (see also Ljung and Box, 1980). However, it is the range $-0.5 \leq \emptyset \leq 0.5$ which is of considerable practical interest (Pierce, 1971) and, here, we found the normal approximation adequate. For example, we simulated the probabilities $P(T \leq h|H_0)$ for numerous symmetric and skew populations, h being the upper 100 $(1-\delta)$% point of the normal N(0,1) distribution; the simulated values (based on $80000/n_1$ Monte Carlo runs) of these probabilities are given in Table 1. It is clear that the null distribution of T is robust to parent populations, and is closely approximated by the normal N(0,1)distribution in the range $-0.5 \leq \emptyset \leq 0.5$.

At this point, one would perhaps like to know about the robustness properties of the analogous Student's t statistic based on the sample means and variances, that is, the statistic T with $r_1 = r_2 = 0$. Given below are the simulated values (based on 8000 runs) of the probability $P(t \leq h|H_0)$; $n_1 = n_2 = 10$:

	$\emptyset$					$\emptyset$				
	-0.5	0.0	0.1	0.5	0.7	-0.5	0.0	0.1	0.5	0.7
	Normal					Dexponential				
5	.048	.036	.034	.024	.017	.041	.035	.033	.023	.022
1	.008	.006	.003	.002	.002	.007	.004	.004	.002	.002
	t_2					χ_2^2-2				
5	.040	.033	.030	.020	.016	.041	.034	.032	.023	.020
1	.005	.003	.003	.002	.001	.007	.003	.003	.002	.001

The results for $(n_1,n_2) = (20,10)$ and $(20,20)$ are similar but we do not reproduce them for conciseness. It is clear that the null distribution of t is very sensitive to changes in $\emptyset$.

3. POWER PROPERTIES

The alternatives to H_0 that are of considerable practical interest are $(E(x_{1i}) = \mu_1^*$ and $E(x_{2i}) = \mu_2^*)$ H_1: $d=\mu_1^* - \mu_2^* > 0$; (9)

large values of T lead to the rejection of H_0 in favour of H_1. Note that $E(y_{i+1} - \emptyset y_i) = \mu(1 - \emptyset)$.

Theorem: If the underlying populations are normal, the asymptotic power function of T is given by (assuming that $\hat{\emptyset}$ converges to $\emptyset$ as n tends to infinity)

$$P[z \geqslant h-d/\sigma\sqrt{\{(1/m_1) + (1/m_2)\}}]; \qquad (10)$$

z being a standard normal variate.

Proof: Follows exactly on the same lines as the previous theorem.

The corresponding power function for the Student's t statistic mentioned above is given by

$$P[z \geqslant h-d/\sigma\sqrt{\{(1/n_1) + (1/n_2)\}}]. \qquad (11)$$

If the underlying populations are normal, the asymptotic values of the power of the statistics T and t can be calculated from equations (10) and (11) and it is seen that T is only slightly less powerful than t; see also the following values of the power.

For small samples, we simulated (from 2000 runs) the values of the power of T and t for $(n_1,n_2) = (10,10)$, $(20,10)$ and $(20,20)$ and for numerous populations and found the statistic T considerably more powerful on the whole. For example, we have the following values of the power; $n_1 = n_2 = 10$, $\emptyset = -0.5$, and the significance level for both T and t is approximately 1%:

Distribution		d 0.5	1.0	1.5	2.0	2.5	3.0
Normal	T	.13	.47	.84	.98	1.00	1.00
	t	.13	.48	.85	.99	1.00	1.00
Dexponential	T	.10	.32	.64	.85	.95	.99
	t	.07	.27	.55	.80	.92	.98
t_2	T	.08	.25	.49	.71	.85	.92
	t	.05	.16	.35	.52	.67	.77
(n-1)N(0,1) &	T	.11	.40	.73	.92	.98	1.00
1N(0,3)	t	.10	.36	.64	.85	.95	.99
χ_2^2-2	T	.19	.61	.89	.97	.99	1.00
	t	.16	.53	.84	.96	.99	1.00

Note that the above values of the power were obtained by adding the constant d to the observations x_{1i}, $i = 1,2,\ldots,n_1$.

ACKNOWLEDGEMENT

Thanks are due to NSERC of Canada and McMaster University Research Board for research grants. Thanks are also due to Mrs. Carmela Civitareale for typing the manuscript.

REFERENCES

Chatfield, D. (1975). The Analysis of Time Series: Theory and Practice. Chapman and Hall, London.

Christopeit, N. and Helmes, K. (1980). Strong consistency of least squares estimators in linear regression models. Ann. Statist. 8, 778-88.

Ljung, G.M. and Box, G.E.P. (1980). Analysis of variance with auto-correlated observations. Scandinavian J. Statist. 7, 172-80.

Pierce, D.A. (1971). Least squares estimation in the regression model with autoregressive-moving average errors. Biometrika 58, 229-312.

Stigler, S.M. (1974). Linear functions of order statistics with smooth weight functions. Ann. Statist. 2, 676-99.

Subrahmaniam, K., Subrahmaniam Kathleen and Messori, J.Y. (1975). On the robustness of some tests of significance in sampling from a compound normal population. J. Amer. Statist. Assoc. 70, 435-38.

Tiku, M.L. (1967). Estimating the mean and standard deviation from censored normal samples. Biometrika 54, 155-65.

Tiku, M.L. (1970). Monte Carlo study of some simple estimators in censored normal samples. Biometrika 57, 207-11.
Tiku, M.L. (1971). Student's t distribution under non-normal situations. Aust. J. Statist. 13, 142-48.
Tiku, M.L. (1978). Linear regression model with censored observations. Commun. Statist. A7(13), 1219-32.
Tiku, M.L. (1980). Robustness of MML estimators based on censored samples and robust test statistics. J. Statistical Planning and Inference 4(2), 123-43.
Tiku, M.L. (1981). Testing linear contrasts of means in experimental design without assuming normality and homogeneity of variances. Invited Paper: Presented at the March 22-26, 1981, Biometric Colloquium of the GDR-Region of the Biometric Society (paper to appear in Biometrical Journal).
Tiku, M.L. and Stewart, D. (1977). Estimating and testing group effects from Type 1 censored normal samples in experimental design. Commun. Statist. A6 (15), 1485-1501.
Tiku, M.L. and Singh, M. (1981). Robust test for means when population variances are unequal. Commun. Statist. A10(20), to appear

APPENDIX

Let

$$X_{r+1},\ X_{r+2},\ldots,X_{n-r} \tag{A.1}$$

be a Type 11 censored sample, obtained by arranging n random observations $x_1, x_2,\ldots,x_n$ in ascending order of magnitude and censoring the r smallest and r largest observations; Tiku's (1967, 1978, 1980) MML estimators (defined formally by Tiku and Stewart, 1977) of the location parameter μ and scale parameter σ are given by

$$\hat{\mu} = \{\sum_{i=r+1}^{n-r} X_i + r\beta(X_{r+1} + X_{n-r})\}/m \tag{A.2}$$

and $\hat{\sigma} = \{B + \sqrt{(B^2 + 4AC)}\}/2\sqrt{\{A(A-1)\}}$, (A.3)

where $m = n-2r+2r\beta$, $A = n-2r$, $B = r\alpha(X_{n-r} - X_{r+1})$, and

$$C = \sum_{i=r+1}^{n-r} X_i^2 + r\beta(X_{r+1}^2 + X_{n-r}^2) - m\hat{\mu}^2;$$

the coefficients α and β are given by Tiku (1967, eq.6). For $n \geqslant 10$, however, α and β are obtained from the following simpler equations (Tiku, 1970):

$$\beta = -f(t)\{t-f(t)/q\}/q \text{ and } \alpha = \{f(t)/q\} - \beta t; \qquad (A.4)$$

$q = r/n$, and t is determined by the equation $F(t) = \int_{-\infty}^{t} f(z)dz = 1 - q$ and $f(z) = \{1/\surd(2\pi)\} \exp(-z^2/2)$, $-\infty \leqslant z \leqslant \infty$. Note that $0 < \alpha < 1$ and $0 < \beta < 1$. In the absence of any knowledge about the underlying population, r is always chosen to be the integer value $r = [0.5+0.1n]$; see Tiku (1980) and Tiku and Singh (1981).

The efficiencies of the estimators $\hat{\mu}$ and $\hat{\sigma}$ are investigated by Tiku (1980); $\hat{\mu}$ and $\hat{\sigma}$ turn out to be remarkably efficient (jointly).

Note that for $r = 0$, $\hat{\mu}$ and $\hat{\sigma}^2$ reduce to the sample mean and variance $\bar{x}$ and s^2.

INFERENCE ABOUT THE POINT OF CHANGE IN A REGRESSION MODEL WITH A STATIONARY ERROR PROCESS

A.H. EL-SHAARAWI and S.R. ESTERBY
National Water Research Institute, Burlington, Ontario

ABSTRACT

Inference about the point of change in a regression model with an autoregressive error process of order one is considered. Expressions for the exact and for an approximate joint maximum likelihood function for the point of change n_1 and the autocorrelation coefficient ρ are derived, and iterative procedures for the estimation of n_1 and ρ are given. The two likelihood functions are shown to give approximately the same results when the number of observations is large. For the number of observations odd, a conditional procedure is used to eliminate the dependency between the random variables, and thus, to permit the derivation of a likelihood function for n_1 alone. The performance of the above procedures and the likelihood function obtained assuming no autocorrelation are compared in a Monte Carlo study, where the samples generated mimic a change in mean level of the volume of discharge of a river.

1. INTRODUCTION

The study of the characteristics of a body of water over a period of time frequently involves the question as to whether a characteristic has changed at some point. Although such a situation can occur in many ways, an example is the abrupt change in industrial discharge during the monitoring of a river for various water quality parameters. There exists a number of different statistical procedures designed to detect or estimate a point of change under various situations. For example, Esterby and El-Shaarawi (1981b) obtain likelihood functions to estimate

Reprinted from *Time Series Methods in Hydrosciences,* by A.H. El-Shaarawi and S.R. Esterby (Editors)

the point of change, Hinkley (1970, 1971) derives the distribution of the maximum likelihood estimate, Pettitt (1979) presents a non-parametric approach to detection and Brown et al. (1975) use cumulative sums and recursive residuals. Examples of application to river data, in particular, discharge data of the Nile River, are the use of cumulative sums (Kraus, 1956), an approximation to the conditional distribution of the maximum likelihood estimator (Cobb, 1978) and the relative marginal likelihood function (Esterby and El-Shaarawi, 1981a). All of the above techniques assume that the observations are independent.

In the present paper, likelihood functions are obtained for the estimation of the point of change in a regression model with an autoregressive error process of order one. The performance of a likelihood function assuming independence is also considered. The likelihood functions are derived in Section 2 and they are compared in a Monte Carlo study in Section 3.

Consider the model

$$y_t = \sum_{j=1}^{p} a_{tj}\,\theta_j + u_t \qquad (t = 1,2,\ldots,n_1) \qquad (1.1)$$

$$y_t = \sum_{j=1}^{q} b_{tj}\,\beta_j + u_t \qquad (t = n_1+1, n_1+2, \ldots, n=n_1+n_2)$$

where $u_1,u_2,\ldots,u_n$ are n successive random variables from a stationary process. The coefficients $\{a_{tj}\}$ and $\{b_{tj}\}$ are known but the parameters $\{\theta_j\}$, $\{\beta_j\}$ and the change point parameter n_1, where $p < n_1$ and $p+q < n$, are unkown. Assume that $u_1,u_2,\ldots,u_n$ is a sample from an autoregressive sequence of normal variables, i.e.,

$$u_t = \rho\, u_{t-1} + e_t \qquad (t = \ldots,-2,-1,0,1,2,\ldots)$$

where $|\rho| < 1$ and the increments $\{e_t\}$ are normally and independently distributed with mean 0 and variance σ^2.

2. LIKELIHOOD FUNCTIONS

2.1 Joint likelihood functions for n_1 and ρ

The joint density of $u_0, u_1, \ldots, u_n$ is

$$f(u_0, u_1, \ldots, u_n) = f(u_0) \prod_{t=1}^{n} f(u_t / u_{t-1}) \tag{2.1}$$

$$= f(u_0)\,(2\Pi\sigma^2)^{-n/2} \exp\{-\tfrac{1}{2}\sigma^2 \sum_{t=1}^{n} (u_t - \rho u_{t-1})^2\}$$

where the initial value u_0 is an unobservable random variable. Three main possibilities for $f(u_0)$ (Cox and Hinkley, 1974) are to assume that u_0 is a known constant, $u_0 = u_n$, or $f(u_0)$ has the form which will make $u_0, u_1, \ldots, u_n$ stationary. The assumption that $u_0 = u_n$ will not be considered here.

Assuming $u_0, u_1, \ldots, u_n$ stationary leads to $f(u_0) \sim N(0, \frac{\sigma^2}{1-\rho^2})$.

Putting this into expression (2.1) and integrating over u_0 gives

$$f(u_1, u_2, \ldots, u_n) = (2\Pi\sigma^2)^{-n/2}(1-\rho^2)^{\frac{1}{2}} \exp\{-\tfrac{1}{2}\sigma^2(u_1^2(1-\rho^2) + \sum_{t=2}^{n} (u_t - \rho u_{t-1})^2)\}. \tag{2.2}$$

The joint likelihood function for $\{\theta_j\}$, $\{\beta_j\}$, σ^2, ρ and n_1 is obtained by replacing u_t by $y_t - \sum_{j=1}^{p} a_{tj}\theta_j$ for $t=1,2,\ldots n_1$ and by $y_t - \sum_{j=1}^{q} b_{tj}\beta_j$ for $t=n_1+1, n_1+2, \ldots, n$.

The assumption that $u_0=0$ can be considered an approximation to the case where $u_0 \sim N(0, \frac{\sigma^2}{1-\rho^2})$ and it is used here because it provides a simpler method for calculating the joint likelihood function for n_1 and ρ. For $u_0=0$, the joint density of $u_1, u_2, \ldots, u_n$ is

$$f(u_1, u_2, \ldots, u_n) = (2\Pi\sigma^2)^{-n/2} \exp\{-\tfrac{1}{2}\sigma^2(u_1^2 + \sum_{t=2}^{n}(u_t - \rho u_{t-1})^2)\}. \tag{2.3}$$

Again the joint likelihood function for $\{\theta_j\}$, $\{\beta_j\}$, σ^2, ρ and n_1 is obtained by replacing u_t by $y_t - \sum_{j=1}^{p} a_{tj}\theta_j$ for $t=1,2,\ldots,n_1$

and by $y_t - \sum_{j=1}^{q} b_{tj}\beta_j$ for $t=n_1+1, n_1+2, \ldots, n$.

Under either the assumption that $u_0=0$ or $u_0 \sim N(0, \frac{\sigma^2}{1-\rho^2})$, given n_1 and ρ, the maximum likelihood estimates of $\{\theta_j\}$ and $\{\beta_j\}$, denoted by $\{\hat{\theta}_j\}$ and $\{\hat{\beta}_j\}$, can be obtained by the method of least squares. Then $\hat{u}_t = y_t - \sum_{j=1}^{p} a_{tj}\hat{\theta}_j$ for $t=1,2,\ldots,n_1$ and $\hat{u}_t = y_t - \sum_{j=1}^{q} b_{tj}\hat{\beta}_j$ for $t=n_1+1, n_1+2, \ldots, n$, and the maximum likelihood estimate of σ^2 is $\hat{\sigma}^2 = \sum_{t=1}^{n}(\hat{u}_t - \rho\hat{u}_{t-1})^2/n$. The joint maximum likelihood function for n_1 and ρ is obtained from the joint likelihood function of $\{\theta_j\}$, $\{\beta_j\}$, σ^2, ρ and n_1 by replacing $\{\theta_j\}$, $\{\beta_j\}$ and σ^2 by their maximum likelihood estimates, or equivalently by replacing u_t and σ^2 by $\hat{u}_t$ and $\hat{\sigma}^2$ in $f(u_1, u_2, \ldots, u_n)$. The maximum likelihood estimate of ρ is the solution to $\partial \log f/\partial\rho = 0$. For $u_0 = 0$, this is

$$\hat{\rho} = \sum_{t=1}^{n} \hat{u}_{t-1}\hat{u}_t \Big/ \sum_{t=1}^{n} \hat{u}_{t-1}^2, \tag{2.4}$$

while for $u_0 \sim N(0, \sigma^2/1-\rho^2)$, $\hat{\rho}$ is the root of the cubic equation

$$\left(\sum_{t=2}^{n-1} \hat{u}_t^2\right)\rho^3 - \left(\sum_{t=2}^{n} \hat{u}_{t-1}\hat{u}_t\right)\rho^2 - \left(\sum_{t=2}^{n-1} \hat{u}_t^2 + \hat{\sigma}^2\right)\rho + \sum_{t=2}^{n} \hat{u}_{t-1}\hat{u}_t = 0 \tag{2.5}$$

for which $|\rho| < 1$.

Denote the maximum likelihood estimate of σ^2, calculated for particular values of n_1 and ρ, by $\hat{\sigma}^2(n_1, \rho)$. The joint maximum likelihood function for n_1 and ρ, assuming $u_0 \sim N(0, \sigma^2/1-\rho^2)$, is

$$L_e(n_1, \rho) = \{\hat{\sigma}^2(n_1, \rho)\}^{-n/2}(1-\rho^2)^{\frac{1}{2}} \tag{2.6}$$

and assuming $u_0 = 0$, is

$$L_a(n_1, \rho) = \{\hat{\sigma}^2(n_1, \rho)\}^{-n/2} \tag{2.7}$$

with the relative maximum likelihood function given by

$R(n_1, \rho) = L(n_1, \rho)/L(\hat{n}_1, \hat{\rho})$

and in each case $\hat{\sigma}^2(n_1,\rho)$ is calculated according to the appropriate procedure described above.

2.2 Maximum likelihood function for n_1

A conditional procedure used to eliminate the dependence between the random variables (Plackett, 1960) can be used with model (1.1). For the number of observations odd, n=2k+1, the conditional distribution of $\underline{u}_2 = (u_2,u_4,\dots,u_{2k})$ given $\underline{u}_1 = (u_1,u_3,\dots u_{2k+1})$ is multivariate normal with mean, variance and covariances

$$E(u_{2i}) = \rho(1+\rho^2)^{-1}(u_{2i-1} + u_{2i+1})$$

$$\text{var}\,(u_{2i}) = \sigma^2(1-\rho^2)/(1+\rho^2) \qquad (2.8)$$

and cov $(u_{2i},u_{2s}) = 0$ for $i \neq s$, where $i = 1,2,\dots,k$.

Replacing u_t by $y_t - \sum_{j=1}^{p} a_{tj}\theta_j$ for $t = 1,2,\dots,n_1$ and by $y_t - \sum_{j=1}^{q} b_{tj}\,\beta_j$ for $t = n_1+1,n_1+2,\dots,n$ and writing λ for $\rho/(1+\rho^2)$, the mean becomes, for even values of n_1,

$$E(y_{2t}) = \sum_{j=1}^{p} a_{2t,j}\theta_j + \lambda(y_{2t-1} + y_{2t+1}) - \lambda\sum_{j=1}^{p}(a_{2t-1,j}+a_{2t+1,j})\theta_j,$$

$$\text{for } t = 1,2,\dots,\frac{n_1}{2} - 1, \qquad (2.9)$$

$$E(y_{n_1}) = \sum_{j=1}^{p} a_{n_1,j}\theta_j + \lambda(y_{n_1-1}+ y_{n_1+1}) - \lambda(\sum_{j=1}^{p} a_{n_1-1,j}\theta_j + \sum_{j=1}^{q} b_{n_1+1,j}\beta_j)$$

and

$$E(y_{2t}) = \sum_{j=1}^{q} b_{2t,j}\beta_j + \lambda(y_{2t-1}+y_{2t+1}) - \lambda\sum_{j=1}^{q}(b_{2t-1,j} + b_{2t+1,j})\beta_j,$$

for $t = \frac{n_1}{2} + 1,\dots,k$.

For odd values of n_1

$$E(y_{2t}) = \sum_{j=1}^{p} a_{2t,j}\theta_j + \lambda(y_{2t-1}+y_{2t+1}) - \lambda\sum_{j=1}^{p}(a_{2t-1,j} + a_{2t+1,j})\theta_j,$$

for $t = 1,2,\ldots,\frac{n_1-1}{2}$,

$$E(y_{n_1+1}) = \sum_{j=1}^{q} b_{n_1+1,j}\beta_j + \lambda(y_{n_1}+y_{n_1+2}) - \lambda(\sum_{j=1}^{p} a_{n_1,j}\theta_j + \sum_{j=1}^{q} b_{n_1,j}\beta_j)$$

and (2.10)

$$E(y_{2t}) = \sum_{j=1}^{p} b_{2t,j}\beta_j + \lambda(y_{2t-1}+y_{2t+1}) - \lambda\sum_{j=1}^{q}(b_{2t-1,j}+b_{2t+1,j})\beta_j$$

for $t = \frac{n_1+3}{2},\ldots,n$, with var $(y_{2t}) = \sigma^2(1-\rho^2)/(1+\rho^2)$. By writing Ψ_{1j} for $\lambda\theta_j$ $(j=1,2,\ldots,p)$ and Ψ_{2j} for $\lambda\beta_j$ $(j=1,2,\ldots q)$, then (2.9) and (2.10) are in the form of the usual linear model when n_1 is known. Let $\{\hat{\theta}_j\}$, $\{\hat{\Psi}_{1j}\}$, $\{\hat{\Psi}_{2j}\}$, $\{\hat{\beta}_j\}$, $\hat{\lambda}$ and $\hat{\sigma}^2$ be the usual least squares estimates for the parameters of the model. Substituting these estimates in the conditional distribution of $\underline{u}_2$ given $\underline{u}_1$ produces a likelihood function for n_1 alone. Inferences about n_1 can then be made by examining the relative maximum likelihood function which is

$$R(n_1) = f_{n_1}(u_2,u_4,\ldots,u_{2k}|u_1,u_3,\ldots,u_{2k+1}) / f_{\hat{n}_1}(u_2,u_4,\ldots,u_{2k}|u_1,u_3,\ldots,u_{2k+1})$$

$$= \{\hat{\sigma}^2(\hat{n}_1)/\hat{\sigma}^2(n_1)\}^{k/2}$$

$$= \{\hat{\sigma}^2(n_1)/\hat{\sigma}^2(\hat{n}_1)\}^{\frac{-n-1}{4}} \quad (2.11)$$

where $f_{n_1}(u_2,\ldots,u_{2k}|u_1,\ldots,u_{2k+1})$ is the conditional distribution of $(u_2,\ldots,u_{2k})$ given $(u_1,\ldots,u_{2k+1})$ with the parameters $\{\theta_j\}$, $\{\Psi_{1j}\}$, $\{\beta_j\}$, $\{\Psi_{2j}\}$, σ^2, and λ replaced by their estimates.

2.3 Maximum likelihood function for n_1 assuming $\rho=0$

Under the assumption of independence, substitution of the maximum likelihood estimates of $\{\theta_j\}$, $\{\beta_j\}$ and σ^2 into the joint likelihood function for $\{\theta_j\}$, $\{\beta_j\}$, σ^2 and n_1 gives the maximum likelihood function for n_1,

$$L_m(n_1) = (\hat{\sigma}^2)^{-n/2}$$

where $\hat{\sigma}^2 = \sum_{t=1}^{n_1} (y_t - \sum_{j=1}^{p} a_{tj}\hat{\theta}_j)^2 + \sum_{t=n_1+1}^{n} (y_t - \sum_{j=1}^{q} b_{tj}\hat{\beta}_j)^2 /n.$

Under the assumption of $\rho=0$, the marginal or conditional likelihood function might be preferred (Esterby and El-Shaarawi, 1981b). However, since maximum likelihood functions have been used above, the maximum likelihood function will be used here to make results more comparable.

2.4 Computational considerations

In the joint likelihood functions for n_1 and ρ, the maximum likelihood estimates of $\{\theta_j\}$, $\{\beta_j\}$ and σ^2 are obtained using the method of least squares when ρ and n_1 are given. This involves minimization of the term in the exponent in both (2.2) and (2.3). Write model (1.1) in the form

$$y_t = \underline{a}'_t \, \underline{\alpha} + u_t$$

where $\underline{a}'_t = (a_{t1}, a_{t2}, \ldots, a_{tp}, b_{t1}, b_{t2}, \ldots, b_{tq})$

and $\underline{\alpha}' = (\theta_1, \theta_2, \ldots, \theta_p, \beta_1, \beta_2, \ldots, \beta_q)$. Thus $u_t - \rho u_{t-1}$ can be written as

$(y_t - \underline{a}'_t\underline{\alpha}) - \rho(y_{t-1} - \underline{a}'_{t-1}\underline{\alpha}) = (y_t - \rho y_{t-1}) - (\underline{a}'_t - \rho\underline{a}'_{t-1})\underline{\alpha}$. Let $z_t = y_t - \rho y_{t-1}$ and $\underline{x}'_t = \underline{a}'_t - \rho \underline{a}'_{t-1}$ and thus the maximum likelihood estimates $\hat{\underline{\alpha}}$ are obtained as $\hat{\underline{\alpha}} = (X'X)^{-1}X'\underline{z}$, where $\underline{z}' = (z_1, z_2, \ldots, z_n)$ and X is the matrix with the tth row equal to $\underline{x}'_t$.

The joint likelihood functions for n_1 and ρ can be obtained by an iterative procedure.

1. Assume an initial value for ρ.
2. For the particular value of ρ, form $\underline{z}$.
3. Let $n_1 = 1, 2, \ldots, n-1$ and for each value of n_1, form X, obtain $\hat{\underline{\alpha}}$ and $\hat{\sigma}^2 = (\underline{z}'\underline{z} - \hat{\underline{\alpha}}\, X'\underline{z})/n$.
4. Choose the maximum likelihood estimate of n_1, $\hat{n}_1$, as the value of n_1 which maximizes the likelihood function, either $L_e(n_1, \rho)$ or $L_a(n_1, \rho)$.

5. For $n_1 = \hat{n}_1$, use $\hat{\underline{\alpha}}$ obtained assuming this value of n_1 and calculate $\{\hat{u}_t\}$ where $\hat{u}_t = y_t - \underline{a}'_t \hat{\underline{\alpha}}$. From $\hat{u}_t$ obtain $\hat{\rho}$ using either equation (2.4) or (2.5), depending upon which likelihood function is being calculated.
6. If this is the first iteration, assume $\rho = \hat{\rho}$ and return to step 2. If it is not the first iteration, test if the difference between the previous and present estimate of ρ is less than some specified value and return to 2 if it is not.
7. Using the final value of $\hat{\rho}$, calculate $R(n_1, \rho)$ using the appropriate expression for $L(n_1, \rho)$, either (2.6) or (2.7).

The likelihood functions for n_1 alone do not involve iteration. For each allowable value of n_1, the parameters of the model are estimated by least squares and $\hat{\sigma}^2(n_1)$ obtained. Next, $\hat{n}_1$ is chosen as the value of n_1 for which $\sigma^2(n_1)$ is the minimum. In the case of the conditional procedure, a vector $\underline{z} = (y_2, y_4, \ldots, y_{2k})$ is formed and for each value of n_1 a matrix X must be formed in a manner similar to that used for the joint likelihood functions.

3. MONTE CARLO STUDY

The behaviour of the four likelihood functions described above was studied by generating data based upon a simple version of model (1.1). It was assumed that the change consisted of a change from one mean level to another, i.e.

$$y_t = \theta_1 + u_t \qquad (t = 1, 2, \ldots, n_1)$$
$$y_t = \beta_1 + u_t \qquad (t = n_1+1, \ldots, n).$$

The values of the parameters were taken as $\theta_1 = 1098$, $\beta_1 = 850$, $\sigma^2 = 128^2$ (the estimates obtained for the Nile River discharge data, Esterby and El-Shaarawi, 1981a), $n=100$, $n_1=50$ and $\rho=-0.90$, -0.75, -0.50, -0.25, 0, 0.25, 0.50, 0.75, 0.90. For each value of ρ, two hundred samples were generated with u_0 and $\{e_t\}$ obtained as pseudo-random normal deviates generated by the polar method. Initial values of $\rho=0$ were used in the computation of both the exact and approximate joint likelihood functions. For each of the two hundred samples, only one real root satisfying $|\rho| < 1$ was found in the computation of the exact joint likelihood functions. See step 5 in the section Computational considerations. This was ascertained by a provision in the computer program to print a message if in fact more than one such real root was found.

The mean of $\hat{n}_1$, the number of times out of 200 for which $\hat{n}_1 \neq 50$, where $n_1 = 50$ was used to generate the samples, and the mean of $\hat{\rho}$ are given for all the likelihood functions in Table 1. All likelihood

TABLE 1
Comparison of the estimates of n_1 and ρ from the four likelihood functions based on 200 samples

	Method	Value of ρ								
		-0.90	-0.75	-0.50	-0.25	0.00	0.25	0.50	0.75	0.90
Mean of n_1	exact	50	50	50	50	50	50	49	49	51
	approximate	50	50	50	50	50	50	49	49	47
	assume ρ=0	50	50	50	50	50	50	50	50	50
	conditional	50	50	50	51	50	49	49	49	47
Number with $\hat{n}_1 \neq n_1$	exact	43	38	52	62	77	91	104	133	152
	approximate	44	38	52	62	77	91	106	136	157
	assume ρ=0	81	71	66	66	76	92	118	148	171
	conditional	141	143	153	149	167	184	189	187	185
Mean of ρ	exact	-0.89	-0.74	-0.51	-0.26	-0.02	0.21	0.45	0.68	0.83
	approximate	-0.89	-0.74	-0.51	-0.26	-0.02	0.21	0.45	0.69	0.84

functions gave better estimates of n_1 for samples generated with $\rho < 0$ and, over all, the joint likelihood functions, exact and approximate, performed about the same but better than the other two likelihood functions. The likelihood function and the frequency distribution of

TABLE 2
Average of the exact relative maximum likelihood function $R_e(n_1, \rho)$ based on the 200 samples

n_1	Value of ρ -.90	-.75	-.50	-.25	.00	.25	.50	.75	.90
1									
					.000	.000	[.001,.016]		
					n_1=30	n_1=27	n_1=30		
42				.000	.006	.019	[.01,.04]	[.01,.08]	[.03,.06]
43				.001	.007	.015			
44				.001	.007	.024	.046		
45	.000		.000	.002	.008	.041	.073		
46	.001	.000	.001	.006	.025	.057	.064		
47	.005	.001	.010	.031	.049	.088	.105	.100	
48	.008	.015	.055	.103	.112	.133	.133	.103	
49	.175	.201	.252	.247	.246	.275	.192	.087	
50	.860	.892	.849	.821	.754	.719	.661	.482	.365
51	.217	.216	.246	.298	.279	.233	.167	.102	
52	.024	.019	.044	.065	.106	.157	.140	.076	
53	.001	.003	.010	.023	.066	.078	.094		
54	.000	.000	.003	.017	.038	.057	.075		
55			.000	.003	.011	.035	.064		
56				.001	.004	.022	.070		
57				.000	.001	.016	.040	[.02,.08]	[.03,.06]
58				.000	.001	.016			
59				.001	.001	.016	[.005,.030]		
60				.000	.000	.011			
61				.000	.003	.012			
					n_1=63	n_1=72	n_1=75		
					.000	.000	[.001,.007]		
99									

↑ indicates that the value remains constant at the last recorded number.

↕ indicates that values are within the closed interval given by the pair of numbers in the brackets [,].

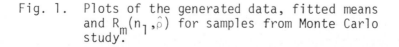

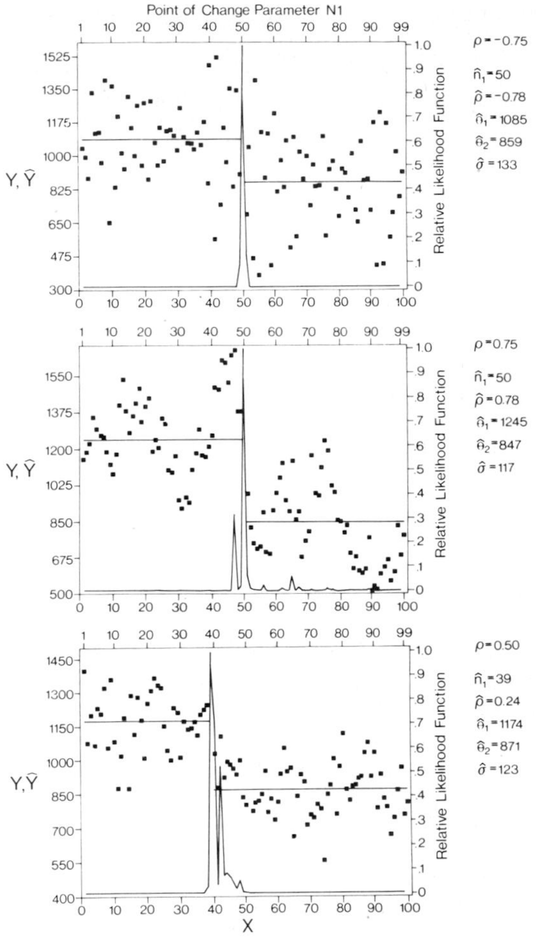

Fig. 1. Plots of the generated data, fitted means and $R_m(n_1, \hat{\rho})$ for samples from Monte Carlo study.

$\hat{n}_1$ are examined in more detail for the exact likelihood function in Tables 2 and 3. Both these tables show that as ρ increases the number of values of n_1 which are plausible also increases. Or in terms of the frequency distribution, given in Table 3, all estimates of n_1 were larger than 45 but no larger than 52 when $\rho = -0.90$ was used to generate the samples, but 12% of the samples had estimates $\leqslant 10$ or $\geqslant 90$ for $\rho = 0.90$.

Examples of the samples generated and the relative maximum likelihood function of n_1 at $\rho = \hat{\rho}$, $R_e(n_1, \hat{\rho})$, are given in Figure 1. The generated data, the estimated constant lines and the relative likelihood function are shown on each plot and the value of ρ used to generate the sample and the estimates of the parameters are given beside the plot. The first two plots are examples for which the estimates of

TABLE 3
Frequency distribution of $\hat{n}_1$ obtained from the exact likelihood function and the 200 samples

$\hat{n}_1 \leqslant c_1$ or $\hat{n}_1 \geqslant c_2$	Relative frequency of $\hat{n}_1 \geqslant c_1$ or $\hat{n}_1 \leqslant c_2$ for ρ =								
	-0.90	-0.75	-0.50	-0.25	0.00	0.25	0.50	0.75	0.90
$\leqslant$ 10	0	0	0	0	0	0	0.02	0.04	0.06
$\leqslant$ 20	0	0	0	0	0	0	0.02	0.08	0.10
$\leqslant$ 30	0	0	0	0	0	0	0.02	0.13	0.19
$\leqslant$ 40	0	0	0	0	0	0.01	0.07	0.19	0.26
$\leqslant$ 45	0	0	0	0	0.02	0.02	0.11	0.25	0.31
$\leqslant$ 48	0.01	0	0.02	0.06	0.08	0.13	0.22	0.45	0.36
= 50	0.79	0.81	0.74	0.69	0.62	0.55	0.48	0.34	0.24
$\geqslant$ 52	0.01	0	0.01	0.04	0.09	0.12	0.18	0.31	0.38
$\geqslant$ 55	0	0	0	0	0	0.03	0.11	0.26	0.35
$\geqslant$ 60	0	0	0	0	0	0.01	0.05	0.20	0.28
$\geqslant$ 70	0	0	0	0	0	0	0.03	0.10	0.19
$\geqslant$ 80	0	0	0	0	0	0	0	0.05	0.11
$\geqslant$ 90	0	0	0	0	0	0	0	0.03	0.06

the parameters were close to the values used to generate the data. The third plot illustrates the type of sample which can be generated with positive ρ and which results in estimates of n_1 and ρ far from the values used to generate the data.

In summary, the Monte Carlo study shows that for negative ρ or moderate positive ρ the joint likelihood functions, exact and approximate, and the likelihood function assuming $\rho=0$ can be used to estimate n_1. The likelihood function for n_1 based upon conditioning did not perform well even though the sample size used was 100. Further work is required to explain the properties of the likelihood functions suggested by the Monte Carlo study.

REFERENCES

Brown, R.L., Durbin, J. and Evans, J.M., 1975. Techniques for testing the constancy of regression relationships over time (with Discussion). J.R. Statist. Soc. B, 37:149-192.

Cobb, G.W., 1978. The problem of the Nile: conditional solution to a changepoint problem. Biometrika, 65(2):243-251.

Cox, D.R. and Hinkley, D.V., 1974. Theoretical statistics. Chapman and Hall, London, 511 pp.

Esterby, S.R. and El-Shaarawi, A.H., 1981a. Likelihood inference about the point of change in a regression regime. J. Hydrology, 53:17-30.

Esterby, S.R. and El-Shaarawi, A.H., 1981b. Inference about the point of change in a regression model. Appl. Statist., 30:277-285.

Hinkley, D.V., 1970. Inference about the change-point in a sequence of random variables. Biometrika, 57(1):1-17.

Hinkley, D.V., 1971. Inference about the change-point from cumulative sum tests. Biometrika, 58(3): 509-523.

Kraus, E.B., 1956. Graphs of cumulative residuals. Q.J.R. Meteorol. Soc., 82:96-98.

Pettitt, A.N., 1979. A non-parametric approach to the change-point problem. Appl. Statist., 28(2):126-135.

Plackett, R.L., 1960. Principles of regression analysis. Oxford: Clarendon Press, London, 173 pp.

THE CHANGE-POINT PROBLEM FOR A SEQUENCE OF BINOMIAL RANDOM VARIABLES

A.H. EL-SHAARAWI AND L.D. DELORME

National Water Research Institute

ABSTRACT

Three statistics are presented for detecting a change in a sequence of ordered binomial random variables. The first two are based on the conditional distribution of the random variables given their sum, while the third statistic is based on the empircal logistic transform approach. The use of these statistics is illustrated using data from a lake-sediment core.

INTRODUCTION

The change-point problem has received considerable attention recently in the field of statistics. Roughly speaking, this problem is concerned with developing procedures for testing the hypothesis that the parameters of the distribution of a sequence of ordered random variables are not constant. If the hypothesis of change is accepted then the problem is how to make inferences about both the position of change and the magnitude of change. Quandt (1958,1960,1972), Quandt and Ramsey (1978), Hinkley (1969), Brown et al (1975), Feder (1975a,b), Ferreira (1975) and Esterby and El-Shaarawi (1981a) have discussed the change-point problem for linear regression models. The case of binary random variables was considered by Pettitt (1979), Hinkley (1970), McGilchrist and Woodyer (1975) and Page (1955,1957).

In water quality studies, the change-point problem has many applica tions. For example, in the study of lake sediment cores one might be interested in determining if a change has occurred in the relative or absolute abundance of a particular biological indicator and of determining the pattern and the depth at which the change has occurred.

Reprinted from *Time Series Methods in Hydrosciences*, by A.H. El-Shaarawi and S.R. Esterby (Editors)

Applications to water quality problems have been considered by Esterby and El-Shaarawi (1981a, 1981b).

In this paper, a number of approaches are presented that can be used for the analysis of the change-point problem in a sequence of ordered binomial random variables. These techniques are then applied to the study of a data set from a lake-sediment core.

THE MODEL

Suppose that we have a sequence of n ordered independent binomial random variables $x_1 \ldots, x_n$. Assume that x_i (for i = 1,2,...,k) has the distribution

$\binom{m_i}{x_i} \theta_1^{x_i} (1-\theta_1)^{m_i - x_i}$, and x_{k+i} (for i = 1,2,...,n-k) has the distribution

$\binom{m_i}{x_i} \theta_2^{x_i} (1-\theta_2)^{m_i - x_i}$.

The integer k (the change-point) is unknown. The problem is to test any of the hypothesis

$H: \theta_1 \neq \theta_2$, $H^+: \theta_1 < \theta_2$ and $H^-: \theta_1 > \theta_2$.

If any of these hypothesis is rejected, the problem is then how to estimate k and to make inferences about its values.

In the binary case, where $m_i = 1$ (i = 1,2,...,n) and x_i (i = 1,2,..., n) takes only the values 0 and 1, McGilchrist and Woodyer (1975) have developed statistics for testing H, H^+, H^- when n is even. These statistics have been generalized by Pettitt (1979) for all values of n. Following Pettitt the statistics

$$U = \max_{1 \leqslant k > n} |U_{k,n}|$$

$$U^+ = \max_{1 \leqslant k > n} U_{k,n}$$

$$U^- = -\min_{1 \leqslant k > n} U_{k,n}$$

could be used for testing H, H^+ and H^- respectively, where $U_{k,n} = S_k - k\bar{x}$ and $S_k = x_1 + \ldots + x_k$. The exact significance levels associated with H, H^+ and H^- can be obtained using the Kolmogorov-Smirnov two sample statistics (Pettitt (1979)).

In the non-binary case, the significance levels associated with the above statistics are not exactly known, however, a conservative value for the significance level is given by Pettitt (1979). For the binomial case, $U_{k,n}$ takes the form

$$U_{k,n} = S_k - TP_k,$$

where $P_k = (m_1 + \ldots + m_k)/M$, $M = m_1 + \ldots + M_n$ and $T = n\bar{x}$.

If the m_i's are moderately large, then it can be shown that the conditional distribution of $U_{1,n}, \ldots, \ldots, U_{(n-1),n}$ given $\bar{x}$ under the assumption $\Theta_1 = \Theta_2$ converges in distribution to the multivariate normal with mean $\underline{0}$ and variance covariance matrix V, when the covariance between S_i and S_j (for $i \neq j$) is $-\frac{T(M-T)}{M-1} P_i P_j$, the variance of S_i is $\frac{T(M-T)}{M-1} P_i (1-P_i)$. An approximate significance level for U can be obtained using Sidak (1968) inequality

$$P_r(|S_i| \leqslant c, i - 1,2,\ldots,n-1) = P_r(U \leqslant c)$$

$$\geqslant \prod_{i=1}^{n-1} P_r (|S_i| \leqslant c).$$

Similarly

$$P_r(U^+ \leqslant c) \geqslant \prod_{i=1}^{n-1} P(S_i \leqslant c)$$

Another statistic for testing H can be constructed by transforming the variables $S_1, \ldots, S_{n-1}$ to the set of uncorrelated random variables $y_1, y_2, \ldots, y_{n-1}$ where (for $k = 1,2,\ldots,n-1$)

$$y_k = (P_{k+1}S_k - P_kS_{k+1})/\sqrt{T(M-T)P_kP_{k+1}(P_{k+1}-P_k)}.$$

The distribution of y_i ($i = 1,2,\ldots,n-1$) is asympotically normal, when $\Theta_1 = \Theta_2$, with mean 0 and unit variance.

Define the random walk

$W_k = W_{k-1} + y_k$, for $(k = 1,2,\ldots,n-1)$

where $W_0 = 0$. The variances and the covariances are given respectively by $\mathrm{Var}(W_k) = k$ and $\mathrm{Cov}(W_k,W_s) = \min(k,s)$.

Approximating $\{W_k\}$ by the corresponding Gaussian process with the same mean and covariance matrix, Brown et al (1975) have shown the probability that $\{W_k\}$ crosses the lines defined by the two points $(0_1 \pm a\sqrt{n-1})$, $(n-1, \pm 3a\sqrt{n-1})$ is $\alpha = .01, .05, .10$ for $a = 1.143, .948, .850$ respectively.

The empircal logistic transform (Cox 1970) is another method for testing H, H^+ and H^-. To do this suppose that the linear logistic model specifies that for $i = 1,2,\ldots,k$

$\lambda_1 = \log[\Theta_1/(1-\Theta_1)]$ and for $i = k+1,\ldots,n$

$\lambda_2 = \log[\Theta_2/(1-\Theta_2)]$

The corresponding empircal logistic transforms are

$Z_k = \log(S_k+1/2)/(M_k-S_k+1/2)$ and $Z'_k = \log(S'_k+1/2)/(M'_k-S'_k+1/2)$,

where $M_k = m_1 + \ldots + m_k$, $S'_k = T-S_k$ and $M'_k = M-M_k$.

The large-sample variances of Z_k and Z'_k are given respectively by

$$V_k = \frac{(M_k+1)(M_k+2)}{M_k(S_k+1)(M_k-S_k+1)} \quad \text{and} \quad V'_k = \frac{(M'_k+1)(M'_k+2)}{M'_k(S'_k+1)(M'_k-S'_k+1)}$$

Under the assumption $\Theta_1 = \Theta_2$, Z'_k-Z_k is an estimate of the logistic difference $\lambda_2-\lambda_1$ and has a standard error approximately $\sqrt{V_k+V'_k}$ For testing H, we use the statistic

$$D = \max_{1\leqslant k>n} \frac{|Z'_k-Z_k|}{\sqrt{V_k+V'_k}}$$

and for testing H^+ and H^- we can similarly define D^+ and D^-. The approximate significance levels associated with D, D^+, and D^- can be calculated in the same manner as that used for the statistic U.

ESTIMATION OF THE CHANGE-POINT

Pettitt (1979) suggested as an estimate for k the value $\hat{k}$ which maximizes $|U_{k,n}|$. This estimate of the change point is justified by noting that the conditional expectation of $U_{j,n}$ given $\bar{x}$ under H is

$$E(U_{j,n}|\bar{x}) = \begin{cases} j(\mu-\bar{x}) & \text{for } j \leqslant k \\ k(n-j)\ (\mu-\bar{x})/n-k & \text{for } j > k \end{cases}$$

$$\text{where} \quad \mu = E(x_r|\bar{x}) \qquad \text{for } r \leqslant k.$$

This shows that the graph of $|U_{j,n}|\bar{x}|$ consists of two straight lines. The first, for $r \leqslant k$, has the positive hope $|\mu-\bar{x}|$, and the second for $k > h$ has the negative slope $-\frac{n}{n-k}|\mu-\bar{x}|$.

Another way of estimating the change-point by talking the value $\hat{k}$ which maximizes the statistic D. Further the plot of W_i against i, $(i = 1,2,\ldots,n-1)$ can be used to estimate k graphically by taking $\hat{k}$ as the point where the process $\{W_i\}$ will start to follow a systematic drift from zero.

APPLICATIONS

In sediment cores, the distribution of fossils is a random variable which is a function of the chemical, physical, and biological controls when the biological unit was alive. In cores where a substantial change in the number of ostracode shells (seed shrimp) has occurred at some point in time, observations of a corresponding change of the relative concentration in a core provides one means of identifying the depth, hence the time, at which a significant change occurred. This is important because it allows one to determine the type of change in the habitat which controlled the faunal element.

A core collected from Echo Lake in the Qu'Appelle valley of southern Saskatchewan contains several ostracode species, one of which is Cyclocpris ampla Furtos, 1933. This species, although not the most abundant, has a characteristic profile with depth down the core. The species appears to diminish in the top third of the profile.

FIGURE 1

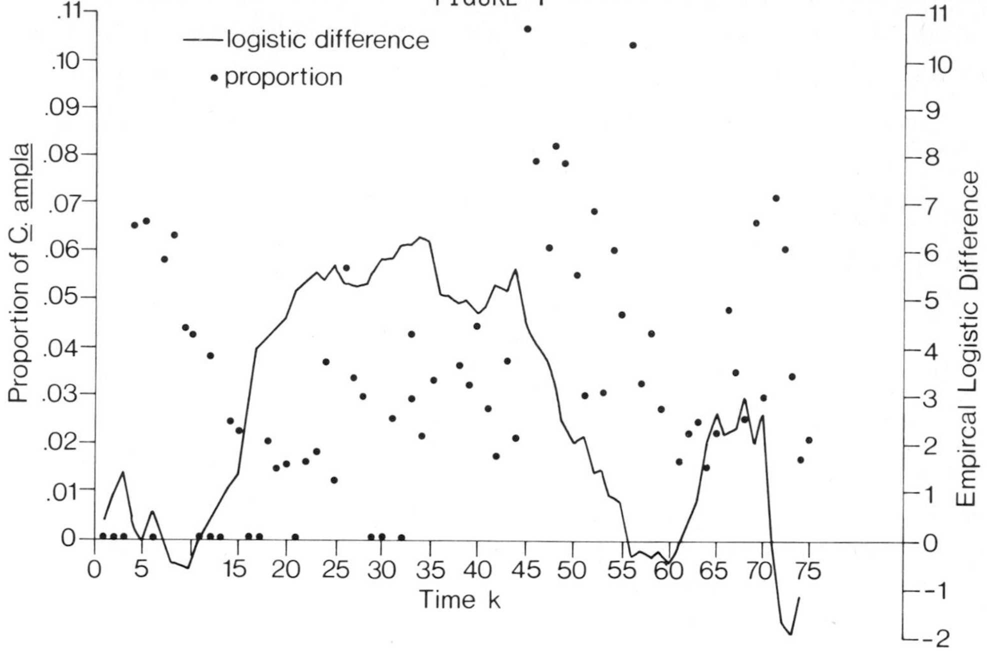

FIGURE 2

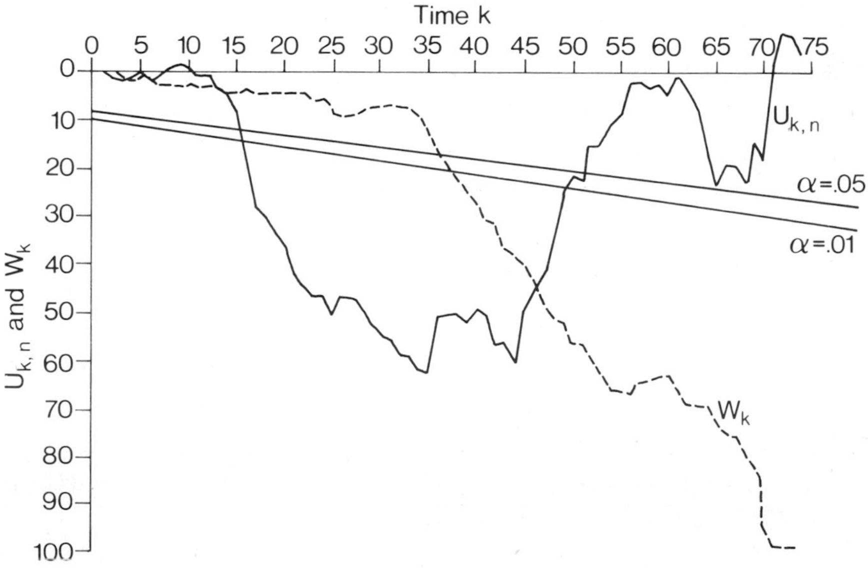

Fig. 1 presents the plot of the proportion of C. ampla shells to the total number of ostracode shells found in each slice of the core. The values of $U_{k,n}$ and W_k are plotted in Figure 2. The maximum of $|U_{k,n}|$ is U = 61.93 and is at depth k = 35 cm. The significance level α associated with U is approximately zero indicating a very high evidence of change. The lines showing the 5% and 1% boundaries for the process $\{W_k\}$ are plotted on Figure 2 which show that the path of $\{W_k\}$ crosses these lines at k = 35. Hence, the hypothesis of change is accepted and as an estimate for the change-point we take 35. On Fig. 1 the empircal logistic transform $(Z'_k - Z_k)/\sqrt{V_k + V'_k}$ is shown. This indicates that D is 6.2563 and occurs at k = 34. The significance level associated with D is almost zero, which indicates that the data shows strong evidence for change. The estimate for k using the logistic transform is k = 34.

Based on an interpretation of the chemical and physical habitat (Delorme et al, 1977), from the fossil ostracodes in the core, a reason for the change in the species population can be seen for the depth of 34 to 35 cm. The depth interval 32 to 37 cm shows an interpreted decrease in major ions and pH. Simultaneously, there is an increase in the annual precipitation. The additional rain had the effect of diluting the ions in the lake water. Although the count of shells for Cyclocypris ampla parallels the total shell population, the interpreted chemistry does not. The decrease in shell population in the top third of the core reflects more than a change in the habitat. Species, such as C. ampla, which had a marginal existence prior to the change in the habitat, did not regain their former status. Some changes may place more stress on the species so that it does not fully recover.

REFERENCES

Brown, R.L. Durbin, J. and Evans, J.M.(1975). Techniques for testing the constancy of regression relationships overtime (with discussion) J.R. Statist. Soc. B, 37:149-192.

Delorme, L.D., Zoltai, S.C., and Kalas, L.L. (1977). Freshwater shelled invertebrate indicators of paleoclimate in northwestern Canada during late glacial times. Can. J. of Earth Sciences, 14(9):2029-2046.
Esterby, S.R. and El-Shaarawi, A.H. (1981a). Likelihood inference about the point of change in a regression regime. J. Hydrology, 53:17-30.
Esterby, S.R. and El-Shaarawi, A.H. (1981b). Inference about the point of change in a regression model. Appl. Statist., 30:277-285.
Feder, P.L. (1975a). On asymptotic distribution theory in segmented regression problems - Identified case. Annals of Statistics, 3:49-83.
Feder, P.L. (1975b). The log likelihood ratio on segmented regression. Annals of statistics, 3:84-97.
Ferreira, P.E. (1975). A Bayesian analysis of a switching regression model: known number of regimes. J. of the American Statistical Association, 70:370-374.
Hinkley, D.V. (1969). Inference about the intersection in two-phase regression. Biometrika, 56:495-504.
Hinkley, D.V. (1970). Inference about the change-point in a sequence of random variables. Biometrika, 57:1-17.
McGilchrist, C.A. and Woodyer, K.D. (1975). Note on a distribution-free Cusum technique, Technometrics, 17:321-325.
Page, E.S. (1955). A test for a change in a parameter occurring at an unknown point. Biometrika, 42:523-527.
Page, E.S. (1957). On problems in which a change in a parameter occurs at an unknown point. Biometrika, 44:248-252.
Pettitt, A.N. (1979). A non-parametric approach to the change-point problem. Appl. Statist., 28(2):126-135.
Quandt, R.E. (1958). The estimation of the parameters of a linear regression system obeying two separate regimes. J. of the American Statistical Association, 53:873-880.
Quandt, R.E. (1960). Test of the hypothesis that a linear regression obeys two separate regimes. J. of the American Statistical Assoc., 55:324:330.
Quandt, R.E. (1972). New approach to estimating switching regressions, J. of the American Statistical Association, 67:306-310.

EXPLORATION OF AN EXTREME VALUE PARTIAL TIME SERIES MODEL IN HYDROSCIENCE

ASHKAR, F., EL-JABI, N. and ROUSSELLE, J.
Ecole Polytechnique de Montréal

ABSTRACT

This paper presents and discusses the characteristics of a probabilistic model suitable for the description of the time and space joint distribution of partial duration series flood flow data.

An important functional of the magnitude of exceedances and their time of occurrence, commonly used in the design of hydraulic structures is the design exceedance corresponding to a given return period. When building a project based on this design exceedance a certain risk resulting from the estimation of the models' parameters is encountered. This risk is studied for different periods of record and different return periods.

In addition to exceedances, the continuous river flow process produces two features essential in several areas of flood control and flood damage analysis. These are namely flood duration and flood volume. The partial duration series approach is shown to offer convenience in the definition, calculation and statistical analysis of quantitative aspects of these two features.

Relative to the planning of water resources and the management of river banks subject to flooding, a methodology is presented in which technical, economic and physical considerations are integrated.

1.0 INTRODUCTION

Flood flow forecasting is undoubtedly one of the most important considerations in the process of water resources management, given the economic damage and the loss of human lives that may be caused by floods. According to Gray (1972), the choice of a forecasting method depends on several factors:

Reprinted from *Time Series Methods in Hydrosciences*, by A.H. El-Shaarawi and S.R. Esterby (Editors)

i - the flood features relevant to the project under consideration; i.e. peak flow, flood intensity and flood volume, flood duration, sedimentation problems, etc...

ii - the available data and the search for the best method for extracting information from these data.

iii- the effect of the basin's topographic and hydrographic characteristics on runoff.

iv - the importance of the project under consideration from the economic, as well as from the social points of view and the span of time covered by this project.

The approach with the longest history, used in flood flow forecasting is the one which uses empirical formulas that express flood discharge as a function of the physical characteristics of the basin, with a special emphasis on the basin area. The application of these formulas requires a profound knowledge about the topographic and hydrographic properties of the catchment under consideration.

A second approach used to foresee the future behavior of floods is statistical and is based on the assumption that a probability measure can be imposed on the available streamflow data. The goal here is to obtain relations capable of describing the interaction between the laws of randomness governing the bahavior of variables. The choice among available statistical techniques is governed by the physical characteristics of the phenomenon, the mathematical properties of the data and by intuition and the prevalent engineering practice.

Credit for the construction of a probabilistic base, suitable for flood analysis goes to Todorovic and Yevjevich (1967, 1969), Todorovic (1967, 1968, 1970), Todorovic and Zelenhasic (1970), Todorovic and Rousselle (1971) and Todorovic (1978). These models are at the base of the current study. As in (Kirby, 1969), they treat flood hydrographs mainly as a sequence of flood peaks, seperated by relatively long periods of low discharge.

Both deterministic and stochastic approaches have gained support from several investigators, and a comparison between the two approaches was done by Yevjevich (1974). The deterministic school is strong on interpretations and developments of a physical nature, but limited on the regional level. The probabilistic school, on the other hand, is more flexible and offers a variety of readily available techniques, often found useful in desperate situations (Tiercelin, 1973).

2.0 THEORETICAL CONSIDERATIONS

A model based on the theory of extreme values and on the theory of random number of random variables was developed at the Engineering research center of Colorado State University, and was presented by Todorovic (1970). Our flood-plain management methodology, to be presented towards the end of this paper, is based on this stochastic model.

2.1 Analysis of the flood phenomenon

Consider a hydrograph representing the instantaneous river flow at a given station, within an interval of time (0,t], (figure 1). Let us consider as floods only those hydrograph peaks exceeding a certain base level (figure 2).

$$\xi_\nu = \begin{cases} 0 & ; \quad Q_\nu \leqslant Q_b \\ Q_\nu - Q_b & ; \quad Q_\nu > Q_b \end{cases} \tag{1}$$

where Q_b is the base flow (often taken as the discharge when the water starts to overflow to spread over the banks of the river).

A discrete and non negative stochastic process representing exceedances occurring in the interval (0,t] is obtained. Let ξ_ν be the νth exceedance occurring at time $\tau(\nu)$ (figure 3). The flood phenomenon may therefore be represented by the following sequence

$$\{\xi_\nu \;;\; \nu = 0,1,2,\ldots,\; t \geqslant 0\} \tag{2}$$

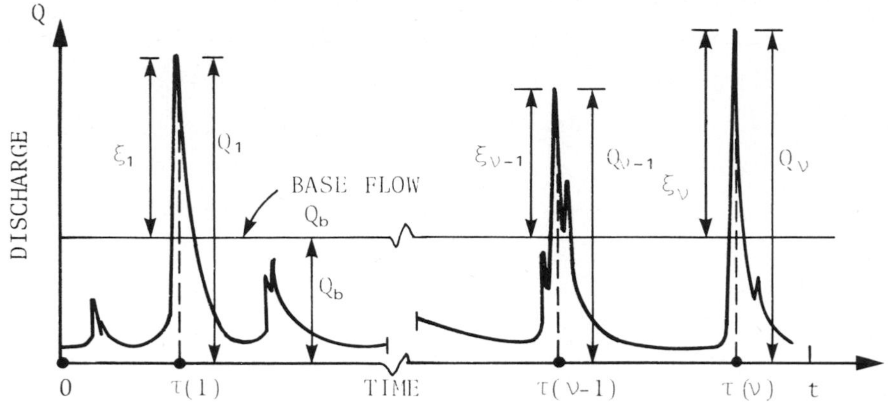

FIGURE 1 - Instantaneous river flow at a given station

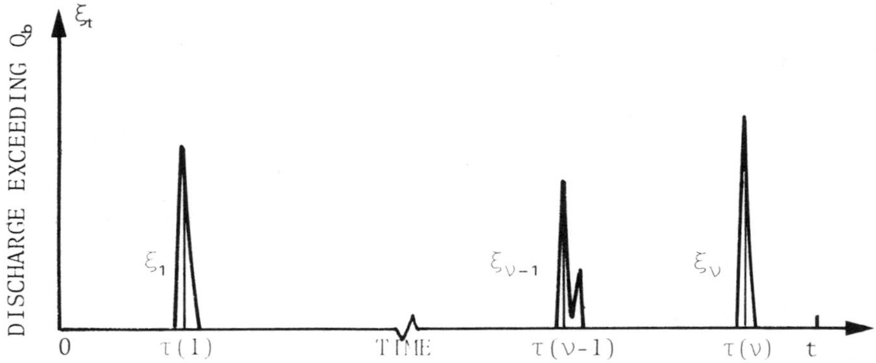

FIGURE 2 - Flood hydrographs

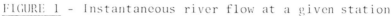

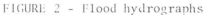

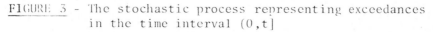

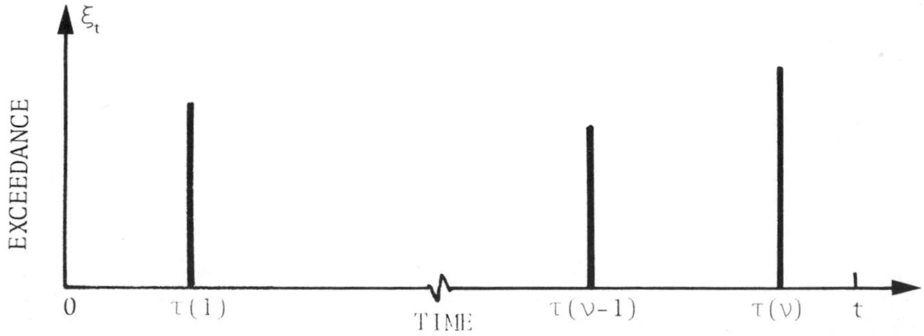

FIGURE 3 - The stochastic process representing exceedances in the time interval $(0,t]$

The existing probabilistic models used in flood forecasting are essentially devoted to the treatment of the two sequences (ξ_ν) and $(\tau(\nu))$. In the case of a multiple-peaked hydrograph, such as the one occurring at time $\tau(\nu - 1)$ in figure 2, only the highest peak is considered. This is done to justify the needed assumption of mutual independence between exceedances ξ_ν. The number of exceedances in $(0,t]$, the times of occurrence of these exceedances, as well as the magnitudes of exceedances are random variables.

2.2 Distribution function of the number of exceedances

The distribution of the number of exceedances is important because design interest usually lies in the multivariate distribution of the number of exceedances in $(0,t]$ and the magnitudes of these exceedances

Denote by $\eta(t)$ the number of exceedances in $(0,t]$, i.e.

$$\eta(t) = \mathrm{Sup}\,\{\nu\ ;\ \tau(\nu) \leqslant t\} \tag{3}$$

$\eta(t)$ may therefore take any non negative integer value for any $t > 0$.

Define E_ν^t by $E_\nu^t = \{\eta(t) = \nu\} = \{\tau(\nu) \leqslant t < \tau(\nu + 1)\}$ (4)

In other terms, E_ν^t is the event that there are exactly ν exceedances in $(0,t]$.

The events E_ν^t form a partition of the sample space, i.e. $E_i^t \cap E_j^t = \theta$ and $\bigcup_{\nu=0}^{\infty} E_\nu^t = \Omega$ for all $i \neq j = 0,1...$ where θ stands for the impossible event and Ω stands for the certain event.

Denote by $\Lambda(t)$ the expected value of $\eta(t)$, so that we have

$$\Lambda(t) = \sum_{\nu=1}^{\infty} \nu\, P(E_\nu^t) \tag{5}$$

Mainly due to the seasonal variations of streamflow, $\Lambda(t)$ is a non linear function in time.

Denote by $F_\nu(t)$ the distribution function of the time of occurrence $\tau(\nu)$ of the νth exceedance,

$$F_\nu(t) = P\{\tau(\nu) \leqslant t\} \tag{6}$$

Equation (4) can, therefore, be written in the following form:

$$P(E_\nu^t) = P\{\tau(\nu) \leqslant t\} - P\{\tau(\nu + 1) \leqslant t\} = F_\nu(t) - F_{\nu+1}(t) \tag{7}$$

By following the same procedure used in arriving at equation (7) we obtain

$$P(E^t_{\nu+1}) = F_{\nu+1}(t) - F_{\nu+2}(t) \tag{8}$$

$$P(E^t_{\nu+2}) = F_{\nu+2}(t) - F_{\nu+3}(t) \tag{9}$$

$$\vdots$$

Summing equations (7), (8) and (9) yields

$$F_\nu(t) = P(E^t_\nu) + P(E^t_{\nu+1}) + P(E^t_{\nu+2}) + \ldots \tag{10}$$

which may be put in the following form

$$F_\nu(t) = \sum_{j=\nu}^{\infty} P(E^t_j) \tag{11}$$

Remark that $\tau(\nu)$ are continuous random variables and for all $\nu = 1,2,\ldots;\ 0 < \tau(\nu) < \tau(\nu + 1);\ \tau(\nu) \to \infty$ if $\nu \to \infty$ and

$$P(E^t_\nu) > 0 \quad \forall\ t > 0 \text{ and } \nu = 0,1,\ldots \tag{12}$$

Suppose that for any "very small" time interval $(t, t + \Delta t]$ on the time scale (by "very small" we mean in the limit when $\Delta t \to 0$) only one of the following two events can occur (Cox and Miller, 1965); *i* - no exceedance occurs in this interval; or *ii* - one and only one exceedance occurs in this interval.

Denote these two events, respectively, by

$$\begin{aligned} E_o^{t+\Delta t} &= \{\eta(t + \Delta t) - \eta(t) = 0\} \\ E_1^{t+\Delta t} &= \{\eta(t + \Delta t) - \eta(t) = 1\} \end{aligned} \tag{13}$$

Ignoring events whose probability of occurrence tends to zero as $\Delta t \to 0$ (for the sake of not entering deep into complex mathematical details) we may write the following equality

$$P(E_\nu^{t+\Delta t}) = P(E^t_\nu)\ P(E_0^{t+\Delta t} | E^t_{\nu-1}) + P(E^t_{\nu-1})\ P(E_1^{t+\Delta t} | E^t_{\nu-1}) \tag{14}$$

Dividing this equation by Δt yields

$$\begin{aligned} \frac{P(E_\nu^{t+\Delta t})}{\Delta t} &= \frac{P(E_\nu)\ P(E_0^{t+\Delta t} | E^t_\nu)}{\Delta t} + \frac{P(E^t_{\nu-1})\ P(E_1^{t+\Delta t} | E^t_{\nu-1})}{\Delta t} \\ &= \frac{P(E^t_\nu)[1 - P(E_1^{t+\Delta t} | E^t_\nu)]}{\Delta t} + \frac{P(E^t_{\nu-1})\ P(E_1^{t+\Delta t} | E^t_{\nu-1})}{\Delta t} \end{aligned} \tag{15}$$

$$\frac{P(E_\nu^{t+\Delta t})-P(E_\nu^t)}{\Delta t} = -\frac{P(E_\nu^t)\ P(E_1^{t+\Delta t}|E_\nu^t)}{\Delta t} + \frac{P(E_{\nu-1}^t)\ P(E_1^{t+\Delta t}|E_{\nu-1}^t)}{\Delta t} \tag{16}$$

and taking the limit of equation (16) as $\Delta t \to 0$, we get

$$\lim_{\Delta t \to 0} \frac{P(E_\nu^{t+\Delta t})-P(E_\nu^t)}{\Delta t} = \lim_{\Delta t \to 0} P(E_{\nu-1}^t)\ \frac{P(E_1^{t+\Delta t}|E_{\nu-1}^t)}{\Delta t} - \lim_{\Delta t \to 0} P(E_\nu^t)\ \frac{P(E_1^{t+\Delta t}|E_\nu^t)}{\Delta t} \tag{17}$$

Using the following notation : $\lambda_\nu(t) = \lim_{\Delta t \to 0} \frac{P(E_1^{t+\Delta t}|E_\nu^t)}{\Delta t}$, equation (17) becomes

$$\frac{dP(E_\nu^t)}{dt} = \lambda_{\nu-1}(t)\ P(E_{\nu-1}^t) - \lambda_\nu(t)\ P(E_\nu^t) \tag{18}$$

with
$$\frac{dP(E_0^t)}{dt} = -\lambda_0(t)\ P(E_0^t) \quad \text{for} \quad \nu = 0 \tag{19}$$

The solution of the system of differential equations (18) and (19) is given by (Zelenhasic, 1970):

$$P(E_0^t) = \exp\{-\int_0^t \lambda_0(s)\ ds\} \tag{20}$$

$$\begin{aligned} P(E_\nu^t) &= \exp\{-\int_0^t \lambda_\nu(s)\ ds\} \cdot \int_0^t \lambda_{\nu-1}(s)\ ds \\ &\cdot \exp\{\int_0^{t_1} [\lambda_\nu(s) - \lambda_{\nu-1}(s)]ds\} \cdot \int_0^{t_1} \ldots \int_0^{t_{\nu-1}} \lambda_0\ (t_\nu) \\ &\cdot \exp\{\int_0^{t_\nu} [\lambda_1(s) - \lambda_0(s)]ds\}\ dt_\nu\ dt_{\nu-1} \ldots dt_1 \end{aligned} \tag{21}$$

Under the following hypothesis : $\lambda_\nu(t) \equiv \lambda(t)$ equation (21) takes the form

$$P(E_\nu^t) = \{\int_0^t \lambda(s)ds\}^\nu \exp\{-\int_0^t \lambda(s)ds\}/\nu! \tag{22}$$

When the following expression for $\Lambda(t)$: $\Lambda(t) = \sum_{\nu=0}^{\infty} \nu\ P(E_\nu^t)$ is taken along with equation (22), we obtain

$$\begin{aligned} \Lambda(t) &= \sum_{\nu=0}^{\infty} \nu\ e^{-\int_0^t \lambda(s)ds} \frac{[\int_0^t \lambda(s)ds]^\nu}{\nu!} \\ &= e^{-\int_0^t \lambda(s)ds} \sum_{\nu=0}^{\infty} \frac{[\int_0^t \lambda(s)ds]^\nu}{(\nu-1)!} \\ &= e^{-\int_0^t \lambda(s)ds} \cdot \int_0^t \lambda(s)ds \cdot e^{\int_0^t \lambda(s)ds} \end{aligned} \tag{23}$$

$$\Lambda(t) = \int_o^t \lambda(s)\, ds \tag{24}$$

$\Lambda(t)$ depends, therefore, on $\lambda(s)$ which is a non negative function representing the rate of occurrence of exceedances ξ_ν at different points in time. This function is effected by climatic variations occurring between seasons and/or between years. Equations (22) and (24) finally yield

$$P(E_\nu^t) = \exp\{-\Lambda(t)\} \frac{\{\Lambda\{t\}^\nu}{\nu!} \tag{25}$$

2.3 Distribution function of the largest exceedance

Among the set of all exceedances ξ_ν within a time interval $(0,t]$, probably the most important variable for design purposes is the largest exceedance $\chi(t)$, defined as $\chi(t) = \sup_{\tau(\nu) \leqslant t} \xi_\nu$; $\chi(t)$ is a non decreasing stochastic process (figure 4).

Let $F_t(x)$ be the distribution function of $\chi(t)$, i.e. $F_t(x) = P\{\chi(t) \leqslant x\}$; A mathematical expression for $F_t(x)$ was derived by Todorovic (1970) in terms of the following conditional probability

$$P\{\sup_{\tau(\nu) \leqslant t} \xi_\nu \leqslant x | \eta(t)\} \tag{26}$$

and under the hypothesis of a mutually independent sequence (ξ_ν), the following equalities were deduced

$$F_t(x) = \sum_{K=0}^{\infty} P[\bigcap_{\nu=0}^{K} \{\xi_\nu \leqslant x\} \cap E_K^t] \tag{27}$$

$$= P(E_o^t) + \sum_{K=1}^{\infty} P[\bigcap_{\nu=0}^{K} \{\xi_\nu \leqslant x\} \cap E_K^t] \tag{28}$$

A schematic representation of the distribution function of $\chi(t)$ is shown in figure 5. The hypothesis of mutual independance of the two sequences (ξ_ν) and $\tau(\nu)$ along with equation (28) leads to the following expression

$$F_t(x) = P(E_0^t) + \sum_{K=1}^{\infty} \{[H(x)]^K \cdot P(E_K^t)\} \tag{29}$$

where

$$H(x) = P\{\xi_\nu \leqslant x\} \tag{30}$$

$H(x)$ is the common distribution function of the exceedances ξ_ν occurring in the invertal $(0,t]$. Making use of equations (25) and (29)

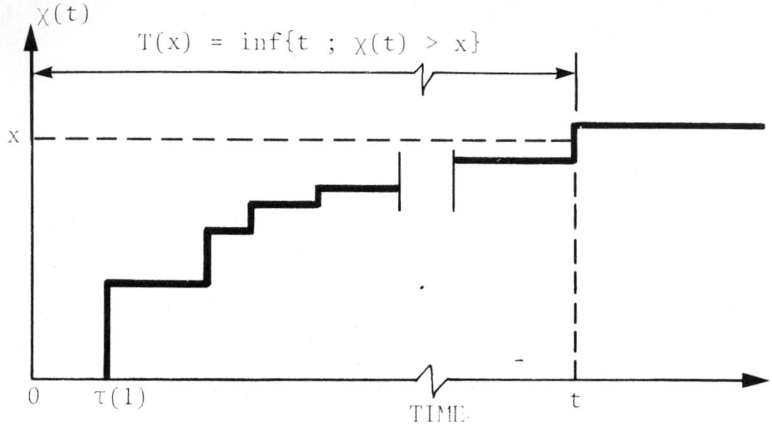

FIGURE 4 - Schematic representation of the stochastic process $\chi(t)$

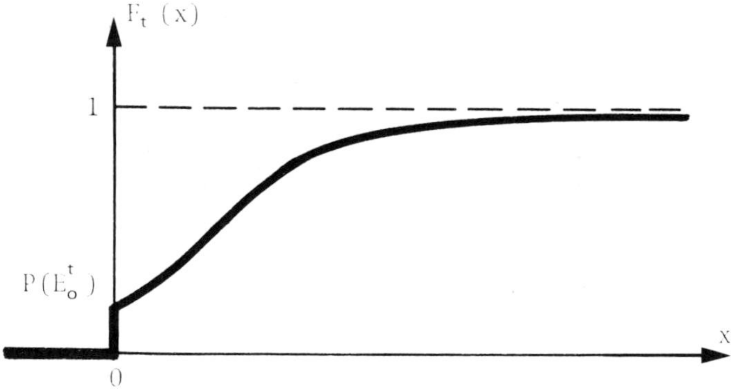

FIGURE 5 - The distribution function of $\chi(t)$

we obtain

$$F_t(x) = \sum_{K=0}^{\infty} P(E_K^t)\ [H(x)]^K = \sum_{K=0}^{\infty} \frac{[\Lambda(t)]^K}{K!} e^{-\Lambda(t)} [H(x)]^K \tag{31}$$

$$= e^{-\Lambda(t)} \sum_{K=0}^{\infty} \frac{[\Lambda(t)H(x)]}{K!} = e^{-\Lambda(t)}\ e^{\Lambda(t)H(x)} \tag{32}$$

$$F_t(x) = e^{-\Lambda(t)[1-H(x)]} \tag{33}$$

In the foregoing analysis it is supposed that all exceedances ξ_ν occurring during a year are mutually independent and identically distributed, and that the flood process is stationary (i.e. there is no change in the probabilistic behavior of floods from year to year). If we wish to divide the one-year interval into a certain number k of "seasons" in the following manner: $(0,T_1]$, $(T_1,T_2]$,..., $(T_{K-1},T_K]$; then the largest exceedance $\chi(t)$ takes the form

$$\chi(t) = \text{Sup}\ \{\chi_1(0,\ T_1),\ \chi_2(T_1,\ T_2),\ldots,\ \chi_K(T_{K-1},\ t)\} \tag{34}$$

where $\chi_K(T_{K-1},\ T_K)$, $k = 1,2,\ldots,$ $T_o \equiv 0$ is the largest exceedance in the interval $(T_{K-1},\ T_K]$. In this case, the distribution function of the largest exceedance in the interval of time $(0,t]$ would be given by

$$F_{t\,K}(x) = P\{\chi_1(0,T_1)\leqslant x, \chi_2(T_1,T_2)\leqslant x,\ldots,\chi_K(T_{K-1},t)\leqslant x\},\ 0<T_1<T_2<\ldots \tag{35}$$

Following the same procedure that lead to equation (33), Rousselle (1972) showed that for an interval $(0,t]$, the distribution function $F_{t\,K}(x)$ takes the form

$$F_{t_K}(x) = \exp\{-\sum_{n=1}^{K-1} [\Lambda(T_n) - \Lambda(T_{n-1})][1 - H_n(x)]$$
$$- [\Lambda(t) - \Lambda(T_{K-1})][1 - H_K(x)] \;\forall\; K = 1,2,\ldots \text{ and } t\varepsilon(T_{K-1},T_K] \tag{36}$$

where $H_i(x)$ stands for the distribution function of exceedances in the ith season. Rousselle (1972) also found that the exponential distribution function fits adequately well the data on some rivers in the United States,

$$H(x) = 1 - e^{-\beta x} \ , \ \beta > 0 \quad \text{with} \quad \beta = \{E(\xi_\nu)\}^{-1} \tag{37}$$

Rousselle and El-Jabi (1976) and Ashkar (1980) showed also that the exponential distribution gives good results when applied to some Canadian rivers.

Equations (37), (33) and (36) lead to the distribution function $F_t(x)$ of the largest exceedance during a year in the case when

i- all exceedances ξ_ν occurring within a year are independent and identically distributed

$$F_t(x) = \exp\{-\Lambda(t)\ e^{-\beta x}\} \tag{38}$$

ii- all exceedances ξ_ν are mutually independent, but are identically distributed only on a seasonal basis.

$$F_{t_4}(x) = \exp\{-\Lambda(T_1)\ e^{-\beta_1 x} - [\Lambda(T_2) - \Lambda(T_1)]\ e^{-\beta_2 x} - [\Lambda(T_3) - \Lambda(T_2)]\ e^{-\beta_3 x} - [\Lambda(T_4) - \Lambda(T_3)]\ e^{-\beta_4 x}\} \tag{39}$$

where $\Lambda(T_1)$, $[\Lambda(T_2) - \Lambda(T_1)]$, $[\Lambda(T_3) - \Lambda(T_2)]$ and $[\Lambda(T_4) - \Lambda(T_3)]$ are the mean numbers of exceedances for the four different seasons (assuming that the year is divided into four seasons), β_1, β_2, β_3 and β_4 are the parameters of the exponential distribution (37) for the four seasons.

3.0 DESIGN CONSIDERATIONS

3.1 The return period

In the process of flood control planning, the choice of the capacity of a hydraulic structure is often hampered by many difficulties. Whatever this choice may be, a certain risk has to be tolerated, and it is desirable in practice to have a quantitative measure of this risk. A widely used measure is the return period T defined as the expected value of N_x, the number of years required to obtain an exceedance greater than the level x for the first time. The level x is called the "design flow" associated with the return period T. Equations (40) through (44) put the notions of N_x and T into mathematical terms.

$$N_x = \text{Inf}\{\nu \; ; \; \chi_\nu > x\} \quad ; \quad x > 0 \tag{40}$$

$$P(N_x = n) = P(\chi_1 \leqslant x, \ldots \chi_{n-1} \leqslant x, \chi_n > x) \; ; \; n = 1,2,\ldots \tag{41}$$

If (χ_ν) is assumed to be a sequence of mutually independent and identically distributed random variables having the distribution function: $F_t(x) = P(\chi_\nu \leqslant x)$, then we obtain

$$P(N_x = n) = [F_t(x)]^{n-1} \; [1 - F_t(x)] \tag{42}$$

from which we can deduce the mean of N_x :

$$E[N_x] = \sum_{n=1}^{\infty} n[F_t(x)]^{n-1} \; [1 - F_t(x)] = \frac{1}{1 - F_t(x)} \tag{43}$$

where $F_t(x)$ is given either by equation (38) or equation (39) depending on whether seasonality is taken into account or not.

Since by definition we have $T = E(N_x)$, we hence obtain

$$T = \frac{1}{1 - F_t(x)} \tag{44}$$

3.2 The first passage time

Define $T(x)$ as the time necessary for the process $\chi(t)$ to exceed the level x (figure 4), i.e. $T(x) = \text{Inf}\{t \; ; \; \chi(t) > x\}$; $T(x)$ is a continuous and non decreasing stochastic process, since $0 \leqslant x_1 \leqslant x_2$ implies that $T(x_1) \leqslant T(x_2)$. Let $Q_x(t)$ be the distribution function of $T(x)$

$$Q_x(t) = P[T(x) \leqslant t] \tag{45}$$

The relation between $\chi(t)$ and $T(x)$ leads to the following equations

$$\{\chi(t) \leqslant x\} \equiv \{T(x) > t\} \tag{46}$$

$$P\{\chi(t) \leqslant x\} \equiv P\{T(x) > t\} \tag{47}$$

which in turn yield the following relation between $Q_x(t)$ and $F_t(x)$

$$Q_x(t) = 1 - F_t(x) \tag{48}$$

4.0 PARAMETER SENSITIVITY ANALYSIS AND ITS RELATION TO DESIGN

From a sample of flood data, design flows X_t corresponding to different return periods T can be calculated, either by using equations (38) and (44) (in the case when seasonal effects on flood flow are not considered important) and solving for x in terms of T to obtain

$$X_T \equiv x = \frac{c + \ell n \lambda}{\beta} \tag{49}$$

where $C = -\ell n(\ell n \frac{T}{T - 1})$; $\lambda = \Lambda(t^*)$; t^* = one-year period or by using equations (39) and (44) (when seasonal variations should be taken into account) and solving numerically for x in terms of T, λ_1, λ_2, λ_3, λ_4, β_1 β_2 β_3 and β_4 (λ_i being the mean number of floods in the ith season).

Because design flows X_T are calculated from the available record, which cannot produce all the information contained in the set of all possible floods, these design flows should be expected to vary if a different flood record was in hand. In other words, a design flow is a random variable whose distribution function can be obtained. This distribution function depends on both the design model used and the length of record.

The randomness inherent in the design flow is an essential element in the calculation of the economic risk to be encountered when building a dam of any size. This randomness is due to the random nature of the parameter estimates $\hat{\lambda}$ and $\hat{\beta}$ used in practice to replace λ and β in equation (49) (or due to the random nature of $\hat{\lambda}_i$, $\hat{\beta}_i$, i = 1,2, 3,4 in the presence of seasonality). It was studied by Hindié (1974) and Ashkar (1980) for the extreme value model presented in the current study. Parameter estimation was done by Bayesian Techniques (Hindié, 1974) and by maximum likelihood (Ashkar, 1980). A comparison between the two approaches was done by Ashkar (1980).

The distribution function of the design flood X_T can be taken along with economic information about the flood-prone region, put in the form of an economic "loss function", to derive the economic risk associated with different possible sizes of the hydraulic structure being planned.

5.0 FLOOD MAGNITUDE, DURATION AND VOLUME

Three important components of flood hydrographs shown in Figure 2 are the exceedance ξ, the flood duration D and the flood volume V.

Ashkar (1980) showed that on some Canadian rivers, it is possible to make the hypothesis that the flood hydrograph is triangular in shape.

Considering the nature of flood flow we may say that the three components ξ, D and V are random variables.

Ashkar (1980) divided the flood duration D into two sub-components D_r and D_f that he called "rise duration" and "fall duration" respectively. He showed that in practice D_r and D_f can be considered as statistically independent and that they can be put in the form of a sum of a deterministic component S, and a non-deterministic component Σ, i.e. $D_r = S_r + \Sigma_r$ and $D_f = S_f + \Sigma_f$. He showed that on some Canadian rivers, the exponential distribution fits adequately well the non deterministic components $\Sigma_{r,\xi}$ and $\Sigma_{f,\xi}$ associated with a given flood magnitude ξ, D and $V = \xi D/2$ for floods in any season.

6.0 FLOOD PLAIN SYSTEM ANALYSIS

This system may be regarded as a complex set containing such elements as streams, floodable lands and the human population living on them. Several decisional constraints complicate the analysis of this system. These constraints may be political, hydrotechnical, environmental, economic, sociological or physical in nature. They form what is called a decisional space which covers the flood phenomenon in its entirety.

Some of these constraints are not amenable to a quantitative treatment due to their piculiar character; nevertheless, they often play a key role in the final implementation of policies, especially when these policies come into conflict with the public's interests.

By a hydroeconomic system, we mean that set of elements or components required in the economic quantification process of flood damage or in the determination of the economic value of a given damage-reduction measure. It is this sub-system within which all economic aspects of a flood occurring on a particular basin at a particular points in time, interact. These economic aspects are composed of physical (topography, slope, area, geology,...) and economic (residential damage, reconstruction costs,...) components.

El-Jabi (1980) showed that the hydroeconomic system may be described by the following set of relations usable in a numerical analysis of the systems: *i*- the depth-discharge relation; *ii*- the depth-damage relation; *iii*- the discharge-frequency relation; and *iv*- the damage-frequency relation.

The pooling of these relations produces a distribution function $\varphi_t(d)$ of flood damage that can be expressed as follows

$$\varphi_t(d) = F_t\{g^{-1}[\theta^{-1}(K,d)]\}$$

d : a vector representing different types of physical and non-physical damage;

k : a vector representing the physical capital (residential property, commerces and industries, stock,...) along with the associated activities (production, services,...);

θ : a discharge-damage functional transfer relation;

g : a depth-discharge functional transfer relation;

F_t : a distribution function of hydrologic characteristics considered in an interval of time (0,t] given by equation (44).

This approach to the flood phenomenon and to its impact on flood plains opens the road to the following fundamental objectives

i- analyzing flood-damage transfer relations to obtain a probability distribution function of flood damage;

ii- finding an estimate of this damage without having to make use of an after-flood survey;

iii- comparing the effectiveness of different flood-control measures.

7.0 CONCLUSIONS

The objective of this study was to present an extreme value model designed for flood analysis. Although no numerical applications were given, several of them may be found in publications listed in the references. This model has the following characteristics:

i- it has no restrictions on the choice of the distribution function of flood magnitude;

ii- the introduction of a base flow makes the analysis of flood duration and flood volume more practical as compared to other flood models in use. It also increases the amount of information on floods occurring in any interval of time;

iii- when using the model, attention may be restricted to intervals of time as short as one day and as long as desired. In practice this is very important for the study of seasonal effects on streamflow in a particular region;

iv- unlike other models in use, this model permits the study of the multivariate distribution of the number of floods occurring in a given interval of time, and the magnitudes of these floods.

REFERENCES

ASHKAR, F., (1980). Partial Duration Series Models for Flood Analysis, Ph.D. Thesis, Ecole Polytechnique de Montréal, 172 p.

COX, D.R. and MILLER, H.D., (1965). The Theory of Stochastic Processes, Chapman and Hall Ltd, 398 p.

EL-JABI, N., (1980). Approche Systématique pour l'Aménagement des Plaines Inondables, Ph.D. Thesis, Ecole Polytechnique de Montréal, 263 p.

GRAY, D.M., (1972). Manuel des Principes d'Hydrologie, CNRC, Canada, 13 chapters.

HINDIE, F.S., (1974). Approche Bayesienne pour l'Estimation de l'Erreur due à l'Echantillonnage dans l'Evaluation des Débits de Crues, Masters Thesis, Ecole Polytechnique de Montréal, 78 p.

KIRBY, W., (1969). On the Random Occurrence of Major Floods, Water Resour. Res., vol. 5, no 4, pp. 778-784.

ROUSSELLE, J., (1972). On Some Problems of Flood Analysis, Ph.D. Thesis, Colorado State Univ., 226 p.

ROUSSELLE, J. and EL-JABI, N., (1976). Rivière des Prairies, Représentation Stochastique des Crues, Eau du Québec, vol. 9, no 1, pp. 23-28.

TIERCELIN, J.R., (1973). Modèles Probabilistes en Hydrologie, La Houille Blanche, no 7, pp. 547-552.

TODOROVIC, P., (1967). A Stochastic Process of Monotonous Sample Functions, Math. Inst. of the Republic of Serbia, vol. 4, no 19, pp. 149-158.

TODOROVIC, P., (1968). A Mathematical Study of Precipitation Phenomena, Report no CET67 - 68PT65, Colorado State Univ., 123 p.

TODOROVIC, P., (1970). On Some Problems Involving Random Number of Random Variables, Ann. Math. Statist., vol. 41, no 3, pp. 1059-1063.

TODOROVIC, P., (1978). Stochastic Models of Floods, Water Resour. Res., vol. 14, no 2, pp. 345-356.

TODOROVIC, P. and ROUSSELLE, J., (1971). Some Problems of Flood Analysis, Water Resour. Res., vol. 7, no 5, pp. 1144-1150.

TODOROVIC, P. and YEVJEVICH, V., (1967). A Particular Stochastic Process as Applied to Hydrology, Proc. Int. Hydrol. Symposium, Colorado, pp. 298-305.

TODOROVIC, P. and YEVJEVICH, V., (1969). Stochastic Process of Precipitation, Colorado State Univ., Hydrol. paper no 35, 61 p.

TODOROVIC, P. and ZELENHASIC, E., (1970). A Stochastic Model for Flood Analysis, Water Resour. Res., vol. 6, no 6, pp. 1641-1648.

YEVJEVICH, V., (1974). Systematization of Flood Control Measures, ASCE, HY11, pp. 1537-1548.

ZELENHASIC, E., (1970). Theoretical Probability Distributions for Flood Peaks, Colorado State Univ.; Hydrol. Paper no 42, 35 p.

A COMPARATIVE STUDY ON ESTIMATION OF PARAMETERS OF A MARKOVIAN PROCESS - I

A.A. ABD-ALLA AND A.M. ABOUAMMOH
University of Riyadh, Saudi Arabia

ABSTRACT

This paper considers the stationary autoregressive process of order one $X_t = \alpha X_{t-1} + \varepsilon_t, t \in I = \{\ldots,-2,-1,0,1,2,\ldots\}$ where ε_t is a sequence of independent normally distributed random variables with zero mean and variance σ_ε^2. The conditional maximum likelihood estimates for σ_ε^2 and α are obtained. In addition these estimates are compared with the empirical estimate of σ_ε^2 and the Bayes' estimate for α. The parameter α is taken to be priori distributed as uniform, standard normal and non-standard normal respectively, where the stationarity condition is $-1 < \alpha < 1$.

INTRODUCTION

Consider the stationary autoregressive process of order one (AR(1))

$$X_t = \alpha X_{t-1} + \varepsilon_t \tag{1.1}$$

where $\{\varepsilon_t\}, t \in I$ is a sequence of independent normally distributed with zero mean and variance σ_ε^2. The process $\{X_t\}, t \in I$ is normally distributed with zero mean and variance $\sigma_X^2 = \sigma_\varepsilon^2/(1-\alpha^2)$. It is known that such process is stationary if the parameter α satisfies the condition $|\alpha| < 1$. The main statistical problem in (1.1) is estimating the unknown parameters σ_ε^2 and α. The problem of estimating these parameters using different methods such as the Yule-Walker's, the maximum likelihood and the conditional maximum likelihood (C.M.L.) is discussed by Box and Jenkins (1976).

The likelihood function of σ_ε^2 and α is given by Abd-Alla (1980) as

$$f(\alpha,\sigma_\varepsilon^2)=(2\pi\ \sigma_\varepsilon^2)^{-n/2}(1-\alpha^2)^{\frac{1}{2}} \exp\left[-(2\sigma_\varepsilon^2)^{-1}\{(1-\alpha^2)x_1^2+\sum_{i=2}^{n}(x_i-\alpha x_{i-1})^2\}\right] \tag{1.2}$$

Reprinted from *Time Series Methods in Hydrosciences*, by A.H. El-Shaarawi and S.R. Esterby (Editors)

This shows that the solutions of the likelihood equations are not simple.

THE CONDITIONAL LIKELIHOOD

In order to simplify the solutions of the likelihood function, we assume in (1.2) that x_1 is fixed. This gives

$$f(\alpha,\sigma^2_\varepsilon/x_1)=(2\pi\ \sigma^2_\varepsilon)^{-(n-1)/2}\ \exp\{-\ (2\sigma^2_\varepsilon)^{-1}\sum_{i=2}^{n}(x_i-\alpha x_{i-1})^2\} \tag{2.1}$$

By solving the C.M.L. equations $\partial L/\partial\alpha = 0$ and $\partial L/\partial\sigma^2_\varepsilon = 0$, where $L = \log f(\alpha,\sigma^2_\varepsilon/x_1)$ to get the C.M.L. estimators

$$\hat{\sigma}^2_\varepsilon = (n-1)^{-1}\sum_{i=2}^{n} x_i^2-2\{(\sum_{i=2}^{n} x_i x_{i-1})/\sum_{i=2}^{n} x_{i-1}^2\}\sum_{i=2}^{n} x_i x_{i-1} + \{(\sum_{i=2}^{n} x_i x_{i-1})/\sum_{i=2}^{n} x_{i-1}^2\}^2\sum_{i=2}^{n} x_{i-1}^2 \tag{2.2}$$

$$\hat{\alpha}_{(CL)} = \sum_{i=2}^{n} x_i x_{i-1}/\sum_{i=2}^{n} x_{i-1}^2 \tag{2.3}$$

for σ^2_ε and α respectively.

THE BAYESIAN APPROACH

It can be easily shown that the empirical estimate $\hat{\sigma}^2_{\varepsilon(E)}$ of σ^2_ε is

$$\hat{\sigma}^2_{\varepsilon(E)} = \{\frac{1}{n}\sum_{t=1}^{n} x_t^2-(\frac{1}{n-1}\sum_{t=1}^{n-1} x_{t+1}^2)(\frac{n}{n-1}\sum_{t=2}^{n} x_{t-1}^2/\sum_{t=1}^{n} x_t^2)\}/$$

$$\{1-\frac{n}{n-1}\sum_{i=2}^{n} x_{t-1}^2/\sum_{i=1}^{n} x_t^2\}. \tag{3.1}$$

Let the prior knowledge about α be represented by the $g(\alpha)$. The other information of sample data are expressed in the conditional likelihood function $f(\alpha,\sigma^2_\varepsilon/x_1)$. The posterior density of α is then

$g(\alpha|x_1) = f(\alpha,\sigma_\varepsilon^2/x_1)\ g(\alpha)/\int_\alpha f(\alpha,\sigma_\varepsilon^2/x_1)\ g(\alpha)\ d\alpha$

and the Bayes point estimate $\hat{\alpha}$ of α is taken as the mean of the posterior distribution

$$\hat{\alpha} = \int \alpha g\ (\alpha|x_1)\ d\alpha \tag{3.2}$$

The following three special cases for the prior distribution of α are considered: (i) the density of α is uniform on (-1,1); (ii) the density of α is normal with mean 0 and unit variance; and (iii) the density of α is normal with mean μ_α and variance σ_α^2. In the three cases, the mean of the posterior distribution is calculated using numerical integration methods (Davis and Rabinowitz (1975)). We shall denote to the Bayes estimates by $\hat{\alpha}_{(U)}$, $\hat{\alpha}_{(SN)}$ and $\hat{\alpha}_{(N)}$ for the uniform, standard normal and the normal prior distributions respectively. The calculations of $\hat{\alpha}_{(U)}$ and $\hat{\alpha}_{(SN)}$ are straight forward but the calculation of $\hat{\alpha}_{(N)}$ requires the knowledge of σ_ε^2, μ_α and σ_α^2.

To find empirical estimates for these values we note the following:

$$E[X_t^2] = (\mu_\alpha^2 + \sigma_\alpha^2)\ E\ X_{t-1}^2 + \sigma_\varepsilon^2 \tag{3.3 a}$$

$$E[X_{t-1}^2] = (\mu_\alpha^2 + \sigma_\alpha^2)\ E\ X_{t-2}^2 + \sigma_\varepsilon^2 \tag{3.3 b}$$

$$E[X_t\ X_{t+1}] = E\ \alpha^3\ E\ X_{t-1}^2 + \mu_\alpha \sigma_\varepsilon^2 \tag{3.3 c}$$

$$E[X_t\ X_{t-1}] = E\ \alpha^3\ E\ X_{t-2}^2 + \mu_\alpha \sigma_\varepsilon^2 \tag{3.3 d}$$

From (3.3 a) and (3.3 b) one can obtain:

$$\sigma_\varepsilon^2 = \{(E\ X_{t-2}^2)(E\ X_t^2) - E\ X_{t-1}^2)^2\}/\{E\ X_{t-2}^2 - E\ X_{t-1}^2\} \tag{3.4}$$

From (3.3 c) and (3.3 d), the quantity $E\ \alpha^3$ can be omitted as follows:

$$\sigma_\varepsilon^2\ \mu_\alpha = \{(E\ X_tX_{t+1})(E\ X_{t-2}^2)-(E\ X_{t-1}^2)(E\ X_tX_{t-1})\}/\{E\ X_{t-2}^2-(E\ X_{t-1}^2)\} \tag{3.5}$$

Using (3.4) and (3.5), we obtain

$$\mu_\alpha = \{(E\ X_{t-1}^2)(E\ X_tX_{t-1})^2 - (E\ X_tX_{t-1})(E\ X_{t-2}^2)\}/\{(E\ X_{t-2}^2)\ (E\ X_t^2) - (E\ X_{t-1}^2)^2\} \tag{3.6}$$

Using (3.3 a) and (3.6) the value of σ_α^2 becomes

$$\sigma_\alpha^2 = [(E\,X_t^2 - \{E\,X_{t-2}^2\;E\,X_t^2 - (E\,X_{t-1}^2)^2\}/\{E\,X_{t-2}^2 - E\,X_{t-1}^2\})/E\,X_{t-1}^2] - \mu_\alpha^2 \qquad (3.7)$$

Then the value of $\alpha_{(N)}$ can be calculated by a numerical integration after replacing the values of $\sigma_\varepsilon^2, \mu_\alpha$ and σ_α^2 by their empirical estimates. This method reduces the undesirable effect of the prior distribution, (Barnett (1973), Lindley (1965)).

RESULTS

The results are summarized in tables (a) and (b). Table (a) shows the conditional maximum likelihood estimate $\hat{\alpha}_{(CL)}$, the Bayes estimates $\hat{\alpha}_{(U)}$, $\hat{\alpha}_{(SN)}$ and $\hat{\alpha}_{(N)}$. Table (b) gives the conditional maximum likelihood estimate $\sigma_{\varepsilon(CL)}^2$ and the empirical estimate $\hat{\sigma}_{\varepsilon(E)}^2$ of σ_ε^2.

From Table (a), it can be concluded for samples of size 100,150,200, that $|\hat{\alpha}_{(CL)} - \alpha| < |\hat{\alpha}_{(U)} - \alpha|$ which shows that $\hat{\alpha}_{(CL)}$ is closer to the true value α than $\hat{\alpha}_{(U)}$, $|\hat{\alpha}_{(CL)} - \alpha| < |\hat{\alpha}_{(N)} - \alpha|$ is true except when $n = 100$ and $\hat{\alpha} = 0.1.0.3,0.4,0.5$, $n = 150$ and $\alpha = 0.1,0.3,0.4,0.5.0.7$, $n = 200$ and $\alpha = 0.1,-0.4,0.4,0.5,0.7$. Also, $|\hat{\alpha}_{(N)} - \alpha| < |\hat{\alpha}_{(CL)} - \alpha|$ is true when mostly, α and n increase in particular for $n = 100$ and $\alpha = 0.1,0.3,0.4,0.5$, $n = 150$ and $\alpha = 0.1,0.3,0.4,0.5,0.7$ and $n = 200$ $\alpha = 0.3,0.4,0.5,0.7$. Generally, $\hat{\alpha}_{(CL)}$ is preferable for negative values of α whereas $\hat{\alpha}_{(N)}$ is better than other estimates, mostly, for all positive values of α. Table (b) shows that the mean square error (MSE) of $\hat{\sigma}_{\varepsilon\,(CL)}^2$ is less than the MSE of $\hat{\sigma}_{\varepsilon(E)}^2$.

ACKNOWLEDGEMENT

The authors are grateful to prof. K. Alam of Clemson University for his encouragement and useful discussion.

Table (a)

n	$\hat{\alpha}$ \ α	-0.7	-0.5	-0.4	-0.3	-0.1	0.1	0.3	0.4	0.5	0.7
100	$\hat{\alpha}$(CL)	-0.755	-0.543	-0.436	-0.328	-0.113	0.189	0.219	0.284	0.579	0.743
	$\hat{\alpha}$(U)	-0.760	-0.557	-0.446	-0.337	-0.116	0.193	0.215	0.278	0.566	0.745
	$\hat{\alpha}$(SN)	-0.636	-0.462	-0.447	-0.333	-0.174	0.101	0.312	0.438	0.524	0.610
	$\hat{\alpha}$(N)	-0.644	-0.552	-0.441	-0.333	-0.115	0.108	0.335	0.448	0.559	0.631
150	$\hat{\alpha}$(CL)	-0.735	-0.521	-0.416	-0.312	-0.103	0.017	0.225	0.373	0.424	0.745
	$\hat{\alpha}$(U)	-0.738	-0.527	-0.422	-0.317	-0.105	0.017	0.220	0.366	0.418	0.763
	$\hat{\alpha}$(SN)	-0.674	-0.440	-0.416	-0.420	-0.163	0.101	0.302	0.404	0.506	0.709
	$\hat{\alpha}$(N)	-0.720	-0.523	-0.419	-0.316	-0.105	0.106	0.315	0.419	0.521	0.716
200	$\hat{\alpha}$(CL)	-0.728	-0.540	-0.422	-0.346	-0.149	0.081	0.306	0.293	0.408	0.609
	$\hat{\alpha}$(U)	-0.732	-0.545	-0.447	-0.350	-0.151	0.880	0.306	0.297	0.401	0.600
	$\hat{\alpha}$(SN)	-0.648	-0.420	-0.416	-0.368	-0.103	0.055	0.261	0.369	0.479	0.700
	$\hat{\alpha}$(N)	-0.736	-0.512	-0.445	-0.348	-0.150	0.056	0.271	0.381	0.493	0.710

Table (b)

σ_ε^2	n	Method \ α	0.1	0.3	0.4	0.5	0.7
$\sigma_\varepsilon^2 = 1$	100	$\hat{\sigma}^2_{\varepsilon}(E)$	0.94074	1.10103	1.17386	1.25807	1.54349
		$\hat{\sigma}^2_{\varepsilon}(CL)$	1.01633	1.01548	1.01451	1.01308	1.00921
	150	$\hat{\sigma}^2_{\varepsilon}(E)$	1.08237	1.22856	1.29405	1.36664	1.59721
		$\hat{\sigma}^2_{\varepsilon}(CL)$	1.04618	1.04601	1.04582	1.04550	1.04446
	200	$\hat{\sigma}^2_{\varepsilon}(E)$	1.01127	1.12622	1.7537	1.23152	1.427
		$\hat{\sigma}^2_{\varepsilon}(CL)$	1.00936	1.01033	1.01089	1.01130	1.01103
$\sigma_\varepsilon^2 = 2$	100	$\hat{\sigma}^2_{\varepsilon}(E)$	1.86904	1.18252	2.32612	2.4955	3.05623
		$\hat{\sigma}^2_{\varepsilon}(CL)$	2.03257	2.03070	2.02874	2.02591	2.01847
	150	$\hat{\sigma}^2_{\varepsilon}(E)$	2.15476	2.44326	2.57290	2.71689	3.01733
		$\hat{\sigma}^2_{\varepsilon}(CL)$	2.09231	2.09191	2.09152	2.09089	3.08903
	200	$\hat{\sigma}^2_{\varepsilon}(E)$	2.01540	2.24116	2.33960	2.4509	2.83880
		$\hat{\sigma}^2_{\varepsilon}(CL)$	2.01892	2.02073	2.02177	2.02256	2.02213
$\sigma_\varepsilon^2 = 3$	100	$\hat{\sigma}^2_{\varepsilon}(E)$	2.80123	3.26742	3.48178	3.73048	4.57235
		$\hat{\sigma}^2_{\varepsilon}(CL)$	3.04873	3.04584	3.04288	3.03869	3.02787
	150	$\hat{\sigma}^2_{\varepsilon}(E)$	3.22945	3.66011	3.85411	4.06932	4.75156
		$\hat{\sigma}^2_{\varepsilon}(CL)$	3.13840	3.13781	3.13722	3.13634	3.13363
	200	$\hat{\sigma}^2_{\varepsilon}(E)$	3.02137	3.35998	3.50550	3.67207	4.25218
		$\hat{\sigma}^2_{\varepsilon}(CL)$	3.02854	3.03119	3.03272	3.03392	3.03326
$\sigma_\varepsilon^2 = 4$	100	$\hat{\sigma}^2_{\varepsilon}(E)$	3.73543	4.35412	4.63924	4.97021	6.09018
		$\hat{\sigma}^2_{\varepsilon}(CL)$	4.06492	4.06097	4.05704	4.05152	4.03720
	150	$\hat{\sigma}^2_{\varepsilon}(E)$	4.30535	4.87809	5.13635	5.42298	6.33088
		$\hat{\sigma}^2_{\varepsilon}(CL)$	4.18451	4.18371	4.18298	4.18179	4.17820
	200	$\hat{\sigma}^2_{\varepsilon}(E)$	4.02829	4.47858	4.67229	4.89418	6.66631
		$\hat{\sigma}^2_{\varepsilon}(CL)$	4.03820	4.04160	4.04370	4.04525	4.04435

REFERENCES:

Abd-Alla, A.A. (1980); The estimation of the parameters of the first and the second order autoregressive process with zero level. In the proc. of the 6th Saudi National comp. conf. p.p. 247-259.

Barnett, V. (1973): Comparative Statistical Inference. Wiley.

Box, G.E.P. and Jenkins, G.M. (1976): Time series analysis, Forecasting and Control. Holden-Day.

Davis, P.J. and Rabinowitz, P. (1975): Numerical Integration. Blaisdell.

Lindley, D.V. (1965): Introduction to Probability and Statistics from a Bayesian Viewpoint. Part 2 Inference. Cambridge Univ. Press.

GENERALIZED LEAST SQUARES PROCEDURE FOR REGRESSION WITH AUTOCORRELATED ERRORS

U.L. GOURANGA RAO

Dalhousie University, Halifax, Canada

ABSTRACT

In this paper the generalized least squares procedure is used to estimate the parameters of the standard linear regression model where the errors follow a first-order autoregressive process. This procedure is similar to the full maximum likelihood procedure of Beach and MacKinnon (1978). Asymptotically the two procedures are equivalent. Results of sampling experiments suggest that the generalized least squares procedure is marginally superior to the full maximum likelihood procedure.

1. INTRODUCTION

Consider the standard linear regression model where the errors follow a first-order autoregressive process,

$$y = X\beta + u \tag{1}$$

$$u_t = \rho u_{t-1} + \varepsilon_t, \quad |\rho| < 1, \qquad t = 1, 2, \ldots, T$$

$$E(\varepsilon_t) = 0 \text{ and } Var(\varepsilon_t) = \sigma^2 \qquad t = 1, 2, \ldots, T.$$

Given the assumptions regarding ε_t, the following results are valid for the error vector u:

$$E(u) = 0 \text{ and } Cov(u) = \frac{\sigma^2}{(1-\rho^2)} V \tag{2}$$

where

$$V = \begin{bmatrix} 1 & \rho & \rho^2 & \cdot & \cdot & \cdot & \rho^{T-1} \\ \rho & 1 & \rho & & & & \rho^{T-2} \\ \cdot & & & & & & \cdot \\ \cdot & & & & & & \cdot \\ \rho^{T-1} & \rho^{T-2} & & & & & 1 \end{bmatrix}$$

For $\rho \varepsilon (-1, +1)$, V is positive definite. The parameters of the model to be estimated are: β, ρ and σ^2.

Reprinted from *Time Series Methods in Hydrosciences*, by A.H. El-Shaarawi and S.R. Esterby (Editors)

Given the above set-up, V^{-1} can be written as:

$$V^{-1} = Q'Q \tag{3}$$

where

$$Q = \begin{bmatrix} (1-\rho^2)^{\frac{1}{2}} & 0 & 0 & . & . & . & 0 \\ -\rho & 1 & 0 & . & . & . & 0 \\ . & & & & & & . \\ . & & & & & & . \\ 0 & 0 & 0 & & & -\rho & 1 \end{bmatrix}$$

The typical Cochrane and Orcutt (1949) estimator is a generalized least squares (GLS) estimator which ignores the information contained in u_1 and minimizes the generalized sum of squares of residuals $u'V_1^{-1}u$ where $V_1 = Q_1'Q_1$ and that Q_1 is of order (T-1)xT which is obtained by deleting the first row of Q. Gurland (1954), Kadiyala (1968), and Rao and Grilliches (1969) have studied the relative efficiency of the GLS estimators based on V_1 and V and concluded that using V_1 instead of V could, in certain cases, result in a substantial loss of efficiency. In this paper, the GLS procedure is used to obtain estimates of β, ρ and σ^2.

2. THE GENERALIZED LEAST SQUARES PROCEDURE

The generalized least squares estimator $\hat{\beta}$ of β is obtained by minimizing the generalized sum of squares of residuals

$$\begin{aligned} S(\beta,\rho) &= (y - X\beta)'V^{-1}(y - X\beta) \\ &= (y^*-X^*\beta)'V^{-1}(y^*-X^*\beta) \end{aligned} \tag{4}$$

where $y^* = Qy$ and $X^* = QX$.

If ρ is known, the efficient Aitken's (1935) estimator is given by

$$\hat{\beta}(\rho) = (X^{*}{}'X^*)^{-1}X^{*}{}'y^* = (X'V^{-1}X)^{-1}X'V^{-1}y \tag{5}$$

If ρ is not known, $\hat{\beta}(\rho)$ represents partial minimization of $S(\beta,\rho)$ with respect to β. Substituting (5) in (4), we have

$$S(\rho) = \hat{u}'(\rho)V^{-1}\hat{u}'(\rho) \tag{6}$$

where $\hat{u}(\rho) = y - X\hat{\beta}(\rho)$ and that $S(\rho)$ has to be minimized with respect to ρ. Noting that S, $\hat{u}$, and $\hat{\beta}$ are all functions of ρ, the argument ρ in these functions will henceforth be dropped for convenience.

It has been pointed by many researchers including Dhrymes (1971, p. 68) that the expression in (6) is highly nonlinear in ρ which precludes a direct approach to the solution of $dS/d\rho = 0$. The non-linearity of the expression (6) led Hildreth and Lu (1960) to suggest a grid-search procedure for estimating ρ. Fortunately, the solution of $dS/d\rho = 0$ is not difficult as we will demonstrate below.

Consider $S = \hat{u}'V^{-1}\hat{u}$.

$$\begin{aligned}\frac{dS}{d\rho} &= \frac{d\hat{u}'}{d\rho}V^{-1}\hat{u} + \hat{u}'\frac{dV^{-1}}{d\rho}\hat{u} + \hat{u}'V^{-1}\frac{d\hat{u}}{d\rho}\\ &= 2\frac{d\hat{u}'}{d\rho}V^{-1}\hat{u} + \hat{u}'\frac{dV^{-1}}{d\rho}\hat{u}\\ &=-2\frac{d\hat{\beta}'}{d\rho}X'V^{-1}\hat{u} + \hat{u}'\frac{dV^{-1}}{d\rho}\hat{u}\\ &= \hat{u}'\frac{dV^{-1}}{d\rho}\hat{u} \end{aligned} \tag{7}$$

The expression in (7) is obtained using the following results:

$$\frac{d\hat{u}}{d\rho} = \frac{d\hat{u}}{d\hat{\beta}}\frac{d\hat{\beta}}{d\rho} = -X\frac{d\hat{\beta}}{d\rho} \tag{8.1}$$

$$X'V^{-1}\hat{u} = X'V^{-1}(y - X\hat{\beta}) = 0. \tag{8.2}$$

Setting $\frac{dS}{d\rho} = \hat{u}'\frac{dV^{-1}}{d\rho}\hat{u} = 0$

yields the unique solution

$$\hat{\rho} = \sum_{2}^{T}\hat{u}_t\hat{u}_{t-1} \Big/ \sum_{2}^{T-1}\hat{u}_t^2 . \tag{9}$$

The second derivative of S with respect to ρ is

$$\begin{aligned}\frac{d^2S}{d\rho^2} &= \frac{d(\hat{u}'\frac{dV^{-1}}{d\rho}\hat{u})}{d\rho}\\ &= \hat{u}'\frac{d^2V^{-1}}{d\rho^2}\hat{u} + 2\frac{d\hat{u}'}{d\rho}\frac{dV^{-1}}{d\rho}\hat{u}\\ &= \hat{u}'\frac{d^2V^{-1}}{d\rho^2}\hat{u} - 2\frac{d\hat{\beta}'}{d\rho}X'\frac{dV^{-1}}{d\rho}\hat{u} . \end{aligned} \tag{10}$$

The derivative of (8.2) with respect to ρ implies

$$X'\frac{dV^{-1}}{d\rho}\hat{u} = -X'V^{-1}\frac{d\hat{u}}{d\rho} \tag{11}$$

and using (11), we can express (10) as

$$\frac{d^2 S}{d\rho^2} = \hat{u}' \frac{d^2 V^{-1}}{d\rho^2} \hat{u} - 2 \frac{d\hat{u}'}{d\rho} V^{-1} \frac{d\hat{u}}{d\rho} . \qquad (12)$$

The second derivative of V^{-1} with respect to ρ in (12) is

$$\frac{d^2 V^{-1}}{d\rho^2} = 2I^* \qquad (13)$$

where I^* is modified unit matrix with zero in the left and right hand corner elements. Using (13), we can express (12) as

$$\frac{d^2 S}{d\rho^2} = 2 \sum_{2}^{T-1} \hat{u}_t^2 - 2 \frac{d\hat{u}'}{d\rho} V^{-1} \frac{d\hat{u}}{d\rho} . \qquad (14)$$

Thus $d^2S/d\rho^2$ is the difference between two quadratic forms which are positive definite involving the GLS estimated residuals $\hat{u}_t$ and hence data-relative. Assuming that the second order condition is satisfied, we have a minimum for S.

We may now summarize the GLS estimators of the parameters of the model as follows:

$$\hat{\beta}(\hat{\rho}) = (X'\hat{V}^{-1}X)^{-1}X'\hat{V}^{-1}y \qquad (15.1)$$

$$\hat{\rho} = \sum_{2}^{T} \hat{u}_t \hat{u}_{t-1} \Big/ \sum_{2}^{T-1} \hat{u}_t^2 \qquad (15.2)$$

$$\hat{\sigma}^2(\hat{\rho}) = S(\hat{\rho})/ T \qquad (15.3)$$

where $\hat{V}^{-1}$ is the matrix V^{-1} with $\hat{\rho}$ substituted for ρ.

The consistency of $\hat{\rho}$ was proved by Hildreth and Lu (1960) and Dhrymes (1971, pp. 91-96) and hence $\hat{\beta}(\hat{\rho})$ and $\hat{\sigma}^2(\hat{\rho})$ are also consistent. An estimate of the asymptotic covariance matrix of $\hat{\beta}(\hat{\rho})$ is given by

$$\mathrm{Cov}(\hat{\beta}(\hat{\rho})) = (X'\hat{V}^{-1}X)^{-1}\hat{\sigma}^2(\hat{\rho}) . \qquad (16)$$

So far we made no assumptions regarding the probability distribution of ε_t except that the first and second moments exist. If we assume that ε_t's are NID$(0,\sigma^2)$, then we can apply the maximum likelihood procedure. Following Beach and MacKinnon (1978) and Dhrymes (1971, p. 70), the concentrated log-likelihood function can be written as

$$L^*(\rho) = \text{Constant} - \frac{T}{2} \log(\sigma^2(\rho) \,/\, (1 - \rho^2)^{1/T}). \qquad (17)$$

It is clear from (17) that maximizing $L^*(\rho)$ with respect to ρ is equivalent to minimizing

$$\sigma^2(\rho) \;/\; (1 - \rho^2)^{1/T} \tag{18}$$

which yields the full maximum likelihood (FML) estimates of β, ρ and σ^2. In GLS we minimize $\sigma^2(\rho)$ whereas in FML we minimize $\sigma^2(\rho) / (1 - \rho^2)^{1/T}$. Asymptotically the two procedures are equivalent because the limit of $(1 - \rho^2)^{1/T}$ is unity as $T \to \infty$.

3. AN ALGORITHM FOR MINIMIZING $S(\beta,\rho)$

Equations (5) and (9) constitute the two components of an algorithm for computing the GLS estimates of the parameters of the model by minimizing $S(\beta,\rho)$. The procedure consists in alternatively minimizing $S(\beta,\rho)$ with respect to β, ρ held constant, and minimizing $S(\beta,\rho)$ with respect to ρ, β held constant. Usually, we start with $\rho = 0$ and compute $\hat{\beta}$ and $\hat{\rho}$. Since S reaches its minimum value at $\rho = \hat{\rho}$, it follows that S is an increasing function of ρ for $\rho > \hat{\rho}$, and a decreasing function of ρ for $\rho < \hat{\rho}$. Accordingly, if $\rho = 0 > \hat{\rho}$, i.e., $\sum_2^T \hat{u}_t\hat{u}_{t-1} < 0$, then S decreases for values of $\rho < 0$. Since our objective is to minimize S, we choose $\rho = \hat{\rho}$ for the next iteration. If, on the other hand, $\rho = 0 < \hat{\rho}$, i.e., $\sum_2^T \hat{u}_t\hat{u}_{t-1} > 0$, then S decreases for values of $\rho > 0$. Since our objective is to minimize S, we choose $\rho = \hat{\rho} > 0$ for the next iteration. Thus, in either case, the computed $\hat{\rho}$ will be the value to be used for the next iteration. When two $\hat{\rho}$-values in successive iterations are close, the iterative procedure is stopped.

Iterative procedures of the type described above raise two questions. The first is concerned with the convergence of S to a stationary value and the second is concerned with the attainment of a global minimum as opposed to a local mimimum. $\hat{\rho}$ being an estimator of the correlation coefficient, it follows from equation (9) that the stationarity condition $|\hat{\rho}| < 1$ may not always be satisfied and the convergence of S may be jeopardized. But the results of sampling experiments we have conducted reveal that this is not a serious drawback of the algorithm. The question of the attainment of global minimum was investigated by Sargan (1964) who concluded that there was no indication of the occurrence of multiple minima.

4. MONTE CARLO RESULTS

The GLS procedure is similar to the FML procedure of Beach and MacKinnon (1978). The advantage of GLS is that it is distribution-free. In order to compare the relative performances of these two estimators in small sample situations, we conducted three sampling experiments. The model-structure used for purposes of experimentation is the same as of Beach and MacKinnon (1978):

$$y_t = \beta_1 + \beta_2 x_t + u_t, \; u_t = \rho u_{t-1} + \varepsilon_t, \; \varepsilon_t \sim NID(0,.0036).$$

The x_t's contain a large trend component and are computed from the equation $x_t = \exp(.04t) + w_t$, $w_t \sim NID(0,.0009)$. The values of β_1 and β_2 are set equal to unity. In each of the three experiments, three values of ρ were considered and they are $\rho = .60$, .80, and .99. Experiments 1 and 2 differ only in the sample size used; in experiment 1 the sample size is 20 and in experiment 2 the sample size is 50. In experiment 3 the sample size is 20, and the ε_t's are realizations on a χ^2 random variable with one degree of freedom adjusted for zero mean. The variance of ε_t is, of course, 2.0. Thus experiment 3 permits us to examine the effect of a specification error (nonnormal errors) on the FML estimator and this might shed light on the robustness of the FML estimator. Each experiment involved one thousand replications. On each replication, FML and GLS estimators are used to compute point estimates of β_1, β_2 and ρ for a given realization of the ε's. It may be mentioned here that in order to ensure that the error process was stationary, u_1 was generated as $(1/(1-\rho^2)^{\frac{1}{2}})\varepsilon_1$. Since FML and GLS point estimates have to be computed by iterative procedures, we started off in each case with the initial value of $\rho = 0$ and stopped iterating when two successive estimated values of ρ differed by less than 10^{-5}.

In Table 1, the mean biases, root mean square errors (RMSE) and the number of GLS estimates that were closer to the corresponding true values than the FML estimates (in the 'NO' column) are presented by experiments. In Table 2, the ratios of the mean square error of FML to the mean square error of GLS are presented by experiments.

The first element in each cell represents the FML estimator and is followed by the corresponding statistic for the GLS estimator. Since experiments 1 and 2 are identical to those of Beach and MacKinnon (1978) except for the number of replications, their format is borrowed for presenting the results to facilitate easy comparisons of our results with theirs.

The biases of both estimators for β_1 and β_2 are not significant at the .05 level. The biases of both estimators for ρ are negative and significant at the .01 level. Each estimator produced biases of β_1 and β_2 which are opposite in sign. The biases are invariably smaller in the case of the GLS estimator for ρ. This is also evident from the frequency of the GLS estimates of ρ which came closer to the true value of ρ than the FML estimator. In experiments 1 and 2, the biases of the GLS estimator are very close to the corresponding biases of the FML estimator. Though the assumption of normal errors is violated, yet, the FML estimator proved to be slightly superior to the GLS estimator by generating smaller biases for β_1 and β_2 in experiment 3. Most often, the frequency of the GLS estimates which came closer to the true values of the parameters is significantly larger at the .05 level. The large sample bionomial test is used for this purpose. See the frequencies marked with an asterisk in Table 1. The RMSE's of the GLS estimator are most often smaller than the corresponding RMSE's of the FML estimator. This is expected of the GLS estimator in view of its minimum variance property. The relative efficiencies of the GLS estimator with respect to the FML estimator are most often greater than unity.

The number of iterations in each replication are either equal or the GLS estimator required just one more iteration. Computational efficiency is the same for both estimators. The average number of iterations in each experiment for the two estimators are displayed in Table 3. In this context one remark is in order. Equation (9) which defines the $\hat{\rho}$ for each iteration may not satisfy the stationarity condition $|\hat{\rho}| < 1$. In such an event, that sample was discarded and a new sample was added. The number of samples thus discarded for want

TABLE 1

BIAS AND ROOT MEAN SQUARE ERROR OF FML AND GLS ESTIMATORS[a]

Parameter	True ρ	Experiment 1 Bias	RMSE	NO.	Experiment 2 Bias	RMSE	NO.	Experiment 3 Bias	RMSE	NO.
ρ	.60	-.2385	.3235	792*	-.0985	.1588	732*	-.2254	.3005	811*
		-.2171	.3177		-.0818	.1552		-.2028	.2933	
	.80	-.3050	.3736	910*	-.1154	.1644	816*	-.2905	.3507	914*
		-.2751	.3582		-.1006	.1565		-.2593	.3334	
	.99	-.4177	.4674	988*	-.1704	.2006	1000*	-.4137	.4618	990*
		-.3816	.4425		-.1516	.1886		-.3770	.4372	
β_1	.60	.0006	.1269	520	.0004	.0410	499	.0201	2.9127	574*
		.0006	.1271		.0004	.0410		.0221	2.9315	
	.80	.0014	.1940	537*	.0010	.0743	524	-.0303	4.3450	583*
		.0014	.1932		.0010	.0743		-.0339	4.3380	
	.99	.0001	.5293	592*	.0057	.4422	528	-.0135	9.4827	596*
		-.0002	.5259		.0056	.4412		-.0353	9.4440	
β_2	.60	-.0002	.0771	508	-.0000	.0106	499	-.0041	1.7657	547*
		-.0003	.0773		-.0000	.0106		-.0017	1.7741	
	.80	-.0004	.1155	560*	-.0001	.0184	513	.0301	2.5911	568*
		-.0004	.1150		-.0002	.0185		.0390	2.5803	
	.99	.0016	.2095	631*	-.0005	.0566	609*	.0575	4.9334	613*
		.0019	.2061		-.0005	.0557		.0795	4.8583	

[a]The first figure in each cell refers to the FML estimator. The second figure in each cell refers to the GLS estimator. The frequency of GLS estimates which came closer to the true parameter value than the corresponding FML estimates and marked with an asterisk are significant at the .05 level.

TABLE 2

RATIOS OF ROOT MEAN SQUARE ERRORS

Parameter	True ρ	Experiment 1	Experiment 2	Experiment 3
ρ	.60	1.0185	1.0234	1.0245
	.80	1.0432	1.0505	1.0517
	.99	1.0562	1.0753	1.0562
β_1	.60	.9981	.9999	.9936
	.80	1.0046	.9996	1.0016
	.99	1.0064	1.0023	1.0040
β_2	.60	.9983	1.0001	.9953
	.80	1.0038	.9999	1.0042
	.99	1.0166	1.0167	1.0155

TABLE 3

AVERAGE NUMBER OF ITERATIONS REQUIRED

	True ρ	FML	GLS
Experiment 1:	.60	4.557	4.735
	.80	5.036	5.365
	.99	5.509	6.211
Experiment 2:	.60	3.597	3.636
	.80	3.963	4.064
	.99	4.728	5.233
Experiment 3:	.60	4.617	4.832
	.80	5.129	5.525
	.99	5.664	6.396

TABLE 4

NUMBER OF SAMPLES DISCARDED

True ρ	Experiment 1	Experiment 2	Experiment 3
.60	0	0	9
.80	2	0	12
.99	10	27	16

of convergence constitutes a very small percentage as is indicated by the figures in Table 4. A similar problem could very well arise in the case of the Cochrane and Orcutt (1949) estimator. See Rao and Grilliches (1969, footnote 6, p. 256).

5. CONCLUSION

The GLS estimator considered in this paper is slightly more efficient than the FML estimator as revealed by the results of the sampling experimets. Except for a few cases wherein convergence is not achieved, the performance of the GLS estimator compares very favourably with the FML estimator. In empirical work it is recommended that both the estimators be used so that the merits of one over the other could be exploited to benefit the researcher.

REFERENCES

Aitken, A. C., 1935. On least squares and linear combinations of observations. Proceedings of the Royal Society of Edinburg, 55: 42-48.

Beach, C.M. and MacKinnon, J. G., 1978. A maximum likelihood procedure for regression with autocorrelated errors. Econometrica, 46: 51-58.

Cochrane, D. and Orcutt, G. H., 1949. Applications of least squares regression to relationships containing autocorrelated error terms. Journal of the American Statistical Association, 44: 32-61.

Dhrymes, P. J., 1971. Distributed Lags. Holden Day, San Francisco.

Gurland, J., 1954. An example of autocorrelated disturbances in linear regression. Econometrica, 22: 218-227.

Hildreth, C. and Lu, J. Y., 1960. Demand relations with autocorrelated disturbances. Research bulletin 276, Michigan State University Agricultural Experiment Station.

Kadiyala, K. R., 1968. A transformation used to circumvent the problem of autocorrelation. Econometrica, 36: 93-96.

Rao, P. and Grilliches, Z., 1969. Small-sample properties of several two-stage regression methods in the context of auto-correlated errors. Journal of the American Statistical Association, 64: 253-272.

Sargan, J. D., 1964. Wages and prices in the United Kingdom: A study in Econometric methodology. In: P.E.Hart, G.Mills and J.K.Whitaker (Editors), Econometric Analysis for National Economic Planning. Butterworth and Co. Ltd., London, pp. 25-54.

FITTING DYNAMIC MODELS TO HYDROLOGICAL TIME SERIES

KEITH W. HIPEL*, A. IAN MCLEOD† and DONALD J. NOAKES*

ABSTRACT

Based upon the physical properties of the phenomena being modelled and valid statistical principles, techniques are presented for fitting dynamic models to hydrological time series. Procedures are devised for properly incorporating one or more covariate series into a dynamic model. Furthermore, when there are missing data points in the series, these can be estimated by including a special type of intervention component in the dynamic model. The efficacy of the model building techniques is clearly demonstrated by designing a transfer function-noise model to describe the dynamic relationships connecting a monthly river flow series in Canada to precipitation and temperature covariate series.

1.1 INTRODUCTION

In order to properly design and operate water resources projects, it is necessary to measure over time and space pertinent hydrological phenomena which may include river flows, precipitation, and temperature. Time series analysis therefore constitutes a very important tool for use in solving water resources problems. In particular, hydrologists often require stochastic models which realistically describe the dynamic relationships connecting a single output series, such as seasonal river flows, to one or more covariate or input series such as precipitation and temperature. Consequently, the purpose of this paper is to clearly demonstrate how both a sound physical understanding of the problem and comprehensive statistical procedures can be employed for developing a comprehensive dynamic model to fit to a set of time series.

* Department of Systems Design Engineering, University of Waterloo, Waterloo, Ontario, Canada
† Department of Statistical and Acturial Sciences, The University of Western Ontario, London, Ontario, Canada

Reprinted from *Time Series Methods in Hydrosciences*, by A.H. El-Shaarawi and S.R. Esterby (Editors)

The specific dynamic model employed in this paper is the transfer function-noise model which is described by Box and Jenkins (1970). By adhering to the identification, estimation and diagnostic checking stages of model construction, a transfer function-noise model is developed for linking a monthly river flow series to precipitation and temperature data sets. The best dynamic model is then selected using the Akaike Information criterion (AIC) (Akaike, 1974). The AIC provides a combined measure of model parsimony and good statistical fit. The model which has the minimum AIC value should be selected when there are several competing models.

In the process of selecting the most appropriate model to fit to the series, a number of useful modelling procedures are suggested. Often there are more than one precipitation and temperature series and a statistical procedure is presented for creating a single sequence to represent the precipitation or temperature series. This approach can be utilized in place of the rather ad hoc methods such as the Isohyetal and Thiessen polygon techniques (see Viessman et al. (1977) for a description of the Isohyetal and Thiessen polygon methods). To decide upon which series to include in the dynamic model and also design the form of the transfer function connecting a covariate series to the output, cross-correlation analyses of the residuals of models fitted to the series can be employed. It is also explained how a dynamic model can be used for estimating missing data points in the output or covariate series and for modelling the effects of one or more external interventions upon the mean level of the output series. The transfer function-noise model which is ultimately chosen by following contemporary modelling procedures can be used for applications such as forecasting and simulation and also providing insight into the physical characteristics of the phenomena being examined.

1.2 THE GENERAL INTERVENTION TRANSFER FUNCTION-NOISE MODEL

Details of the mathematical theory underlying transfer function-noise and intervention models are given in a number of papers (see for example Box and Tiao (1975), Hipel et al. (1975), Hipel et al. (1977b) and Hipel and McLeod (1982)). To give the reader some perspective of the model, the basic development is now presented.

The general intervention transfer function-noise model may be written in the form

$$y_t = f(\underline{k},\underline{x},\underline{\xi},t) + N_t \qquad (1)$$

where t is discrete time; y_t is the response variable; N_t is the stochastic noise component which may be autocorrelated; and $f(\underline{k},\underline{x},\underline{\xi},t)$ is the dynamic component of y_t. The dynamic term includes a set of parameters $\underline{k}$, a group of covariate series $\underline{x}$, and the set of intervention series $\underline{\xi}$ which are required when modelling the effects of external interventions. When required, both y_t and x_t may be transformed using a suitable Box-Cox transformation (Box and Cox, 1964). The reasons for transforming the series include stabilizing the variance and improving the normality assumption of the white noise series which is included in N_t. In hydrologic applications, Box-Cox transformations, in particular the logarithmic transformation, often rectify problems associated with the untransformed series. It should also be noted that the same Box-Cox transformation need not be applied to all of the series.

Once a given series has been transformed, it is then necessary to remove any trends or seasonality in the data. A common procedure employed in hydrology for monthly sequences is to deseasonalize the series by subtracting the estimated monthly mean and dividing by the estimated monthly standard deviation for each data point.

Included in the dynamic component of the model are the effects of all input covariate series and all external interventions. In general, if there are I_1 input covariate series and I_2 interventions, the dynamic component of the model is given by

$$f(\underline{k},\underline{x},\underline{\xi},t) = \sum_{i=1}^{I_1} \frac{\omega_i(B)}{\delta_i(B)} B^{b_i} x_{ti} + \sum_{j=I_1+1}^{I_1+I_2} \frac{\omega_j(B)}{\delta_j(B)} B^{b_j} \xi_{tj} \qquad (2)$$

where x_{ti} is the ith transformed and deseasonalized input covariate series; ξ_{tj} is the jth intervention series consisting of 0's and 1's to indicate the nonoccurrence and occurrence, respectively, of the jth intervention; and B is the backward shift operator such that $Bx_t = x_{t-1}$ and $B^s x_t = x_{t-s}$ where s is a positive integer. The term $\omega_j(B)B^{b_j}/\delta_j(B) = \nu_j(B)$ is the transfer function of the jth intervention or input series and b_j is the delay time for ξ_{tj} or x_{tj} to effect y_t. The transfer function, $\nu_j(B)$, has the form

$$\nu_j(B) = \frac{(\omega_{oj} - \omega_{1j}B - \omega_{2j}B^2 - \ldots - \omega_{u_j j}B^{u_j})}{(1 - \delta_{1j}B - \delta_{2j}B^2 - \ldots - \delta_{r_j j}B^{r_j})} B^{b_j} \tag{3}$$

In practice, u_j and r_j are usually 0 or 1 (see for example Hipel et al. (1975), Hipel et al. (1977b), D'Astous and Hipel (1979), Baracos et al. (1981)). When there are no input covariate series in (2) and hence $I_1 = 0$, the model in (1) is referred to as an intervention model. On the other hand, if there are no external interventions in (2) and therefore $I_2 = 0$, the model in (1) is called a transfer function-noise model.

The noise component of the general intervention transfer function-noise model is defined by

$$N_t = y_t - f(\underline{k},\underline{x},\underline{\xi},t) \tag{4}$$

That is, the noise term of the model is simply the difference between the response variable, y_t, and the dynamic component. The form of the noise term, N_t, is not restricted to any particular form, but a commonly employed family of models is the autoregressive moving-average (ARMA) models.

MODEL CONSTRUCTION

2.1 INTRODUCTION

No matter what types of stochastic models are being considered for fitting to a specified data set, it is recommended to follow the identification, estimation and diagnostic check stages of model construction. At the identification stage, tentative model forms are identified by employing various statistical techniques which are usually easiest to interpret graphically (Box and Jenkins, 1970; Hipel et al., 1977a; McLeod et al., 1977; Hipel et al., 1981; Hipel and McLeod, 1982). For the case of the intervention transfer function-noise model in (1), both the dynamic and noise components must be identified. Following this, maximum likelihood estimates (MLE's) of the model parameters are obtained (McLeod, 1977). Finally, diagnostic checks are employed to insure that the key modelling assumptions are satisfied for a given model.

2.2 IDENTIFYING THE DYNAMIC COMPONENT

Often it is known in advance whether or not a covariate series, x_t, causes the output series, y_t. For instance, precipitation obviously causes river flow. If, however, it is not certain whether one series causes another, procedures are available for determining the type of relationship which may exist between the two series (see for example Granger (1969), Pierce and Haugh (1977), Hipel et al. (1981), and Hipel and McLeod (1982)). In particular, the form of the cross-correlation function (CCF) of the residuals from stochastic models fitted to the two transformed and deseasonalized series, x_t and y_t, can be examined.

In general, the ARMA models fitted to the two series, x_t and y_t, may be symbolically written as

$$\phi_x(B)x_t = \theta_x(B)u_t \qquad (5)$$

and

$$\phi_y(B)y_t = \theta_y(B)v_y \qquad (6)$$

where u_t is NID$(0,\sigma_u^2)$; v_t is NID$(0,\sigma_v^2)$; and $\phi(B)$ and $\theta(B)$ are the AR and MA operators respectively. When there are n observations in each series the residual CCF at lag k between the residuals for the two models can then be estimated by

$$r_{\hat{u}\hat{v}}(k) = \frac{c_{\hat{u}\hat{v}}(k)}{[c_{\hat{u}}(0)c_{\hat{v}}(0)]^{1/2}} \tag{7}$$

where

$$c_{\hat{u}\hat{v}}(k) = \begin{cases} n^{-1} \sum_{t=1}^{n-k} \hat{u}_t \hat{v}_{t+k} , & k \geq 0 \\ n^{-1} \sum_{t=1-k}^{n} \hat{u}_t \hat{v}_{t+k} , & k < 0 \end{cases} \tag{8}$$

is the estimated cross-covariance function at lag k between the estimated residual series $\hat{u}_t$ and $\hat{v}_t$; $c_{\hat{u}}(0)$ is the estimated variance of the $\hat{u}_t$ sequence; and $c_{\hat{v}}(0)$ is the estimated variance of the $\hat{v}_t$ series.

Approximate confidence intervals for the CCF may be obtained by assuming x_t and y_t are independent (Haugh, 1972, 1976). However, McLeod (1979) obtained the asymptotic distribution of the residual CCF for the general case where the x_t and y_t series do not have to be independent of each other, and consequently more accurate confidence limits can be obtained by utilizing his results.

Causal relationships between x_t and y_t can be detected or confirmed by examining the form of the residual CCF (Pierce and Haugh, 1977). For example, if x_t causes y_t, which is the case when the x_t series is precipitation and the y_t sequence is made up of river flows, then for $k \geq 0$ there will be at least one value of $r_{\hat{u}\hat{v}}(k)$ which is significantly different from zero. However, all values of the residual CCF for $k < 0$ would not be significantly different from zero. When x_t causes y_t instantaneously a significant value of the sample CCF will exist at lag zero. A description of the various types of causal relationships is provided by authors such as Piere and Haugh (1977) and Hipel et al. (1981).

To identify the form of the dynamic and noise components when interventions aren't present, Haugh and Box (1977) provide a technique which is based upon the residual CCF. By utilizing a physical understanding of the problem plus statistical procedures, Hipel et al. (1981) suggest an empirical approach for identifying transfer functions plus the noise term. The empirical approach for identifying the noise term is described in the next section.

2.3 IDENTIFYING THE NOISE COMPONENT

After the dynamic component has been designed, the noise component is tentatively assumed to be white noise and consequently the transfer function-noise model in (1) has the form

$$y_t = f(\underline{k},\underline{x},\underline{\xi},t) + a_t \tag{9}$$

In practice, the noise term is usually correlated. Therefore, after obtaining the estimated residual series, $\hat{a}_t$, for the model in (9), the type of ARMA model to fit to this series can be ascertained by following the usual three stages of model construction for ARMA models. Subsequently, the identified form of the ARMA model can be used for N_t in (1) and MLE's for all the model parameters can be simultaneously estimated. Diagnostic checks can then be employed to insure that the residual assumptions are satisfied. If more than one dynamic model passes diagnostic tests, the AIC can be employed to assist in selecting the best model.

2.4 ESTIMATION OF MISSING DATA

The model used for estimating missing data is a special case of the general intervention transfer function-noise model given in (1) (Baracos et al., 1981; D'Astous and Hipel, 1979). When dealing with monthly data the first step in the estimation procedure is to substitute the appropriate monthly means for all of the missing data. This substitution is necessary to achieve a zero value for each missing data point when the series is deseasonalized by subtracting out the estimated monthly means and dividing by the estimated monthly standard deviation for each observation. To demonstrate how an intervention model can be employed to estimate a single missing observation, consider the case where there is a single missing data point at time t = T. The model used to estimate the missing observation for a single series such as y_t is

$$y_t = \omega_0 \xi_t + N_t \tag{10}$$

where y_t is set to zero when the series is deseasonalized; and ξ_t is a pulse intervention defined by

$$\xi_t = \begin{cases} 1 & , t = T \\ 0 & , \text{otherwise} \end{cases} \tag{11}$$

The noise term, N_t, in (10) can be identified by fitting an ARMA model to the y_t series without the intervention component. Once the form of the model is identified, the parameter estimates of the noise model are estimated simultaneously with the estimate of ω_0. At time $t = T$ the model reduces to

$$-\omega_0 = N_t \tag{12}$$

Thus, the MLE of $-\omega_0$ constitutes an estimate of the missing observation at time T. By considering the standard error of estimate for ω_0, a confidence interval can be obtained for the estimated missing observation represented by $-\hat{\omega}_0$. Furthermore, because the estimate of the missing data depends on the noise component, N_t, the correlation structure of the series is preserved. To obtain the estimate of the missing observation for the original series, the inverse deseasonalization transformation followed by the inverse Box-Cox transformation can be invoked.

If more than one missing observation are to be estimated, the model is simply extended by adding the appropriate number of intervention terms. For example, if I_3 missing data are to be estimated, the model is given by

$$y_t = \sum_{i=1}^{I_3} \omega_{0i} \xi_{ti} + N_t \tag{13}$$

where y_t is the transformed and deseasonalized series, and

$$\xi_{ti} = \begin{cases} 1 & , \text{ if } t = T_i \\ 0 & , \text{ otherwise} \end{cases} \qquad (14)$$

The term T_i in this case refers to the time or location of the ith missing data in the series.

2.5 COMBINING MULTIPLE TIME SERIES

Often more than one covariate series is available to the analyst. In hydrological studies, data from several precipitation and temperature stations within or near the basin may be available. A common procedure employed by hydrologists to reduce model complexity is to combine similar types of series to form a single input covariate series. In the case of precipitation data, the records from the various stations are often combined to provide a single series of mean area precipitation. Two common methods of combining precipitation series are the Isohyetal and the Thiessen polygon techniques (Viessman et al., 1977). These procedures are essentially graphical methods and require a skilled analyst to obtain reasonable and consistent results. In an effort to automate procedures for combining similar types of series and provide more consistent results, a technique based on combining transfer function coefficients is presented.

Consider the case where two input covariate series, x_{t1} and x_{t2}, are to be combined to form a single input covariate series x_t. If x_{ti} causes y_t instantaneously, then the transfer function-noise models for the two series would be

$$y_t = \omega_{01} x_{t1} + N_{t1} \qquad (15)$$

and

$$y_t = \omega_{02} x_{t2} + N_{t2} \qquad (16)$$

where ω_{01} and ω_{02} are the transfer function parameters for the series x_{t1} and x_{t2}, respectively. In this case the two series x_{t1} and x_{t2} would be combined using the relative ratio of the transfer function coefficients such that

$$\left(\frac{\omega_{01}}{\omega_{01} + \omega_{02}}\right) x_{t1} + \left(\frac{\omega_{02}}{\omega_{01} + \omega_{02}}\right) x_{t2} = x_t \tag{17}$$

If more than two input series are available, this procedure could simply be extended to combine all of the available data into one input covariate series.

MODELLING HYDROMETEOROLOGICAL DATA

3.1 INTRODUCTION

The general procedure in any modelling problem is to start with a simple model and then increase the model complexity until an acceptable description of the phenomenon is achieved or until further improvements in the model cannot be obtained by increasing the model complexity. In order to demonstrate the model building techniques discussed in the previous sections, different transfer function-noise models are considered and the most appropriate model is eventually selected. The output for each transfer function-noise model always represents the average monthly flows of the Saugeen River at Walkerton, Ontario, while the covariate series consist of either precipitation or temperature data sets, or both types of series. The type, length and location of measurement for the data sets entertained are shown in Table 1 for the single river flow sequence, the two precipitation and the two temperature series. The river flow data are obtained from Environment Canada (1980a) and the precipitation and temperature data are provided by the Atmospheric Environment Service in Downsview Ontario (Environment Canada, 1980b).

3.2 ESTIMATING MISSING DATA

Prior to performing the cross-correlation analyses and fitting the transfer function-noise models, any missing data must be estimated. The only missing data in this study are ten precipitation and corresponding temperature data points for the Lucknow station where the dates of these missing data are given in Table 2. When considering data

TABLE 1. Available Monthly Data

Type	Location	Period
Riverflows	Saugeen River at Walkerton, Ontario.	1963-1979
Precipitation	Paisley, Ontario.	1963-1979
Precipitation	Lucknow, Ontario.	1950-1979
Temperature	Paisley, Ontario.	1963-1979
Temperature	Lucknow, Ontario.	1950-1979

sets in the residual cross-correlation analyses or for use an input or output series in the dynamic model, the data are only considered for the time period during which all of the series overlap. However, when estimating missing values in a single sequence, the entire time series is utilized in order to take full advantage of all the available information and thereby obtain better estimates of the missing data.

TABLE 2. Estimates of Missing Temperature Data at Lucknow

Date	Monthly Mean (C°)	Estimate (C°) (Standard Error)
February 1953	-6.48	-6.57 (2.32)
May 1968	11.99	11.81 (1.78)
September 1968	15.17	15.34 (1.16)
October 1973	9.60	9.78 (1.67)
August 1975	18.89	18.59 (1.20)
September 1975	15.17	15.29 (1.16)
July 1976	19.68	19.47 (1.09)
September 1978	15.17	15.30 (1.16)
October 1978	9.60	8.30 (1.70)
August 1979	18.89	18.56 (1.18)

For the case of the precipitation and temperature series at Lucknow a separate dynamic model is fitted to each of the series where there are ten intervention components for estimating the ten missing observations plus a correlated noise term. The most appropriate noise model for the Lucknow temperature data is found to be an ARMA(0,4) model with the second and third MA parameters constrained to zero. Using this form of the noise model, MLE's of the missing data are obtained and are shown in Table 2. In all cases, the estimates are within one standard deviation of the monthly mean. The best noise model for the Lucknow precipitation data is found to be a white noise model. Thus, the estimates of the missing data are simply taken as the appropriate estimated mean monthly values.

3.3 IDENTIFYING THE DYNAMIC COMPONENT

Once estimates of the missing data are obtained, a residual cross-correlation analysis can be performed to identify the forms of possible transfer functions for linking precipitation or temperature to the river flow output. Appropriate ARMA models are first fitted to each of the five transformed and deseasonalized series and the model residuals are estimated. The CCF between the residuals from the model fitted to the Saugeen River flows and each of the other four residual series are then calculated.

The results of the cross-correlation analyses shown a positive significant relationship at lag zero for each of the two precipitation series. For instance, the plot of the CCF for the Lucknow precipitation and Saugeen River flows is displayed in Figure 1 along with the estimated 95% confidence interval. The value of the CCF at lag zero in Figure 1 is 0.448 whereas for the Paisley precipitation the estimated value of 0.365 is slightly smaller. Although the CCF plot for the Paisley precipitation is not shown, it is indeed similar in form to Figure 1. The characteristics of the CCF's for the two precipitation series makes intuitive sense from a physical point of view since for monthly data, most of the precipitation for a particular month will result in runoff in the same month.

The results of the cross-correlation analyses for the two temperature series and the Saugeen River flows are somewhat different.

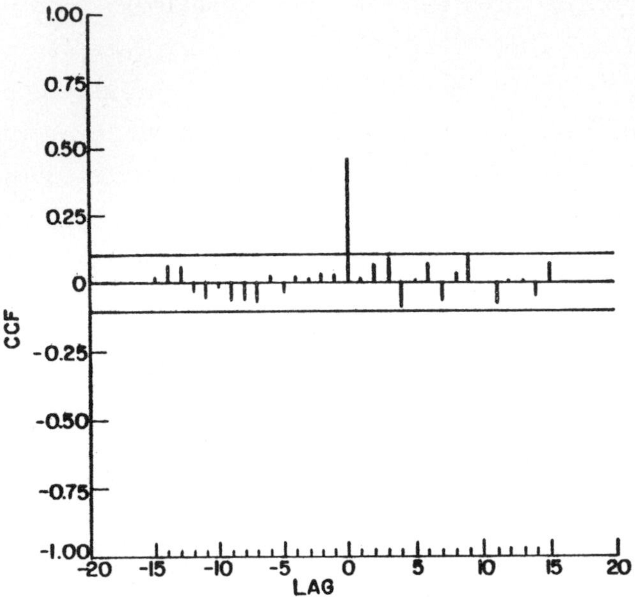

Fig. 1. CCF for the Lucknow Precipitation and Saugeen River Flows.

In these cases there are no significant cross-correlations at any lag. However, from a physical view point, one might expect that above average temperatures during the winter season would increase snow melt and thus river flow. For this reason, the temperature series are considered in the model building. The temperature series are assumed to have a significant contribution at lag zero and as will be shown later, this assumption is found to be justifiable.

3.4 THE TRANSFER FUNCTION-NOISE MODELS

The various transfer function-noise models examined and their associated AIC values are presented in Table 3. A decrease in the value of the AIC indicates that the additional model complexity is probably warranted since a better statistical fit is obtained. As expected, each increase in model complexity leads to a corresponding decrease in the value of the AIC. Thus a 'better' description of the phenomena results with each addition of available information. Details of each of the models are now discussed.

TABLE 3. Transfer Function-Noise Models Fitted to the Data

Output and Covariate Time Series	Parameter Estimates (Standard Errors)	AIC
Saugeen Flows	$\hat{\phi}_1$ = 0.407 (0.064)	963.121
Saugeen Flows Paisley Precipitation (ω_0)	$\hat{\phi}_1$ = 0.405 (0.064) $\hat{\omega}_0$ = 0.310 (0.055)	936.129
Saugeen Flows Lucknow Precipitation (ω_0)	$\hat{\phi}_1$ = 0.429 (0.063) $\hat{\omega}_0$ = 0.350 (0.053)	925.606
Saugeen Flows Combined Precipitation (ω_0)	$\hat{\phi}_1$ = 0.418 (0.064) $\hat{\omega}_0$ = 0.350 (0.054)	926.340
Saugeen Flows Summer Precipitation (ω_{01}) Winter Precipitation (ω_{02})	$\hat{\phi}_1$ = 0.407 (0.064) $\hat{\omega}_{01}$ = 0.444 (0.065) $\hat{\omega}_{02}$ = 0.183 (0.092)	922.270
Saugeen Flows Summer Precipitation (ω_{01}) Accumulated Snow (ω_{02})	$\hat{\phi}_1$ = 0.408 (0.064) $\hat{\omega}_{01}$ = 0.448 (0.066) $\hat{\omega}_{02}$ = -0.162 (0.109)	924.037
Saugeen Flows Paisley Temperature (ω_0)	$\hat{\phi}_1$ = 0.414 (0.064) $\hat{\omega}_0$ = 0.073 (0.061)	963.700
Saugeen Flows Lucknow Temperature (ω_0)	$\hat{\phi}_1$ = 0.419 (0.064) $\hat{\omega}_0$ = 0.112 (0.062)	961.860
Saugeen Flows Summer Precipitation (ω_{01}) Winter Precipitation (ω_{02}) Lucknow Temperature (ω_{03})	$\hat{\phi}_1$ = 0.422 (0.064) $\hat{\omega}_{01}$ = 0.453 (0.064) $\hat{\omega}_{02}$ = 0.194 (0.090) $\hat{\omega}_{03}$ = 0.144 (0.055)	917.558
Saugeen Flows Summer Precipitation (ω_{01}) Winter Precipitation (ω_{02}) Combined Temperature (ω_{03})	$\hat{\phi}_1$ = 0.420 (0.064) $\hat{\omega}_{01}$ = 0.453 (0.064) $\hat{\omega}_{02}$ = 0.194 (0.090) $\hat{\omega}_{03}$ = 0.133 (0.055)	918.586

The first model considers only the Saugeen river flows. The time series is first transformed by taking natural logarithms of the data. This series is then deseasonalized by subtracting the estimated monthly mean and dividing by the estimated monthly standard deviation for each observation. An ARMA(1,0) model is then fitted to these deseasonalized flows. The value of the AIC is 963.121 and this value is used as a basis for comparing improvements in each of the models.

3.4.1 Precipitation Series as Inputs

As suggested by the results of the cross-correlation analyses, each of the precipitation series is used as an input covariate series. Prior to fitting the transfer function-noise model, each of the series is first deseasonalized. Each series is then used independently as an input covariate series in a transfer function-noise model. As shown in Table 3, the transfer function parameter, ω_0, for Paisley and Lucknow are estimated as 0.310 and 0.350, respectively. Note that the AIC value for the Lucknow precipitation series is significantly less than the AIC value for Paisley. This may suggest that the pattern of the overall precipitation which falls on the Saugeen River basin upstream from Walkerton, is more similar to the precipitation at Lucknow than the precipitation at Paisley, even though Paisley is closer to Walkerton than Lucknow.

Using the procedure outlined in the section entitled "Combining Input Covariate Series", the two precipitation series are combined to form a single input covariate series. In this study, the Lucknow and Paisley precipitation series are combined in the ratio 53:47, respectively. This combined precipitation series is then deseasonalized and used as an input series for the transfer function-noise model. The resulting AIC value is only slightly larger than the AIC value obtained when only the Lucknow precipitation series is employed. Since the difference is small, either model would be satisfactory and for the balance of this paper the combined precipitation series is employed.

In the previous models the precipitation series are input as a single series having a single transfer function parameter. In these cases it is therefore assumed that the contribution of precipitation is the same throughout the year. Physically, however, it makes sense that the contribution of precipitation during the winter months would be less than the contribution during the warmer periods of the year since the precipitation accumulates on the ground in the form of snow during the cold season. In an effort to better reflect reality, the single precipitation series formed by combining the Lucknow and Paisley data, is divided into two separate seasons.

In dividing the precipitation into two seasons, the winter season is taken as those months where the mean monthly temperature is below zero degrees celcius. For both the Paisley and the Lucknow temperature series, December, January February and March have mean monthly temperatures below freezing. Therefore, the winter precipitation series consists of the deseasonalized precipitations for these four months and zeros for the other eight months of the year. Conversely, the summer precipitation series has zeros for the four winter months and the deseasonalized precipitations for the remaining entries. These two series are input as separate covariate series with separate transfer function parameters. The resulting transfer function-noise model is

$$y_t = 0.444x_{t1} + 0.183x_{t2} + \frac{a_t}{(1 - 0.407B)} \qquad (18)$$

where y_t is the deseasonalized logarithmic flow at time t; x_{t1} is the combined deseasonalized summer precipitation series; and x_{t2} is the combined deseasonalized winter precipitation series. As expected, the transfer function coefficient for the summer precipitation is larger than the transfer function parameter for the winter precipitation. It is also reassuring to note that the better representation of the physics of the system also leads to an improved statistical fit as indicated by a lower AIC value.

A second type of dynamic model aimed at modelling the spring runoff resulting from snowmelt is also considered. In this model, the summer precipitation is the same as the model in (18). However, the snow accumulated during the winter months from December to March is represented as a single pulse input in April where the temperature is above zero for the first time and hence spring runoff occurs. For the other eleven months of the year this series has zero entries.

This type of dynamic model has been shown to work well for river systems located in areas that experience Arctic climate (Baracos et al., 1981) and there are rarely any thaws during the winter months. However, the climatic conditions in the Saugeen River basin during the winter are not extremely cold and several midwinter melts result in a significant reduction in the accumulated snow cover on the ground. For this reason, the transfer function parameter for the accumulated winter precipitation is not significantly different from zero.

3.4.2 Temperature Series as Inputs

Although the cross-correlation analyses indicate no significant relationships between temperature and river flow, the two temperature series are used as input covariate series in transfer function-noise models. As before, both series are first deseasonalized by subtracting out the estimated monthly means and dividing by the estimated monthly standard deviations for each observation. These series are then input independently as covariate series in transfer function-noise models. The resulting models and their associated AIC values are shown in Table 3. Because 1.96 times the standard error for each parameter is larger than the parameter estimate, neither transfer function parameter is significant at the five percent significance level. Recall that the CCF for each temperature series and the Saugeen River flows, also suggests that there may not be a marked relationship between the temperatures and river flows. However, the Lucknow temperature parameter is significantly different from zero at the ten percent significance level. As a result, the Lucknow temperature series is included in the transfer function-noise models where both the temperature and precipitation series are included.

3.4.3 Precipitation and Temperature Series as Inputs

In an effort to combine all the available information, both the temperature and the precipitation data are used as input covariate series to the transfer function-noise model. In the first model of this type in Table 3 the combined precipitation is deseasonalized and split into two seasons as is done in (18). The deseasonalized Lucknow temperature data is used as another input covariate series. The resulting model is given by

$$y_t = 0.453x_{t1} + 0.194x_{t2} + 0.144x_{t3} + \frac{a_t}{(1 - 0.422B)} \qquad (19)$$

where y_t is the deseasonalized logarithmic flow at time t; x_{t1} is the deseasonalized summer precipitation; x_{t2} is the deseasonalized winter precipitation; x_{t3} is the deseasonalized Lucknow temperature data; and a_t is the white noise term. The model and its associated AIC are also shown in Table 3. This model provides a significant improvement over any of the models previously employed with a decrease of almost five in the AIC when compared to the model in (18). Also, the transfer function parameter for the temperature series is significantly different from zero in this case. Recall from before that the transfer function parameter for either temperature series is not significantly different from zero at the five percent significance level. However, when the precipitation series is included in the model the temperature series provides a significant contribution. This point illustrates the need for more research in identifying the dynamic component of transfer function-noise models when more than one input covariate series is available.

The last model fitted to the data employs the combined precipitation and the combined temperature data. The temperature series are combined in the same fashion as the precipitation series but is not divided into two separate seasons. The resulting model and its associated AIC are shown in Table 3. Note that the AIC value is only marginally larger than that of the previous model where the Lucknow temperature data is employed instead of the combined temperature series. In this case either of these last two models could be employed as the most appropriate model for the available data.

CONCLUSIONS

As exemplified by the application in this paper, both physical justifications and flexible statistical methods can be employed to design a suitable transfer function-noise model to fit to hydrological time series. When the covariate series possess missing data, these can be estimated by using a separate intervention model for each series. The covariate series can then be linked with the output river flow series in an overall transfer function-noise model. If there are more than one precipitation or temperature series, a procedure is available for obtaining a single precipitation or temperature series. For the case where snow accumulates during the winter time, the precipitation series can be incorporated into the dynamic model in specified manners so that the model makes sense from a physical point of view. For the case of the best Saugeen River dynamic model, the precipitation series was split into a winter and summer series, and a separate transfer function was designed for each of the series. The residual CCF is useful for statistically identifying a transfer function to link a covariate series with the output while standard identification procedures can be used to design the form of the noise term in (1). If the output series has been affected by natural or man induced interventions, an intervention component could be built into the model as shown in (2). Likewise, intervention components can be built into the dynamic model containing the covariate series to estimate missing data in the output series. An automatic selection procedure such as the AIC plus diagnostic checks can be employed for choosing the best overall transfer function-noise model which can then be used for practical applications.

REFERENCES

Akaike, H., 1974. A new look at the statistical model identification. IEEE Transactions on Automatic Control, 19: 716-723.

Baracos, P.C., Hipel, K.W. and McLeod, A.I., 1981. Modelling hydrologic time series from the Arctic. Water Resources Bulletin, 17(3).

Box, G.E.P. and Cox, D.R., 1964. An analysis of transformations. Journal of the Royal Statistical Society, Series B, 26: 211-252.

Box, G.E.P. and Jenkins, G.M., 1970. Time Series Analysis: Forecasting and Control. Holden-Day, San Francisco.

Box, G.E.P. and Tiao, G.C., 1975. Intervention analysis with applications to economic and environmental problems. Journal of the American Statistical Association, 70(349): 70-79.

D'Astous, F. and Hipel, K.W., 1979. Analyzing environmental time series. Journal of the Environmental Engineering Division, American Society of Civil Engineers, 105(EES): 979-992.

Environment Canada, 1980a. Historical Stremaflow Summary, Ontario to 1979. Inland Waters Directorate, Water Resources Branch, Water Survey of Canada, Ottawa, Ontario, Canada.

Environment Canada, 1980b. Monthly Meteorological Summary, Ontario to 1979. Meteorological Branch, Environment Canada, Ottawa, Ontario, Canada.

Granger, C.W.J., 1969. Investigating causal relations by econometric models and cross-spectral methods. Econometrica, 37(3): 424-438.

Haugh, L.D., 1972. The Identification of Time Series Interrelationships with Special Reference to Dynamic Regression. Ph.D. Thesis, Department of Statistics, University of Wisconsin, Madison, Wisconsin, U.S.A.

Haugh. L.D., 1976. Checking the independence of two covariance-stationary time series: a univariate residual cross-correlation approach. Journal of the American Statistical Association, 71(354): 378-385.

Haugh, L.D. and Box, G.E.P., 1977. Identification of dynamic regression (distributed lag) models connecting two time series. Journal of the American Statistical Association, 72(357): 121-130.

Hipel, K.W., Lennox, W.C., Unny, T.E. and McLeod, A.I., 1975. Intervention analysis in water resources. Water Resources Research, 11(6): 855-861.

Hipel, K.W., Li, W.K. and McLeod, A.I., 1981. Causal and Dynamic Relationships between Natural Phenomena. Technical Report No. TR-81-05, Dept. of Statistical and Actuarial Sciences, The University of Western Ontario, London, Ontario, Canada.

Hipel, K.W. and McLeod, A.I., 1982. Time Series Modelling for Water Resources and Environmental Engineers. Elsevier, Amsterdam, in press.

Hipel, K.W., McLeod, A.I. and Lennox, W.C., 1977a. Advances in Box-Jenkins modelling, 1, model construction. Water Resources Research, 13(3): 567-575.

Hipel, K.W., McLeod, A.I. and McBean, E.A., 1977b. Stochastic modelling of the effects of reservoir operation. Journal of Hydrology, 32: 97-113.

McLeod, A.I., 1977. Improved Box-Jenkins estimators. Biometrika, 64(3): 531-534.

McLeod, A.I., 1979. Distribution of the residual cross-correlation in univariate ARMA time series models. Journal of the American Statistical Association, 74(368): 849-855.

McLeod, A.I., Hipel, K.W. and Lennox, W.C., 1977. Advances in Box-Jenkins modelling, 2, applications. Water Resources Research, 13(3): 577-586.

Peirce, D.A. and Haugh, L.D., 1977. Causality in temporal systems. Journal of Econometrics, 5: 265-293.

Viessman, W., Jr., Knapp, J.W., Lewis, G.L. and Harbaugh, T.E., 1977. Introduction to Hydrology, 2nd Edition, Harper and Row, New York.

SOME ASPECTS OF NON-STATIONARY BEHAVIOUR IN HYDROLOGY

N.T. KOTTEGODA

Department of Civil Engineering, University of Birmingham,
Birmingham, B15 2TT, England, U.K.

ABSTRACT

Non-stationary behaviour in hydrology is investigated by paying particular attention to modelling. Application is made to 12 series of annual rainfall and river flows from Northern Utah and to the Nile flows at Aswan Dam. Different aspects of non-randomness are investigated by fitting linear autoregressive type of models and examining the residuals by means of parametric and non-parametric tests. The alternative approach adopted herein is to investigate evolutionary changes through estimated spectral functions of overlapping sequences. Non-stationarities are then quantified by relative mean values of chi-squared in particular frequency bands.

GENERAL APPROACH TO NON-STATIONARITY

In hydrologic time series non-stationary behaviour may be seen to occur in a number of different forms. Random shifts in the mean constitute a basic example of departures from stationarity. Comparatively deterministic in type are cyclical changes such as the annual seasons and the diurnal periodicities, brought about by external stimuli. More complicated series have non-stationarity in the mean and variance and possibly in the higher moments, caused by climatic and environmental changes. In general, we can view non-stationary series as ones that are being disturbed stochastically by transient effects of one or more kinds with different causative factors.

In the analysis of time series, shifts in the mean may be investigated by differencing of the first order. Second order differencing can be used to separate trends. Alternatively, we may take moving averages or simple $(\frac{1}{2},\frac{1}{2})$ weights of adjacent values.

Reprinted from *Time Series Methods in Hydrosciences,* by A.H. El-Shaarawi and S.R. Esterby (Editors)

These techniques belong to the category of methods called local filtering. Fitting of a polynomial or the use of a general global model constitute another approach. An original study of trends and jumps of various kinds applicable to hydrology was made by Yevjevich and Jeng (1969). If there is strong physical reasoning for a change in the mean and if the point in time when this change occurs can be approximately identified, then a non-stationary series may be modelled by intervention analysis (Box and Tiao, 1965).

Of greater concern and more difficult to assess practically are evolutionary changes in the structure of a time series. For instance, slow movements may be induced by the generating process, which can be simulated by changing the parameters of a model over different time spans. These modifications are caused primarily by gradual movements in climate such as increasing or decreasing rainfalls over finite time horizons. Urbanization, variations in catchment characteristics, which may be natural or man-made, gradual silting of reservoirs and changes in evaporation can also have significant effects.

Prior to investigation, errors of measurement and inconsistencies in the data should be eliminated, if they exist. Then if a long series is available, a simplistic approach is to divide the series into non-overlapping segments, compute suitable statistics or empirical functions such as serial correlograms from each segment and test these for significant differences. There is however subjectivity in the choice of sub-samples and hence some bias is often introduced.

OBJECTIVES AND SCOPE OF STUDY

This study aims to find objective procedures for investigating non-stationary behaviour. Because nearly all hydrologic series possess some degree of non-randomness we commence by initially fitting linear stochastic models (Box and Jenkins, 1976). Then the residuals are analysed by different methods. These include in addition to the Box-Pierce portmanteau lack-of-fit test, three non-parametric tests, the Von Neumann ratio test for normal data and rescaled range. More importantly, a likelihood ratio test is used to test for a shift in

Table 1 Tests on Independent Residuals

No.	Station	Type*	Length of Record in years	Model Fitted	Q_ν Box Pearce Portmanteau Statistic			Test Statistic for Comparison with Standard Normal Variates				Other Parametric Test Statistics		
								Non-parametric			Parametric			
					Q_{10}	Q_{15}	Q_N	Runs Above or Below Average	Mann-Whitney	Wald-Wolfowitz	Neumann Ratio	Kolm.Smir. Test for Normality	r_N^{**} rescaled range	Likelihood Ratio W Test
1	Jordan River, Lehi 1914 to 1976	R	64	AR(1)	4.83	7.69		0.64	-0.89	-0.89	-0.62	1.12	11.88	2.50
2	Bear River, Collingston 1890 to 1977	R	88	AR(2)	15.03	24.20		-0.43	-0.43	-0.43	-0.40	1.00	13.21	2.67
3	Weber River, Plain City 1906 to 1977	R	72	AR(1)	10.56	11.25		-0.12	-0.83	-0.83	-1.32	0.94	12.21	3.62
4	Blacksmith Fork, Hyrum 1914 to 1977	R	64	AR(1)	4.93	8.26		0.13	1.14	1.14	-1.22	0.90	8.97	1.74
5	Logan River, Logan 1901 to 1976	R	76	AR(1)	9.35	10.69		1.28	1.65	1.65	-1.46	0.83	13.88	3.22
6	Weber River, Oakley 1905 to 1977	R	73	AR(1)	7.05	10.15		-0.24	0.47	1.40	-1.07	0.58	13.46	4.04
7	Farmington 1890 to 1977	P	88	AR(1)	9.83	16.26		0.75	1.40	-0.57	-1.73	0.38	10.05	2.32
8	Kelton 1897 to 1929	P	53	AR(1)	5.84	10.06		-0.57	-0.57	0.45	-1.04	1.02	8.13	2.01
9	Delp 1897 to 1929	P	81	AR(1)	2.92	3.24		-0.68	0.45	0.10	-2.17	0.51	9.73	2.61
10	Corrine 1871 to 1972	P	102	AR(1)	6.34	7.74	N=20 11.56	-0.90	0.10	0.99	-1.86	0.47	15.26	3.03
11	Ogden 1871 to 1977	P	107	AR(3)	7.83	9.65	N=20 18.09	-0.79	0.99	1.25	-0.82	0.58	13.35	2.55
12	Tree Ring Index Rex Peak, Bear Lake Drainage, 1698 to 1977	T	280	AR(1)	9.10	12.54	N=20 31.80	0.30	1.25	0.47	-1.07	0.52	22.82	3.45

* P : precipitation R : Flow Recording T : Tree Ring Index

** 95 percent values of chi-squared are 16 for, degrees of freedom, ν = 10; 25 for ν = 15; 31.4 for ν = 30 and 67.5 for ν = 50.

the mean without à priori knowledge of the point in time at which it occurs. In the second phase of the study evolutionary spectral densities are estimated from overlapping sections and mean chi-squared values are compared over certain frequency bands. This is applied to historical data, and repeated using synthetic data initially with model parameters estimated from a full record and then from different sections of it.

APPLICATION TO NORTHERN UTAH

Application is made to series of annual riverflows and precipitation in Northern Utah some of which extend in time over 100 years. A tree ring index series from the same region is also investigated. Optimum use of scarce water resources in this area has necessitated the use of realitic and adequate models for time series, so that proper assessment can be made of future supplies. The hydrological series studied are listed in Table 1. Comparison is made with a 172 year segment of the Nile flows at Aswan.

RESIDUALS FROM LINEAR STOCHASTIC MODELS

Figure 1 shows traces of the six time series of river flows and the precipitation series are shown in Figure 2 with the tree ring index. In general, no significant trends or periodicities are evident. There are varying degrees of serial correlation in nearly all the series. Figure 3 shows the serial correlograms, r_k, and estimated partial autocorrelograms, $\hat{\phi}_{k,k}$, of five typical series. The serial correlation r_k for lag k is computed from observations $(x_1, x_2, \ldots, x_N)$ with mean μ and estimated $\bar{x}$, as follows.

$$r_k = \left[\sum_{t=1}^{N-k} \{(x_t - \bar{x})(x_{t+k} - \bar{x})\}\right] / \; (x_t - \bar{x})^2. \tag{1}$$

Initially on the assumption of stationarity, appropriate Box-Jenkins models are fitted from the general ARMA (p,q) type given by

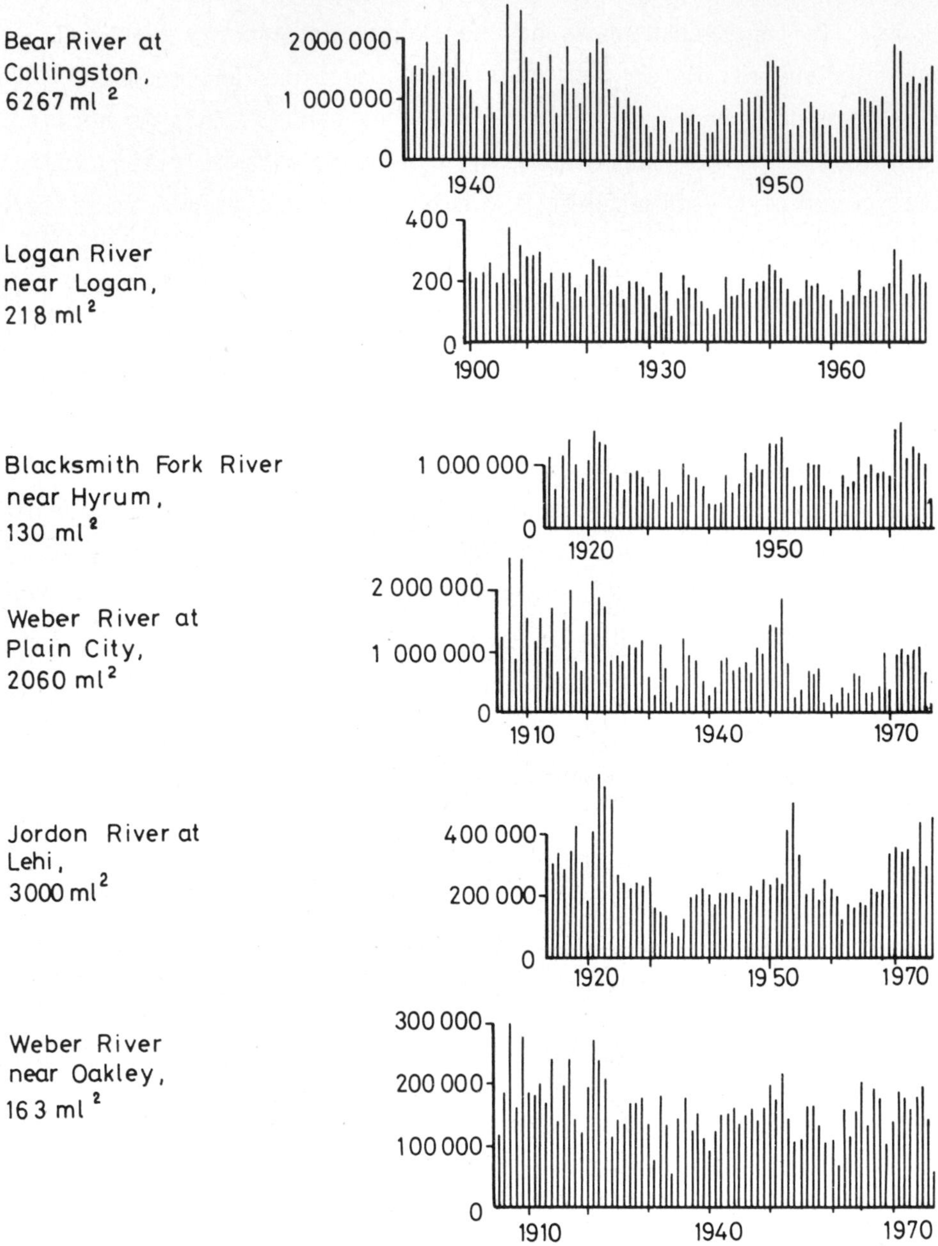

Fig. 1. Six River Flow Series From Northern Utah.

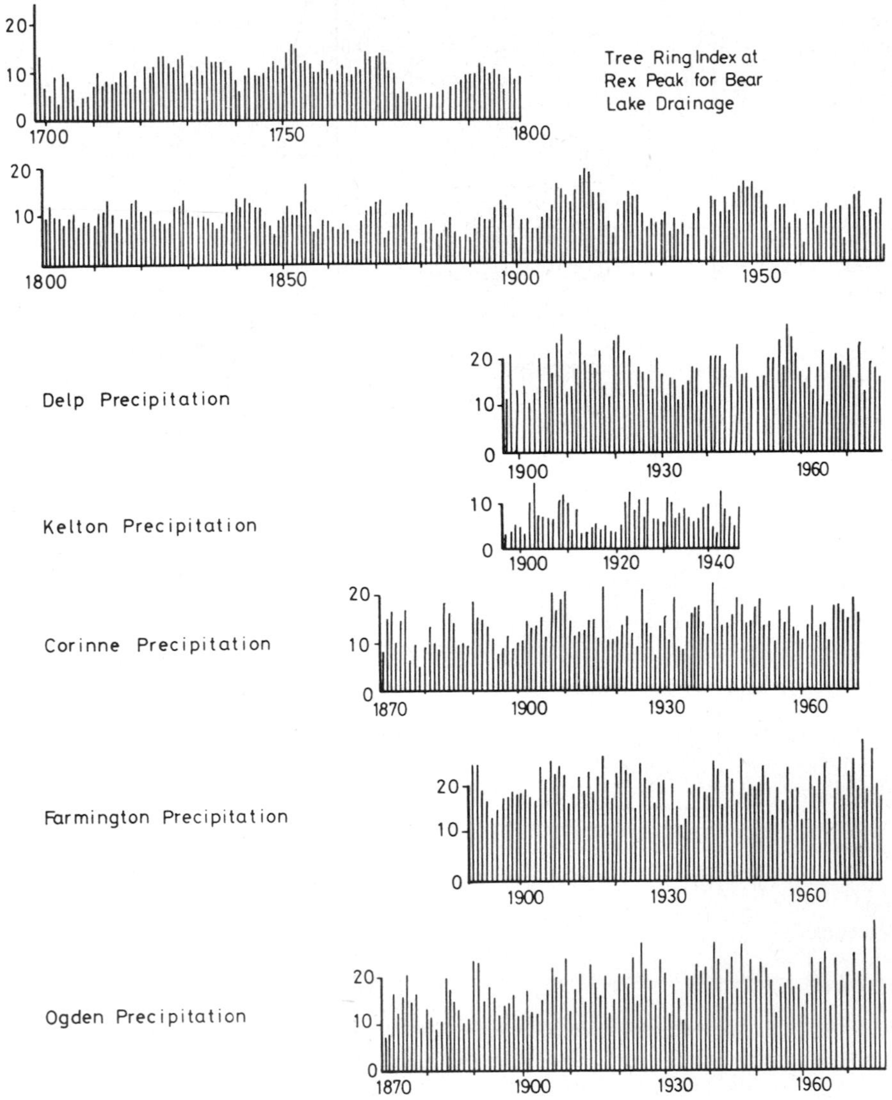

Fig.2. Annual Tree Ring Index and 5 Precipitation Series From Northern Utah.

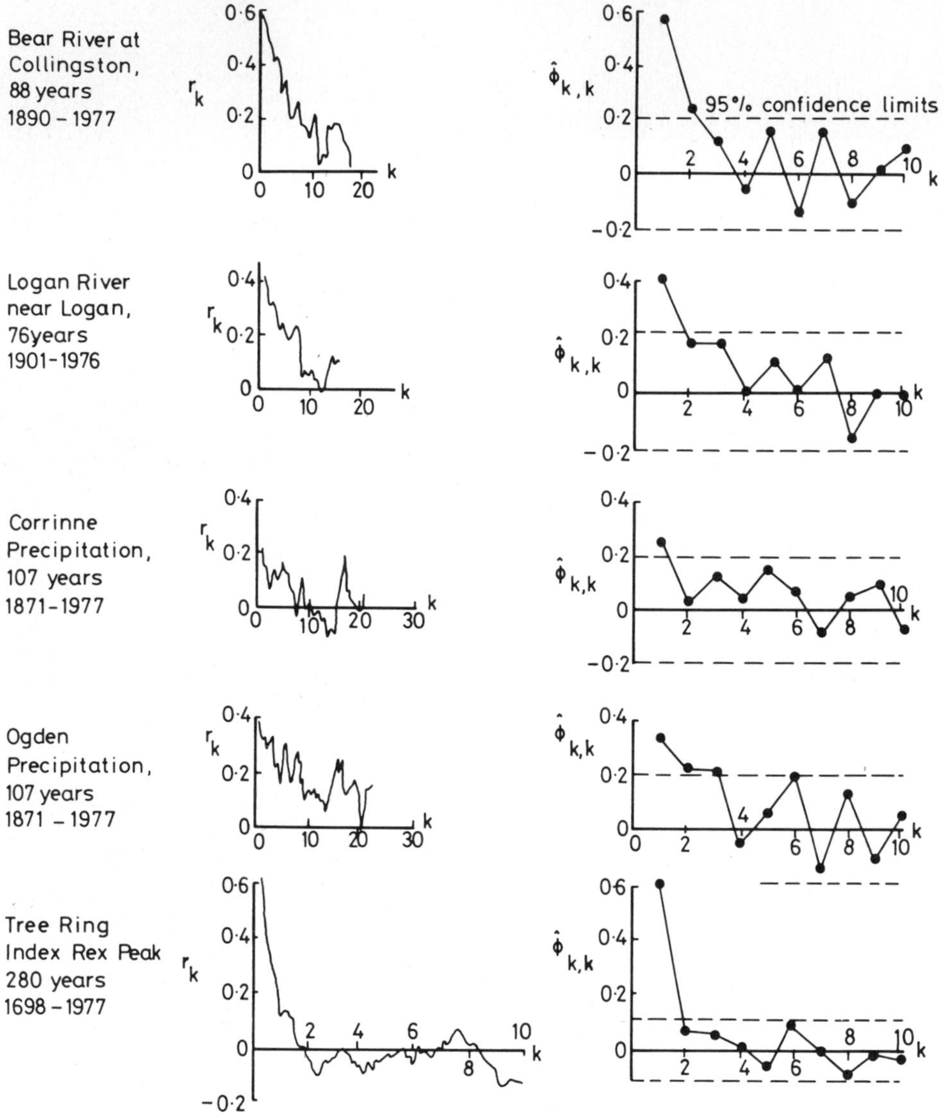

Fig. 3. Serial Correlograms and Partial Autocorrelograms of 5 Annual Hydrological Series From Northern Utah.

$$X_t = \mu + \sum_{i=1}^{p} \phi_i (X_{t-i} - \mu) + Z_t + \sum_{i=1}^{q} \theta_i Z_{t-i} \qquad (2)$$

in which the ϕ_i and θ_i are parameters and Z_t is an independent normal variate with zero mean. The fitted models are of the autoregressive type: models AR(1), AR(2) and AR(3). Spectral densities are calculated, for a truncation point M, as follows

$$s(\omega) = \{r_o + 2 \sum_{k=1}^{M-1} r_k \cos(\omega k/M) + r_M \cos(\omega)\}/\pi \qquad (3)$$

where ω = frequency in radians per unit time (year), and $\hat{s}(\omega)$ is calculated at $\omega = \pi/M, 2\pi/M, \ldots, \pi$. The $\hat{s}_k$ are smoothed using the Tukey window which has degrees of freedom given by $\nu = 2.67\ N/M$, if N observations are used for estimation.

Figure 4 shows on the left estimated spectral densities for 5 of the series using two values of M in each case. These are plotted on semi-logarithmic paper. Also shown are the theoretical spectra for the appropriate model from equation (2) as calculated from

$$\Gamma(\omega) = \sigma_Z^2 \left| \frac{1 + \theta_1 e^{-j\omega} + \theta_2 e^{-2j\omega} + \ldots + \theta_q e^{-qj\omega}}{1 - \phi_1 e^{-j\omega} - \phi_2 e^{-j\omega} - \ldots - \phi_p e^{-pj\omega}} \right|^2 \qquad (4)$$

where σ_Z^2 is the variance of the Z series and $j = \sqrt{-1}$.

Shown on the right of Figure 4 are the spectral densities of the residuals from each model using the corresponding two values of M. Also given are the 95 percent confidence limits $\nu\bar{s}(\omega)/\chi^2_{\nu,0.975}$ and $\nu\bar{s}(\omega)/\chi^2_{\nu,0.025}$ where $\bar{s}(\omega)$ is the mean spectral density for white noise. Results of the Box-Pierce Portmanteau test [Box-Jenkins (1976)] are shown as Q_ν values in Table 1. From tables of the chi-squared distribution (see footnote of Table 1) these are not significant for the appropriate degrees of freedom ν. However, it is well known that this test for white noise has low power.

Seven additional tests were carried out in order to assess the nature of the residuals; the purpose and scope of each are set out below. The results are given in Table 1. It will of course be noted

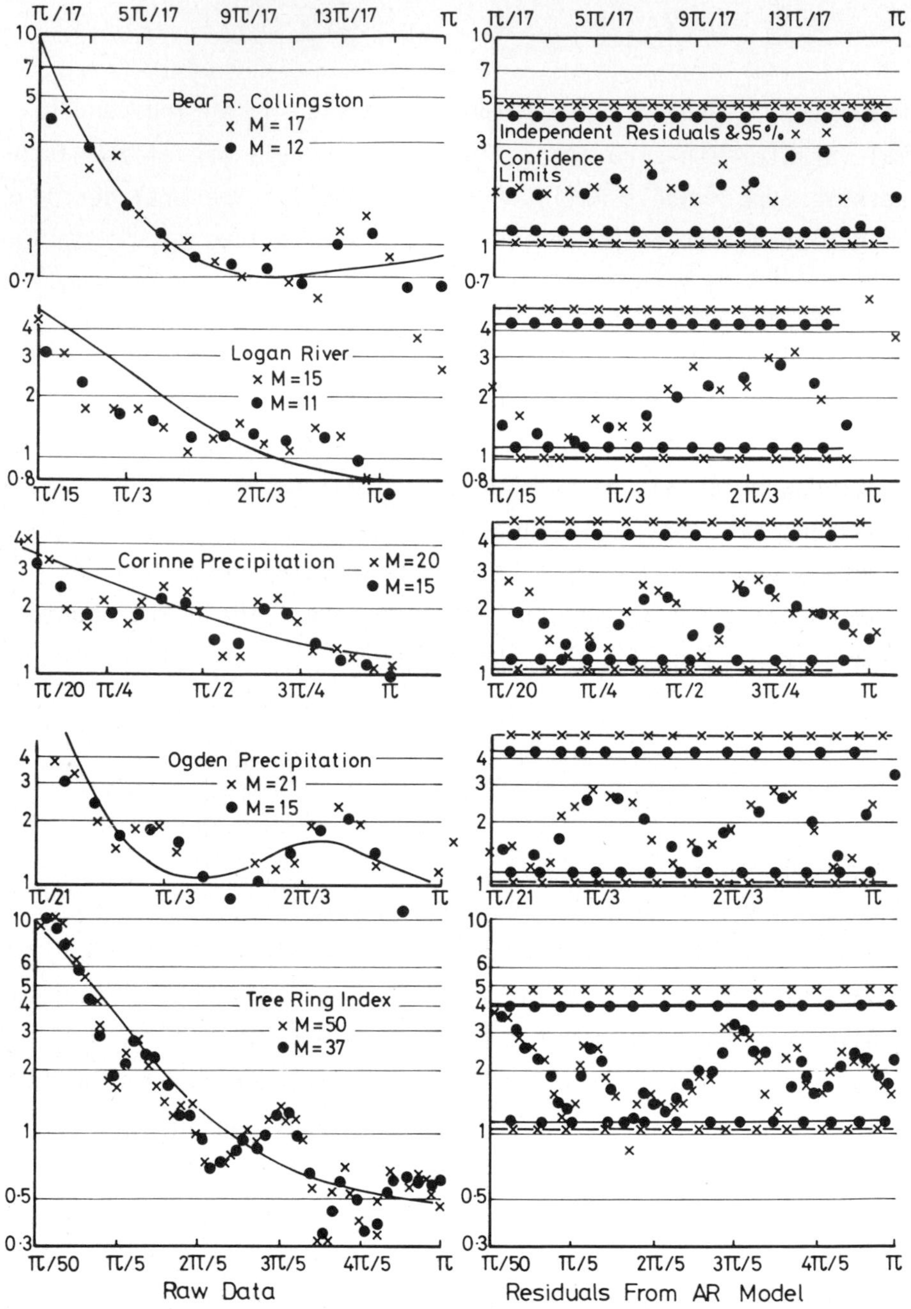

Fig. 4. Estimated Spectra of 5 Annual Hydrological Series From Northern Utah.
Curves Show Theoretical Spectra of AR Models.
Also Estimated Spectra of Independent Residuals are Shown on the Right

that some bias is introduced because the estimated residuals are investigated and not the true residuals.

NON-PARAMETRIC TESTS

Runs Test In this basic one sample non-parametric test, runs above and below the estimated median of the series are counted. The total number of runs is standardised using the mean and standard deviation from the normal distribution of the number of runs of a random sequence as given for instance by Conover (1971) and Siegel (1956). Values are well within the range -1.96 to 1.96, so they are not significant at the 5-percent level of significance. If they are significant it implies non-randomness on account of serial correlation, periodic movement or trend. The test is not a powerful one but is nevertheless useful for preliminary purposes.

Man-Whitney and Wald-Wolfowitz Tests For these well-known two sample tests (Conover, 1971), the full sequence of residuals is divided into two samples of equal length. The Mann-Whitney test is to determine whether there is a significant difference in the locations of the two samples. On the other hand significance in the Wald-Wolfowitz test can mean differences in general distribution properties such as location, dispersion and skewness. The test statistics are, as in the runs test, standardised by using the respective means and standard deviations of the large sample normal distributions of the statistics expected in a random population. Again the results do not show any significance. All of the aforementioned tests are distribution-free.

Kolmogorov-Smirnov Goodness of Fit Test The distributions of the residuals are then tested using the Kolmogorov-Smirnov Goodness of Fit Test. As shown in Table 1 the statistics are much less than the critical value of 1.36 which is the 95 percent value for large samples. On account of the normality of the residuals, it is possible to carry out three additional tests of importance as follows.

PARAMETRIC TESTS

Von-Neumann Ratio Test This test of randomness can be used against unspecified alternatives. It covers inconsistencies such as jumps. One needs to calculate the ratio of the mean square successive differences to the variance of the observations, given by

$$V = (N/N - 1)) \sum_{i=1}^{N-1} (x_i - x_{i+1})^2 / \left[\sum_{i=1}^{N} x_i^2 - (1/N)\left(\sum_{i=1}^{N} x_i\right)^2 \right] \quad (5)$$

For $N > 50$, V may be assumed to be normally distributed with mean $2N(N-1)$ and variance $4(N-2)/(N-1)^2$. Again as shown in Table 1, nearly all standardised values of V are within the limits of ± 1.96 and hence non-randomness is not indicated.

Rescaled Range Test The adjusted rescaled range of the observations $(x_1, x_2, \ldots, x_N)$ is obtained from

$$r_N^{**} = \left[\max_{1 \leq i \leq N} (x_1 + x_2 + \ldots + x_i - i\bar{x}) - \min_{1 \leq i \leq N} (x_1 + x_2 + \ldots + x_i - i\bar{x}) \right] / s_N \quad (6)$$

in which $\bar{x}$ is the estimated mean and s_N is the estimated standard deviation from the observations.

In order to ascertain the small sample distribution of the random variable, R_N^{**}, 100,000 sets of independent standard normal variates x_i, $i = 1,2,\ldots,N$, were generated, the r_N^{**} were then calculated from each set and the empirical percentage points obtained for N = 50,60, ...,110. These are shown by the curves on the left of Figure 5. Also given are the estimates r_N^{**} for each of the residual series. Two of the series show significant departures from randomness.

Likelihood Ratio Test Following the work of Worsley (1979) a likelihood test can be used to test for a shift in location at unknown time in a normal population; to test for a change in variance one may apply the Bayesian method of Menzefricke (1981) and, alternatively, the non-parametric approach of Petitt (1979) determines whether there is a significant change in distribution at an unknown point. The Worsely test is based on a statistic W which is calculated as follows.

Let $X_1, X_2, \ldots, X_N$ be a sequence of independent random variables with distributions $X_i \sim N(\mu_i, \sigma^2)$ $i = 1,2,\ldots,N$. Consider the hypothesis

$$H_o: \ \mu_i = \mu, \ i = 1,2,\ldots,N$$

against the alternative hypothesis

$$H_1: \ \mu_i = \mu, \ i = 1,2,\ldots,k$$
$$= \mu', \ i = k+1, k+2,\ldots,N$$

where σ^2, μ, μ' and k are unknown. If we denote the mean of the first k observations by $\bar{x}_k$ and the mean of the remaining N-K observations by $\bar{x}'_k$, the within groups sum of squares of the observations is

$$S_k^2 = \sum_{i=1}^{k} (x_i - \bar{x}_k)^2 + \sum_{i=k+1}^{N} (x_i - \bar{x}'_k)^2 \tag{7}$$

Also, the normalised between-groups sum of squares is given by

$$T_k^2 = (k(N-k)/N)(\bar{x}_k - \bar{x}'_k)^2, \ k = 1,\ldots,N-1 \tag{8}$$

Then the likelihood ratio test is based on the statistic

$$W = \max_{1 \leqslant k \leqslant N} (N-2)^{\frac{1}{2}} \mid T_k \mid / S_k \tag{9}$$

As in the case of R_N^{**}, the empirical distribution of W was obtained by simulation using independent standard normal variates.

It is found that for sample sizes greater than N = 50 the form of the distribution can be approximated by a single curve as shown on the right of Figure 5. The estimates of W for each of the residual series are given on the same figure. About 4 of the series are significant at the five percent level showing shifts in location. Also shown on the right is the step function for the 12 data series. It is seen that there is a significant difference in distribution in terms of the Kolmogorov-Smirnov statistic between this and the smooth curve.

TESTS BASED ON THE SPECTRUM

The longer series of data were tested for evolutionary changes by means of the spectrum and overlapping samples. The theory of the spectrum is given by Jenkins and Watts (1967). There is empirical work in relation to climate by W.M.O. (1966). Priestley (1965)

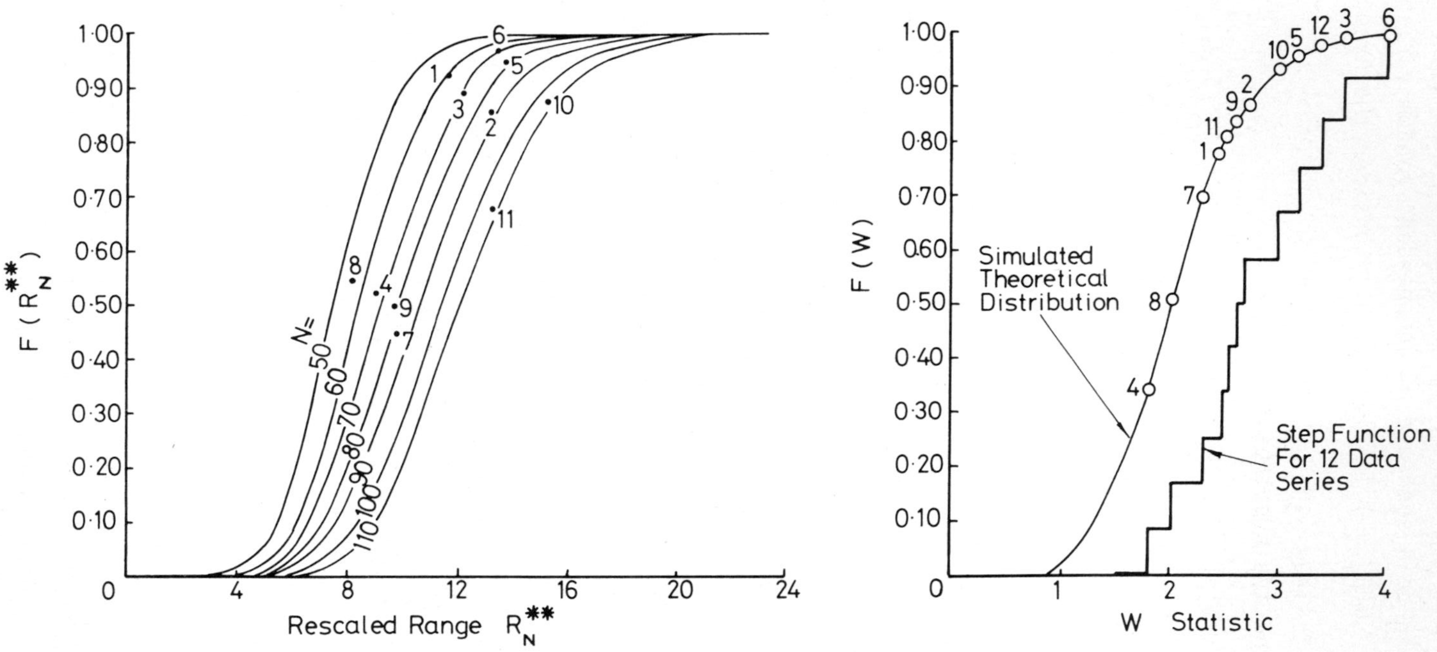

Fig. 5. Emperical Probability Distribution of Rescaled Range and W Statistic of Independent Standard Normal Variates and Estimates From Samples of Residuals of Hydrological Data, With Step Function of W Statistic for 12 Series.

originated analytical work on evolutionary spectra and non-stationary processes.

Results for the 107 year record of precipitation at Ogden are given in Figure 6. Against the background of the theoretical spectrum based on equation 4 for an AR(3) model, the estimated spectral density functions of ten overlapping sequences are shown here. The first of these span the initial 53 years. Using the same sample size, a second sample is formed starting six years after the commencement of the first. Then the third is lagged by a further six years and the procedure is repeated to cover the full 107 year period. Figure 7 shows a repetition of the method but in this case it is applied to a generated 107 sample of data based on the AR(3) model applicable to the series. Although results are affected by the choice of the truncation point M for which a value of M = 17 is adopted, the dispersion in the historical spectra in Figure 6 seems to be greater.

In order to evaluate the variability of the spectra in time, a theoretical spectrum $s(\omega)$ based on equation 4 was fitted to the first 53 years after estimating the best model and its estimates. Then the spectra $\hat{s}(\omega)$ estimated from each historical sequence of 53 years and shown in Figure 6 were used to calculate values of $\nu\hat{s}(\omega)/s(\omega)$ which have the chi-squared distribution. Results given in Table 2 show the spectral variations in time at different frequencies. Rapidly increasing or decreasing values of chi-squared at some frequencies indicate non-stationarities of some form. For purposes of comparison some percentage points from tables of the chi-squared distribution are given below Table 2.

The following procedure is used to quantify the non-stationary behaviour. Initially a frequency band was selected. For the Ogden precipitation frequencies 10 to 17 seem to be the ones having consistently higher than the median value of chi-squared and showing the greatest changes in time. It is seen that chi-squared values tend to increase in time; this implies that the Ogden precipitation has become more random in recent years, although differences may not be significant. For each sequence these chi-squared values are averaged

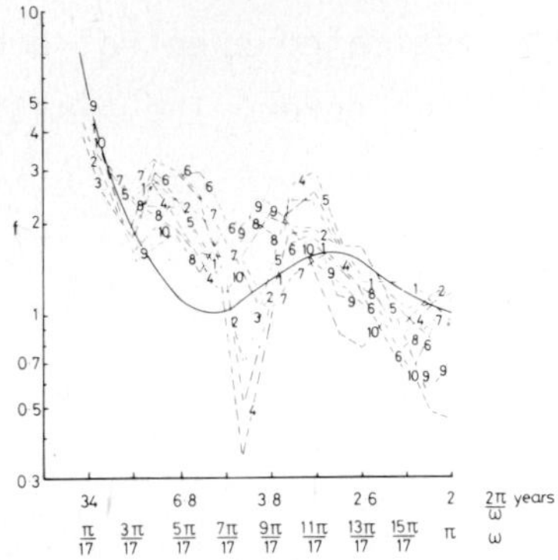

Fig 6 Ogden Annual Precipitation - 107 years - 1901 to 1977
Evolutionary Spectra - Historical
Overlapping Sequences of 53 years.
Smooth Curve Shows Theoretical Function for AR(3) Model

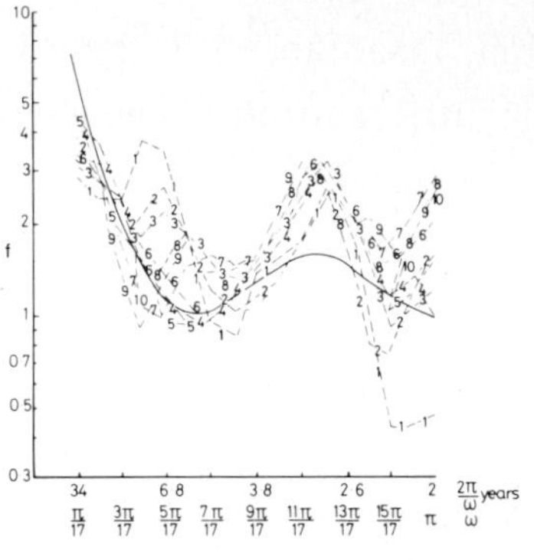

Fig 7 Ogden Annual Precipitation - 107 years Synthetic
Evolutionary Spectra From One Sequence From AR(3) Model
10 Overlapping Sequences of 53 years
Smooth Curve Shows Theoretical Function for AR (3) Model

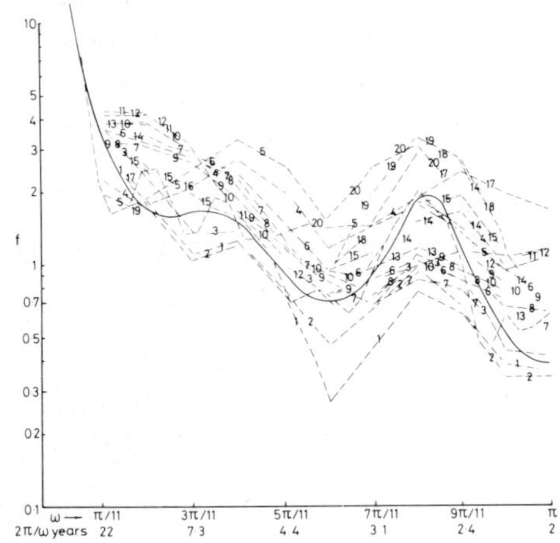

Fig. 8 Nile at Aswan Dam - 172 Annual Flows - 1701 to 1872
Evolutionary Spectra - Historical.
20 Overlapping Sequences of 58 years
Smooth Curve Shows Theoretical Function for AR(5) Model

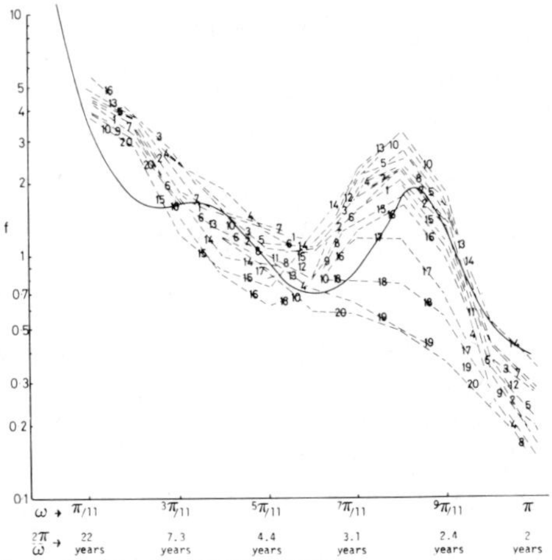

Figure 9 Nile at Aswan Dam - 172 Synthetic Annual Flows.
Evolutionary Spectra from One Sequence using AR(5) model.
20 Overlapping Sequences of 58 years.
Smooth curve shows theoretical function for AR(5) model.

Table 2 Ogden Precipitation : 107 years Values of $\nu\hat{s}(\omega)/s(\omega)$ from Historical Data

	1	2	3	4	5	6	7	8	9	10	11	12	13	14	15	16	17	K see footnote*
ω	$\pi/17$	$2\pi/17$	$3\pi/17$	$4\pi/17$	$5\pi/17$	$6\pi/17$	$7\pi/17$	$8\pi/17$	$9\pi/17$	$10\pi/17$	$11\pi/17$	$12\pi/17$	$13\pi/17$	$14\pi/17$	$15\pi/17$	$16\pi/17$	π	
$2\pi/\omega$	34	17	11.3	8.5	6.8	5.7	4.9	4.3	3.8	3.4	3.1	2.8	2.6	2.4	2.3	2.1		
1	6.5	5.9	6.5	11.1	11.8	6.9	4.1	4.1	7.3	8.7	11.1	17.4	16.1	7.3	3.3	3.5	3.8	0.32
2	8.7	7.3	4.7	6.3	8.7	6.8	5.2	5.0	6.0	7.9	11.7	16.8	14.1	6.0	5.7	10.8	13.1	0.39
3	7.0	6.5	4.8	5.5	7.3	7.0	6.5	6.2	7.1	10.0	15.5	21.4	19.5	11.0	7.1	8.4	9.4	0.46
4	9.2	9.0	6.4	4.3	4.0	3.7	4.1	5.5	8.3	9.9	14.1	20.9	19.5	11.8	8.8	10.6	12.2	0.54
5	11.0	9.3	5.0	3.2	3.4	3.5	4.2	5.4	6.0	7.8	15.7	20.4	17.4	12.2	8.0	9.6	12.5	0.52
6	7.4	7.5	6.4	5.2	4.6	4.1	4.4	5.0	5.4	8.2	17.0	21.2	18.0	14.6	10.7	12.3	16.2	0.54
7	7.6	7.2	5.2	3.3	3.3	5.6	6.7	6.9	7.1	9.0	17.7	21.8	18.0	14.7	11.3	16.2	22.8	0.60
8	8.7	8.2	5.6	3.8	4.9	7.2	7.0	5.6	6.5	11.1	19.5	18.2	12.7	12.0	10.2	15.4	21.4	0.55
9	8.9	7.4	4.1	2.7	4.3	6.9	6.3	5.8	7.8	12.9	21.8	19.4	13.5	15.6	12.2	14.3	20.7	0.59
10	10.5	8.2	4.6	3.2	4.4	7.2	5.9	5.1	7.9	9.9	14.5	19.0	17.8	12.3	7.8	14.2	21.5	0.53

Notation

$Pr(\chi^2 \le \chi^2_{\nu,\alpha})$ =

$\chi^2_{17,.005}$	$\chi^2_{17,.025}$	$\chi^2_{17,.05}$	$\chi^2_{17,.10}$	$\chi^2_{17,.50}$	$\chi^2_{17,.90}$	$\chi^2_{17,.95}$	$\chi^2_{17,.975}$	$\chi^2_{17,.995}$
5.7	7.6	8.7	10.1	16.3	24.8	27.6	30.2	35.7

* Value of K is obtained by dividing the mean value of $\nu\hat{s}(\omega)/s(\omega)$ from columns 10 to 17 by $\chi^2_{17,.95}$(= 27.6)

Table 3 Ogden Precipitation Values of $\nu\hat{s}(\omega)/s(\omega)$ from a 107 year synthetic sequence using an AR(3) model with parameters estimated from historical data.

	1	2	3	4	5	6	7	8	9	10	11	12	13	14	15	16	17	K from one sequence*	K from five sequences
ω	$\pi/17$	$2\pi/17$	$3\pi/17$	$4\pi/17$	$5\pi/17$	$6\pi/17$	$7\pi/17$	$8\pi/17$	$9\pi/17$	$10\pi/17$	$11\pi/17$	$12\pi/17$	$13\pi/17$	$14\pi/17$	$15\pi/17$	$16\pi/17$	π		
$2\pi/\omega$	34	17	11.3	8.5	6.8	5.7	4.9	4.3	3.8	3.4	3.1	2.8	2.6	2.4	2.3	2.1	2		
1	3.8	12.0	9.0	5.6	2.2	1.2	3.2	6.7	4.6	3.0	6.0	7.6	8.3	12.3	17.4	15.3	11.4	0.37	0.34
2	5.4	12.0	7.6	5.5	3.3	2.8	5.4	8.0	4.2	2.6	6.2	7.1	7.0	10.1	14.6	14.4	12.6	0.34	0.33
3	5.7	12.2	8.0	6.0	3.1	2.0	6.0	9.8	5.3	2.7	4.9	5.5	7.1	9.5	9.8	12.1	15.0	0.30	0.30
4	7.9	14.2	9.5	7.0	3.8	2.6	6.6	11.1	6.3	1.6	2.4	3.9	7.4	8.4	5.0	7.7	12.5	0.22	0.31
5	7.7	14.3	9.8	7.0	3.7	2.7	6.9	11.4	6.4	1.7	2.7	4.1	6.9	7.4	4.8	8.7	13.8	0.22	0.34
6	7.1	14.6	9.9	5.8	3.6	4.0	7.5	11.0	6.5	1.7	2.4	4.5	6.8	6.9	5.6	8.3	11.8	0.22	0.31
7	7.3	14.5	8.9	4.7	3.2	3.7	6.4	8.8	7.6	4.6	5.9	7.3	6.7	5.3	5.4	9.1	12.1	0.26	0.32
8	3.7	12.7	10.2	8.8	8.2	5.9	6.3	10.2	8.9	5.5	5.5	5.3	4.4	4.4	5.7	8.1	9.3	0.22	0.29
9	4.9	14.3	10.7	6.2	4.2	4.1	6.2	8.9	6.6	4.3	6.8	6.5	4.8	6.8	10.6	13.4	14.0	0.30	0.32
10	7.5	16.8	11.4	6.2	4.6	3.1	3.4	6.2	6.8	5.1	7.2	8.1	6.4	8.3	11.5	11.7	10.4	0.31	0.32

Notation

$Pr(\chi^2 \le \chi^2_{\nu,\alpha}) = \alpha$

$\chi^2_{17,.005}$	$\chi^2_{17,.025}$	$\chi^2_{17,.05}$	$\chi^2_{17,.10}$	$\chi^2_{17,.50}$	$\chi^2_{17,.90}$	$\chi^2_{17,.95}$	$\chi^2_{17,.975}$	$\chi^2_{17,.995}$
5.7	7.6	8.7	10.1	16.3	24.8	27.6	30.2	35.7

* The values of K are calculated as in Table 2 except that synthetic data are used.
In the last column K is averaged over five randomly selected sequences of 107 years.

Table 4 Ogden Precipitation Values $\nu\hat{s}(\omega)/s(\omega)$ from a 107 year synthetic sequence using an AR(1) model for the first half with parameter estimated from historical data and a random normal sequence for the second half.

	1	2	3	4	5	6	7	8	9	10	11	12	13	14	15	16	17	K from one sequence*	K from five sequences
ω	$\pi/17$	$2\pi/17$	$3\pi/17$	$4\pi/17$	$5\pi/17$	$6\pi/17$	$7\pi/17$	$8\pi/17$	$9\pi/17$	$10\pi/17$	$11\pi/17$	$12\pi/17$	$13\pi/17$	$14\pi/17$	$15\pi/17$	$16\pi/17$	π		
$2\pi/\omega$	34	17	11.3	8.5	6.8	5.7	4.9	4.3	3.8	3.4	3.1	2.8	2.6	2.4	2.3	2.1	2		
1	9.8	11.2	13.3	10.2	4.6	2.6	5.0	9.6	6.5	2.8	4.7	5.6	6.4	10.9	16.4	14.4	10.7	0.33	0.29
2	9.6	10.5	11.7	9.3	5.8	4.7	7.8	11.8	6.1	2.3	5.0	5.3	5.3	8.5	13.3	14.4	13.4	0.31	0.32
3	9.6	10.5	11.8	9.4	5.0	3.4	8.9	14.8	8.2	2.5	3.9	4.2	5.6	8.1	9.3	13.9	18.2	0.30	0.31
4	0.5	11.4	12.5	9.8	5.6	4.0	9.6	17.0	10.0	1.7	1.9	3.0	6.2	7.5	5.1	10.2	16.7	0.24	0.35
5	10.0	11.2	12.7	10.0	5.8	4.5	10.4	17.3	9.6	1.7	2.2	3.3	5.7	6.1	4.8	10.8	17.0	0.24	0.38
6	9.6	11.6	12.4	8.1	5.5	6.3	10.6	16.3	10.3	2.2	1.8	3.4	5.7	6.6	6.4	11.0	15.4	0.24	0.37
7	0.9	11.5	9.9	6.1	5.0	5.8	9.6	15.1	11.1	5.3	6.0	6.9	6.3	5.1	6.2	12.0	16.0	0.29	0.39
8	6.8	8.3	9.7	10.2	10.4	8.1	9.5	15.1	12.3	6.4	5.7	5.4	4.8	5.1	7.2	11.5	13.6	0.27	0.39
9	6.5	7.6	8.3	7.1	6.1	6.2	9.7	13.8	10.0	5.0	7.1	7.0	5.8	9.7	16.2	20.4	21.0	0.42	0.45
10	7.8	9.3	9.1	7.5	6.9	5.3	6.2	10.3	10.0	6.0	7.4	8.6	7.7	11.5	17.7	19.0	17.3	0.43	0.47

Notation

$Pr(\chi^2 \leq \chi^2_{\nu,\alpha}) = \alpha$

$\chi^2_{17,.005}$	$\chi^2_{17,.025}$	$\chi^2_{17,.05}$	$\chi^2_{17,.10}$	$\chi^2_{17,.50}$	$\chi^2_{17,.90}$	$\chi^2_{17,.95}$	$\chi^2_{17,.975}$	$\chi^2_{17,.995}$
5.7	7.6	8.7	10.1	16.3	24.8	27.6	30.2	35.7

and divided by 27.6 (which is the corresponding 95% value of chi-squared) and entered as the variable K on the right of Table 2. The rate at which K approaches and exceeds unity is taken as an index of the lack of reliability of a hydrological record for modelling future events. In another situation in which serial correlation tends to increase with time, or in general when chi-squared values become very low, K ought to be obtained by dividing the 5% value of chi-squared by the estimated mean value over the frequency band. There is also the possibility of seeing both types of behaviour over a long period, in which case the 97.5% and 2.5% values of chi-squared should be used.

The shortness of records in addition to the common divisor of $s(\omega)$ leads to the highly correlated chi-squared values in the table and this precludes more stringent statistical tests. For purposes of comparison Table 3 shows the corresponding chi-squared values obtained from a synthetic sequence using an AR(3) model. Given in the penultimate column are K values for this sequence. These are uniformly distributed as shown by the K values averaged over 5 sequences which are given in the last column.

To complete this aspect of the study, ARMA models were fitted separately to the first and second halves of the Ogden precipitation record of 107 years. This meant that the first half was generated using an AR(1) model with $\hat{\phi}_1 = 0.30$ followed by a completely random sequence for the second half. Then the same procedure as in Table 3 was repeated and evolutionary spectra were obtained from the resulting synthetic non-stationary sequence. Table 4 shows the chi-squared values with mean values of K in the penultimate column and the K values averaged over 5 sequences at the extreme right. The differences in these mean values of K from Tables 3 and 4 is a measure of the non-stationarity. Results shown in Table 4 are comparable with those from the historical data, shown in Table 2.

NILE FLOWS

By way of comparison the method is applied to a 172 year record of the Nile annual flows at Aswan Dam from 1701 to 1872. Some aspects of

Table 5 Nile River Flows : 172 years Values of $\nu\hat{s}(\omega)/s(\omega)$ from Historical Data

	1	2	3	4	5	6	7	8	9	10	11	K
ω	$\pi/11$	$2\pi/11$	$3\pi/11$	$4\pi/11$	$5\pi/11$	$6\pi/11$	$7\pi/11$	$8\pi/11$	$9\pi/11$	$10\pi/11$	π	see footnote*
$2\pi/\omega$	22	11	7.3	5.5	4.4	3.7	3.1	2.8	2.4	2.2	2	
1	5.7	8.0	8.8	16.0	13.8	6.3	12.9	25.6	23.3	15.3	14.2	0.45
2	5.7	7.7	8.2	15.3	14.7	11.1	19.2	32.1	25.2	13.5	13.4	0.52
3	7.3	10.5	10.5	17.7	20.4	15.7	19.7	36.6	32.6	17.4	16.2	0.61
4	4.4	7.8	16.5	34.2	35.5	27.6	41.1	62.0	51.8	40.1	46.3	1.20
5	4.0	8.2	19.6	40.8	46.1	33.7	40.9	60.8	49.3	37.3	42.9	1.15
6	7.3	14.2	22.3	29.6	27.5	22.0	25.5	34.8	32.4	28.4	34.1	0.77
7	5.8	13.2	23.7	28.2	23.2	17.6	20.0	28.9	29.2	22.1	22.2	0.61
8	6.2	13.4	22.6	26.8	22.1	17.2	21.7	33.4	35.0	26.2	25.2	0.71
9	6.4	13.8	21.6	23.4	21.9	19.5	20.1	33.4	41.4	32.4	27.4	0.77
10	7.6	16.7	23.8	20.7	19.6	21.7	25.2	32.4	35.4	30.9	30.0	0.77
11	8.9	18.4	23.7	21.4	22.6	21.2	21.9	30.6	41.7	42.6	41.5	0.89
12	8.3	18.4	25.0	20.1	17.9	18.3	24.2	29.6	36.5	43.1	44.5	0.89
13	7.7	17.0	24.2	20.1	18.1	19.7	28.5	39.0	36.1	25.8	21.0	0.75
14	6.4	14.9	23.9	20.7	17.4	17.8	23.6	50.2	60.4	37.8	25.3	0.98
15	6.1	11.1	15.7	20.1	21.8	23.1	32.3	60.0	70.9	44.4	26.8	1.17
16	6.7	11.5	16.3	22.0	21.5	20.1	27.8	56.7	64.9	40.9	30.1	1.10
17	5.4	9.4	15.4	21.3	17.8	16.0	25.0	62.4	89.3	77.8	67.5	1.61
18	5.3	9.6	16.4	20.3	15.4	20.2	43.0	93.7	103.5	46.6	18.3	1.52
19	3.5	7.6	17.3	25.1	20.3	21.2	56.1	111.6	102.4	49.2	28.7	1.74
20	3.4	8.1	16.6	22.3	24.7	35.7	70.7	103.2	78.0	38.7	28.3	1.59

Notation

$\Pr(\chi^2 \leq \chi^2_{\nu,\alpha}) = \alpha$

$\chi^2_{27,.005}$	$\chi^2_{27,.025}$	$\chi^2_{27,.05}$	$\chi^2_{27,.10}$	$\chi^2_{27,.50}$	$\chi^2_{27,.90}$	$\chi^2_{27,.95}$	$\chi^2_{27,.975}$	$\chi^2_{27,.995}$
11.8	14.6	16.2	18.1	26.3	36.7	40.1	43.2	49.6

* Value of K obtained is the mean value of $\nu\hat{s}(\omega)/\hat{s}(\omega)$ from columns 7 to 11 divided by $\chi^2_{27,.95}$ (= 40.1).

this record have been studied by Reihl and Meitin (1979); it is thought that there are no errors of measurement and that non-stationarities are less than in the period after 1872. However, it is found that three non-overlapping sub-samples from the chosen record have contrasting features with some evidence of non-stationarity particularly in the period after 1766. Investigations also show that first and second differencing and weighting do not appear to be helpful for modelling purposes.

Figure 8 which shows historical evolutionary spectra for the Nile flows corresponds to Figure 6 for the Ogden precipitation. The base period is taken as 58 years and the lagging of consecutive overlapping samples is six years as before giving 20 sequences. Then, the procedure is repeated for a randomly selected 172 year synthetic sequence using an AR(5) model and the resulting spectra are shown in Figure 9.

Table 5 gives the chi-squared values for the Nile annual flows following the same procedure as for Table 2. The frequency band studied is the high frequency section numbered 6 to 11. Here the 95% value of chi-squared is 40.1 and variable K commences with 0.45 for the first sequence and is seen to exceed 1 after a few lags of six years. This indicates a larger degree of non-stationarity than for the Ogden precipitation. It also means that in applications such as this modelling becomes a less reliable procedure.

CONCLUSION

The results of this study show that a variety of techniques should be adopted when testing the validity of stochastic models in hydrology. This is partly on account of the low power of some statistical tests. Some form of non-stationary behaviour is evident in several series as seen here. It is important to gauge the nature, magnitude and extent of the variations. The outcomes can be considered to be a measure of the resilience one needs to build into planned water resource projects to account for future uncertainties.

ACKNOWLEDGEMENTS

The work commenced whilst the author was a Visiting Professor at Utah State University. Thanks are due to Professors L.D. James and D.S. Bowles and to Dr. W.R. James of the Water Research Laboratory, Logan. The Nile data was received from Dr. J. Meitin of the University of Colorado at Boulder.

REFERENCES

Box, G.E.P. and Jenkins, G.M. Time Series: Forecasting and Control. revised edn., Holden Day, San Francisco, 1976.

Box, G.E.P. and Tiao, G.C. A change in level of non-stationary time series. Biometrika, 52, pp. 181-192, 1965.

Conover, W.J. Practical Nonparametric Statistics. Wiley, New York, 1971.

Jenkins, G.M. and Watts, D.G. Spectral Analysis and its Applications. Holden-Day, San Francisco, 1968.

Menzefricke, U.A. Bayesian analysis of a change in the precision of a sequence of independent normal random variables at an unknown time point. Applied Statistics, vol. 30, pp. 141-146, 1981.

Pettitt, A.N. A non-parametric approach to the change-point problem. Applied Statistics, vol. 28, pp. 126-135, 1979.

Priestley, M.B. Evolutionary spectra and non-stationary processes. J.R. Statist. Soc., B, 19, 1-12, 1965.

Riehl, H. and Meitin, J. Discharge of the Nile River. A Barometer of Short Period Climate Variation. Science, vol. 206, pp. 1178-1179, 1979.

Siegel, S. Nonparametric statistics for the behavioural sciences. McGraw Hill, New York, 1956.

Worsley, K.J. On the Likelihood Ratio Test for a Shift in Location of Normal Populations. Jour. Am. Stat. Assoc., vol. 74, pp. 365-367, 1979.

World Meteorological Organisation. Climatic Change. World Meteor. Organ., Geneva, Tech. Note No. 79, 1966.

Yevjevich, V. and Jeng, R.I. Properties of Non-Homogeneous Hydrologic Series. Hydrology Papers, no. 32, Colorado State University, Fort Collins, 1969.

PERSISTENCE ESTIMATION FROM A TIME-SERIES CONTAINING OCCASIONAL MISSING DATA

DANIEL A. CLUIS AND PIERRE BOUCHER

Universite du Quebec, Quebec, Canada

ABSTRACT

In a real life situation, such as that encountered with weekly sampled water-quality parameters, measured time-series are often incomplete. The missing data complicate the estimation of the lag-one autocorrelation coefficient, a most important parameter in the evaluation of serial persistence necessary for efficient sampling strategies. In the case of occasional missing data, we have investigated the biases and sampling variances resulting from the introduction of simple but non-optimal estimators (e.g. the "general mean" estimator, the "local mean" estimator and the "combined mean" estimator) or from compressing the time-series by simply neglecting the missing data. These studies were carried out in two ways:

1) via theoretical developments, in the case of large samples and
2) via simulation of Markovian series in the case of smaller samples where a classical short sample correction was combined with the theoretical biases derived from the large samples case.

The efficiencies of the estimators (residual biases and sampling variances), which depend on the sample lengths and on the persistence parameter of the parent population, have been evaluated when 5%, 10%, 15% and 20% of the data were missing. The proposed approach (combined mean estimator) leads, in many cases, to better results than would be obtained using classical techniques derived from Fisher's Z transformation.

INTRODUCTION AND PREVIOUS WORK

The estimation of the lag 1 autocorrelation coefficient of a stationary time-series constitutes an important step in the evaluation of the short-term persistence of this series. This knowled-

Reprinted from *Time Series Methods in Hydrosciences*, by A.H. El-Shaarawi and S.R. Esterby (Editors)

ge is useful both for the understanding of the phenomenon represented by the series and also to the determination of the optimal sampling rate of future data acquisition programs. It happens quite often that in order to assess this autocorrelation coefficient one has at his disposal a sample of equispaced, systematically sampled data, but containing either occasionnal missing data or abnormal values which could distort considerably the results and bias any interpretation. In this situation, two types of techniques have been described to assess the values of the autocorrelation.

The first one consists, on each continuous section of data, in the calculation of an autocorrelation coefficient that may be highly biased. Given the relatively short length of each section, the Fisher-z-transformation is performed to normalize their distribution and the transformed values are combined according to the relative length of the data. Finally, the inverse z-transformation on the composite values is obtained, giving an estimation of the global autocorrelation coefficient.

The second technique consists in the evaluation of missing data using a model of variable complexity, established with all the valid data. This procedure gives the missing data values maintaining the basic characteristics of the sample: mean, variance and persistence. Given a sufficient number of data, the BOX and JENKINS approach, followed by a forward and backward residual generation procedure, provides an estimation of each missing value. Such a technique which should be applied in a stepwise and interactive manner for each missing data is quite time-consuming; the procedure can even eventually diverge if the number of missing data is large and if their distance is too small to eliminate the possible interactions.

In the recent literature, numerous papers deal with the problem of estimating missing data in univariate and multivariate cases; it is essentially a problem of data interpolation using the local trend of the data. WILKINSON (1958) and PREECE (1971) have applied this kind of technique to experimental data, but in our case, its value is limited as no attempt is made to preserve the autocorrelation structure. More recently BRUBACHER and WILSON (1976) have developed a predictive approach based on the least squares technique to estimate the effect of a national holiday on the electricity demand using previous and posterior normal consumptions; the n interpolated values are obtained via a method of forecasting and backforecasting which minimizes the residual serials; the ratio of the actual demand to the historical demand during holidays allows the evaluation of the future demand for similar holidays. The technique seems well adapted to the problem, even if numerous missing data are to be recreated simultaneously. The very nature of electricity demand explains why such an approach was succesfull and why such good results are not to be expected for any time-series. A model of the BOX and JENKINS (1970) type can be identified for the whole series from a short sequence of data; weekly fluctuations possessing a much larger variability than yearly fluctuations, a few weeks of data allow sufficient identification for the global model and is not significantly disturbed by the holidays. D'ASTOUS and HIPEL (1979) have presented an Intervention Analysis model derived from the BOX and JENKINS approach which permits, on a very general way, to reconstitute missing-data. The simulations performed by these authors on homogeneous and non-homogeneous series of yearly streamflow and monthly phosphorous data seem very conclusive.

PRESENT APPROACH

In this paper, we suggest the use of some non-optimal, but very simple estimators of the missing data; they introduce biases in the calculation of the autocorrelation coefficients which are easy to estimate and thus can be corrected.

The first estimator r_0 is built in replacing the missing values by the general mean, or zero if the sample has been centered: such an estimator generally underestimates the true value r of the autocorrelation of the sample. The second estimator r_i is built in replacing each missing value by the local mean, i.e. the arithmetical mean between the preceeding and the following values: such an estimator generally overestimates the true value r of the autocorrelation of the sample. A combined value $r_c = f(r_0, r_i)$ possessing a large sample unbiasedness property is discussed. The third estimator studied, r_n, is obtained as the series is simply compressed of its missing values, shortening its length; this method of dealing with missing data is used in practice quite often and is not without danger. For these cases, the biases and the amplification factors of the sampling variance are derived theoretically and compared in the large sample case.

In the case of a short sample, WALLIS and O'CONNELL (1972) have demonstrated the efficiency of the KENDALL (1954) circular length correction to relate the sampled correlation value with that of the parent population which is the parameter to be assessed. Given the two corrections, a large-sample estimator correction and a length correction, a simulation programm using the Monte-Carlo technique was devised to assess the order in which they should be applied and to compare the efficiencies of the estimators with that of the classical FISHER's z-transformation technique. One should note that the previous estimators are making no hypothesis about the generating processes of the samples which is consistent with the fact that a same short trace may originate from different parent populations.

HYPOTHESIS AND NOTATIONS

Let's consider a sample X_i (i = 1, ... N) from a stationary time-series, of length N in which n values are missing; we will develop here only the case of occasional missing data i.e. their number is relatively small n ≪ N; each missing data is isolated and followed by non-isolated section Z_i of valid data, so that the sample may be written:

$$Z_1 \; ? \; Z_2 \; ? \; .... \; ? \; Z_n \; ? \; Z_{n+1} .$$

Evidence presented PARZEN (1964) and RODRIGUEZ-ITURBE (1971) has lead us to consider the unbiased definition of variance and covariance to be specially well suited for relatively short samples:

$$\text{var} = \sum_{i=1}^{N} (X_i - m)^2 \, / \, (N-1)$$

$$\text{cov} = \sum_{i=1}^{N-1} (X_i - m) \, (X_{i+1} - m)/(N-1)$$

and

$$m = \sum_{i=1}^{N} X_i \, / \, N$$

The corresponding expression of the lag 1 autocorrelation coefficient is that of JENKINS and WATTS (1968) and of BOX and JENKINS (1970):

$$\text{var} = \sum_{i=1}^{N} X_i^2 \, / \, (N-1) \qquad\qquad \text{cov} = \sum_{i=1}^{N-1} X_i X_{i+1} \, / \, (N-1)$$

and

$$r = \sum_{i=1}^{N-1} X_i X_{i+1} \, / \, \sum_{i=1}^{N} X_i^2 \tag{1}$$

The expected values of the products involving missing values are:

$$E[X_i X_{i+1}] = \text{cov}$$

$$E[X_i^2] = [(N-1) \, / \, N] \, \text{var}$$

THE LARGE SAMPLE CASE

This section deals with the case where, with regard to the sampling error, the KENDALL (1954) circular length correction is negligible, i.e. when the length of the sample containing missing values is larger than 100.

Bias related to the estimator r_o

Using the general mean as an estimation of missing data leads to the disappearance in the numerator of equation (1) of two products and of one square at the denominator for each of the n missing data. Taking the expected values, one gets:

$$E[r_o] = \frac{N-2n-1}{N-n} \frac{N}{N-1} r \tag{2}$$

which can be inverted as:

$$r = \frac{(1-\frac{1}{N})(1-\frac{n}{N})}{1 - \frac{2n+1}{N}} E[r_o] \tag{3}$$

This expression can be expanded to order N^{-2}:

$$r \simeq [1 + \frac{n}{N} + \frac{2n(n+1)}{N^2}] E[r_o]$$

The replacement of $E[r_o]$ by its sample estimate r_o yields the adequate bias correction:

$$r - r_o = \frac{n}{N} (1 + \frac{2n(n+1)}{N}) r_o \tag{4}$$

if we suppose a sufficient number of missing data ($n \gg 1$) and by putting $\frac{n}{N} = \alpha \ll 1$, then the preceeding equations reduce to:

$$r = (1 + \alpha + 2\alpha^2) r_o \tag{5}$$

$$\overline{r-r_o} = \alpha (1+2\alpha) r_o \tag{6}$$

One can note that for positive values of r_o, which is the most common case in hydrosciences, the bias is positive; the estimator underestimates the true values of r.

Bias related to the estimator r_i

Using the local mean as an estimation of missing data, any missing value X_{im} is estimated by: $X_{im} = \frac{X_{i+1} + X_{i-1}}{2}$ so that each product and each square containing X_{im} are estimated by:

$$X_{im}X_{i\pm1} = \frac{X_{i\pm1}^2}{2} + \frac{X_{i+1}X_{i-1}}{2}$$

$$X_{im}^2 = \frac{X_{i+1}^2}{4} + \frac{X_{i-1}^2}{4} + \frac{X_{i+1}X_{i-1}}{2}$$

Introducing those expressions in equation (1) and approximating the expected values $E[X_{i+1}\ X_{i-1}]$ by $E[X_{i+1}\ X_i]$ gives;

$$E[r_i] = \frac{2N(N-n-1)r + 2n(N-1)}{nNr + (2N-n)\ (N-1)} \qquad (7)$$

which can be inverted as:

$$r = \frac{(1-\frac{n}{2N})\ (1-\frac{1}{N})\ E\ [r_i] - \frac{n}{N}\ (1-\frac{1}{N})}{(1-\frac{n+1}{N}) - \frac{n}{N}\ \frac{E[r_i]}{2}} \qquad (8)$$

This expression can be expanded to order N^{-2}:

$$r \simeq E[r_i] - \left(\frac{E[r_i]}{2} + 1\right)\ (1- E[r_i])\left[\frac{n}{N} + \frac{n^2}{N^2}\left(\frac{E[r_i]}{2} + 1 + \frac{1}{n}\right)\right]$$

The replacement of $E[r_i]$ by its sample estimate r_i yields an adequate bias correction:

$$\overline{r - r_i} = -\left(\frac{r_i}{2} + 1\right)\ (1-r_i)\ \left[\frac{n}{N} + \frac{n^2}{N^2}\left(\frac{r_i}{2} + 1 + \frac{1}{n}\right)\right] \qquad (9)$$

If we suppose a sufficient number of missing data ($n \gg 1$) and by putting $\frac{n}{N} = \alpha \ll 1$, then the preceeding equations reduce to:

$$r = r_i - \alpha\ \left(1 - \frac{r_i}{2} - \frac{r_i^2}{2}\right) - \alpha^2\ \left(1 - \frac{3}{4} r_i^2 - \frac{1}{4} r_i^3\right) \qquad (10)$$

$$\overline{r - r_i} = \alpha\ \left(1 - \frac{r_i}{2} - \frac{r_i^2}{2}\right) - \alpha^2\ \left(1 - \frac{3}{4} r_i^2 - \frac{1}{4} r_i^3\right) \qquad (11)$$

One can note that for all values of r_i, the bias is always negative; the estimator overestimates the true value of r.

Definition of an almost unbiased combined estimator

In the preceeding sections, we have developed the calculation of the biases of two very simple estimators as a function of the percentage of missing values α, one overestimating, the other underestimating the true value of r. In this section we define a combined estimator r_c as a combination of r and r_i which reduces the bias to order α^2, making bias correction unnecessary.

Taking the linear part of equations (5) and (10) and eliminating α between them gives the definition of r_c:

$$r_c = \frac{r_o(1+r_i)(2-r_i)}{2r_o + (1-r_i)(2+r_i)} \tag{12}$$

Using the same technique as previously, the bias can be approximated by:

$$\overline{r - r}_c \simeq - 2\alpha^2 \left(\frac{2r_c}{1+r_c} - r_c^2 \right) \tag{13}$$

which is a negative and small value of order α^2 and vanishes for $r_c = 0$ and $r_c = 1$. The Table 1 presents the estimator r_c as a function of the biased sample estimators r_o and r_i. The Figure 1, which can be used as a nomograph, shows the range of application of equation (12) and the relative location of the three estimations:

$$1 \geqslant r_i \geqslant r_c \geqslant r_o \geqslant 0$$

$$1 \geqslant r_i \geqslant 0 \geqslant r_o \geqslant r_c \geqslant 1 \qquad \text{and}$$

$$0 \geqslant r_i \geqslant r_o \geqslant r_c \geqslant -1.$$

This almost unbiased estimator r_c is very easy to apply, but presents a drawback for values of r_i greater that 0.9, as the singular point around $r_i = 1$ and the sampling errors make the definition of r_c less accurate.

Bias related to the estimator r_D

Compressing the time-series by simply neglecting the occasional missing values gives way to the estimator r_D; doing so, the length of the sample is reduced to (N-n) and each missing data leads in equation (1) to the suppression of two products $X_{im}X_{i+1} + X_{im}X_{i-1}$ at the numerator and at the same time a new product $X_{i-1}X_{i+1}$ is created; at the denominator, one of the squares X_{im}^2 disappears. If we suppose an exponential law for the decrease of the autocorrelation with the lag, a case of pure persistence fairly common in in hydrometeorology and which includes the often-used first-order Markov model, the equation (1) becomes:

TABLE 1: Values of r_c for different values of r_o and r_i

r_i \ r_o	-1	-0.9	-0.8	-0.7	-0.6	-0.5	-0.4	-0.3	-0.2	-0.1	0	0.1	0.2	0.3	0.4	0.5	0.6	0.7	0.8	0.9	1
-1	-1.00																				
-0.9		-0.90																			
-0.8			-0.80																		
-0.7				-0.70																	
-0.6				-0.87	-0.60																
-0.5					-0.71	-0.50															
-0.4					-0.83	-0.58	-0.40														
-0.3					-0.96	-0.67	-0.45	-0.30													
-0.2						-0.76	-0.52	-0.34	-0.20												
-0.1						-0.87	-0.59	-0.38	-0.22	-0.10											
0						-1.00	-0.67	-0.43	-0.25	-0.11	0.00										
0.1							-0.77	-0.49	-0.28	-0.12	0.00	0.10									
0.2							-0.90	-0.56	-0.32	-0.14	0.00	0.11	0.20								
0.3								-0.65	-0.37	-0.16	0.00	0.12	0.22	0.30							
0.4								-0.80	-0.43	-0.18	0.00	0.14	0.24	0.33	0.40						
0.5									-0.53	-0.21	0.00	0.15	0.27	0.36	0.44	0.50					
0.6									-0.70	-0.27	0.00	0.18	0.31	0.41	0.49	0.55	0.60				
0.7										-0.36	0.00	0.22	0.37	0.47	0.55	0.61	0.66	0.70			
0.8										-0.60	0.00	0.28	0.45	0.56	0.63	0.69	0.74	0.77	0.80		
0.9											0.00	0.43	0.61	0.70	0.77	0.81	0.84	0.87	0.88	0.90	
1												1.00	1.00	1.00	1.00	1.00	1.00	1.00	1.00	1.00	1.00

TABLE 2: Theoretical biases $\overline{\rho - r}$ and sampling variances var r for short samples derived from Markovian parent populations

N	100		80		60		40		20	
ρ	$\overline{\rho - r}$	var r	$\overline{\rho - r}$	var r	$\overline{\rho - r}$	var r	$\overline{\rho - r}$	var r	$\overline{\rho - r}$	var r
0	0.010	0.010	0.013	0.013	0.017	0.017	0.025	0.025	0.050	0.050
0.1	0.014	0.010	0.018	0.012	0.023	0.017	0.035	0.025	0.070	0.049
0.2	0.018	0.009	0.023	0.012	0.030	0.016	0.045	0.024	0.090	0.048
0.3	0.022	0.009	0.028	0.011	0.037	0.015	0.055	0.023	0.110	0.045
0.4	0.026	0.008	0.033	0.010	0.043	0.014	0.065	0.021	0.130	0.042
0.5	0.030	0.007	0.038	0.009	0.050	0.013	0.075	0.019	0.150	0.038
0.6	0.034	0.006	0.043	0.008	0.057	0.011	0.085	0.016	0.170	0.032
0.7	0.038	0.005	0.048	0.006	0.063	0.008	0.095	0.013	0.210	0.018
0.8	0.042	0.004	0.053	0.004	0.070	0.006	0.105	0.009	0.210	0.018
0.9	0.046	0.003	0.058	0.002	0.077	0.003	0.115	0.005	0.230	0.010
1	0.050	0	0.063	0	0.083	0	0.125	0	0.250	0

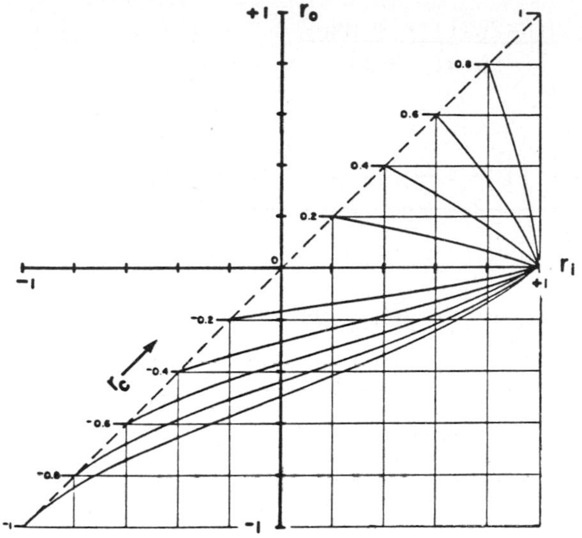

FIGURE 1: Nomograph relating r_c to r_o and r_i.

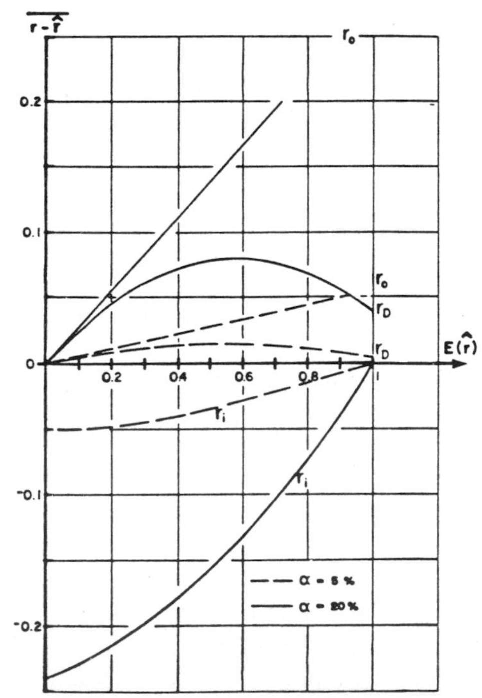

FIGURE 2: Theoretical bias corrections related to various estimators $\hat{r}$ in the large sample case.

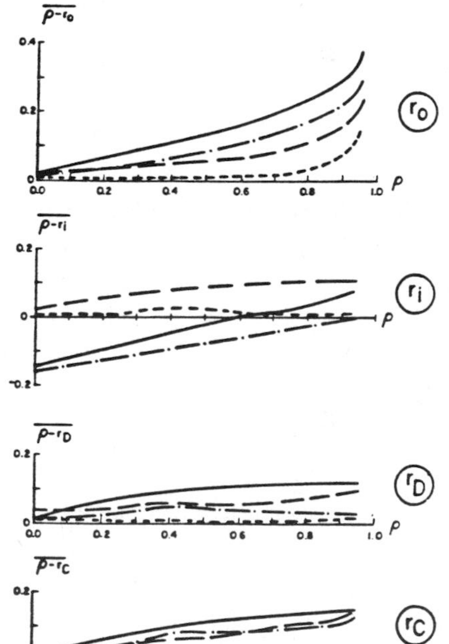

FIGURE 3: Compared efficiencies in the bias corrections (N= 60, n= 9) based on 2000 entries.

$$E(r_D) = \frac{N(N-2n-1)\ r + nNr^2}{(N-n)\ (N-1)} \tag{14}$$

which can be inverted as:

$$r = \frac{-(1-\frac{2n+1}{N}) + \left[1 - (\frac{2n+1}{N})\right]^2 + 4\frac{n}{N}(1-\frac{n}{N})(1-\frac{1}{N})\ E[r_D]^{\frac{1}{2}}}{2\ \frac{n}{N}} \tag{15}$$

This expression can be expanded to order N^{-2}:

$$r \simeq E\ [r_D]\ (1-\frac{2n+1}{N})\ (1 + \frac{n(3-E[r_D])+1}{N} + \frac{n^2\ (8-3E[r_D])+n(7-3E[r_D])+1)}{N^2})$$

One can verify that if n =0, $E(r_D)$ = r, if $E[r_D]$ = 0, r=0 independently of n and N, and if $E[r_D]$ = 1, r=1 to the order N^{-2}. The replacement of $E[r_D]$ by its sample estimate r yields an adequate bias correction:

$$\overline{r - r_D} = r_D[\ \frac{n}{N}\ (1-r_D) + \frac{n^2}{N^2}\ (2-r_D) + \frac{2n}{N^2}\ (1-r_D)] \tag{16}$$

If we suppose a sufficient number of missing data ($n \gg 1$) and by putting $\frac{n}{N} = \alpha \ll 1$, then the preceeding equations reduce to:

$$r = r_D\ [1 + \alpha\ (1-r_D) + \alpha^2\ (2-r_D)] \tag{17}$$

$$r - r_D\ = r_D\ [\alpha\ (1-r_D) + \alpha^2\ (2-r_D)] \tag{18}$$

One can note that for all values of r_D, the bias is always positive; the estimator r_D underestimates the true value of r as r_0 did.

Comparison of the biases

The biases of r_0, r_i and r_D, given by equations (6), (11) and (18) are compared on the Figure 2, for α = 0.05 and 0.20. We suggest the use of r_0 for sampled values lower than 0.5 and of r_i for sampled values higher than 0.5, which minimizes the influence of the sampling error on the estimator's bias corrections. Even if the bias on r_D are smaller, we prefer r_0 and r_i as no hypothesis is made on the generating process of the parent population.

The biases of r_c being much smaller cannot be represented on this figure.

5. The short sample case

The complete sample (no missing value)

More interesting that an unbiased evaluation of the autocorrelation r of a sample is the estimation of the autocorrelation ρ of the parent population; SOPER et al, (1918), ORCUTT (1948) and SASTRY (1951) worked on the relationships between autocorrelations estimated from a few traces of relatively short ($N \leqslant 100$) samples and that of their parent population; in a review of previous works, MARIOTT and POPE (1954) have shown that bias may arise from two sources: if the mean value of the parent population is known, autocorrelations estimated from short samples are generally biased toward zero; if the mean of the parent population is not known and has to be estimated from the sample, it induces another bias (except for $\rho = 0$), which is always negative for a long series. The combinaison of the two biases can either compensate or reinforce each other; as they are not independent, they cannot be investigated separately. For that reason, WALLIS and O'CONNELL (1972) have used the simulation technique to evaluate the biases resulting from samples of various lengths drawn from parent population for known auto-correlation ρ; they have shown that if the parent population is generated by a Markovian process of order one, the theoretical bias correction due to KENDALL (1954) and derived from a circular series assumption is equally valid for the case of classical open series. This length correction can be written as:

$$\rho - r = \frac{1}{N}(1 + 4\rho) = \frac{4E[r] + 1}{N-4} \qquad (19)$$

For the same type of parent population, BOX and JENKINS (1970) have developped the sampling variance for short samples:

$$\mathrm{var}(r) = \frac{1-\rho^2}{N} \qquad (20)$$

As we are going to use this case as a reference, the theoretical biases and sampling variances for complete samples of length N = 100, 80, 60, 40 and 20 and parent populations of autocorrelation $\rho = 0,(0.1),1$ have been tabulated in the Table 2.

The incomplete sample (occasional missing values)

Given the short sample bias correction, equation (19), and the estimator's bias corrections, equations (3), (8) and (15), we questioned whether successive application of those two corrections could improve significantly the bias in the autocorrelation in the case of short samples with occasional missing values. Bearing in mind that the sampling variance is generally increased by a bias correction and that one generally disposes of a single sample, we should verify that the final sampling variance was kept in reasonable limits to really benefit from an almost unbiased estimation; in this regard, one can note that the equations (19) and (3) being linear, the variance amplification factor is independent of the order in which the corrections are carried out; this is not the case for the equations (8) and (15), so that anterior and posterior length corrections yield different sampling variances. The analytical expressions derived from the combinations of the theoretical corrections as well as their numerical values for the cases considered here are given in CLUIS and BOUCHER (1981).

SIMULATIONS AND RESULTS

To test the efficiencies of the proposed theoretical bias corrections pertaining to the various estimators, we used a Monte Carlo technique by generating 2000 synthetic sequences of length 100, 80, 60, 40 and 20 from a Markovian parent population of known parameter ρ; those series were created using the recurring formula:

$$X_i = \rho X_{i-1} + (1-\rho^2)^{\frac{1}{2}}\varepsilon_i$$

In this expression, ε_i are NIP (0,1), provided by the algorithm of BOX and MULLER (1958). At first, the missing values were introduced in appropriate number by suppressing at random some elements of the sequence in each of the 2000 samples; as found also by KNOKE (1979) in a study on some statistics of the lag-one autocorrelation distribution, the results were remarkably unaffected by the location of the missing data when the number of missing data did not exceed 20% of the sample length: the means and the variances calculated with 2000 series were stable within 10^{-3}; so, it was decided to give missing values fixed positions in each of the 2000 samples, making them approximately equispaced.

Five series of results obtained by simulation and concerning residual biases and sampling variances will be discussed here:

<u>series 1</u>: on complete samples (no missing value) to serve as reference for the efficiencies of the estimators;

<u>series 2</u>: on incomplete samples for which no correction has been applied;

<u>series 3</u>: on incomplete samples for which only the length correction has been applied;

<u>series 4</u>: on incomplete samples for which only the estimator's correction has been applied;

series 5: on incomplete samples for which both corrections have been applied.

The results of the series 1 shown on the Table 2 prove the high efficiency of the KENDALL length bias correction and exhibit a large increase of the sampling variances for very short samples (N=40, N=20), but also very similar results to the theoretical values of the Table 1 for larger samples (N=100, N=80), especially, if ρ is close to 1.

For the series 2, 3, 4 and 5, typical results concerning the cases N=100, n=10, N=60, n=15 and N=20, n=4 are presented on the Tables 4, 5, 6, 7; the Table 4 displays the magnitude of the biases without any correction applied to the estimators; all the estimators yield poor results, but r_c and r_D are relatively more efficient for the larger series; the Tables 5 and 6 presents the results of the series 3 and 4 when only one correction has been applied; the residual biases are still very important, and except in one case, always positive; one can note that if one has to apply only one correction, the length correction is generally more efficient than the estimator's correction. For the study of the series 5, the two corrections can be applied in two possible permutations in the case of r_o, r_i and r_D and in five possible permutations in the case or r_c; the better results both for the correction of the biases and the reduction of the sampling variance were consistently obtained when the length correction is applied first and are displayed on the Table 7; the estimator r_o proves to be the worst for large values of ρ and the best for small values of ρ; the estimator r_i is the best estimator for small values of ρ; both r_D and r_c are very efficient in the whole range of values of ρ; there is an advantage in using r_c as its theoretical bias correction makes no hypothesis about the generating process of the parent population which is not the case for r_D; there is also a disadvantage related to its poor definition around $\rho = 1$ which can be seen for very short samples (N = 20).

The Figure 3 shows the progress toward unbiasedness when not using or when using one or two corrections in the case of a sample of length 60 containing 9 missing values. In parallel to the bias corrections, the sampling variances increased slightly, with no noteworthy difference between the estimators, the sampling variances corresponding to the ultimate bias correction being multiplied by a factor 2 or 3 compared with the ones obtained on the Table 3 for complete samples.

COMPARISON WITH THE FISHER'S TRANSFORMATION

When two populations are correlated, the distribution of their correlation coefficients is neither Gaussian nor even symetrical,

TABLE 3: Residual biases and sampling variances for length-corrected autocorrelations of complete samples

N	100		80		60		40		20	
ρ	$\overline{\rho - r}$	var r	$\overline{\rho - r}$	var r	$\overline{\rho - r}$	var r	$\overline{\rho - r}$	var r	$\overline{\rho - r}$	var r
0	0.003	0.010	0.003	0.013	0.003	0.018	0.005	0.028	0.000	0.068
0.1	-0.001	0.011	0.001	0.014	0.003	0.018	0.005	0.028	-0.006	0.069
0.2	-0.003	0.010	-0.004	0.013	-0.003	0.018	-0.003	0.028	-0.003	0.064
0.3	-0.000	0.010	0.002	0.012	0.001	0.017	-0.002	0.028	-0.003	0.066
0.4	0.003	0.009	0.005	0.012	0.005	0.017	0.006	0.027	0.003	0.067
0.5	-0.000	0.008	0.001	0.011	-0.001	0.015	-0.001	0.025	0.005	0.065
0.6	0.005	0.008	0.003	0.010	0.004	0.014	0.005	0.022	0.009	0.060
0.7	0.000	0.006	0.000	0.008	0.002	0.012	0.007	0.021	0.020	0.057
0.8	0.004	0.005	0.007	0.007	0.008	0.010	0.017	0.018	0.027	0.054
0.9	0.006	0.004	0.009	0.005	0.016	0.008	0.024	0.015	0.044	0.046
0.95	0.008	0.003	0.011	0.004	0.018	0.006	0.032	0.013	0.066	0.044

Based on 2000 entries

TABLE 4: Original biases induced by the estimators (no correction applied)

	N=100 n=10				N=60 n=9				N=20 n=4			
ρ	r_o	r_i	r_D	r_c	r_o	r_i	r_D	r_c	r_o	r_i	r_D	r_c
0	0.011	-0.092	0.013	0.010	0.021	-0.136	0.025	0.020	0.051	-0.157	0.065	0.050
0.1	0.023	-0.077	0.025	0.026	0.043	-0.110	0.046	0.044	0.084	-0.119	0.097	0.082
0.2	0.038	-0.061	0.035	0.044	0.061	-0.091	0.057	0.063	0.135	-0.070	0.141	0.133
0.3	0.057	-0.042	0.048	0.063	0.092	-0.062	0.080	0.092	0.178	-0.028	0.174	0.174
0.4	0.072	-0.028	0.056	0.077	0.119	-0.036	0.096	0.116	0.233	0.019	0.213	0.225
0.5	0.087	-0.016	0.061	0.086	0.136	-0.022	0.099	0.125	0.279	0.054	0.242	0.263
0.6	0.106	0.001	0.068	0.098	0.167	0.002	0.110	0.144	0.330	0.092	0.265	0.304
0.7	0.120	0.010	0.065	0.097	0.196	0.020	0.111	0.152	0.388	0.133	0.290	0.347
0.8	0.140	0.025	0.068	0.099	0.231	0.044	0.117	0.162	0.463	0.174	0.314	0.401
0.9	0.170	0.040	0.068	0.096	0.287	0.069	0.121	0.174	0.569	0.218	0.341	0.486
0.95	0.205	0.047	0.067	0.097	0.356	0.080	0.121	0.193	0.697	0.250	0.366	0.629

TABLE 5: Residuals biases after the application of the lenth correction only

	N=100 n=10				N=60 n=9				N=20 n=4			
ρ	r_o	r_i	r_D	r_c	r_o	r_i	r_D	r_c	r_o	r_i	r_D	r_c
0	0.001	-0.106	0.002	0.000	0.005	-0.164	0.006	0.003	0.001	-0.259	0.002	-0.000
0.1	0.009	-0.095	0.009	0.013	0.021	-0.142	0.020	0.022	0.017	-0.236	0.013	0.015
0.2	0.021	-0.082	0.016	0.027	0.033	-0.129	0.023	0.035	0.056	-0.200	0.038	0.054
0.3	0.036	-0.067	0.025	0.043	0.059	-0.106	0.040	0.059	0.084	-0.173	0.048	0.080
0.4	0.048	-0.056	0.029	0.053	0.082	-0.085	0.049	0.078	0.129	-0.139	0.068	0.119
0.5	0.059	-0.048	0.029	0.058	0.092	-0.077	0.043	0.080	0.161	-0.120	0.072	0.141
0.6	0.075	-0.034	0.032	0.067	0.119	-0.058	0.047	0.093	0.200	-0.098	0.071	0.167
0.7	0.085	-0.030	0.024	0.062	0.142	-0.046	0.040	0.095	0.247	-0.071	0.070	0.197
0.8	0.102	-0.018	0.023	0.059	0.173	-0.028	0.038	0.098	0.316	-0.045	0.069	0.239
0.9	0.129	-0.007	0.017	0.052	0.226	-0.008	0.034	0.104	0.423	-0.015	0.071	0.320
0.95	0.163	-0.001	0.015	0.052	0.295	0.000	0.029	0.121	0.571	0.012	0.088	0.487

TABLE 6: Residuals biases after the application of the estimator's correction only

	N=100 n=10				N=60 n=9				N=20 n=4			
ρ	r_o	r_i	r_D	r_c	r_o	r_i	r_D	r_c	r_o	r_i	r_D	r_c
0	0.013	0.106	0.016	0.014	0.026	0.027	0.037	0.030	0.070	0.069	0.134	0.094
0.1	0.013	0.023	0.017	0.015	0.031	0.046	0.040	0.035	0.078	0.096	0.137	0.098
0.2	0.018	0.032	0.020	0.021	0.030	0.054	0.036	0.037	0.110	0.136	0.154	0.130
0.3	0.026	0.044	0.027	0.031	0.045	0.073	0.050	0.054	0.131	0.167	0.164	0.154
0.4	0.031	0.050	0.031	0.038	0.057	0.087	0.059	0.069	0.170	0.203	0.190	0.194
0.5	0.033	0.052	0.034	0.042	0.054	0.087	0.056	0.069	0.194	0.223	0.203	0.218
0.6	0.043	0.059	0.041	0.052	0.071	0.097	0.007	0.085	0.227	0.245	0.215	0.249
0.7	0.046	0.056	0.040	0.052	0.083	0.097	0.070	0.092	0.269	0.270	0.334	0.287
0.8	0.055	0.060	0.047	0.059	0.105	0.103	0.081	0.106	0.334	0.293	0.255	0.339
0.9	0.076	0.061	0.052	0.065	0.151	0.110	0.092	0.125	0.442	0.319	0.282	0.430
0.95	0.109	0.062	0.056	0.072	0.223	0.112	0.097	0.151	0.600	0.345	0.307	0.608

based on 2000 entries

TABLE 7: Ultimate biases (both corrections applied)

	N=100 n=10				N=60 n=9				N=20 n=4			
ρ	r_o	r_i	r_D	r_c	r_o	r_i	r_D	r_c	r_o	r_i	r_D	r_c
0	0.002	-0.001	0.004	0.003	0.006	-0.004	0.014	0.011	0.001	-0.054	0.056	0.044
0.1	-0.002	0.004	0.001	0.000	-0.004	0.009	0.010	0.008	-0.014	-0.048	0.037	0.013
0.2	-0.002	0.009	-0.001	0.002	-0.004	0.010	-0.001	0.002	0.001	-0.026	0.038	0.021
0.3	0.002	0.017	0.003	0.007	0.005	0.022	0.007	0.013	0.002	-0.016	0.028	0.022
0.4	0.003	0.019	0.002	0.010	0.011	0.030	0.010	0.020	0.026	-0.000	0.036	0.039
0.5	0.003	0.016	0.002	0.009	0.001	0.021	0.000	0.012	0.031	-0.004	0.030	0.032
0.6	0.008	0.019	0.005	0.015	0.011	0.023	0.005	0.019	0.047	-0.007	0.023	0.033
0.7	0.007	0.012	0.000	0.010	0.017	0.015	0.003	0.017	0.075	-0.006	0.025	0.039
0.8	0.013	0.010	0.004	0.011	0.033	0.014	0.008	0.020	0.131	-0.009	0.031	0.047
0.9	0.030	0.007	0.006	0.012	0.075	0.012	0.014	0.026	0.241	-0.007	0.044	0.155
0.95	0.062	0.006	0.008	0.013	0.145	0.009	0.017	0.033	0.426	-0.011	0.064	0.303

based on 2000 entries

TABLE 8: Residual biases and sampling variances using Fisher's transformation (N=100)

	no length correction				with length correction			
	n=5		n=10		n=5		n=10	
ρ	$\overline{\rho - r_f}$	var r_f	$\overline{\rho - r_f}$	var r_f	$\overline{\rho - r_f}$	var r_f	$\overline{\rho - r_f}$	var r_f
0	0.067	0.009	0.135	0.009	0.001	0.019	0.000	0.064
0.1	0.088	0.009	0.171	0.008	-0.011	0.019	-0.070	0.058
0.2	0.106	0.010	0.210	0.009	-0.028	0.021	-0.119	0.058
0.3	0.130	0.009	0.255	0.009	-0.039	0.020	-0.143	0.048
0.4	0.156	0.009	0.303	0.009	-0.048	0.021	-0.159	0.037
0.5	0.175	0.009	0.347	0.008	-0.069	0.019	-0.165	0.026
0.6	0.207	0.008	0.401	0.009	-0.068	0.017	-0.135	0.021
0.7	0.231	0.008	0.448	0.008	-0.068	0.012	-0.102	0.012
0.8	0.271	0.008	0.510	0.008	-0.035	0.009	-0.040	0.009
0.9	0.311	0.006	0.571	0.007	-0.013	0.005	0.027	0.006
0.95	0.336	0.006	0.607	0.007	-0.045	0.003	0.005	0.005

based on 2000 entries.

which makes impossible the tests of hypotheses; FISHER (1921) gave a solution to this problem by transforming sampled values of r into a quantity Z almost normally distributed with a standard deviation σ_Z practically independent of the correlation level:

$$Z = \tfrac{1}{2} \left[\mathrm{Ln}(1 + r_f) - \mathrm{Ln}\,(1-r_f)\right] = \tanh^{-1} r_f \quad \text{or} \tag{21}$$

$$r_f = \tanh Z$$

Dealing with autocorrelations, we are in the case of "intra-class" or "fraternal" correlation with regard to the parent population; the standard deviation of the transformed variate is then $\sigma_Z = (m - \frac{3}{2})^{-\frac{1}{2}}$ where m is the length of the sequence for which the transformed value Z is calculated. Making use of this property, we derived an estimator r_f in the case of occasional missing values; on each sequence k of uninterupted data, an autocorrelation coefficient r_k is calculated and eventually corrected for length; the corresponding transformed values z_k are then compounded into a global value Z:

$$z_k = \tanh^{-1} r_k \tag{22}$$

$$Z = \frac{\sum_1^k z_k \left(m_k - \frac{3}{2}\right)}{\left(\sum_1^k m_k\right) - \frac{3}{2}} \qquad {}^{-1} r_f \text{ or} \tag{23}$$

$$r_f = \tanh Z$$

One should note that the application of the Fisher's transformation is less restrictive than the previously designed estimators as it does not require for the missing values to be occasional; the residual bias is always negative and small; for equidistant missing data, the expansion of the previous expression permits to figure its value:

$$\overline{r - r_f} = - \frac{r_f\,(1-r_f)^2}{2N\,(n+1)} \left(1 + \frac{n+2}{2N}\right) \tag{24}$$

In our case this expression was kept below 10^{-3}.

To test the efficiency of such an estimator, we used the same simulation program as previously; 2000 samples of length 100 were generated, containing 5 and 10 missing values approximately equispaced, which created sequences of about 16 and 8 data. For such short samples, it seems necessary to apply the length correction given by equation (9), but we found in the literature no report of such a use with the Fisher's transformation; this creates a difficulty as, for very short samples, some corrected values of r_k become larger than one, especially for large values of ρ; so we constrained r_k between - 0.98 and + 0.98. The results are shown on the Table 8; the improvement, due to the length correction is quite visible, but even with it, the efficiency of the estimator derived from the Fisher's transformation is largely worse than other estimator efficiencies given by the Table 7. One can also suggest that the values presented in Table 8 are possibly too optimistic, due to the constraints imposed upon the values of r_k.

CONCLUSIONS

Various estimators have been tested in order to estimate the lag-one autocorrelation coefficient of a time-series containing occasional missing values; the corresponding bias corrections have been first established theoretically in the case of large samples, then combined with a length correction to yield a procedure valid for short samples; it has been demonstrated that the compression of a time-series, a common practice in this situation, should not be performed without applying the proper corrections. The results tested by a Monte-Carlo technique with Markovian series seemed to be more effective than the classical Z-transformation and should be useful for a whole range of real-life hydrometeorological time-series whose persistence structure is approximately Markovian.

REFERENCES

ANDERSON, R.L. (1942).
Distribution of the serial correlation coefficient. Ann. Math. Stat., B: 1-13.

BOX, J.P. and G.M. JENKINS (1970).
Time series analysis, forecasting and control. Holden-Day, San Francisco, Calif., 553 p.

BOX, G.E.P. and M.E. MULLER (1958).
A note on the generation of random normal deviates. Ann. Math. Statist., 29: 610-611.

BRUBACHER, S.R. and G.T. WILSON (1976).
Interpolating time series with applications to the estimation of holiday effects on electricity demand. Journal of the Royal Statistical Society, London, England, Series C (Applied Statistics), 25 (2): 107-116.

CLUIS, D. and P. BOUCHER (1981).
Estimation de l'autocorrélation d'ordre 1 d'un échantillon court comportant des valeurs manquantes occasionnelles. Technical Report No 138. INRS-Eau, Université du Québec, Qué., Canada.

D'ASTOUS, F. and K.W. HIPEL (1979).
Analysing environmental time series. ASCE, Env. Eng. Div., 105 (EE5): 979-992.

FISHER, R.A. (1921).
On the probable error of a coefficient of correlation deduced from a small sample. Metron., 1(4): 3-32.
JENKINS, M.G and D.G. WATTS (1968).
Spectral analysis and its applications. Holden-Day, San Francisco, Calif., 525 p.
KENDALL, M.G. (1954).
Note on bias in the estimation of autocorrelation. Biometrika, 41: 403-404.
KNOKE, J.D. (1979).
Normal approximations for serial correlation statistics. Biometrics, 35: 491-495.
MARRIOTT, F.H.C. and J.A. POPE (1954).
Bias in the estimation of autocorrelation, Biometrika, 42: 390-402.
ORCUTT, G.H. (1948).
A study of the autoregressive nature of the time series used for Tinbergen's model of the economic system of the United States, 1919-1932. J.R. Statist. Soc., B.10: 1-54.
PARZEN, E. (1964).
An approach to empirical time series analysis. Radio-Science, 68(9): 937-951.
PREECE, D.A. (1971).
Iterative procedures for missing values in experiments. Technometrics, 13 (4):743-753.
RODRIGUEZ-ITURBE, I. (1971).
Structural analysis of hydrological sequences. Proceeding Warsaw Symposium. In: Mathematical models in hydrology, 3: 1157-1165.
SASTRY, A.S.R. (1951).
Bias in estimation of serial coefficients. Sankhya, 11: 281-296.
SOPER, H.E., A.W. YOUNG, B.M. CAVE, A. LEE and K. PEARSON (1916).
On the distribution of the correlation coefficient in small samples - a cooperative study. Biometrika, 11: 328-413.
WALLIS, J.R. and N.C. MATALAS (1971).
Correlogram analysis revisited. Wat. Resour. Res., 7(6): 1448-1459.
WALLIS, J.R. and P.E. O'CONNELL (1972).
Small sample estimation of ρ_1. Water Resour. Res., 8(3): 707-712.
WILKINSON, G.N. (1958).
Estimation of missing values for the analysis of incomplete data. Biometrika, 14 (2): 257-286.

TIDAL ANALYSIS - A RETROSPECT

D.E. CARTWRIGHT

Institute of Oceanographic Sciences, Bidston, UK

LIMITATIONS TO PROGRESS

I first became involved with tidal analysis - as an oceanographer as distinct from one committed to producing tide-tables - in the late 1950's when primitive electronic digital computers first became accessible to the general scientist. At that time, tide-tables were still handwritten by operators sitting in front of tide-predicting machines, and tidal data were analysed with the aid of sheets of squared paper, stencils and mechanical desk-adders. Twenty-odd years later, such methods may appear laughable, but in fact their tide predictions were very good, the methods having been evolved by expert mathematicians who regarded them as having reached a plateau of practical perfection. Therefore, although modern computers opened up a new range of analytical techniques which were previously unthinkable in terms of labour, their improvement in accuracy of prediction was at best only marginal.

To be more explicit, a data series of sea surface elevation z(t) may be expressed as

$$z(t) = \zeta(t) + R(t)$$

where t is the time (preferably 'Universal' or Greenwich Mean Time), ζ is the part of the signal which is directly related to the tide-generating potential of the Moon and Sun, and R is the residual which is uncorrelated with ζ and which depends on meteorological and other non-tidal influences. At a typical estuarine site in the UK the total variance of z for a certain period was $2.420m^2$. Of this, $2.361m^2$ was within the possible spectral bands of the

Reprinted from *Time Series Methods in Hydrosciences*, by A.H. El-Shaarawi and S.R. Esterby (Editors)

tides, making 2.361 an upper bound to the variance of ζ and 0.059 a lower bound for the variance of R. Comparison of the same data with three different tidal syntheses (theoretical approximations to ζ) gave the following residual variances:

Elementary harmonic method	(60 terms)	0.125m^2
'Extended' harmonic method	(114 terms)	0.114m^2
Advanced response method	(47 terms)	0.088m^2

It is clear that the most advanced methods of tidal analysis developed during the computer age can only improve prediction variance (i.e. reduce the residual) by a very small fraction of the total variance. To reduce it more effectively one has to investigage methods of predicting R(t) through weather-forecast models, and this is outside the scope of this survey. In terms of representation of $\zeta(t)$ however the modern methods are very good indeed, although poorer cases are found in regions strongly affected by shallow water and variable river run-off. For those interested in the tides 'per se', there have been considerable advances in understanding their physical nature in relation to their generating forces and their propagation in the ocean basins, (Cartwright, 1977). The principal techniques in time series analysis which have made this possible have been the use of spectral analysis for exploring the nature of both $\zeta(t)$ and R(t), and the use of directly computed time series of the tide-generating potential.

In the following sections I shall review briefly some of the uses which have been made of these and other techniques, and also point out some areas where there is still need for improvement.

INITIAL DATA SCANNING

Rising analytical precision has emphasised the need to check data series carefully before any serious analysis. Digital tide gauges and automated chart readers have reduced the frequency of some types of error, but many sea level records are still recorded by moving pen and digitised by a human reader, and these may contain a host of errors of various origin. The simplest error detectors which have been widely used are the Wiener predictor (1) and the interpolator (2),

$$z'(t) = \sum_{1}^{k} w'_k \; z(t-k\delta) \tag{1}$$

$$z''(t) = \sum_{1}^{k} w''_k \; [z(t-k\delta)] \tag{2}$$

designed so that $z'-z$ or $z''-z$ has a variance very much less than that of z itself. (δ is the time interval of the series). On scanning in t, any values z' or z'' differing from z by more than a pre-assigned quantity is flagged, and the data checked, if possible back to its original source. Some favour the Wiener predictor (1) because if $|z'(t) - z(t)|$ is large but all previous k values are small it is a fair indication that $z(t)$ itself is suspect, whereas (2) gives a symmetrical pattern of anomalous differences before and after each true error. However, (2) always allows a lower threshold of detectability. Several forms of (2) are discussed by Karunaratne (1980), including the use of a 25h interval.

Another error-detecting method which we have found very effective at Bidston in recent years is the automated plotting of the residual series after subtraction of a fair tidal synthesis. All forms of error are revealed at a glance and the method is especially effective in identifying all-too-common timing errors (Pugh & Vassie, 1978: Graff & Karunaratne, 1980).

SPECTRAL ANALYSIS

Spectral analysis is a vital tool for understanding the nature of tidal data. It has transformed thinking from the older textbooks which treat tides as if they were an isolated line-spectral process, to the modern concept of a weakly nonlinear signal highly correlated with its source function embedded in a continuous noise background. The source function is of course the tide-generating potential, or more usefully, the time-varying part of its leading spherical harmonics which divide naturally into the tidal 'species' 0,1,2,3.... representing long-period, diurnal, semi-diurnal tides etc.

It is important to preserve spectral phase. Spectral analysis through the standard method employing cross-correlation with the source function is not in my experience very fruitful, because of the very high power density in the M_2 line and others, which spill over into neighbouring filter bands, masking the finer detail one seeks there. The normal approach is through fourier analysis of synodic periods of data, that is periods for which most of the major lines come near to an integral number of cycles, important examples being 29,59,355,738 days. Unfortunately, each of these contains a large prime factor, so the 'fast fourier transform' technique is inapplicable in its ideal form involving powers of 2, or if adapted to take large prime factors it loses much of its advantage in speed. Besides this, one does not usually require the complete transform. Franco & Rock (1971), however, have adapted the fast fourier transform to the 'harmonic method' of tidal analysis.

I have a personal preference for a basic spectral unit consisting of a fourier transform of a 59d span of data with a 'hanning' window:

$$C_s(t) = 2N^{-1} \sum_{r=-N/2}^{N/2} z(t+r\delta)(1+\cos 2\pi r/N) \exp(2\pi i s r/N) \qquad (3)$$

where for δ = 1 hour, N = 59x24 = 1416, and s takes all integral values from 0 to about 60P where P is the highest tidal species of interest. The transform is effected by a common algorithm such as 'Watt-Iteration', adapted to make use of the fact that t is usually stepped sequentially in steps of T days where T is typically in the range 5-15. (The 'hanning', represented by the middle bracket, is omitted from the first stage of computation and brought in at the end by applying (½,1,½) smoothing to the sequence of harmonics).

The spectral filter (3) neatly separates the tidal 'group centred on

p cycles/lunar day + q cycles/month

($0 \leqslant p \leqslant 12$, $-4 \leqslant q \leqslant 4$),

into the spectral elements $s = 57p + 2q$, while the overspill is so small that outside the species-bands the elements C_s may be fairly taken as measures of the non-tidal continuum. A set of C_s for a single value of t has some limited use, but the most useful applications stem from a sequence of t extending over say one or several years. These applications are of three sorts:

(i) The complex sequences $C_s(t)$ for selected values of s of tidal interest may be further fourier analysed at higher resolution to reveal the tidal structure within each 'group'.

(ii) A parallel sequence $C_s'(t)$ may be generated from a related time series such as the tidal potential or some meteorological function (for the non-tidal values of s) and the mean transfer function and coherence evaluated at each frequency (s/59 cycles d^{-1}) from the relations (cf Munk & Cartwright, 1966)

$$Z_s \sim \langle C_s' C_s^* \rangle / \langle C_s' C_s'^* \rangle, \quad \gamma_s^2 \sim (\langle |C_s' C_s^*| \rangle)^2 / \langle |C_s'|^2 \rangle \langle |C_s|^2 \rangle \qquad (4)$$

where * denote the complex conjugate and < > an ensemble average over all values of t.

(iii) For tidal groups s possibly containing two distinct tidal elements, (eg linear and nonlinear, gravitational and radiational), cross spectra can be made with two parallel series for each s, and the distinct elements separated by inverting a cross-correlation matrix.

Examples of (i) are the harmonic development of the tide potentials from spectral analysis of 18y sequences (Cartwright & Taylor, 1971; Cartwright & Edden, 1973), the resolution of an unexpected term at exactly 1 cycle/lunar day (Cartwright, 1975, 1976), a detailed examination of the structure of the non-gravitational solar tides (Cartwright & Edden, 1977), and several other examinations of oceanic tidal records.

Procedure (ii) was applied at cy^{-1} resolution ($N\delta$ = 710d) in the analyses of long tidal series by Munk & Cartwright (1966), and, as presented here, to an alalysis of surges and surge-tide interaction round Britain by Cartwright (1968) and in numerous cases where an admittance at cm^{-1} resolution or a variance spectrum covering tidal and non-tidal frequencies are required.

Procedure (iii) has had some applications to separating admittances to tidal effects with close or overlapping frequency structure mentioned under (i), and for a quick examination of the different characteristics of say S_2 and K_2 (radiational effects) or $2N_2$ and μ_2 (nonlinear effects).

THE NOISE CONTINUUM

It is worth commenting on the main features of the continuum because although well known to specialists they are little used by practitioners. By definition, the continuum represents a random variation. It may be partly related to weather parameters but it must usually be regarded as unpredictable noise, possibly with seasonal variations in general level. Its spectrum, like that of most geophsical variables, rises monotonically towards the lowest measurable frequencies, and presents a threshold for the detectability of tidal components in a record of given duration.

Because of this frequency-dependence, the diurnal tides in places like the Atlantic Ocean where they are weak are usually less reliably estimated than the very weak ter-diurnal tides with frequencies near 2.9 cd^{-1} where the continuum is typically an order of magnitude lower. The most notable casualties are of course the tides of long period (species 0). The continuum density at the monthly frequency Mm, say, is typically $3cm^2$ $(cycle/year)^{-1}$ so the noise variance of a spectral element derived from N years' data is about 3/N cm^2. A typical amplitude of the Mm tide is expected to be 1 cm or less, with a variance of at most 0.5 cm^2. If we take as criterion for reliability of an estimate, that its signal variance should be ten times the noise variance, then Mm would typically require N = 60 years for a proper estimation. (Despite this, one still sees lists of harmonic constants from ly analyses which religiously quote amplitudes and phases for Mm and as if they had some predictive value.)

Some improvement in the low frequency continuum level can be effected by subtracting a convolution of the atmospheric pressure field (Cartwright, p.45, 1968), but in general the only long-period tides which can be reliably estimated from a few years' data are the seasonal yearly (Sa) and possibly the half-yearly (Ssa) components. The most appropriate data sets for these are the long series of monthly sea levels available from the International Permanent Service for Mean Sea Level.

Another awkward property of the continuum is its invariable tendency to rise in the neighbourhood of the strong tidal bands, even converging in cusp-like fashion on individual strong lines like M_2. This property was first discovered by Munk, Zetler & Groves (1965), but it is also examined in Munk & Cartwright, (1966) and Cartwright (1968). Its cause has been vaguely ascribed to internal tides or to some modulation of the tides by the weather continuum, but in my experience of dealing with common tide gauge records the cusp-like rise is in practice more likely to be due to mediocre instrumental maintenance, notably in time-keeping.

Tidal cusps make it impossible to estimate the reliability of tidal constants from the inter-species spectral noise level, otherwise fairly easy to compute. It would be most useful if all estimated sets of tidal constants were accompanied by species-band variances of the original data and of the residuals associated with the listed constants. The latter gives a reliable measure of the true cusp-variance within the species concerned, against which the likely variation of the individual terms may be assessed. The total residual variance is not sufficient because it includes much low-frequency and inter-species noise.

EXTENDED HARMONIC METHODS

For its compromise between accuracy and simplicity the 'harmonic method' remains as it was a century ago, the best principle for routine practical dealing with weakly nonlinear tidal systems. As is well-known, it expresses the tidal part of sea level at a given place in the form

$$\zeta(t) = \sum_{n=1}^{n'} f_n(t)\, H_n \cos\,[s_n t + x_n(o) + u_n(t) - G_n] \qquad (5)$$

where s_n is a set of known frequencies, not differing by less than about 1 cy^{-1}, and $x_n(o)$ are known initial phases at the contemporary epoch t=0; f_n, u_n are slow modulating functions, mostly with the period 18.6y of the lunar node, derived from the corresponding terms in the tide-generating potential, and H_n, G_n are arbitrary constants to be assigned to the place in question.

The ingenious but noise-leaking filters with integral multipliers used to extract H_n, G_n in the old hand computations (most neatly summarised in the context of more recent techniques by Godin (1972)), have long since been superseded by the superior techniques made possible by automatic computers such as 'least-squares' (Horn, 1960) and FFT (Franco & Rock 1971).

One of the vaguest aspects of 'harmonic' practice is how to assign the frequencies s_n and their total number n' out of the many hundred possible terms appearing in the tidal potential and its cross-products. The old procedures were limited by the number of terms which the largest tide-predicting machines could handle, in the range 45-60. By studying the power spectra of residuals, Zetler & Cummings (1967) and Rossiter & Lennon (1968) independently arrived at n' = 114 as a suitable number, although their choice of individual terms differed. Some of the new terms chosen required considerable contortion of the original harmonic concept. For example, MNK2S, implying a quintic interaction between the primary constituents M_2, N_2, S_2, and K_2, to produce a new term of species 2, was found to have significant amplitude at the three shallow water stations considered. This and other odd-order interactions identified are almost certainly symptoms of strong friction, as distinct from even-order interactions which arise from advective terms in the dynamics of shallow water propagation.

Identification of a certain type of interactive term because it stands alone in frequency, dyamically implies the existence of similar and perhaps stronger interactions between all similar terms. However, many of these are hidden by the fact that their combined frequencies coincide with the frequencies of primary linear terms or with lower-order interactions which are already taken into account. For example, the presence of $2SM_2$ must imply the presence of $2MS_2$ and $2MN_2$, but because the latter coincide in their central frequencies with μ_2 and L_2 respectively, they are ignored as individual effects. In practice, this works up to a point, but involves inaccuracy when (f,u) factors are assigned, because these are quite different in the two cases.

Amin (1976) seriously tackled this difficulty by harmonically analysing nearly 19 years of data from Southend. This process resolved the line structure of the spectrum to 'nodal splitting' level, so that the true (f,u) modulation of each constituent could be estimated directly without recourse to the known modulation of the potential.

Amin was thereby able to assess the respective contributions of linear and interactive terms (provided no more than two influences were present) and assign a more accurate harmonic formulation than is usually possible. Amin also identified the seasonal modulations to M_2, first studied by Corkan (1934) who named them Ma_2 and MA_2*, since recognised as widespread in the North Sea and probably elsewhere (e.g. Pugh & Vassie, 1976). A 19-year analysis of the tides at the long-established station at Brest, France is presented by Simon (1980); this includes the curiosity of a distinct harmonic term generated by a fault in the tide-gauge mechanism, identified by Desnoës (1977).

If the harmonic method is to be stretched to its theoretical limit, one should do away with the (f,u) factors in (5), except in cases of simple oceanic tides, and represent the full set of pure harmonic constants down to nodal-splitting, with six parameters to denote the frequency. Amin (1976) records 326 terms up to species 6, without including the 20 terms (without nodal-splitting) belonging to higher species listed by Rossiter & Lennon (1968) for the same place. No doubt, more terms would be required for a port with stronger diurnal inequality. However, the mere number of terms presents no deterrent to a modern computer, and they may be easier to deal with than the rather awkward (f,u) factors. Analysis of course would require 18-19 years of data, and it would be all the more important to specify the threshold of the noise continuum.

* Because of the frequent need for computerised typesetting, I have recommended the notation MB_2 for the higher frequency seasonal modulation.

RESPONSE METHODS

The 'response' method of tidal analysis was introduced by Munk & Cartwright (1966) as a research tool rather than as a means of bettering the accuracy of predictions. Although comparisons of its prediction accuracy with that of harmonic methods (e.g. Zetler, Cartwright & Berkman, 1979) has always shown the 'response' predictions to be slightly better, the margin of improvement is not enough to justify replacing familiar routine procedures by the rather bulky and unfamiliar programs involved in the 'response' formalism. There is no space here for a full discussion of the method, but I should like to summarise points of improvement which have been made since the publication of the original paper.

The aim is to escape from the enslavement to a multitude of independent time-harmonic terms by expressing in a few parameters the response functions of the measured tide to the leading spherical harmonics of the tide-generating potential. (In the earliest work, the response functions were referred to the 'equilibrium tide' at the place of measurement, equivalent to using phase lags in the 'Kappa' notation, but this was soon found to be an irrelevant complication, so all response functions are now referred to the same time-variable part of the potential at the Greenwich meridian, equivalent to phase lags in the G-notation.)

The gravitational potential $V(\theta,\lambda,t)$ due to the Moon and the Sun on a sphere with the Earth's equatorial radius can be computed at time t in the form

$$g^{-1}V(\theta,\lambda,t) = \sum_{n=2}^{3} \sum_{m=o}^{n} [a_n^m(t)U_n^m(\theta,\lambda) + b_n^m(t)V_n^m(\theta,\lambda)] \qquad (6)$$

where θ is north colatitude, λ is east longitude, and

$$U_n^m + iV_n^m = (-1)^m \left[\frac{2n+1}{4\pi}\right]^{\frac{1}{2}} \left[\frac{(n-m)!}{(n+m)!}\right]^{\frac{1}{2}} P_n^m (\cos\theta)\, e^{im\lambda} \tag{7}$$

is the spherical harmonic of order m degree n in a standard normalisation. m is identical with the tidal 'species'. The four terms with degree 3 are much smaller than the three terms of degree 2 but they produce tidal effects which are detectable. Terms with degree higher than 3 can be entirely neglected.

The scheme is to express that part of the tidal sea level which is related to the harmonic m,n by a relation of the type (dropping the suffices m,n for convenience):

$$\zeta_n^m(t) = \sum_{s=-S}^{S} [u_s a(t-s\Delta t) + v_s b(t-s\Delta t)] \tag{8}$$

where $w_s = u_s + iv_s$ is a set of arbitrary 'response weights' for the system. Munk & Cartwright (1966) justify an invariable choice of 2 days for the incremental time lag Δt, and the use of negative as well as positive lags which appears to violate physical laws. They also suggested that S=3 is a suitable maximum limit to the summation, but later investigations (Cartwright, (1968) showed that S=3 tends to 'overfit' the data, and I have found in general that S=2 (5 complex response weights) gives as good a representation as noise levels will permit. For the weaker harmonics with degree n=3, s=1 or even 0 is adequate. See also Zetler & Munk, (1975).

Individually, the response weights w_s mean very little, but taken as a complete group they define the admittance of the response system at frequency f,

$$Z(f) = \sum_{s=-S}^{S} w_s \exp(-2\pi i f\Delta t) \tag{9}$$

which is the most physically meaningful quantity which emerges from the analysis. In conjunction with the known time-harmonic development of the potential coefficients $a_n^m(t)$, (Cartwright & Tayler, 1971; Cartwright & Edden, 1973), Z(f) may be used to derive any harmonic amplitude of ζ, and its phase Lag,

$$G = \pi m - \text{Arg}\,[Z(\sigma/2\pi)], \qquad (10)$$

equivalently to and more accurately than the direct 'harmonic' method. The essential difference between the two methods is in fact that the response method assumes the existence of a reasonably smooth admittance function Z(f), whereas the harmonic method makes no use of the physical reality of the system. Munk & Cartwright's "credo of smoothness" (of the admittance) has been well vindicated in numerous testing examples.

Groves & Reynolds (1976) pointed out an inherent clumsiness of the response representation (8) in that, while a(t) and b(t) are mutually orthogonal functions, and so in the long term are the corresponding functions with differing (m,n), the elements of (8) with different lags $s\Delta t$ are rather strongly correlated, resulting in lack of convergence and instability in the weights w_s. In order to remedy this, they replaced the functions a,b in (8) by linear combinations of a,b which they computed to form a completely orthogonal set of functions which they called 'orthotides'. Use of these orthotides does indeed restore convergence and stability to their response weights, but since these are linearly related to w_s and the orthogonal properties of the orthotides themselves require 18.6 year averages in their cross-products, their use does not add any computational advantages for the formalism 8, (Alcock & Cartwright, 1978).

Complications are required to deal with the slightly anomalous behaviour of the solar tides and with nonlinear terms. The former is allowed for by the addition of a 'radiational' potential analogous to (6), derived from the Sun's position and requiring additional response weights, unlagged in time. The time-harmonics of the radiational potential are listed in the Table 6 of Cartwright & Tayler (1971). They are effective, but rather variable from year to year. Detailed analysis has suggested that the anomalies in species 2 are more closely related to the atmospheric tide than to the radiational potential, but the subtle differences are close to the tidal noise level (Cartwright & Edden 1977).

Treatment of nonlinear effects as 'response' processes has proved to be the least satisfactory aspect of the scheme, although adequate as an approximation. The present procedure is to add further 'potential' functions derived from products of the primary potentials a_n^m, b_n^m. If these are expressed as complex variables,

$$c_n^m(t) = a_n^m(t) + ib_n^m(t), \tag{11}$$

then $c_n^m c_n^{m'}$ is a suitable reference function containing terms only with the sum of the frequencies of the primary terms of species m,m'; that is, it defines a new function of species (m+m') of nonlinear origin. Similarly, $c_n^m c_n^{m'*}$ defines a new function of species (m-m'). In practice, however, it is reasonable to suppose that nonlinear terms, being locally generated, are more closely related to the products of the local tide itself than to products of the potential. Accordingly, we compute linear terms similar to (8), approximating to the observed tides of species 1 and 2, and form products analogous to $c_n^m c_n^{m'}$ from them. Response weights are then assigned to these products (which include triple and higher order interactions) along with the linear terms for the gravitational and radiational potentials in a least-squares evaluation process.

Where the nonlinear terms are weak there are no problems and the described nonlinear formalism significantly improves the predictable variance. However, I have experienced some cases where even the relatively simple (2+2) interaction does not satisfactorily account for all the observed species 4 variance, even with the addition of some time-lagged terms. In one case at least it appears that the assumption that the nonlinearity is generated locally is only partially valid. Again, the important triple interaction (2+2-2), which embodies the major friction effect, seems to require more subtlety of definition. Probably, a closer modelling of the relevant dynamical process is needed.

In a current version of the 'response' analysis package, provision is made for up to 71 complex response weights, including 10 radiational terms, some annual modulations and nonlinear forms up to order 5. It is highly improbable that a case could arise where all these terms are needed simultaneously. Many variables are mutually correlated, resulting in an unstable normal matrix and large, unrealistic response weights. Considerable pre-selection is required, and this can only be done as a result of experience and preliminary spectral analysis. I prefer to regard a response analysis as a final means of optimising the definition of a set of admittance of terms whose presence and approximate magnitude are already known by other techniques or by knowledge of the characteristics of the local sea area.

COMPUTING THE TIDAL POTENTIAL

Finally, I should like to comment on methods of computing the tide-generating potential, on which much of the modern research is based. Accuracy depends on the computation of the lunar and solar positions, and the standards used by geophysicists and astronomers vary enormously. The lunar formulae used in Munk & Cartwright (1966) do not compare well with astronomical ephemerides although they evidently give passable results at the level of accuracy required by tidal analysis. My own programs use a selection of some 280 of the 'Brown' terms and other refinements used in modern ephemerides to maintain a consistent accuracy of 2" in latitude and longitude and 10^{-5} in parallax (Cartwright & Tayler, 1971). This is certainly excessive, but was done with a view to checking the computations against the published ephemerides and removing all possible doubt about errors from this source.

Some compromise between orbital precision and bulk of computation is desirable for the purposes of good quality tidal research, and I outline below a complete set of formulae which achieves the ideal balance.

APPENDIX - POTENTIAL FORMULAE

Let t be (Ephemeris) time in days counted from 1900 January 0.5 (i.e. December 31 noon) and T=t/36525.

The <u>mean</u> longitudes of the following quantities are given in the form $L = L_o + L_1 T + L_2 T^2$ in revolutions

Name	Symbol	L_o	L_1	L_2
Moon	s	0.751206	1336.855231	-0.000003
M. Perigee	p	0.928693	11.302872	-0.000029
Node	n	0.719954	-5.372617	0.000006
Sun	h	0.776935	100.002136	0.000001
S. Perigee	p'	0.781169	0.004775	0.000001

Also, obliquity of ecliptic, $\varepsilon = 0.409318 - 0.000227T$ radians

and Earth's eccentricity, $e = 0.016751 - 0.000042T$.

For the Sun, latitude $\delta = 0$,

longitude $\rho = h + 2e \sin (h-p') + \frac{5}{4} e^2 \sin 2(h-p'$

parallax $\xi = \bar{\xi} [1 + e \cos(h-p') + e^2 \cos 2(h-p')$

where $\bar{\xi} = 8''.794$.

For the Moon, latitude and longitude (sines) and parallax (cosines) are defined as

$$A_i \begin{matrix}\sin \\ \cos\end{matrix} \left(\sum_{n=1}^{4} k_n^{(i)} B_n \right)$$

where $B_1 = s-p$, $B_2 = h-p'$, $B_3 = s-n$, $B_4 = s-h$, and A_i and $k_n^{(i)}$ are as in the following table, where $\bar{\xi} = 3422''.540$

AMPLITUDES A_i			
$k_n^{(i)}$	Parallax (ξ) x 10^{-5} $\bar{\xi}$	Longitude (ρ) x 10^{-5} rad	Latitude (δ) x 10^{-5} rad
0 0 0 0	100000	0	
0 0 0 1	-29	-61	
0 0 0 2	824	1149	
0 0 2 0	0	-200	
0 1 0 0	-12	-324	
0 1 0-2	56	-80	
1 0 0 2	90	93	
1 0 0 0	5450	10976	
1 0 0-2	1003	-2224	
1 1 0 0	-28	-53	
1 1 0-2	42	-100	
1-1 0 0	34	72	
2 0 0 0	297	373	
2 0 0-2	-9	-103	
0 0 1 0			8950
0 0 1 2			51
0 0-1 2			306
1 0 1 0			491
1 0-1 0			481
1 0 1-2			-78
1 0-1-2			-99

For both Sun and Moon, Right Ascension R and Declination D are given by the formulae

$$\sin D = \cos\delta \sin\rho \sin\varepsilon + \sin\delta \cos\varepsilon$$
$$\cos R \cos D = \cos\delta \cos\rho ,$$
$$\sin R \cos D = \cos\delta \sin\rho \cos\varepsilon - \sin\delta \sin\varepsilon ,$$

and finally, the required time-variables in (6, 11) are given by the sum of the solar and lunar contributions to

$$a_2^0 = (\pi/5)^{\frac{1}{2}} h_2 (\xi/\bar{\xi})^3 (3 \sin^2 D-1),$$
$$c_2^1 = -(6\pi/5)^{\frac{1}{2}} h_2 (\xi/\bar{\xi})^3 \sin 2D \exp [i(R-R_G)],$$
$$c_2^2 = (6\pi/5)^{\frac{1}{2}} h_2 (\xi/\bar{\xi})^3 \cos^2 D \exp [2i(R-R_G)],$$

etc. .....

where h_2 = 0.16458m (Sun), 0.35838m (Moon), and R_G is the Right Ascension of the Greenwich meridian, namely

$$R_G \div h - \pi + 2\pi \text{ (Universal Time in days).}$$

The radiational potential elements are formed similarly, from the Sun's elements alone.

REFERENCES

Alcock, G.A. & Cartwright D.E., 1978. Some experiments with 'orthotides', Geophys.J.R.astr.Soc. 54: 681-696.

Amin, M., 1976. The fine resolution of tidal harmonics. Geophys.J.R. astr.Soc. 44: 293-310.

Cartwright, D.E., 1968. A unified analysis of tides and surges. Phil.Trans.R.Soc. A, 263: 1-55.

Cartwright, D.E., 1975. A subharmonic lunar tide in the seas off Western Europe. Nature, 257: 5524, 277-280.

Cartwright, D.E., 1976. Anomalous M1 tide at Lagos. Nature, 263: 5574, 217-218.

Cartwright, D.E., 1977. Ocean tides. Rep.Prog.Phys. 40: 665-708.

Cartwright, D.E. & Edden, A.C., 1973. Corrected tables of tidal harmonics. Geophys.J.R.astr.Soc. 33: 253-264.

Cartwright, D.E. & Edden, A.C., 1977. Spectroscopy of the tide-generating potentials. Ann.Geophys. 33: 179-182.

Cartwright D.E. & Tayler, R.J., 1971. New computations of the tide generating potential. Geophys.J.R.astr.Soc. 23: 45-74.

Corkan, R.H., 1934. An annual perturbation in the range of the tide Proc.R.Soc.London A, 100: 305-329.

Desnoës, M.Y., 1977. Le bruit dans les analyses de marée. Ann.Hydrog. 5(2): 31-46.

Franco, A.S. & Rock, N.J., 1971. The FFT and its application to tidal oscillations. Bol.Inst.Oceanog.Univ.S.Paulo, 20: 1-56.

Godin, G., 1972. The analysis of tides. Univ. Toronto 264pp.

Graff, J. & Karunaratne, A., 1980. Accurate reduction of sea level records. Int.Hydrog.Rev. 57: 2, 151-166.

Groves, G.W. & Reynolds, R.W., 1975. An orthogonalised convolution method of tide prediction. J.Geophys.Rev. 80: 30, 4131-4138.

Horn, W., 1960. Some recent approaches to tidal problems. Int. Hydrog.Rev. 37: 2, 65-84.

Karunaratne, D.A., 1980. An improved method for smoothing and interpolating hourly sea level data. Int.Hydrog.Rev. 57: 1, 135-148.

Munk, W.H. & Cartwright, D.E., 1966. Tidal spectroscopy and prediction. Phil.Trans.R.Soc. A, 259: 533-581.

Munk, W.H., Zetler, B. & Groves, G., 1965. Tidal Cusps. Geophys. J.R.astr.Soc. 10: 211-219.

Pugh, D.T. & Vassie, J.M., 1976. Tide and surge propagation offshore in the Dowsing region of the North Sea. Deut.Hydrog.Zeitsch. 29: 163-213.

Pugh, D.T. & Vassie, J.M., 1978. Extreme sea levels from tide and surge probability. Proc. 16th Conf.Coastal Eng., Hamburg, Ch.52: 911-930.

Simon, B., 1977. Analyse de 19 ans d'observations de marée à Brest. Ann.Hydrog. 8(1): 5-17.

Zetler, B., Cartwright, D. & Berkman, S., 1979. Some comparisons of response and harmonic tide predictions. Int.Hydrog.Rev. 56: 2, 105-115.

Zetler, B.D. & Munk, W.H., 1975. The optimum wiggliness of tidal admittances. J.Mar.Res.Supp. 33: 1-13.

IDENTIFICATION OF INTERNAL TIDES IN TIDAL CURRENT RECORDS FROM THE MIDDLE ESTUARY OF THE ST. LAWRENCE

LANGLEY R. MUIR

Bayfield Laboratory for Marine Science and Surveys, Department of Fisheries and Oceans, Canada Centre for Inland Waters, Burlington, Ontario, Canada

Abstract

Observations of tidal currents are traditionally analysed by means of spectral analysis, to ensure that only tidal frequencies are present and by harmonic analysis, to obtain the amplitude and phases of the tidal constituents and to allow the prediction of future currents.

When this was done for a series of records obtained in the St. Lawrence it was found that 95% of the energy was at the tidal frequencies, and that the calculated "tidal constituents" allowed hindcasting the tidal currents with great accuracy. However, neither the amplitudes nor the phases of the "tidal constituents" could be interpreted as standing or progressive waves over the whole Estuary and these amplitudes and phases seemed to be entirely uncorrelated, even though the observations were separated by relatively short distances when compared to the expected wave lengths of the barotropic waves.

The topography and density field of the Estuary allow the generation and propagation of internal gravity waves at the same frequencies as the barotropic tides but with much shorter wave lengths. These internal tides may interfere linearly with the barotropic tidal currents, causing phase and amplitude variations, depending upon the location in the Estuary and the density field in the Estuary. By means of admittance calculations on very short portions of the observed currents records, it is possible to show that the amplitudes and phases of the "tidal constituents" obtained from the harmonic analyses vary in a way that would be expected if both barotropic and internal tides were present in the Estuary.

1. INTRODUCTION

The barotropic, or surface, tides of the St. Lawrence River have been the subject of an intensive amount of research over the past 15

Reprinted from *Time Series Methods in Hydrosciences*, by A.H. El-Shaarawi and S.R. Esterby (Editors)

years. As a result, the physics of the surface tides are fairly well understood, in the sense that numerical models adequately represent the observed water levels and analytical models adequately predict the observed distortion of the tidal wave as it progresses upstream. Godin (1979) and LeBlond (1978) give extended descriptions of the propagation of the surface tide in the whole system from two different viewpoints.

Godin (1979) shows cotidal charts for two constituents K_1 and M_2, but a brief description of the behaviour of the M_2 constituent in the lower part of the system will give an indication of the tidal propagation. There is an amphidromic point about half-way between Prince Edward Island and Anticosti Island in the Gulf of the St. Lawrence. The M_2 tide then progresses upstream in the Lower Estuary, becoming a Kelvin wave, and continues as a Kelvin wave, past the Saguenay River and through the Middle Estuary, (Fig. 1) although it loses most of the cross-channel height difference as the Estuary narrows. By the time the wave has reached the Ile d'Orleans, the rotational effects have disappeared and the wave crest is virtually horizontal.

The phase of the M_2 surface tide at the Saguenay River is 220° and at Baie-St-Paul, near Ile aux Coudres, the phase is 255.5°, and so the wave takes 35.5x2.07 = 73.5 minutes to make a passage of 95 km. The mean celerity is therefore 77.6 km/hr. Since the mean water depth in the North Channel is approximately 100m, the phase speed should be about 110 km/hr, and so the M_2 surface tide in the North Channel could be considered to be part way between a standing and a progressive wave. The sill stretching from Pte au Pic to the Morin Shoal, which shall be called the Morin Bank, provides an adequate reflector as does the abrupt change in direction of the channel as it goes around Ile aux Coudres. On the south side the phase at Riviere du Loup is 224.8°, while at Pte aux Orignaux the phase is 251.1°, so that the time of passage is 54.4 minutes, which gives an average celerity of about 57.4 km/hr. The mean depth in the South Channel is about 25 m, which gives a wave speed of about 55 km/hr and would indicate that in the South Channel the M_2 surface tide is a simple progressive wave.

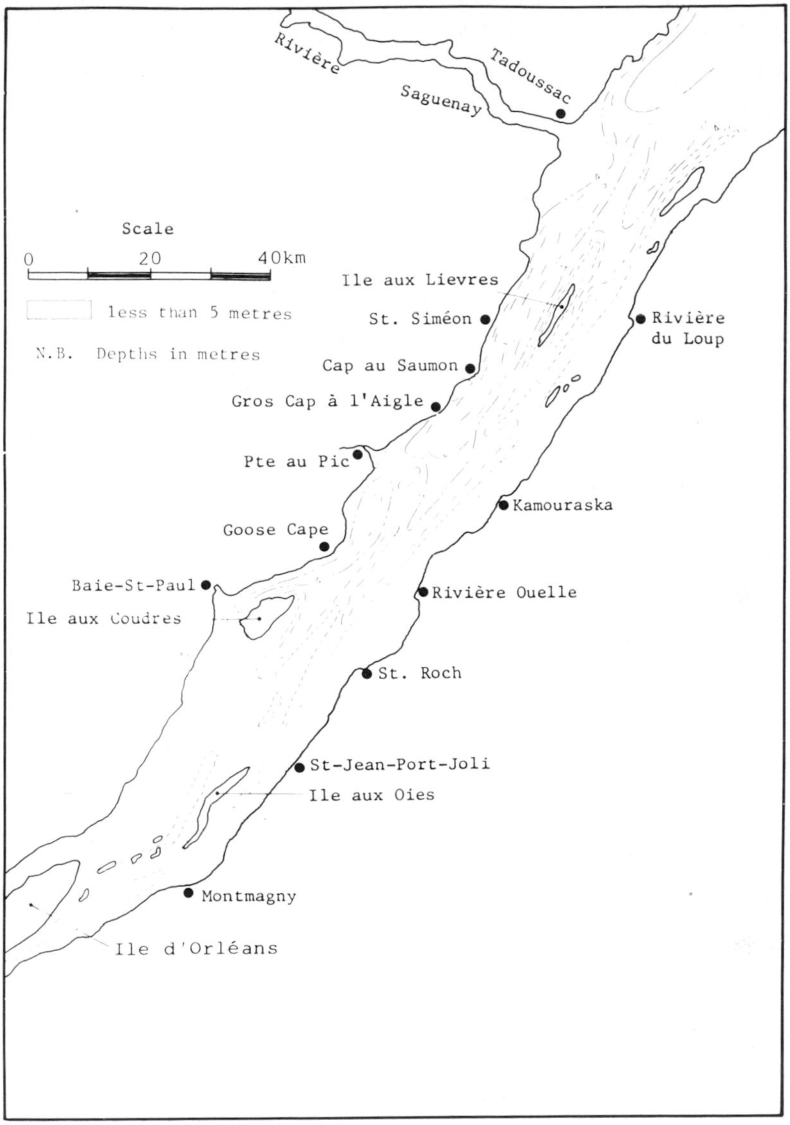

Figure 1 Location diagram. The Middle Estuary stretches from the Saguenay River to Ile aux Coudres

Predictions of the tidal currents in the Middle Estuary have not been very successful. It has been realized for a number of years that the 1939 Tidal Current Atlas is very inaccurate. The verification of the two-dimensional numerical models of Prandle and Crookshank (1974), Ouellet and Cheylus (1971), Aubin et al. (1979), and El Sabh et al. (1979) consisted of matching observed with computed water levels, and little attempt was made to match the observed and computed tidal currents. This is primarily because, until the St. Lawrence Tidal Current Surveys of 1974, 1975, and 1977, there has been an almost total lack of data for verification of the tidal currents. For some of the models, computed trajectories of the currents were shown, but these computed currents bear no more than a passing relation to observed currents and would be completely useless in any operational sense, such as computing the dispersion of oil spills, or for navigational tidal current atlases.

2. HARMONIC ANALYSIS OF THE TIDAL CURRENTS.

A total of 44 current meter records have been obtained in the Middle Estuary in the course of the surveys conducted by the author in 1974, 1975 and 1977. All of the plots and mooring details are to be found in the St. Lawrence Data Reports (Budgell and Muir, 1975; Muir, 1978; and Muir, 1979a) and these details will not be listed here.

The preliminary analysis of the tidal current records was carried out using the standard Canadian Hydrographic Service Tidal Streams Analysis program, which is described by Godin (1972). The data were filtered and reduced to hourly values which were then analysed for the appropriate tidal constituents, with no related constituents used. From the statistics of the analyses, the harmonic analyses of the current records were very good as shown in Table 1. The residual currents were quite small, the coefficient matrices were well-conditioned, and the amount of explained variance was approximately 95%. The constituent lists and standard plots may be found in the appropriate Data Reports.

It was expected that the tidal currents would progress up the river as a simple Kelvin wave in the same fashion as the surface tides, but this is not the case, as can be seen from Figure 2. This Figure gives the phase of the M_2 constituent for the surface tide and for the tidal currents at the location of the measurements. The number in parentheses after the phase gives the depth of the current meter below chart datum in metres, and the arrow shows the direction of rotation of the M_2 ellipse.

If the assumption is made that the tidal currents are driven directly by the barotropic surface tide, then there are many anomalous features shown in Figure 2. First, the phases at any given depth cannot be contoured as a simple Kelvin wave, and, in fact, cannot be contoured in any meaningful way at all. Second, the phase speeds that can be calculated for a wave moving from mooring to mooring vary considerably and do not match with the wave speeds that can be calculated from the water level records. For example, it takes the surface tide about 30 minutes to travel from St. Simeon to Goose Cape, while the tidal current at 5 m depth apparently takes 92 minutes to cover the same distance and, in addition, it arrives on the south shore at Pte. Aux Orignaux about 47 minutes before it arrives at Goose Cape. Third, one would expect that for a barotropic tide, the phase at any given location should be constant throughout the vertical with, perhaps, a slight phase lead for the bottom currents due to the effects of friction (Soulsby, 1978), and this does not happen. The presence of these anomalous features throws considerable doubt on the assumption that the tidal currents are due simply to a barotropic surface tide.

3. ADMITTANCE ANALYSIS OF THE TIDAL CURRENTS.

In order to check that there was no error in the harmonic analysis method and to gain more information from the tidal current records, an admittance analysis was carried out on all of the 1974 current meter records. Since this method is not widely known, a brief description of it will be given here. The only complete description that is available

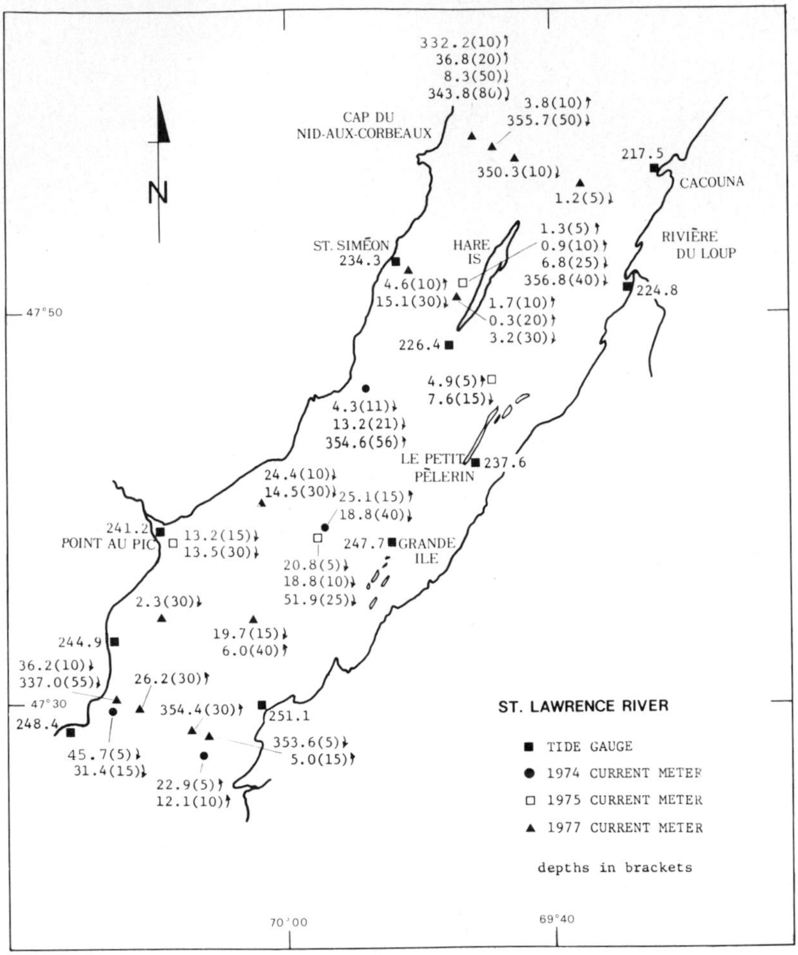

Figure 2 Phase of the M_2 tide.

is to be found in Godin (1976), where it was used with success in the analysis of tidal currents in Robeson Channel.

The basic idea is the same as the tidal spectroscopy methods of Munk and Cartwright (1966), but very much shorter periods of record are used, and instead of using the equilibrium tide as an input function, a well-resolved water level record is used. It is assumed that there is a linear relation between the tidal constituents derived from the water level record and the tidal constituents which make up the tidal current record for each frequency band.

If there are two time series which may be assumed to be linearly related, then the complex admittance (or the gain) between them, at frequency σ, is given by

$$G(\sigma) = C_{xy}(\sigma)/X(\sigma)$$

and the complex coherence is given by

$$\gamma_{xy}(\sigma) = C_{xy}(\sigma)/(X(\sigma)Y(\sigma))^{\frac{1}{2}} \qquad (1)$$

where C_{xy} is the complex cross-spectrum between the two series, x and y, and X and Y are the auto spectra of the two time series.

Given the amplitude and phase of the admittance, a component of the series y may be predicted from the corresponding component of the series x simply by multiplying the amplitude of the component of x by the modulus of the admittance, and by adding the phase of the admittance to the phase of x. Godin (1976) gives the confidence limits for the estimates of the admittance and the phase.

In the case of the analysis of tidal current records, the input function is the water level record for a nearby tide gauging station, and the u(east) and v(north) components of the tidal currents are the independent output records. So long as the input record can supply a well-resolved set of tidal constituents, it is then possible to calculate $G_u(\sigma)$, $\gamma_u(\sigma)$, $G_v(\sigma)$ and $\gamma_v(\sigma)$ for a given bandwidth Δ, and from these to calculate the component constituents for the tidal currents and thence the constituent ellipse parameters. On the St. Lawrence, in the semidiurnal bands, for a bandwidth of 0.005 cycles/hr, the modulus of the coherence between the input tidal elevation and the tidal current components is seldom less than 0.98.

4. APPLICATION TO THE ST. LAWRENCE.

The reference water levels used for the application of the admittance method to the 1974 St. Lawrence data, were the observed water levels at St. Jean Port Joli. This station is somewhat upstream of the Middle Estuary, but has a very long period of tidal record whose tidal constituents are very well resolved. The hourly water levels were used as the input function, while the filtered, hourly current observations in u and v components were used as the output functions. The admittances and coherences were calculated using a power-of-two Fast Fourier Transform. The mean and trend were removed from the data using a linear, least-squares fit and a cosine taper put on the first and last tenths of the data to remove end effects. The raw spectral estimates were frequency smoothed to produce estimates for both 12 and 36 bands. The estimates from the 36 band analyses were used to ensure that the admittance was not changing significantly within the tidal frequency bands, and that the coherences were also constant within the tidal frequency bands.

After the spectral analysis had been done for each of the records, the ellipse parameters (Godin, 1972) were calculated for each of the data sets using the admittances and the 55 tidal constituents from a one-year harmonic analysis of the St.-Jean-Port-Joli water level record. Using these ellipse components, the tidal currents were then predicted for each of the periods of record for the observed currents, and then the residual currents were calculated by subtracting the predicted currents from the observed currents.

Mooring Number	OBSERVED VARIANCE (TOTAL)		HARMONIC RESIDUALS				ADMITTANCE RESIDUALS			
	U	V	R-u	R-v	V-u	V-v	R-u	R-v	V-u	V-v
01Z005	3485.11	3172.22	76.00	219.00	153.61	698.60	116.57	254.69	338.27	1549.29
01Z015	2319.23	7963.10	55.00	98.00	70.41	158.00	79.10	139.12	149.82	412.37
04Z005	4906.00	5852.45	47.00	49.00	52.10	54.22	71.66	78.93	263.83	190.38
04Z010	2761.41	5345.28	55.00	58.00	62.11	69.89	81.37	74.91	203.18	194.55
07Z015	744.66	2697.96	75.00	78.00	56.01	78.05	92.51	110.57	90.96	145.47
07Z040	334.90	1462.88	48.00	80.00	21.92	50.37	58.44	80.59	32.98	85.02
11Z011	246.30	1794.24	41.00	48.00	31.80	48.41	51.10	60.46	59.09	102.09
11Z021	189.74	472.89	23.00	32.00	14.00	20.39	26.91	46.68	22.93	36.80
11Z056	853.78	1586.43	53.00	68.00	27.59	39.89	50.66	60.97	56.79	103.86

TABLE 1 Residual Statistics for 1974 St. Lawrence Moorings. Observed Variance is the variance of the observed tide, for east and north components. R-u and R-v give the range in the residuals for the u and v components. V-u and V-v give the variance in the residuals for the u and v components. Admittance residuals calculated using 55 tidal constituents.

Table 1 shows a comparison between the residuals derived from the harmonic method and the residuals derived from the admittance method. The residuals from the harmonic method are smaller than the residuals from the admittance method, and this would indicate that the harmonic analysis provides a better anaylsis procedure than the admittance method. The reason for this is not difficult to find. Table 2 provides the coherences computed in three of the frequency bands for each record. The coherences are very high in the semi-diurnal band, poorer in the diurnal band, and poorer still in the quarter-diurnal band. Since most of the energy in the tidal signal is in the semi-diurnal band, the signal to noise ratio in the other bands is too much for the admittance method to cope with, and this is why the harmonic method provides a better fit, although both methods reduce the residuals to acceptable levels.

Mooring Number	Diurnal Band		Semi-diurnal Band		Quarter-diurnal Band	
	u	v	u	v	u	v
01Z005	.911	.431	.988	.810	.914	.721
01Z015	.910	.947	.992	.990	.780	.312
04Z005	.961	.915	.988	.997	.859	.776
04Z010	.892	.912	.984	.995	.899	.799
07Z015	.557	.947	.995	.997	.832	.903
07Z040	.805	.952	.995	.997	.974	.958
11Z011	.796	.979	.982	.998	.917	.582
11Z021	.919	.971	.990	.996	.852	.802
11Z056	.961	.934	.995	.995	.726	.757

TABLE 2 Coherences for three frequency bands, u and v components, between the 1974 St. Lawrence moorings and St.-Jean-Port-Joli.

Mooring Number	Harmonic Components M	m	θ	g	Admittance Components M	m	θ	g
01Z005	97.2	5.1	37.6	45.7	90.4	1.9	35.2	48.8
01Z015	136.1	1.6	62.0	31.4	131.4	- .1	61.8	34.8
04Z005	140.3	-9.6	47.3	22.9	133.7	-8.2	47.9	27.1
04Z010	121.4	-8.7	54.0	12.1	115.2	-8.1	55.2	16.6
07Z015	77.4	-5.9	62.0	25.1	75.6	-6.0	63.3	29.6
07Z040	52.6	19.4	73.4	18.8	50.7	18.5	73.6	21.7
11Z011	56.0	3.0	74.5	4.30	54.5	3.0	74.6	10.3
11Z021	31.8	6.1	60.9	13.2	30.7	5.7	59.8	19.0
11Z056	64.6	-2.0	54.2	-5.4	62.7	-1.9	54.2	-1.7

TABLE 3 Comparison of ellipse components for the M_2 derived from the harmonic analysis and the admittance analysis.

Table 3 gives a comparison of the M_2 ellipse parameters for each of the 1974 current meter records, and, since the coherences in the semi-diurnal band are high, these two sets of constituents are quite close. The admittance analysis shows, then, that there is no major fault in the harmonic analysis of the current meter records and that all of the discrepancies mentioned previously require a physical explanation.

The major advantage of the admittance method is that it allows the analysis of relatively short periods of record in which it would be impossible to separate the individual tidal constituents by means of the harmonic method. This is possible since it is assumed that the relationships between the constituents in any given frequency band are known from the analysis of the water level record. Therefore, some reanalysis of the 1974 data was done in order to examine the assumption that the tidal currents are due to a barotropic surface tide.

The first of the reanalyses was performed on the two sixty-day records 74-12C-07Z015 and 74-12C-07Z040 which were located in the middle of the Estuary, just off Pointe Au Pic. These were analysed in two, thirty-day sections and then in four, fifteen-day sections. Tables 4 and 5 give the results of the anaylses for each component along with the associated errors. The errors are in degrees and associated with the error in the estimate of the phase.

It should be noted from these two tables that the amplitudes and phases of none of the bands are stable. This instability does not seem to be related to the coherence estimates. One would expect that low coherences should be associated with extreme estimates of the phase and amplitude, but this does not seem to be the case.

The immediate inference that can be drawn is that the tidal currents of the Middle Estuary are not very predictable. Using the analysed constituents from any particular one of the 15-day analyses, one could not predict the currents in another of the 15-day periods with any degree of accuracy. The main carrier signal, which is the semi-diurnal band, would not be very far out in terms of amplitude, but its phase would certainly be incorrect and the modulations due to other frequency bands would be considerably wrong.

A second observation about the analyses is that there does not seem to be any consistency in the variations in the transfer function from one record to the other. These two records were obtained from the same mooring, over the same time period. One of the meters was at 15 metres depth and the other was at 40 metres depth. If the vertical velocity profile due to a barotropic surface tide is considered to be linear or semi-logarithmic, then the ratios of the amplitudes of the admittance should remain more or less constant and the phase differences should also remain constant throughout the vertical, but neither of these things happen.

The third observation that may be made about the analyses concerns the coherences. If the tidal currents at the two meters are caused by the same barotropic tide, then the signals should be related to each other linearly and the coherence between them should be exactly 1 in the absence of noise. Since the same water level record is used as the input signal for both tables, the coherence over any given period for the u-component, say, should be identical in each table. This does not happen, except possibly in the semi-diurnal band, where the coherences are so high that it is difficult to distinguish between them.

One of the most puzzling features of this first reanalysis of the data is the instability in the amplitudes and phases of the admittances, which implies that the amplitude and phase of the tidal constituents are variable with time. The major assumption of tidal theory is that the amplitude and phase of a tidal constituent must remain constant with time; otherwise, there could be no hope of predicting the tides. A second reanalysis of the 1974 St. Lawrence data was carried out to determine whether or not there was any pattern in the instability of the admittance.

This second reanalysis consisted of analysing three contiguous, non-overlapping segments of 236 hours each, for each of the 1974 current records. The length of 236 hours was chosen because it is exactly 19 M_2 tidal cycles; it contains part of a spring tide and a neap tide cycle, and, with 12 frequency bands for the spectral analysis, retains 19.5 degrees of freedom in the estimates so that they have a reasonable amount of significance. For each of the 1974 records, the first three segments were analysed and the phases of the M_2 constituent calculated. The results are given in Table 6 where the difference in phase between the phase of the segment and the phase of the whole record is given, along with the error in the calculation of the phase.

Mooring Number	Phase Total Record	Phase Difference of segments 1	2	3	Phase Error of segments 1	2	3
01Z005	45.70	1.02	16.46	0.58	5.94	9.70	19.24
01Z015	31.40	3.50	9.62	1.39	3.33	5.60	9.02
04Z005	22.90	2.26	12.68	1.37	5.12	6.44	5.63
04Z010	12.10	0.94	12.00	3.91	4.08	5.86	8.57
07Z015	25.10	4.35	8.10	4.81			
07Z040	18.80	2.45	10.09	1.56			
11Z011	4.30	5.13	9.31	5.00	5.00	7.13	4.65
11Z021	13.20	6.15	6.39	6.36	3.47	4.92	8.06
11Z056	-5.40	3.28	10.72	-.10	7.00	4.49	2.68

TABLE 6 Instabilities of the M_2 phase from the 1974 moorings. Phase difference = Phase Whole Record - Phase segment.

The calculated errors for the phase are quite large for each case, which reflects the short period of record used for each of the analyses. However, there is a pattern which emerges from the phase differences. For all of the records, there is a distinct phase shift, which is always in the same direction, from the first to the second ten-day segment and then another phase shift, back again, from the second to the third ten-day segment. The magnitude of these phase shifts is different for each of the records, and most of them, on a purely statistical basis, are not significant phase shifts. However, the fact that all of them are in the same direction is significant. It would argue that this phase shift is related to the differing portions of the spring-neap cycle in each of the ten-day records.

Comparing these phase shifts to the M_2 phase shifts shown in Tables 4 and 5 for the successive 15-day segments is revealing. In the 15-day segments, there is one complete spring-neap cycle and the phase shifts are always in the same direction. That is, the phase of the M_2 consistently lags from one 15-day segment to the next. In the 10-day segments, there is an unequal portion of the spring-neap cycle in each of the records and the phase of the M_2 lags from the first to the second segment and then returns almost to its original value by the third segment. Taking the two together, it seems that the variation in the phase of the M_2 is not linear and that this variation may have both a fortnightly component in it as well as a longer period component. To obtain reasonable estimates of these variations in the phase of the M_2 or any other constituent would require a large number of very long period tidal current records, and these are not available at the present time.

The fact that the harmonic analysis and the admittance analysis, on the whole, agree, would argue that there is no error in either method of analysis and that the observed inconsistencies in the tidal constituents and the instabilities of these constituents are a real, physical phenomena in the Middle Estuary of the St. Lawrence. It is necessary, therefore, to find an explanation for them.

5. THE PROPAGATION OF THE TIDAL CURRENTS.

The use of the admittance method for the analysis of the 1974 tidal current records has shown that neither the amplitude nor the phase of the tidal constituents at a given location in the Middle Estuary is a constant. Since both of the methods of analysis are essentially least squares, curve fitting exercises, and since neither of the methods makes use of any physical principles, other than those involved in choosing the frequencies for the curve fits, it is not the fault of the analysis methods that they give results which are inconsistent with our understanding of tidal processes. However, this understanding of tidal processes rests on the assumption that the tidal currents in the Middle Estuary result from the propagation of a simple barotropic tide. It is possible to explain the observed phenomena if this assumption is dropped, and the tidal currents are considered to be the result of not only barotropic tides but also baroclinic tides.

There are two major reasons for expecting to have internal tides present in the Middle Estuary. In Muir (1982) it was shown that there is considerable vertical density structure in the Estuary and, if there were internal tides generated, these would be allowed to propagate because of this vertical density structure. The second reason for expecting the presence of internal tides is that Forrester (1974) found them present in the Lower Estuary, where they were shown to be generated by the abrupt change in depth associated with the end of the Laurentian Channel, which is at the downstream end of the Middle Estuary. LeBlond and Mysak (1978), reporting on the work of Rattray and his co-workers (1960 onward), point out that internal waves may be generated by a step-like change in topography and although the work of Rattray is concerned with the seaward propagation of these waves, there is also a landward propagation. It is this landward, or upstream, propagating set of internal waves which would be found in the Middle Estuary. The internal tides affect only the currents and do not affect the water levels. This would explain why it is possible to predict the water levels accurately, but not the currents, by considering only the barotropic constituents.

If we assume a uniform channel with a constant surface density and appropriate vertical density variation, and that a surface tide of frequency, σ, is generated at one end of the semi-infinite channel which generates internal modes of the same frequency, then, at a distance x from the origin, the current will be given by

$$u(x,z,t) = \sum_{i=0}^{m} a_i(z)\cos(\sigma t + k_i x) \qquad (2)$$

where; a_i is the amplitude of the ith vertical mode at depth z; t is time; k_i is the horizontal wave number associated with the ith vertical mode; x is the distance from the generation point and m is the total number of vertical modes, with the zeroth mode being associated with the barotropic tide. Using standard trigonometric identities, equation (2) may be transformed into

$$u(x,z,t) = \text{AMP} \cos(\sigma t + \text{PHASE}) \qquad (3)$$

where;

$$\text{AMP} = \left[\left(\sum_{i=0}^{m} a_i \cos k_i x \right)^2 + \left(\sum_{i=0}^{m} a_i \sin k_i x \right)^2 \right]^{\frac{1}{2}}$$

$$\text{PHASE} = \arctan \left[\sum_{i=0}^{m} a_i \sin k_i x \Big/ \sum_{i=0}^{m} a_i \cos k_i x \right]$$

and so the current at any distance from the source could still be described by a cosine of the appropriate frequency, but both the amplitude and the phase of this cosine would be functions of the amplitude and phase of the modes as well as a function of the distance from the source.

To give an example of how this process could affect an estuary such as the St. Lawrence, assume that a surface tide at the M_2 frequency generates three internal modes at $x = 0$. If the wave properties are given in Table 7, then equation (3) may be used to calculate the amplitude and phase of a 'tidal constituent' at various depths and distances from the source of the internal waves. The values are given in Table 8.

It will be seen that even a very simple example can produce the type of features that are found in the Middle Estuary. The phase speeds of the "tidal constituent" vary considerably and are not consistent with a simple progressive wave, there is considerable variation in the amplitude, and the depth dependence of both the amplitude and the phase are similar to those observed. If a realistic topography and a cross-channel variation, such as due to Poincare waves were introduced, the variations could very well look even more like those observed in the Middle Estuary.

If the St. Lawrence were a uniform channel with a constant density structure, then the amplitudes and phases of the internal waves could be calculated from a number of simultaneous current meter records in a fairly restricted area. This information could then be used to predict the tidal currents at any point in the Estuary from the knowledge of

Mode Number	Amplitude (m/s) Surf.	10m	40m	Wave-number (rad/km)
0	0.50	0.50	0.50	0.00406
1	0.21	0.17	−0.12	0.1093
2	0.15	0.03	−0.02	0.2389
3	0.08	0.04	0.03	0.3570

TABLE 7 Wave parameters for the example in Table 8

Distance (km)	Amplitude Surf.	10m	40m	Phase Surf.	10m	40m	Phase Barotropic
60	0.696	0.647	0.359	31.92	19.65	13.65	13.96
65	0.508	0.647	0.410	27.65	23.35	1.52	15.12
70	0.562	0.584	0.497	21.63	30.07	4.08	16.28
75	0.606	0.543	0.510	30.45	43.93	9.35	17.45
80	0.488	0.506	0.524	39.14	34.86	7.72	18.61
85	0.421	0.374	0.611	40.98	29.89	10.07	19.77

TABLE 8 Amplitude and phase of a "tidal constituent" at various depths and distances from the source, compared to the phase of the barotropic tide. Wave parameters are as in Table 7.

the amplitudes, phases and wave number at the source. However, quite apart from the fact that the St. Lawrence is not a simple channel and has large and important topographic features, it is shown in Muir (1982) that the horizontal and vertical density structure itself is quite complicated. Since the density structure is a complicated function of both space and time, then the wave numbers and the modal structure at the generation point would be a function of time and the propagation properties of the whole water mass would also be a function of both space and time. Hence the 'tidal constituents', as defined by Equation (3), at any given location will be a function of time as well as a function of the density structure which lies between the measuring point and the generation point.

6. DISCUSSION

The purpose of this paper was to draw out some of the information available from the tidal current records that have been collected on the Middle Estuary. The admittance method of analysis of tidal currents has been very useful in revealing the inconsistencies in these records. However, the inconsistencies are due to the assumption that the tidal currents in the St. Lawrence are due to surface, or barotropic, tides alone. It has been shown that the inconsistencies may be explained in a qualitative manner if it is assumed that there are baroclinic, or internal, tides present which also contribute to the observed tidal currents. Unless the internal tides are taken into account, it will be impossible to predict the currents in the Middle Estuary, even though the statistics of the tidal analyses indicate that very good least-squares fits are obtained.

REFERENCES

AUBIN, F., T.S. MURTY, and M.I. EL-SABH, 1979. Numerical Simulation of the Movement and Dispersion of Oil Slicks in the Upper St. Lawrence Estuary: Preliminary Results. Le Naturaliste Canadien, 106: 37-44.

BUDGELL, W.P. and L.R. MUIR, 1975. St. Lawrence River Current Survey 1974 Data Report. Ocean & Aquatic Sciences, Cent. Region, Dept. Fish. Envir., 335 pp.

EL-SABH, M.I., T.S. MURTY, and L. LEVESQUE, 1979. Mouvement des Eauxs Induits Par La Marée et le Vent dans l'Estuaire du Saint-Laurent. Le Naturaliste Canadien, 106: 89-104.

FORRESTER, W.D., 1974. Internal Tides in St. Lawrence Estuary. J. Marine Research, 32: 55-66.

GODIN, G., 1972. The Analysis of Tides. University of Toronto, Toronto, 264 pp.

GODIN, G., 1976. The Reduction of Current Observations with the Help of the Admittance Function. Technical Note 14 - Mar. Envir. Data. Serv. Dept., Fish. Envir., 13 pp.

GODIN, G., 1979. La Marée Dans le Golfe et l'Estuarire du Saint-Laurent. Le Naturaliste Canadien, 106: 105-121.

LEBLOND, P.H., 1978. On Tidal Propagation In Shallow Rivers. J. Geophysical Res. 83: 4717-4721.

LEBLOND, P.H. and L.A. MYSAK, 1978. Waves in the Ocean. Elsevier Scientific Publishing Co., Amsterdam, 602 pp.

MUIR, L.R., 1978. St. Lawrence Current Survey: 1975 Data Report. Ocean and Aquatic Sciences, Cent. Region, Dept. Fisheries and the Environment. 315 pp.

MUIR, L.R., 1979. St. Lawrence River Oceanographic Survey: 1977 Data Report. Ocean and Aquatic Sciences, Cent. Region, Dept. Fisheries and the Environment. Tidal, Met. and Current Meter Data, Vol. 1, 199 p., Profile Data, Vol. 2, 278 pp.

MUIR, L.R., 1982. Variability of Temperature, Salinity and Tidally-Averaged Density in the Middle Estuary of the St. Lawrence, Atmosphere- Ocean, (in the press).

MUNK, W. and D. CARTWRIGHT, 1966. Tidal Spectroscopy and Prediction. Phil. Trans. Royal Soc. London, Ser. A., 259: 533-581.

OUELLET, M.Y. and M.J.F. CHEYLUS, 1971. Étude du Modèle Mathématique du La Propagation Des Marées dans le Fleuve St. Laurent. Report CRE-71-05, Laval University. 37 pp.

PRANDLE, D. and N. CROOKSHANK, 1974. Numerical Model of the St. Lawrence Estuary. J. Hyd. Div. ASCE, 100: 517-529.

SOULSBY, R.L., 1978. The Use of Depth Averaged Current to Estimate Bed Shear Stress. Internal Document 26, IOS, Crossway, Taunton, Somerset.

SIMULATION OF THE LOW FREQUENCY PORTION OF THE SEA LEVEL SIGNAL AT YARMOUTH, NOVA SCOTIA

D.L. DEWOLFE AND R.H. LOUCKS

Bedford Institute of Oceanography,Dartmouth, N.S., and R.H. Loucks Oceanology Ltd., Halifax, N.S., Canada

ABSTRACT

In the course of fisheries oceanography research, it was found necessary to gap-fill an existing tidal record at Yarmouth, N.S. and to hindcast the record for a five-year period prior to the establishment of the tide gauge, specifically for low frequency (periods >12 hours) oscillations.

A "neighbouring" time series, at Saint John, N.B., was determined by cross-spectral analysis to be coherent with Yarmouth at the desired low frequencies. After suitable low pass filtering and decimation of the two time series for the longest period for which both had complete records, the "admittance" between the two was calculated. The resulting gain and phase information was then Fourier Transformed into the impulse response function, with Saint John as input. The Saint John record was then filtered with the impulse-response function weights which produced a synthetic low frequency signal for Yarmouth which differed from the actual by about 5%.

The paper describes in detail the theory and methods used in this novel application, together with the results.

Reprinted from *Time Series Methods in Hydrosciences,* by A.H. El-Shaarawi and S.R. Esterby (Editors)

INTRODUCTION

In the course of fisheries oceanography research into the possible influence of environmental factors on herring year-class strength off Southwest Nova Scotia (Metuzals, et al., 1978), it was noticed that there was a relatively strong correlation between the sea level signal at Halifax and the herring year-class strength. To explore this correlation further, it was considered desirable to investigate the use of the sea level signal at Yarmouth as a possible indicator of the oceanographic situation year to year. However, the period of record available at Yarmouth was relatively short and the record itself contained gaps. It was decided to develop a means of simulating sea levels at Yarmouth, both to fill gaps in the record of observations and to extend the signal beyond the period of record (from 1960 to 1966), thus giving a continuous low frequency signal from 1960 to 1978. The data at Saint John, continuous except for small gaps, would be used as a "leading indicator", due to the similarity and proximity of the two tidal stations.

DATA

The data, obtained from Marine Environmental Data Services (MEDS) in Ottawa, consisted of hourly tidal heights for Saint John, N.B. (1960-1976) and Yarmouth, N.S. (1966-1978).

PROCEDURE

The procedure, in capsule, is to take the cross-spectra between Yarmouth and Saint John as defining the transfer function relating the two signals. The impulse response (time domain) form of the transfer function with Saint John as input yields the simulated Yarmouth signal as output.

Now we can describe the procedure in more detail. A period of record complete in both Yarmouth and Saint John observations was selected. Actually, the longest such period was selected. This was to be the basis on which the cross-spectra were calculated. The period selected was somewhat more than a year in duration and included two freshet seasons. This turned out to be a disadvantage because the freshet of the Saint John River is the one feature which the Yarmouth and Saint John signals will not have in common. In future this will be avoided. For each signal and for the selected period, astronomically predicted values of sea levels were subtracted. These residuals were low-pass filtered and decimated to six-hour intervals from one-hour intervals, which effectively removes the tidal signal. Cross-spectra were then calculated and the resulting gains and phase differences were obtained. These are listed in Table 1.

TABLE 1. Coherence, gain and phase difference, Saint John to Yarmouth. (Negative phase means Yarmouth Leads.)

Freq. Band	Coh.	Gain	Phase	Freq. Band	Coh.	Gain	Phase
0	0.69	0.73	-11.6°	9	0.58	0.44	- 5
1	0.94	1.12	- 4	10	0.63	0.49	-10
2	0.95	1.09	- 8	11	0.74	0.60	3
3	0.93	0.96	1	12	0.47	0.45	-32
4	0.93	0.95	- 8	13	0.51	0.47	-15
5	0.83	0.75	1	14	0.20	0.17	-33
6	0.89	0.82	0	15	0.36	0.21	-19
7	0.86	0.88	1	16	0.30	0.19	17
8	0.72	0.59	-16				

Following Holloway (1959), we can proceed from the definition of the impulse response (time domain) function to the actual algorithm for its computation using as input the values of gain and phase differences (Table 1) in the frequency domain, as follows:

$$w(t) = \int_{-\infty}^{\infty} G(f) \exp [i(\Phi(f) - 2\pi\ ft)]\ df \qquad (1)$$

where $G(f) = G(-f)$ and $\Phi(f) = -\Phi(-f)$

$$\therefore\ w(t) = - \int_{0}^{-\infty} G(f) \exp [i(\Phi(f) - 2\pi\ ft)]\ df$$

$$+ \int_{0}^{\infty} G(f) \exp [i(\Phi(f) - 2\pi\ ft)]\ df \qquad (2)$$

Now substituting $f' = -f$ in (2) gives

$$w(t) = \int_{0}^{\infty} G(f') \exp [-i(\Phi(f') - 2\pi\ f't)]df'$$

$$+ \int_{0}^{\infty} G(f) \exp [i(\Phi(f) - 2\pi\ ft)]\ df$$

$$= 2 \int_{0}^{\infty} G(f) \cos (\Phi(f) - 2\pi\ ft)\ df$$

$$\simeq \frac{1}{2m} [G(o) \cos \Phi(o) + 2 \sum_{q=1}^{m} G(q) \cos \left\{\Phi(q) - \frac{q\pi t}{m}\right\}$$

$$+ G(m) \cos (\Phi(m) + \pi t)] \qquad (3)$$

where $w(t)$ is the impulse response at time t,

$G(f)$ is the gain at frequency f,

$\Phi(f)$ is the phase difference at frequency f,

and $\frac{1}{2m}$ is the bandwidth of the discrete gain and phase estimates.

Starting with the data in Table 1 and utilizing equation (3), the calculated values for the impulse response function extending away from the central value at 6-hour intervals are shown in Table 2.

The filtered and decimated portion of the record at Saint John was then filtered with the impulse response weights shown in Table 2, yielding a simulated record at Yarmouth.

The simulated signal at Yarmouth was then compared with the (suitably lagged) signal at Yarmouth for the period of the test data (1-1/2 years). This comparison showed that the maximum difference between the real signal and simulated signal was 5 cm which met our acceptability criterion.

The hourly data at Saint John was then made continuous for the period 1960-1976 by filling in the small gaps with predicted tides and was then low-passed, decimated and filtered with the impulse

Table 2. Values or weights for the impulse response function, Saint John to Yarmouth.

Time	Weight	Time	Weight	Time	Weight
-16	-0.033	- 5	-0.020	6	-0.013
-15	-0.022	- 4	-0.020	7	-0.032
-14	-0.013	- 3	0.040	8	-0.025
-13	-0.030	- 2	-0.025	9	-0.013
-12	-0.027	- 1	0.239	10	0.005
-11	-0.030	0	0.641	11	-0.043
-10	-0.005	1	0.155	12	-0.000
- 9	0.023	2	0.006	13	-0.013
- 8	0.002	3	-0.017	14	-0.012
- 7	-0.060	4	-0.030	15	0.043
- 6	-0.002	5	-0.003	16	-0.033

response weights. The Yarmouth hourly data was low-passed, filtered and lagged to provide alignment with the simulated Yarmouth data. The two signals were then spliced together, yielding a continuous Yarmouth signal from 1960 to 1978.

DISCUSSION

Noting that the maximum difference between the actual and simulated signals at Yarmouth was only 5 cm over a period of a year and a half, it would appear that this method of simulating one signal from a related one is sound. The resulting Yarmouth signal, mostly observed but partly simulated was then, after further filtering and decimation to 60-hour data, used for exploring the relationship between herring year-class strength and the oceanographic environment. Although far from complete, this exploration done by others indicates very high correlations in the order of 0.9 between the herring year-class strength and the sea level at Yarmouth.

REFERENCES

Holloway, J.L., Jr. 1959. Smoothing and filtering of time series and space fields. Advances in Geophysics, pp. 351-389.

Metuzals, K., Sinclair, M. and Sutcliffe, W. 1978. A preliminary analysis of recruitment variability in 4WK herring. Working paper, Bedford Institute of Oceanography, Dartmouth, N.S., Canada.

THE COMPUTATION OF TIDES FROM IRREGULARLY SAMPLED SEA SURFACE HEIGHT DATA

LUNG-FA KU, CANADIAN HYDROGRAPHIC SERVICE, OTTAWA

I INTRODUCTION

Several investigators have attempted to obtain the geometrical distribution of tides in a region using the sea surface height data obtained from GEOS-3 without any success (Won and Miller, 1979; Brown and Hutchinson, 1980; Maul and Yanaway, 1977; Parke, 1980; and Ku, 1982). Another approach is to divide the ocean into small areas where the spatial distribution of tides can be neglected. The problem is then reduced to merely that of a time series analysis. This paper discusses problems encounteed in the analysis due to the irregularity of the sampling interval and the bias caused by the geoidal height remained in the sea surface height.

II HARMONIC ANALYSIS OF TIDES

Assuming A_k is the complex amplitude of the kth tidal constituent with the frequency σ_k , h_n is the nth sea surface height sampled at t_n , then the least squares fit solution of A_k can be expressed in matrix notaton as

$$[A] = [C]^{-1} [Y] \quad (1)$$

where

$$[C] = [R^*]^t [R] \quad (2)$$

$$[Y] = [R^*]^t [h] \quad (3)$$

The element of [R] and [Y] are

Reprinted from *Time Series Methods in Hydrosciences*, by A.H. El-Shaarawi and S.R. Esterby (Editors)

$$r_{nk} = \exp\,(i\sigma_k\, t_n) \tag{4}$$

$$y_k = \sum_{n=1}^{N} h_n \exp\,(-i\,\sigma_k t_n) \tag{5}$$

which is basically N times the Fourier transform of h_n at the angular frequency σ_k.

Since the sampled sea surface height h_n can be expressed as the product of the height h(t) and a data sampler s_n, y_k can therefore be represented by the convolution between the Fourier transforms of the sea surface height H (σ) and the data sampler S (σ). The data sampler is usually chosen to reduce the aliasing in y_k.

III DATA SAMPLER

The sampling time interval of the sea surface height data obtained from GEOS-3 is primarily determined by the period of the satellite. The satellite's track on earth does not form a closed curve. It is usually described by the offset of the equatorial crossing of the track. For GEOS-3, the equatorial crossing moves westward 25.32° for each revolution. Depending on the longitudinal width of the study area, the data collected may have come in a burst of several passes,followed by a gap of several passes. To investigate the effect of bursting in the data sampler on the harmonic analysis of tides, let us assume that the sampler takes the form shown in Figure 1. The period of the satellite is T, the duration of the burst is KT, and the burst repeats at an interval of LT. The total number of bursts is N.

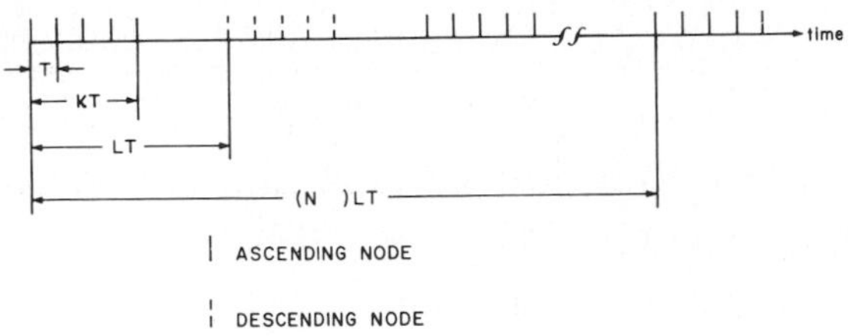

Figure 1. Simplified data sampler (T is the period of the satellite).

The Fourier transform of this sampler at a frequency σ is

$$S(\sigma) = \left(\sum_{k=1}^{K} \exp(i\sigma kt)\right)\left(\sum_{n=1}^{N} \exp(-i\sigma LTn)\right) \qquad (6)$$

The term in the first bracket represents the Fourier transform of the sampler within a burst, and the second one represents that of the sampler burst. Equation (6) can also be expressed as

$$S(\sigma) = \frac{\text{Sin}\,(\sigma KT/2)}{\text{Sin}\,(\sigma T/2)} \, \frac{\text{Sin}\,(\sigma NLT/2)}{\text{Sin}\,(\sigma LT/2)} \exp\,(-i\sigma T(K+1-(N+1)L)/2) \qquad (7)$$

Therefore, S (σ) is the product of two diffraction functions: the first one is due to the sampler within a burst and the second one is due to the repetition of the burst. Since the second diffraction function fluctuates faster than the first, the first function has the effect of modulating the amplitude of the second function. The feature of S (σ) will repeat at 1/LT, 1/kt, and 1/T intervals, with its first zero crossing at 1/NT.

Near the equator, an area with a longitudinal width less than 25°, the burst will disappear, being replaced by one pass about every 7T. In reality, data might not be collected at every pass over the area, or some data must be discarded because of its poor quality. Consequently, altimeter data would possess many gaps.

Table 1 lists the sampling time of the data supplied by NASA in N.E. Pacific Ocean. Some data from other passes are not included due to the possible large error indicated in the analysis of the height difference at crossing points between two satellite tracks. It shows that most of them are the multiple of 14T, where T = 101.8 minutes is the period of the GEOS-3 satellite. Consequently, we define 14T as the median of the sampling interval.

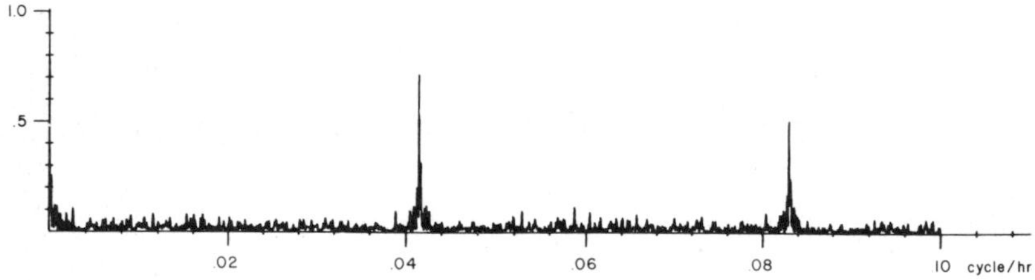

Figure 2. Spectrum of the data sampler.

The spectrum of this data sampler is shown in Figure 2. It has a similar feature as the spectrum of a fixed interval data sampler. The major lobe repeats at an interval of .0414 cycle/hour which is equal to 1/14T. Therefore, the median of the sampling interval is equivalent to the sampling interval for a fixed interval data sampler. The major lobe and several neighboring minor lobes, however, does not decrease to zero, they reach a minimum instead.

Therefore, the resolution of the spectrum of the irregularly sampled data will be poorer than that sampled regularly. The first minimum with a magnitude of .08 occurs at about 1.2×10^{-4} cycle/hr which is about twice $1/T_d$, where T_d is the duration of the observation. All minor lobes which are not adjacent to the major lobes maintain a magnitude of about .05.

IV THE BIAS IN THE ANALYSIS

By ignoring the effect of the minor lobes in $S(\sigma)$, equation (5) can be approximated as

$$y_k = \sum_{j=-\infty}^{\infty} H(\sigma_k + j\sigma_s) \tag{8}$$

Where σ_s is the angular sampling frequency. This is a well known problem which is called aliasing.

For most data, this problem can be overcome by low-pass filtering the data provided that $\sigma_k < .5\ \sigma_s$, A similar remedy is not available for data sampled at irregular intervals shown in Table 1. Therefore, the covariance vector could be biased by the spectral energy of the real sea surface height at other frequencies. Since the sea surface height is the difference between the altimeter height and the satellite height, the spectrum of the sea surface height contains the spectrum of the satellite height which has not been accounted for in the orbital computation. This effect, however, is negligible (Ku, 1982).

Figure 3 plots the track of the satellite passes used in this study. It covers an area with a large change in the geoid as shown in the same figure. To reduce this effect in

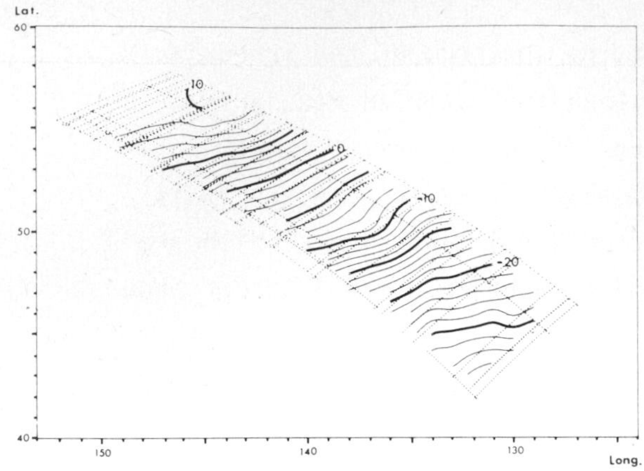

Figure 3. Satellite tracks and the gravimetric geoid (meter).

the computation, the gravimetric geoidal heights supplied by NASA are subtracted from the sea surface heights. The average sea surface height is then computed for each pass. After removing the mean of all the average sea surface heights from these heights, their spectrum is computed and plotted in Figure 4. The six most important tidal components in this area are O1, P1 and K1 in the diurnal tidal frequency band, and N2, M2 and S2 in the semi-diurnal tidal frequency band. Their frequencies have been indicated in the same figure. The spectrum shows a large variation at all frequencies, and there is no significantly large peak at any of the six tidal frequencies. The spectrum is clearly aliased as shown by the numbering of some of the peaks between 0. and .0414 cycle/hr.

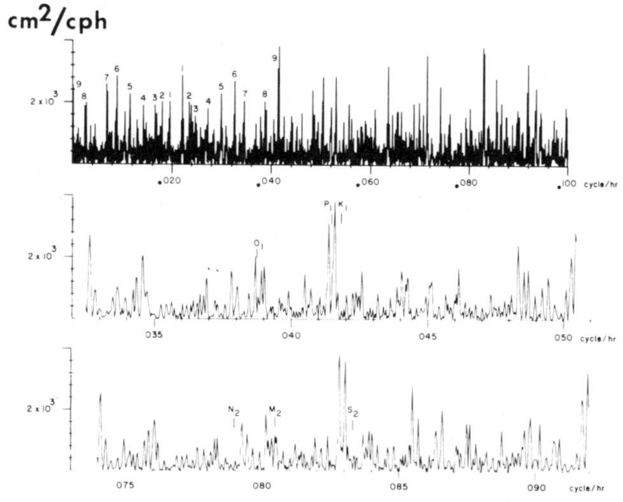

Figure 4. Spectrum of the average sea surface heights.

Since the gravimetric geoidal height provided by NASA might not be identical with the actual geoid in this area, it is possible that some of the variances in the sea surface height data originated from the residues in the geoidal heights. Ku (1982) indicates that the residues could be stronger along the northwest direction. To study its effect in the spectrum of the sea surface height, we assume that the remainder of the geoidal height in the sea surface

Table 1

Sampling Time Intervals

n	k_n	k_n	$k_n/14$	n	k_n	k_n	$k_n/14$
0	300			25	2844	213	15.2
1	357	57	4.1	26	2859	15	1.1
2	599	242	17.3	27	2972	113	8.1
3	855	256	18.3	28	3043	71	5.1
4	1267	412	29.4	29	3072	29	2.1
5	1438	171	12.2	30	3087	15	1.1
6	*1527	89	6.4	31	3115	28	2.0
7	1537	10	.7	32	3229	114	8.1
8	1949	412	29.4	33	3342	113	8.1
9	2220	271	19.4	34	*3745	403	24.8
10	2234	14	1.0	35	*4669	924	66.0
11	2318	84	6.0	36	4921	252	18.0
12	2319	1	.1	37	5049	128	9.1
13	2347	28	2.0	38	5248	199	14.2
14	2361	14	1.0	39	5319	71	5.1
15	2390	29	2.1	40	*5323	4	.3
16	2404	14	1.0	41	5333	10	.7
17	2419	15	1.1	42	5361	28	2.0
18	2447	28	2.0	43	5390	29	2.1
19	2461	14	1.0	44	5405	14	1.0
20	2503	42	3.0	45	5916	512	36.6
21	2504	1	.1	46	5986	70	5.0
22	2560	56	4.0	47	*7882	1896	135.4
23	2617	57	4.1	48	*7910	28	14.0
24	2631	14	1.0				

* indicates ascending pass $k_n = k_n - k_{n-1}$

height data is a linear function of the latitude ϕ and the longitude λ as

$$h_g = \phi - 50 - (\lambda - 220) \tag{9}$$

A synthetic time series is then generated by taking the sample of h_g at the time of the satellite passes shown in Table 1 and at the corresponding centres of the track shown in Figure 3. The spectrum of this time series is shown in Figure 5. Some of the peaks in the figure have been numbered to indicate the effect of aliasing.

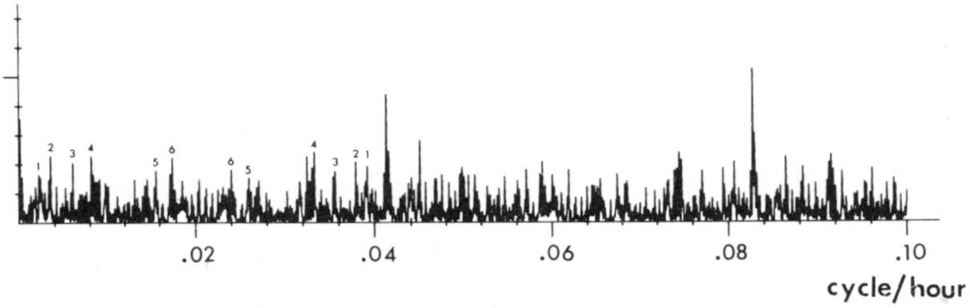

Figure 5. Spectrum of the simulated geoidal height residues.

V THE HARMONIC ANALYSIS OF THE SEA SURFACE HEIGHT

The significant tidal periodic components in this area are O1, P1, K1, N2, M2 and S2. According to the global ocean tidal computation carried out by Schwiderski (1979), and the tidal measurement reported by Rapatz and Huggett (1972), the expected values of the amplitudes and phases of these components are given Table 2. Conventional tide gauges measure the tide as a change in the sea surface height with respect to the sea bottom, while the tide derived from the altimeter height and the satellite height is the change of the

Table 2

Mean tides computed from the harmonic analysis of sea surface heights (cm and degree)

	H	θ	A	$\bar{A}$	G	$\bar{G}$	R
O_1	65(43)	-146	71	25	118	240	86
P_1	108(57)	-106	108	15	244	250	93
K_1	92(54)	-49	97	40	329	260	91
N_2	50(46)	-39	49	15	255	250	34
M_2	55(45)	10	54	80	296	265	44
S_2	35(50)	-46	35	25	314	300	12

$\nu_o = 154$ $\nu_r = 134$ N = 49

A, θ = amplitude and Greenwich phase lag obtained from the harmonic analysis

( ) = standard error

A,G = amplitude and phase after correcting astronomica modulation

bar = expected value

ν_o , ν_r = standard deviation of observed data and its residues

N = number of data points

R = $| A \exp(iG) - \bar{A} \exp(i\bar{G}) |$

sea surface height with respect to the center of the earth. Therefore, the former is called the surface tide and the latter is called the geocentric tide. According to the result of Parke (1978), the difference between these two tides is negligible for the semi-diurnal tide. The phase lag of the diurnal surface tide is about 15° larger than that of the geocentric tide, and its amplitude is 15% larger.

Table 2 shows the results of the harmonic analysis. The standard error of the analysis is largely due to the large standard deviation in the residues, and the ill condition of the covariance matrix. The estimated values of the tidal component differ significantly from the expected values. The discrepancy is about 90 cm in the diurnal frequency band, which is about twice the amplitude of the strongest diurnal component, K1. The discrepancy for the semi-diurnal tide is about 30 cm which is only one-third of that of the diurnal tide, and is about half of the amplitute of the strongest semi-diurnal tide, M2. In general, the discrepancy is well within the estimated error.

VII CONCLUSION

This study concludes that the major lobe of the spectrum of the data sampler with an irregular sampling interval repeats at an interval 1/MT where MT is defined as the median of the sampling intervals. At any frequency outside the major lobe, the spectrum maintains a level of about 5% of the major lobe.

The spectrum of the sea surface height shows the effect of aliasing and, therefore, the covariance vector and the result of the harmonic analysis could be biased. The aliasing is caused by the fact that the data can't be smoothed prior to the analysis to reduce the undesired components in the spectrum. The major sources of energy could be the large residues of the geoidal height in the sea surface height data.

The result of the analysis is unsatisfactory due to the large noise level in the data. The standard error of the estimate is about 50 cm, which is about equal to the amplitude of K1 and about one-half the amplitude of M2 in the area.

REFERENCES

Brown, R.D., and M.K. Hutchinson, 1980: Ocean tide determination from satellite altimetry. Presented at COSPAR/SCOR/IUCRM Symposium on Oceanography from Space, Venice, Italy, May 26-30.

Ku, L.F., 1982: The computation of tides from GEOS-3 altimeter data.

Maul, G.A. and A. Yanaway, 1977: Deep sea tide determination from GEOS-3. NASA CR-141435.

Parke, M.E., 1978: Global numerical models of the open ocean tides M_2, S_2, K_1 on an elastic earth. Ph.D. thesis, Univ. of California, San Diego.

Parke, M.E., 1980: Tides on the Patagonian shelf from the SEASAT radar altimeter. Presented at COSPAR/SCOR/IUCRM Symposium on Oceanography from Space, Venice, Italy, May 26-30.

Rapatz, W.J. and W.S. Huggett, 1975: Pacific Ocean offshore tidal program. Presented at ICGU meeting in Grenoble, France.

Schwiderski, E.W., 1979: Global Ocean tides, Part II: The semi-diurnal principal lunar tide (M_2), Atlas of tidal charts and maps. Naval Surface Weapon Center, NSWC TR 79-414.

Won, I.J. and L.S. Miller, 1978: Oceanic geoid and tides obtained from GEOS-3 satellite data in the Northwestern Atlantic Ocean. NASA CR-156845.

ON STOCHASTIC MODELLING OF HYDROLOGIC DATA

T.E. UNNY
University of Waterloo

INTRODUCTION

Hydrologic data primarily pertain to rainfall, to streamflows and to other variables included in the hydrologic cycle. These data can be roughly divided into three categories: there is the historical data series recorded on a single variable in time; there is the field data series recorded in space; and finally, there are simultaneously measured data series recorded in time on many variables distributed in space.

Most of the historical data collected in the past on hydrologic variables, and particularly on streamflows, are those observed in discrete time. Thereby, time becomes an indexing variable in the historical data. It is in this sense that an ordered set of data in time is called a time series. The discussion below will be specifically concerned with streamflow time series although the points of discussion contained herein are of general applicability to other hydrologic time series.

Hydrologic time series are described depending on the time interval between successive observations as daily, monthly and yearly time series. The properties possessed by these time series are different.

The yearly streamflow time series has been considered in many cases as a sample from a purely random sequence. However, it may be noted here that such a time series is characterised by groups consisting of values corresponding to spells of flood and drought years (good and bad years)(Fig. 1b). The existence of groups indicates the presence of persistence in these time series.

Reprinted from *Time Series Methods in Hydrosciences*, by A.H. El-Shaarawi and S.R. Esterby (Editors)

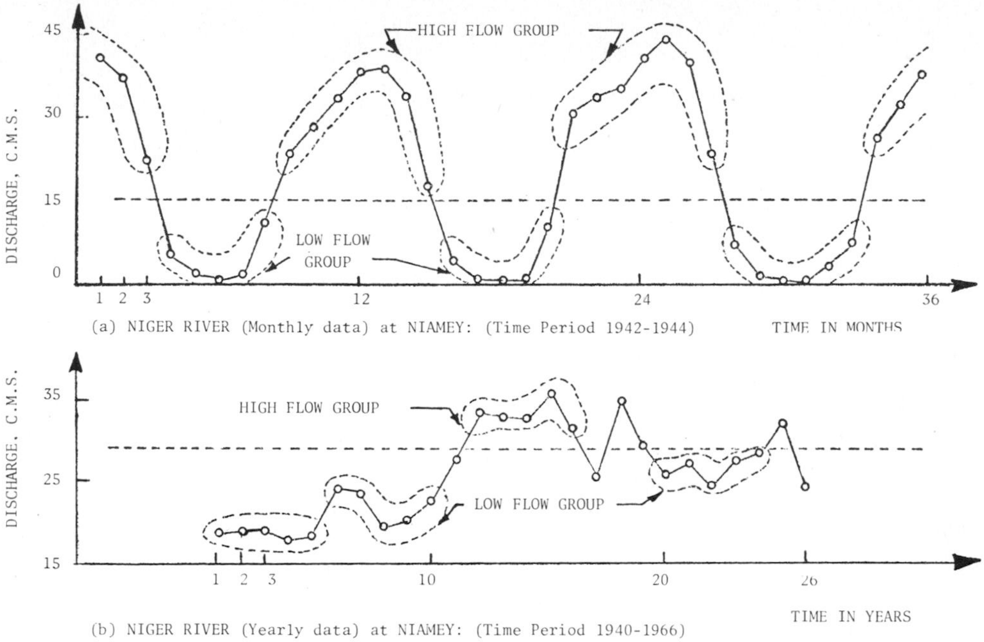

Fig. 1. Group formation in yearly and monthly time series of flow

In contrast, the monthly time series are marked by seasonality according to the geophysical year (Fig. 1a). There is also the presence of characteristic groups of high and low values among the years and also within the year.

The daily time series are different from monthly and yearly time series in the sense that these time series are characterised by the occurrence of sharp peaks and exponential decay. The cause-effect relationship in rainfall-runoff process is stronger in the short interval time series.

The periodicities are induced in the streamflow time series by the geophysical cycle. This is reflected in the occurrence of high precipitation and high runoff during the summer months, and low precipitation and low runoff during the winter months in Northern climates. Means and variances of hydrologic time series are found larger in summer and smaller in winter months. Further, non-homogeneity in time also occurs in the form of trend in data as well as in the form of gradual and sudden variations in the stochastic nature of the data.

The implication arising from the presence of persistence, especially that corresponding to prolonged wet and dry periods, is significant from the point of view of water resources system planning, design and operations. Both short-term persistence and long-term persistence are integral parts of hydrologic time series which thereby become different from those time series found in other disciplines such as stock-market analysis.

An historically recorded streamflow time series can be considered to be derived from a stochastic generating mechanism that evolves according to certain probabilistic laws. If it were possible to decipher these probabilistic laws, then there would be a satisfactory model for the time series. However, this procedure is difficult, if not practically impossible. Because streamflow data represent only a single time series a check as to whether such a series forms part of a stationary process and also an ergodic process is impossible. In spite of this, the following procedure is often carried out in connection with streamflow time series modelling:

Given an historically recorded data series:

i) Assume ergodicity

ii) Assume a stochastic process

iii) Calculate certain parameters from the recorded data and assume that these parameters are the statistics for the process

iv) Assume that the so-defined stochastic process is the generating mechanism from which the recorded data is derived as a sample.

Thereby, and with so many assumptions, a stochastic model is considered to be obtained for the time series. It should be emphasized here that this procedure bears no relationship to the physical phenomena on which the data has been recorded. In addition, this procedure could often become irrelevant when its results are applied in connection with water resources planning and management. Objections can also be raised from heuristic and philosophic points of view. All that is required to complete the above procedure is a few parameters - at the most three of four - determined from the data; otherwise the whole set of data so laboriously collected can be discarded.

An Example

Figure 2 represents a bivariate time series of length 100 units. Some points in the series are shown by open circles and others by solid ones. This will be explained subsequently. No attempt has been made here to generate an ARMA process or any other stochastic process to represent these time series. It was not the purpose for which these series were 'recorded'. They are taken from a recent Ph.D. thesis completed at the University of Waterloo (McInnes, 1981) and it is acknowledged here. The fact remains that, if the data had been printed in a tabular format it would have been easy to extract parameters and develop an ARMA process as a model for the series. This could result in a delusion of the reality.

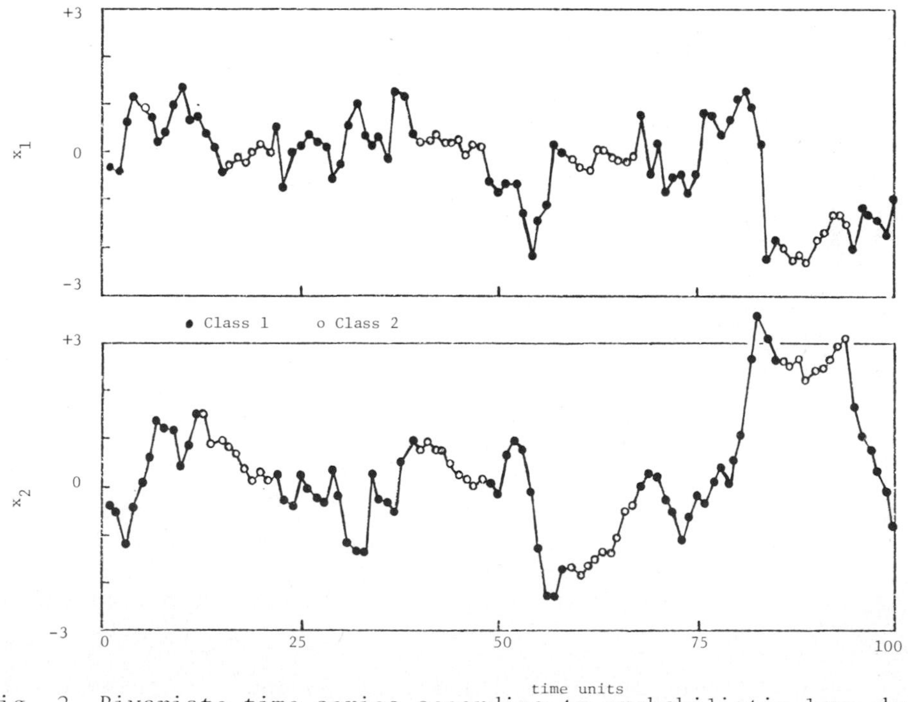

Fig. 2. Bivariate time series according to probabilistic laws described in equations (12) and (13) (from ref: MacInnes, 1980).

In order to illustrate the main point of this discussion, consider the generating mechanism of the series in Fig. 2. These series are obtained as samples of two separate processes. The solid circles are derived from the process:

$$\underline{x}(t) = \begin{bmatrix} 0.7 & -0.3 \\ 0.3 & 0.9 \end{bmatrix} \underline{x}(t-1) + \underline{\varepsilon}(t) \text{ where } E[\underline{x}(t)]=\underline{0}=E[\underline{\varepsilon}(t)] \ , \qquad (1)$$

and

$$E[\underline{\varepsilon}(t)\underline{\varepsilon}^T(t-k)] = \begin{bmatrix} 0.3 & 0 \\ 0 & 0.3 \end{bmatrix} \delta_k \text{ where } \delta_k = \begin{cases} 1 \ , & k=0 \\ 0 \ , & k\neq 0 \end{cases} .$$

The open circles are derived from the random walk model

$$\underline{x}(t) = \begin{bmatrix} 1 & 0 \\ 0 & 1 \end{bmatrix} \underline{x}(t-1) + \underline{\varepsilon}(t) \quad \text{where} \quad \begin{aligned} E[\underline{x}(t)] &= \underline{x}(t^*) \\ E[\underline{\varepsilon}(t)] &= \underline{0} \end{aligned} \qquad (2)$$

and

$$E[\underline{\varepsilon}(t)\underline{\varepsilon}^T(t-k)] = \begin{bmatrix} 0.05 & 0 \\ 0 & 0.05 \end{bmatrix} \delta_k \ .$$

In the above t* is the final time point associated with the regime of the earlier described process immediately preceding the regime associated with the random walk model. The length of run (number of time points) in each regime is generated by an equilibrium discrete renewal process given by:

$$\ell = 1 + B(n,\theta) \qquad (3)$$

where B is a binomial random variable described by

$$P_B(B|N,\theta) = \binom{n}{B} \theta^B (1-\theta)^{n-B} \qquad (4)$$

which is the probability of exactly B occurrences in n independent Bernoulli trials with the probability of an occurrence in any one trial being θ. The values of n and θ are 55 and 0.2, respectively, such that

$$E[\ell] = 1 + n\theta = 12 \;, \quad \text{and} \quad var[\ell] = n\theta(1-\theta) \approx 2.97^2 \quad . \tag{5}$$

Though these series are artificial creations, they do have some practical relevance. In most hydrologic data there are several generating processes at work one after the other at different periods in time and these cannot be averaged into a single generating process. As an example, a bivariate ARMA process derived for the whole series in Fig. 2 would be based on a value for the first order autocorrelation coefficient significantly different from that for the process from which the solid circles are obtained. The solid circles and the open circles represent data with vastly different persistence characteristic and any time series model that does not consider this difference should be treated as unsatisfactory.

On Persistence

Persistence implies a certain deterministic relationship between successive values of data in the time series. Such a relationship may be due to the fact that the cause resulting in the effect persists for a span of time longer than one or more increments between successive datum. The cause being of limited length, the persistence introduced by the cause is also of limited length. Succeeding parts of the time series may exhibit different persistence.

Persistence is often characterised by the correlogram which is a plot of the correlation coefficient against lag. Correlation coefficient at each lag is determined by a scanning procedure across the whole data. This has an averaging effect. The fallacy of using this indicator to denote persistence is apparent in the case of annual hydrologic series. In most annual series the correlation coefficients at lag 1 and at higher lags are found to be insignificantly different from zero, meaning that they lie within the confidence bounds of similar coefficients for an independent series. Thus, in the literature, and in water resources applications as well, annual series are often treated as independent. However, it is seen from Fig. 1 that there are well defined group formations in annual series which is indicative of strong persistence in stretches of the series. It is also valid to note in this connection the recently recorded examples in various parts of the world of 7 bad years, etc.

Furthermore, it is difficult to obtain a reliable and stable estimate for the correlation coefficient from the data series. With regard to the data series in Table 1 of length 59, different sections of this series have been used in cases numbered 1 to 10 for

TABLE 1. Explanation of the Cases of Time Series

Case Number	Data Values Considered For The Case	Remarks
1	1 to 50	Series "A" Data Points. Monthly Discharge in C.M.S.
2	2 to 51	325, 228, 201, 143, 103, 83,
3	3 to 52	
4	4 to 53	151, 300, 251, 640, 511, 208,
5	5 to 54	123, 142, 278, 242, 264, 248,
6	6 to 55	281, 399, 531, 572, 700, 477,
7	7 to 56	409, 313, 202, 111, 149, 128,
8	8 to 57	145, 304, 211, 577, 389, 162,
9	9 to 58	92, 74, 230, 172, 129, 145,
10	10 to 59	152, 306, 213, 579, 391, 164, 94, 76, 828, 795, 834, 827, 885, 1106, 1230, 1467, 898.

the derivation of the correlogram. These are plotted in Fig. 3. There is a large difference in the correlograms indeed. Especially noteworthy is the difference between cases 1 and 10. In case 1, there is a bad year at the beginning. This has been deleted in case 10 and substituted by a good year at the end. The above example is taken from a thesis submitted at the University of Waterloo (Panu, 1978) and it is acknowledged here. The primary reason for the fluctuation in the correlogram is the non-homogeneity caused by the existence of different generating processes in different parts of the data. This also leads to the conclusion that it is most unlikely that the correlogram would tend to a constant "population" value with longer length of data. In effect, a sample from a stationary and ergodic stochastic process cannot form a model for the observed time series.

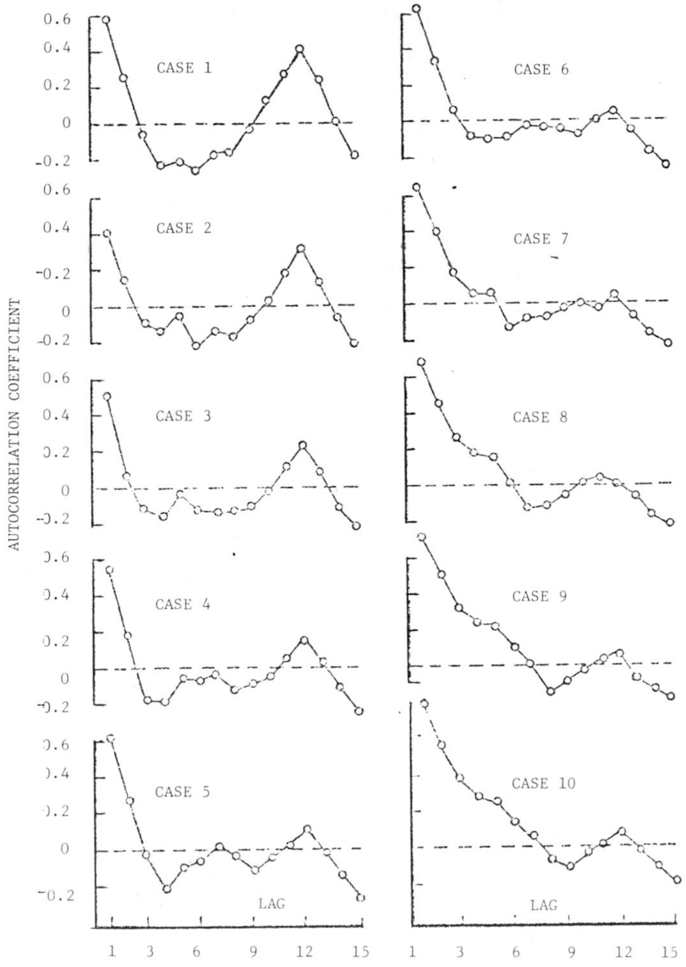

FIG. 3. Correlograms for cases 1 to 10 described in Table 1 from (Panu, 1978)

The presence of persistence in yearly time series, and in streamflow time series in particular, was first brought to the attention of engineers by the pioneering studies of (Hurst, 1951,1956) on long term storage requirements in the Nile Basin.

Through an extensive analysis of geophysical time series (consisting of annual values), including the extremely long time series of the average annual flow on the river Nile, and also of normal independent series generated by various experiments, Hurst derived the relationship

$$R_N/S_N \propto N^H \tag{6}$$

where N is the length of data. In the above, R_N is the adjusted range and it can be expressed for an annual time series $\{x_i, i=1,2,\ldots, N\}$ of length N years, with mean $\bar{x}_N$ and standard deviation S_N, as

$$R_N = \max_{1\leq i\leq N} \left\{ \sum_{i=1}^{N} (x_i - \bar{x}_N) \right\} - \min_{1\leq i\leq N} \left\{ \sum_{i=1}^{N} (x_i - \bar{x}_N) \right\} . \tag{7}$$

The term $R_N/S_N (=\overset{*}{R}_N)$ is the adjusted rescaled range. An estimate of H, denoted by K_H, was defined as

$$K_H = \log(R_N/S_N)/\log(N/2) \quad . \tag{8}$$

Hurst observed that the value of the exponent H in relationship (6) has on the average a value of 0.73 for the geophysical time series, and 0.5 for the normal independent series. In hydrologic literature, the discrepancy in the values of the exponent in hydrological time series and that in all independent series has been called the Hurst phenomenon and the exponent in relationship (6) is now known as the Hurst coefficient. A value of the Hurst coefficient greater than 0.5 is considered to indicate long-term persistence.

Parameter values for the Hurst coefficient derived from historical record is found sensitive to non-homogeneities in data including changes in the probabilistic laws defining the generating mechanism of the data. Klemes (1974) has shown that independent series (persistence of order zero) with fluctuating means exhibited an H-coefficient greater than 0.5. A reference may also be made at this point to Wing (1951) who, while commenting on Hurst's original paper expressed doubt about Hurst's findings and implied that perhaps discontinuities in the record could have caused an H-coefficient greater than 0.5.

Extensive analysis of large assemblage of recorded data by (Hurst, 1951) encompassing many geophysical phenomena, e.g., rainfall, run-off, lake levels, tree rings and mud varves, showed that groups of high and low values tended to occur more frequently in natural events than in purely random events. In addition, Hurst observed with particular reference to annual streamflow time series that groups associated with stretches of floods and droughts occurred without any regularity either in their duration or in the time of occurrence (Fig. 1). This, then, is the fundamental difference between natural streamflow time series and other purely 'man-made' series such as those derived from random processes, autoregressive processes and fractional Gaussian noise sequences.

On Some of the Commonly Used Models for Hydrologic Time Series

Certain stochastic processes have been suggested by hydrologists as models for time series. Specified statistics of the chosen process are adjusted to have numerical values equal to that of equivalent parameters evaluated from the observed series. The term used in hydrologic literature is preservation. Thereby it is meant that all sample functions obtained from the process deliver the same parameter values. Considered significant in this connection are one or more of the following: Mean of the series, variance, correlation coefficients at lag 1 and at higher lags, and Hurst coefficient. Two commonly used models are discussed below.

Mandelbrot and van Ness (1968a) define fractional Brownian motion process (fBm) as:

$$B_H(t) = \frac{1}{\sqrt{H+0.5}} \int_{-\infty}^{t} (t-v)^{H-0.5} dB(v) \ ; \ 0 < H < 1 \qquad (9)$$

where dB(v) is the differential of the Brownian motion process and H is a specified exponent. This process reduces to a Brownian motion process (Wiener-Levi process) for H = 0.5.

The fractional Gaussian noise process (fGn) is defined as the derivative of the above process. The discretised version, the discrete fractional Gaussian noise sequence (dfGn), is defined by Mandelbrot and van Ness (1968a) as follows:

$$x_t^{(f)} = B'_H(t) = [B_H(t) - B_H(t-1)] \quad \text{with}$$

$$B_H(t) = \frac{1}{\sqrt{H+0.5}} \sum_{v=-\infty}^{t} (t-v)^{H-0.5} \, \Delta B(v+1) \tag{10}$$

where t has integer values from $-\infty$ to present. Further $\Delta B(v)$ is the finite difference in the Brownian motion process with $\Delta B(v)=B(v+1+\varepsilon)-B(v+1)$, and $x_t^{(f)}$ is the realized value of the process at time point t.

The fact that dfGn has the asymptotic property that its adjusted range R_N defined in equation (7) is such that $R_N \propto N^H$ is the primary reason for developing this process as a model for geophysical time series. By keeping (preserving) H = the Hurst coefficient derived from the recorded time series, the sample functions of dfGn are made to possess the same Hurst coefficient.

There is absolutely no other similarity whatever between sample functions of dfGn and recorded time series. The proponents of the dfGn models claim that it is capable of providing samples with extremes (highs and lows) that are more severe than that in the historic series (Mandelbrot and Wallis, 1968b). This has not been demonstrated in any convincing manner.

There is no doubt that there is a theoretical beauty in the fractional Brownian motion process itself. The theory and the underlying assumptions of both fBm and dfGn are provided in a series of articles by Mandelbrot and van Ness (1968a) and Mandelbrot and Wallis (1968b, 1969a,b,c). Review articles indicating their relevance to hydrology are also given in Chi, et al (1973), O'Connel (1974) and Lawrence and Kottegoda (1977).

The dfGn involves summation from infinite past to the present. In other words, what happened in the distant past is considered as an influencing factor in the present occurrence. This concept is the antithesis of that of the Markov processes and Markov chains. It should be considered in this connection that Markov chains have not only theoretical elegance but they also have found wide applications in hydraulics and hydrology. These include applications in storage theory (Moran, 1954; Prabhu, 1967; Lloyd, 1967; Klemes, 1981; Soares et al, 1977), and in estimation theory and forecasting (Jazwinski, 1970). For a comprehensive set of articles with applications in hydrology see Chiu (1978) and also Unny (1977).

As a result of the difficulties involved in the infinite summation approximations have been developed for the dfGn. These include the Types I and II approximations of Mandelbrot and Wallis (1969c), the fast fractional Gaussian noise approximation (ffGn) of Mandelbrot, (1971a), and the filtered fGn of Matalas and Wallis (1971b).

The autocovariance function of dfGn is found to tend to zero very slowly. It is primarily the result of non-stationarity in the dfGn. In fact, non-decaying correlograms are considered to indicate non-stationarity in the data according to the procedure adopted in the ARIMA modelling of the time series (Box and Jenkins, 1970). In these instances, the data in the series are successively differenced, if necessary d times, until a decaying correlogram is obtained. Modelling then involves fitting an ARMA model of order (p,q) of the form

$$\Phi(B)(x_t-\overline{x}) = \theta(B)a_t \tag{11}$$

to the differences data. In the above,

$$\Phi(B)=(1-\Phi_1 B-\Phi_2 B^2 \ldots \phi_p B^p) \quad \text{and} \quad \theta(B)=(1-\theta_1 B-\theta_2 B^2 \ldots \theta_q B^q) \tag{12}$$

with B as the backward shift operator and the Φ's and θ's are specified coefficients. Further, a_t is normal independently distributed random variable with zero mean and variance σ_a, and $\overline{x}$ is the mean of the series. The statistics of the ARMA process are functions of the coefficients, as well as the specified values for $\overline{x}$ and σ_a. Assuming ergodicity, the statistics are evaluated from the recorded series, thus enabling the determination of the coefficients in the ARMA (p,q) process. The sample functions preserve the mean, the variance and the correlogram. Apart from this preservation, there is no similarity whatever between sampel functions of the ARMA process and the recorded series.

Since the publication of the book by Box and Jenkins (1970), there has been a flood of articles on the ARMA (p,q) models or, equivalently, on ARIMA (p,d,q) models in the hydrologic context. Seasonal and non-seasonal models and many other infinite variations of these models have been reported. "Best" models have been determined for a given time series using criteria such as the Akaike information criteria (Akaike, 1974). It is surprising that much of the developments in the ARIMA modelling during the last decade has taken place without any concern being expressed as to the objectives of modelling and the application made of these models. The scanned parameters employed in the development of ARIMA models for given streamflow time series are of questionable relevance because of the fact that these natural geophysical time series do not evolve according to simple probabilistic laws. Despite these shortcomings, the ease that accompanies the use of preprogrammed logic has stimulated an acceptance of these models. In many cases model development for a given series has been reduced to the level of a mechanistic procedure carried out on the machine. Inference about streamflow phenomena are being made without any reference to the physical nature of the problem and, in extreme cases, without any consideration other than the data sheet. The only prerequisite to providing an inference has become a capacity to program; in fact much less because the programs are already available on the system.

On the Requirements of Models for Time Series in Hydrology

Models for historically recorded time series in the hydrologic context are required so that such models can be used for extrapolation of data into future times beyond the present. It has become an accepted practice in the last two decades or so to consider these extrapolated data in the design and planning of water resources systems. This is based on the understanding that the past represented by the historical data will never be repeated and that data series employed in water resources applications should be such that they are likely to occur in probabilistic terms in the performance time horizon of the system which lies in the future.

Data extrapolation is required in various formats. Specifically four different formats are discussed below:

a) Generation of unbiased equiprobable samples for use in long term planning and design. The purpose is to provide several and various scenarios on which the efficacy of the proposed design can be tested.

b) Generation of biased equiprobable samples for use in planning and operation of the system in the short term in the immediate future. The purpose now is to obtain different scenarios biased to the present time.

c) Forecasting on a stochastic basis data for several periods ahead. Forecasting involves the determination of the expected value and the probability distribution of the future event on a period by period basis. Such forecasted samples are required as an aid to decision making on the operation of the system for the next few time periods.

d) Deterministic or stochastic forecasting of a single datum on a single step ahead basis. This forecasted value is used in the actual scheduling of the real-time operation of the system.

The general purpose of the extrapolation of data is to provide an understanding at the present time of future events so that certain decisions can be taken based on this understanding. This purpose includes the successful exaggeration of extremes in the historical data as well as generation of extrapolated data with increased information content derived from a priori sources. However, the purpose does not involve prediction with any specified "Degree of Accuracy" the real-time events into the future. Also, then, there is no such thing as a correct model or a "best" model; however, there are appropriate models; and the only justification for the validity of a model is that based on an investigation whether the purpose for which the model has been developed is served by its use. This also leads to the conclusion that, for any given historical data recorded up to the present time, it is necessary to have separate models for extrapolation of data noted in formats a to d above.

Consider the case of data extrapolation, format a, with the purpose of generating equiprobable samples. This is often referred to as data synthesis. As an application the following can be mentioned. The several samples of inflow are routed through a reservoir system and, using an optimization procedure, samples of optimal release policies are determined. This is the implicit stochastic optimization. This procedure results in the development of long term rule curves in system operation. It is clearly seen, then, that a model for data extrapolation should be such that it should be capable of providing samples with extremes of flood and drought sequences, so that the validity of the development of rule curves could be investigated with regard to these samples. A model for format "a" can be considered to generate equiprobable scenarios into the future if the following three conditions are satisfied by the samples derived from the model:

(a) the samples exhibit extreme flood sequences within a range lying on both sides of that found in the historical sample;

(b) the samples provide extreme drought sequences within a range lying on both sides of that found in the historical sample;

(c) the samples provide distribution of data in the state space similar to that embedded in the historical data.

Satisfaction of the above three conditions should be the criteria in justifying a model. These conditions are specific and quantifiable.

Consider the format "d" connected with the forecasting of a single datum on a step ahead basis. This is required in real-time operation which is the scheduling of the system operation for the next time period. At the completion of the time period when the actual measured value is available, it is used to update the system states and the so updated states form the initial conditions for decisions on real-time operation for the succeeding time period based on a new forecasted value.

Again, for emphasis, it should be stated that a comparison of the forecasted value on a step ahead basis with the value occurring in real-time is excluded as a purpose. The following can form criteria for validating the model. Perform physical operation of the system in real-time. Simulation on the computer of the real-time physical operation is an alternative procedure. After having completed the operation for a reasonable horizon of time, evaluate the results. Justification of the model can now be based on any criteria that is an appropriate function of these results. For example, the following questions are valid on a post operation basis. Was the operation, so far carried out, optimal? Was there any failure (withdrawal below targeted or required level) involved in the operation? Could the operation have been improved if a different model had been used for step ahead forecasting?

Some Further Thoughts on Modelling

The severe shortcomings of ARIMA models and dfGn models for data synthesis have been noted previously. Primarily these models neglect the consideration of the distinguishing characteristics of well defined groups in the data record. The existence of groups as postulated by Hurst is evident from Fig. 1. Even a visual examination will indicate the extreme interrelationship between succeeding datum values in each group. There is, then, a need for investigations pertaining to these groups so that the intrarelationship between identifiable groups as well as the interrelationship within each group could be properly considered in time series modelling.

For example, consider the data record in Fig. 2. It is obvious that the open circles represent data that have less variation or perturbation from one another, while the solid circles show data that have a moderately large oscillatory behaviour, strong interdependence and a small negative correlation between x_1 and x_2, perhpas, with a lag. Clearly, the open and solid circles, as seen from the data, represent two easily discernible random behaviour types. In many cases, in most hydrologic cases, this understanding can be enhanced by a priori information concerning the data set. Is there any procedure, then, that would enable us to divide the data into several separate classes?

Models of time series should be based on an analysis of data and its synthesis. Analysis is the process of determining the fundamental components, groups, etc., embodied in the data by separation and isolation. Its purpose is for close scrutiny and examination of the constituent components, as well as for accurate resolution of an overall structure or the nature of the whole or parts of the data set.

Analysis of empirical information or data from the physical world is the result of mapping of this information from one form to another. The mapping should be based on training set of data and supervised learning procedures. The meaning of this latter term often used in connection with pattern recognition and pattern analysis is quite obvious. It involves the inclusion at the analysis stage of any experience based understanding of the analyst, as well as his knowledge of the causative forces that create the progression of data in time (Unny et al, 1981).

Synthesis represents the action of combining various parts or components having different characteristics into one coherent, consistent whole. It is the result of remapping in the original format of the data configuration recognized in the learning phase. It is quite obvious that analysis and synthesis interact with each other. Breaking up into components is not possible without specifying the manner in which the components could be put together.

The two step mapping procedure leading to analysis and synthesis can be repreated a number of times with continued improvements in the learning procedure, provided a basis exists for such improvements. This basis is the understanding built on the interaction of the analyst with the physical world.

Recently a series of articles have appeared that employ concepts of pattern recognition for data synthesis (Panu and Unny, 1980a,b,c and d; Unny et al, 1981). A pattern is a shape representation of sections of the physical world, for example, sections of streamflow time wave form corresponding to geophysical seasons. Patterns result from innumerable causes and a study of patterns is a study of all these causes. A chronological reflection of the causative mechanism is contained in a series of such patterns. What is attempted in the articles noted above is the development of a technique that provides flexibility in data processing, that accepts input from the analyst and that is adaptive to the requirements satisfying various objectives of modelling. The approach occupies an intermediate position between a purely subjective experience based formulation and a totally machine derived alternative. The motivation has been to avoid irrelevant results arising out of the use of preconceived models and preprogrammed logic that impose an external structure upon the otherwise unique behaviour of the time series.

CONCLUDING REMARKS

The main point of discussion contained in this paper can be summarized as follows: The last decade has seen various attempts at refining some of the previously proposed models for time series synthesis. Much of the developments in this regard has taken place without due regard given to the unique nature of the physical problem and without consideration of the origin of data (stock market data versus streamflow data). Perhaps it is the appropriate time to take a fresh look at time series modelling procedures in the hydrologic context.

REFERENCES

Akaike, H., 1974. A New Look at the Statistical Model Identification, IEEE Trans, Automatic Control, 19(6): 716-723.

Box, G.E.P. and Jenkins, G.M., 1973. Time Series Analysis: Forecasting and Control, Holden-Day, San Francisco, California.

Chi, M., Neal, E. and Young, G.K., 1973. Practical Application of Fractional Brownian Motion and Noise to Synthetic Hydrology, Water Resour. Res., 9: 1569-1582.

Chiu, C.L., (Editor), 1978. Applications of Kalman Filter to Hydrology, Hydraulic and Water Resources, Proc. A.G.U., Chapman Conference, University of Pittsburgh.

Hurst, H.E., 1951. Long-term Storage Capacity of Reservoirs, Trans. A.S.C.E., 116: 770-808.

Hurst, H.E., 1956. Methods of Using Long-term Storage in Reservoirs, Proc. Instn. Civil Engrs., 1: 519-543.

Jazwinski, A.H., 1970. Stochastic Processes and Filtering Theory, Academic Press, New York.

Klemes, V., 1974. The Hurst Phenomena — A Puzzle? Water Resour. Res., 10(4): 675-688.

Klemes, V., 1981. Applied Stochastic Theory of Storage in Evolution, Advances in Hydrosciences, 12: 79-141.

Lawrance, A.J. and Kottegoda, N.T., 1977. Stochastic Modelling of Riverflow Time Series, Jour. Royal Statist. Soc. Series A, 140(1): 1-47.

Lloyd, E.H., 1967. Stochastic Reservoir Theory, Advances in Hydrosciences, 4: 281-339.

Mandelbrot, B.B., 1971. A Fast Fractional Gaussian Noise Generator, Water Resour. Res. 76(3): 543-553.

Mandelbrot, B.B. and vanNess, J.W., 1968a. Fractional Brownian Motions, Fractional Noises and Applications, Soc. Ind. Appl. Math. Rev., 10(4): 422-437.

Mandelbrot, B.B. and Wallis, J.R., 1968b. Noah, Joseph and Operational Hydrology, Water Resour. Res., 4(5): 909-918.

Mandelbrot, B.B. and Wallis, J.R., 1969a. Computer Experiments with Fractional Gaussian Noises. Part 1 - Averages and Variances; Part 2 - Rescaled Ranges and Spectra; and Part 3 - Mathematical Appendix, Water Resour. Res., 5(1): 228-267.

Mandelbrot, B.B. and Wallis, J.R., 1969b. Some Long Run Properties of Geophysical Records, Water Resour. Res. 5(2): 321-340.

Mandelbrot, B.B. and Willis, J.R., 1969c. Robustness of the Rescaled Range R/S in the Measurement of Non-cyclic Long-run Statistical Dependence, Water Resour. Res., 5(5): 967-988.

Matalas, N.C. and Wallis, J.R., 1971. Statistical Properties of Multi-variate Fractional Gaussian Noise Processes, Water Resources Research, 7(6): 1460-1668.

MacInnes, C.D., 1981. Multiple Time Series Data Extrapolation in Water Resources Engineering Applications using Pattern Recognition Techniques, Doctoral Dissertation, Department of Civil Engineering, University of Waterloo, Ontario, Canada.

Moran, P.A.P., 1954. A Probability Theory of Dams and Storage Systems, Australian Journal of Applied Science, 5: 116-124.

O'Connell, P.E., 1974. Stochastic Modelling of Long-term Persistence in Streamflow Sequences," Ph.D. Thesis, Univ. of London, England.

Panu, U.S., 1978. Stochastic Synthesis of Monthly Streamflows Based on Pattern Recognition, Doctoral Dissertation, Department of Civil Engineering, University of Waterloo, Ontario.

Panu, U.S. and Unny, T.E., 1980a. Extension and Application of Feature Prediction Model for Synthesis of Hydrologic Records, Water Resources Research, 16(1): 77-96.

Panu, U.S. and Unny, T.E., 1980b. Stochastic Synthesis of Hydrologic Data Based on Concepts of Pattern Recognition I. General Methodology of the Approach, Journal of Hydrology, 46: 5-34.

Panu, U.S. and Unny, T.E., 1980c. Stochastic Synthesis of Hydrologic Data Based on Concepts of Pattern Recognition II. Application of Natural Watersheds, Journal of Hydrology, 46: 197-217.

Panu, U.S. and Unny, T.E., 1980d. Stochastic Synthesis of Hydrologic Data Based on Concepts of Pattern Recognition III. Performance Evaluation of the Methodology, Journal of Hydrology, 46: 219-237.

Prabhu, N.U., 1964. Time Dependent Results in Storage Theory. J. Applied Probability, Vol 1: 1-46.

Soares, E.F., Unny, T.E. and Lennox, W.C., 1977. On a Stochastic Sediment Storage Model for Reservoirs, in Stochastic Processes in Water Resources Engineering. (Eds.) L. Gottschalk, G. Lindh and L. de Marie, Water Resources Publications, Fort Collins, Colorado, 141-166.

Unny, T.E., Panu, U.S., MacInnes, C.D. and Wong, A.K.C., 19 . Pattern Analysis and Synthesis of Time Dependent Hydrologic Data. Advances in Hydrosciences, Vol 12: 195-295.

Unny, T.E. 1977. Transient and non-stationary Random Processes, in Stochastic Processes in Water Resources Engineering, (Eds.) L. Gottschalk, G. Lindh and L. de Marie, Water Resources Publications, Fort Collins, Colorado.

Wing, S.P., 1951. Discussion on Long-term Storage Capacity of Reserviors, by Hurst, H.E., Trans. A.S.C.E., 116: 807-808.

A DYNAMIC-STOCHASTIC APPROACH FOR MODELLING ADVECTION-DISPERSION PROCESSES IN OPEN CHANNELS

W.P. BUDGELL

Bayfield Laboratory for Marine Science and Surveys, Department of Fisheries and Oceans, Canada Centre for Inland Waters, Burlington, Ontario, Canada

ABSTRACT

A combined stochastic-deterministic model has been developed to describe the temporal and spatial distribution of conservative substances in open channel flows. The model consists of a finite difference approximation to the one-dimensional advection-dispersion equation embedded within a stochastic filter. The time and measurement updates of the estimated concentrations and their covariance are carried out through the use of a factored form of the covariance matrix. The resulting filtering algorithm is more computationally stable than the standard Kalman filter approach. The dynamic-stochastic model is shown to perform well when it is applied to simulated observations of salinity in an Arctic estuary. It is shown that this type of modelling approach can be used as a tool in planning field experiments.

1. INTRODUCTION

The time and space distribution of a conservative substance in rivers and estuaries can often be described using the one-dimensional time-dependent advection-dispersion equation (Harleman, 1971; Hann and Young, 1972; Hinwood and Wallace, 1975). If the cross-sectional area, velocity and dispersion coefficient are constants, an analytical solution to the equation can be obtained (Harleman, 1971). For realistic channel geometry and flow conditions, it is necessary to make use of numerical techniques (Roache, 1972). Although numerical advection-dispersion models have often provided reliable results, it should be noted that these models are crude approximations to the actual

Reprinted from *Time Series Methods in Hydrosciences*, by A.H. El-Shaarawi and S.R. Esterby (Editors)

transport processes taking place in open channels (Fischer, 1973, 1976). Errors in the specified cross-sectional area, flow field, dispersion coefficients and boundary conditions and the numerical discretization of the original partial differential equation introduce uncertainty or noise into the modelling process. This modelling error, or system noise, is propagated through time and space by the deterministic numerical model. Because of the dynamic nature of the problem, the variance of the errors in predicted concentrations will increase exponentially with time for the case of constant coefficients in the original equation.

If time series observations of concentration are available from the river or estuary under consideration, the modelling error at each time step can be estimated and the model results can be corrected. By updating the computed concentrations using observations, less error is propagated through the model. A major difficulty associated with this procedure is that the observations will also contain a certain degree of error or measurement noise. Thus, the actual corrections to be applied to the computed concentrations will not be the difference between the observed and computed values, but rather some portion of that difference. The magnitude of the correction will depend upon the reliability of the observations relative to that of the model. A means of computing the optimal corrections to be applied to the computed values at each time step is through the use of the Kalman filter (Kalman, 1960; Kalman and Bucy, 1961). Kalman filter theory has been applied to satellite tracking (Jazwinski, 1970), air pollution monitoring (Desalu, Gould and Schweppe, 1974; Bankoff and Hanzevak, 1975; Koda and Seinfeld, 1978; Fronza, Spirito and Tonielli, 1979), water resources problems (Chiu, 1978) and the estimation of water levels and velocities in tidal estuaries (Budgell, 1981).

DeGuida, Connor and Pearce (1977) have used a Kalman filter to combine observations of concentration with a finite element numerical model of the time-dependent two-dimensional horizontal distribution of estuarine pollution. The model includes advection, dispersion and source-sink terms. However, Koda and Seinfeld (1978) have noted that such a straightforward application of the Kalman and Bucy (1961) filtering algorithm to large scale distributed parameter systems (systems with both time and space dependence) can lead to filter divergence and computational instability. These computational problems are attributable to the covariance matrix associated with the estimated concentrations becoming non-positive definite.

In this paper, stability problems are avoided by implementing a square root form of the Kalman filter in the estimation of the cross-sectionally averaged concentration of a conservative substance in rivers and estuaries. The filter is constructed around an implicit finite difference representation of the time-dependent one-dimensional advection-dispersion equation. A square root formulation for the filter ensures that the covariance matrix remains positive definite and reduces the computational burden from that imposed by the conventional Kalman filter algorithm.

2. THE DETERMINISTIC MODEL

The numerical model combined with the stochastic filter is referred to as a dynamic-stochastic model. The deterministic component of the dynamic-stochastic model is governed by the one-dimensional advection-dispersion equation describing the distribution of a conservative constituent in open channels (Harleman, 1971):

$$\frac{\partial (Ac)}{\partial t} + \frac{\partial (Qc)}{\partial x} - \frac{\partial}{\partial x} \left\{ AE \frac{\partial c}{\partial x} \right\} = 0 \qquad (1)$$

Boundary conditions must be specified at the upper and lower ends of the channel. The upstream boundary conditions are specified as follows:

$$c(0,t) = c_0(t) \quad \text{for } Q(0,t) \geq 0$$

$$\left.\frac{\partial^2 c}{\partial x^2}\right|_{x=0} = 0 \qquad \text{for } Q(0,t) < 0 \tag{2}$$

and the downstream boundary conditions are as follows:

$$c(L,t) = c_L(t) \quad \text{for } Q(L,t) \leq 0$$

$$\left.\frac{\partial^2 c}{\partial x^2}\right|_{x=L} = 0 \qquad \text{for } Q(L,t) > 0 \tag{3}$$

where $c(0,t)$ and $c(L,t)$ are the concentrations at the upstream and downstream ends, respectively, of an open channel and $c_0(t)$ and $c_L(t)$ are specified concentrations at the upstream and downstream ends. These boundary conditions are described in greater detail by Thatcher and Harleman (1972).

Since A, Q and E can be time- and space-dependent parameters, it is necessary to solve (1) to (3) using numerical approximations. In this study the Stone and Brian (1963) six-point finite difference scheme has been used to approximate the time derivative and advective flux terms in (1). The dispersion term is modelled using the Crank-Nicholson (1947) scheme. The resulting finite difference representation possesses second order accuracy in space and time, produces no numerical dispersion and is stable for cell Peclet numbers less than 20 (Lam, 1977). There is no stability restriction on the time step.

When the finite difference equations are applied to the N-2 interior grid points of the discretized open channel and the finite difference approximations of boundary condition equations (2) and (3) are imposed, the result is a set of N linear equations in N unknowns. In matrix form this may be expressed as:

$$\underline{A}(n,n+1)\ \underline{c}(n+1) = \underline{B}(n,n+1)\ \underline{c}(n) + \underline{G}(n,n+1)\ \underline{u}(n+1) \tag{4}$$

If flow is into a boundary from the interior of the computational region, the specified concentration for that boundary condition is not used in the computation of $\underline{c}(n+1)$. This is accomplished by setting the corresponding column of $\underline{G}(n,n+1)$ to zero.

Since $\underline{A}(n,n+1)$ is a tri-diagonal matrix and the right-hand side of (4) constitutes a known vector if the initial conditions are supplied, $\underline{c}(n+1)$ can be obtained in an efficient manner using the well known Thomas algorithm (Roache, 1972, p.349). Equation (4) can be expressed in an alternate manner as:

$$\underline{c}(n+1) = \underline{\Phi}(n,n+1)\ \underline{c}(n) + \underline{\Omega}(n,n+1)\ \underline{u}(n+1) \tag{5}$$

Although the matrices $\underline{\Phi}(n,n+1)$ and $\underline{\Omega}(n,n+1)$ are neither computed nor stored, they serve to represent the sequence of linear operations performed to obtain $\underline{c}(n+1)$ given $\underline{c}(n)$ and $\underline{u}(n+1)$.

3. THE STOCHASTIC FILTER

The process described by (5) is purely deterministic. Given the correct values of $\underline{c}(n)$ and $\underline{u}(n+1)$, $\underline{c}(n+1)$ will be known with certainty. Unfortunately, the distribution of a conservative solute in dynamic open channel flows is far from perfectly described by (1) to (4). Errors associated with the cross-sectional integration of the original three-dimensional mass transport equation, the numerical approximation of a continuous partial differential equation, the specification of the time- and space-dependent parameters Q and E, and the specification of boundary conditions all result in considerable uncertainty being associated with the computed concentration vector, $\underline{c}(n+1)$.

This uncertainty may be considered to result from noise, or error, caused by imperfect modelling of the process under consideration. If the effects of system noise, or modelling error, are included, the system model may be described by the following equation:

$$\underline{c}(n+1) = \underline{\Phi}(n,n+1)\ \underline{c}(n) + \underline{\Omega}(n,n+1)\ \underline{u}(n+1) + \underline{w}(n+1) \tag{6}$$

where $\underline{w}(n+1)$ is a vector of length N containing system noise, or sources of uncertainty in the modelling process. Thus, each grid point of the model has noise associated with it. The system noise is assumed to be Gaussian and uncorrelated with zero mean and covariance $\underline{Q}(n)$.

Taking the expected value of (6) yields the time update equation for concentration:

$$\hat{\underline{c}}(n+1) = \underline{\Phi}(n,n+1)\ \tilde{\underline{c}}(n) + \underline{\Omega}(n,n+1)\ \underline{u}(n+1) \tag{7}$$

where $\hat{\underline{c}}(n+1)$ is the one step ahead prediction, or the expected value of $\underline{c}(n+1)$ conditioned on information up to time $n\Delta t$, and $\tilde{\underline{c}}(n)$ is the filtered estimate, or the expected value of $\underline{c}(n)$ conditioned on information up to time $n\Delta t$.

Subtracting (7) from (6), squaring and taking the expected value gives the covariance time update equation:

$$\hat{\underline{P}}(n+1) = \underline{\Phi}(n,n+1)\ \tilde{\underline{P}}(n)\ \underline{\Phi}^T(n,n+1) + \underline{Q}(n+1) \tag{8}$$

where $\hat{\underline{P}}(n+1)$ and $\tilde{\underline{P}}(n)$ are the covariances associated with $\hat{\underline{c}}(n+1)$ and $\tilde{\underline{c}}(n)$, respectively.

If observations are available, they can be used to improve the accuracy of the estimates of $\underline{c}(n)$. Measurements have error associated with them. If it can be assumed that measurement error $\underline{\zeta}(n)$ is additive noise then the observations $\underline{z}(n)$ are related to the state, or concentration, vector in the following manner:

$$\underline{z}(n) = \underline{H}\ \underline{c}(n) + \underline{\zeta}(n) \tag{9}$$

The measurement noise is assumed to be uncorrelated and Gaussian with zero mean and covariance $\underline{R}(n)$.

If observations are included in the estimation process, the following measurement updates, or filter estimates, can be obtained for the state vector and covariance matrix (Jazwinski, 1970);

$$\underline{\tilde{c}}(n+1) = \underline{\hat{c}}(n+1) + \underline{K}(n+1) [\underline{z}(n+1) - \underline{H}\, \underline{\hat{c}}(n+1)] \quad (10)$$

$$\underline{\tilde{P}}(n+1) = \underline{\hat{P}}(n+1) - \underline{K}(n+1)\, \underline{H}\, \underline{\hat{P}}(n+1) \quad (11)$$

where

$$\underline{K}(n+1) = \underline{\hat{P}}(n+1)\, \underline{H}^T [\underline{H}\, \underline{\hat{P}}(n+1)\, \underline{H}^T + \underline{R}(n+1)]^{-1} \quad (12)$$

is the Kalman gain matrix. Prediction for $t = \ell\Delta t$, $\ell > n$ ($\underline{\hat{c}}(\ell)$, $\underline{\hat{P}}(\ell)$) is accomplished using (7) and (8) with initial condition ($\underline{\tilde{c}}(n)$, $\underline{\tilde{P}}(n)$). Equations (7) through (12) constitute the discrete form of the Kalman-Bucy filter (Kalman, 1960; Kalman and Bucy, 1961).

From (11) it can be seen that the measurement update of the covariance matrix, $\underline{\tilde{P}}(n+1)$, is computed by subtracting the positive definite matrix $\underline{K}(n+1)\underline{H}\, \underline{\hat{P}}(n+1)$ from the positive definite matrix $\underline{\hat{P}}(n+1)$. Because of round-off errors, the resulting matrix may become non-positive definite or weakly positive definite ultimately causing severe computational instability when the matrix inverse in (12) is computed (Koda and Seinfeld, 1978).

One means of avoiding these difficulties is the application of square root filtering theory. Desalu, Gould, and Schweppe (1974),Koda and Seinfeld (1978), and Budgell (1981) have found that applying square root filtering to distributed parameter state estimation problems results in stable algorithms.

The covariance square root filter used here is an algorithm based upon triangular square root factorization of the estimation error covariance matrix. The filter covariance matrix may be factored as follows:

$$\underline{\tilde{P}}(n) = \underline{\tilde{U}}(n)\, \underline{\tilde{D}}(n)\, \underline{\tilde{U}}^T(n) \quad (13)$$

where $\underline{\tilde{U}}(n)$ is a unit upper triangular matrix

and $\underline{\tilde{D}}(n)$ is a diagonal matrix.

Similarly:

$$\underline{\hat{P}}(n) = \underline{\hat{U}}(n)\ \underline{\hat{D}}(n)\ \underline{\hat{U}}^T(n) \tag{14}$$

The time update equations can be obtained from (7) and (8):

$$\underline{\hat{c}}(n+1) = \underline{\Phi}(n,n+1)\ \underline{\tilde{c}}(n) + \underline{\Omega}(n,n+1)\ \underline{u}(n+1) \tag{15}$$

$$\underline{U}(n+1) = \underline{\Phi}(n,n+1)\ \underline{\tilde{U}}(n) \tag{16}$$

$$\underline{\hat{U}}(n+1)\ \underline{\hat{D}}(n+1)\ \underline{\hat{U}}^T(n+1) = \underline{U}(n+1)\ \underline{\tilde{D}}(n)\cdot \underline{U}^T(n+1) + \underline{Q}(n+1) \tag{17}$$

The matrix $\underline{U}(n+1)$ is computed in (16) by applying the tri-diagonal algorithm required to solve (4) to the factor matrix $\underline{\tilde{U}}(n)$. It should be noted that $\underline{\Phi}(n,n+1)$ merely represents the sequence of operations carried out by the equation solver. The matrix $\underline{\Phi}(n,n+1)$ is never actually computed.

The factor matrices $\underline{\hat{U}}(n+1)$ and $\underline{\hat{D}}(n+1)$ are computed from (17) by applying a modified weighted Gram-Schmidt (MWGS) orthogonalization procedure (Bierman, 1977) which is reputed to have accuracy comparable to the Householder algorithm. Unlike the classical procedure, the modified algorithm produces almost orthogonal vectors and pivoting is unnecessary.

The state and covariance measurement updates are accomplished through the equations:

$$\underline{\tilde{c}}(n+1) = \underline{\hat{c}}(n+1) + \sum_{i=1}^{m} \underline{k}_i(n+1)[z_i(n+1) - \underline{h}_i^T\underline{\hat{c}}(n+1)] \tag{18}$$

$$\underline{\tilde{U}}(n+1)\ \underline{\tilde{D}}(n+1)\ \underline{\tilde{U}}^T(n+1) = \underline{\hat{U}}(n+1)\ \underline{\hat{D}}(n+1)\ \underline{\hat{U}}^T(n+1)$$

$$- \sum_{i=1}^{m} [\underline{k}_i(n+1)\ \underline{h}_i\ \underline{\hat{U}}(n+1)\ \underline{\hat{D}}(n+1)\ \underline{\hat{U}}^T(n+1)] \tag{19}$$

where $\underline{h}_i$ is a column vector specifying the location of the i-th measurement sensor such that $\underline{H}^T = [\underline{h}_1,\ldots,\underline{h}_m]$;
$\underline{k}_i(n+1)$ is a column vector specifying the gain associated with the

i-th measurement at time (n+1)Δt such that

$$\underline{K}(n+1) = [\underline{k}_1(n+1),\ldots,\underline{k}_m(n+1)]$$

and $z_i(n+1)$ is the measurement from the i-th sensor at time (n+1)Δt.

The gain vectors $\underline{k}_i(n)$ are obtained one at a time using the $\underline{UDU}^T$ estimate-covariance updating algorithm of Bierman (1977). This updating algorithm is numerically stable since numerical differencing is avoided in the computation of $\underline{D}$. Using the factored form of the covariance matrices for the time and measurement updates results in a computationally stable algorithm that requires fewer arithmetic operations than the conventional Kalman filter (Bierman, 1977).

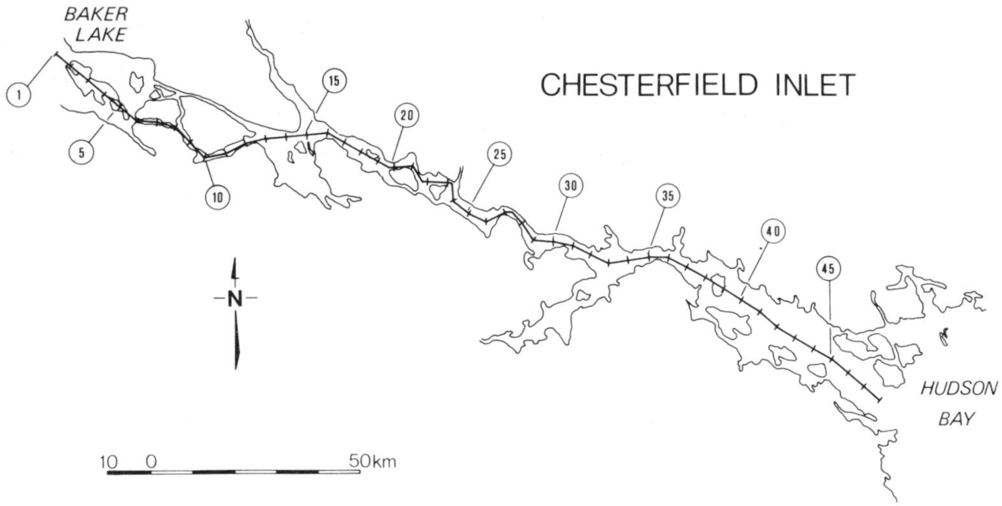

Fig. 1. Model schematization of Chesterfield Inlet.

4. A TEST SIMULATION

The characteristics of the combined dynamic-stochastic model as represented by (12) to (19) are illustrated using simulated observations of salinity concentration from Chesterfield Inlet, situated on the Northwest coast of Hudson Bay in the Canadian Arctic. As shown in Figure 1, the estuary is discretized into 48 grid points with a Δx of 5000 m. The channel depth and topwidth were obtained from a study by Budgell (1976). Dispersion coefficients were obtained from Roff et al (1980) and varied from 500 to 5000 m^2/sec.

The time- and space-dependent flow rates and water levels were obtained using a numerical tidal model developed previously (Budgell, 1976). The boundary conditions consisted of no flow through the upper end of Baker Lake and predicted tidal water surface elevations at the mouth of the estuary. The tidal water level predictions were obtained for the month of September, 1978 using tidal harmonic constituents (Godin, 1972). The predominant tidal constituent in the water level and channel flow time series is the lunar semi-diurnal with a period of 12.42 hours. Typical amplitudes of the cross-sectionally averaged velocity are 0.3 to 1.0 m/sec.

The simulated test data were generated using (6) and (9) on a time step of 20 minutes. The true state vector of salinity concentrations at time $(n+1)\Delta t$ was obtained from (6). Uncorrelated Gaussian noise $\underline{w}(n+1)$ was added to the concentrations computed by the numerical model in (4) to produce the true state vector $\underline{c}(n+1)$. The variance of the system noise was specified as being 10 percent of the variance attributable to tidal fluctuations in salinity as computed with the deterministic numerical model (4). The system noise variance varied from zero to 0.5 ppt^2 throughout the estuary.

In essence, then, salinities are computed by running the deterministic model for one time step. A small quantity of Gaussian uncorrelated system noise is added to these computed values to produce the true salinity concentrations. These salinities then constitute the initial condition for the next time level. The numerical model is then run for another time step. As before, a vector of system noise values is added to the computed concentrations to create true salinities at the next time level. This process is repeated to obtain state (salinity) vectors for the desired length of record. In this manner, system noise added to the numerical model at each time step propagates through space and time altering future salinity values throughout the estuary.

The boundary conditions used in the creation of the data are a salinity of zero ppt at the upstream end at ebb tide and a salinity of 32 ppt plus a random noise component at the downstream end during flood tide. Otherwise, a condition of $\partial^2 c/\partial^2 x = 0$ was specified at the boundaries. The random perturbation applied to the downstream (ocean) end of the estuary at flood tide has a variance of 0.04 ppt^2. This perturbation simulates error in the specification of the boundary condition.

In order to obtain initial conditions, the deterministic numerical model was used to compute salinities for a 15 day period. The salinities averaged over the final tidal cycle were used as the initial values in the creation of the test data.

Measurements were simulated by adding uncorrelated Gaussian noise with a variance of 0.04 ppt^2 to the "true" salinity values at specified measurement sensor locations. The measurement noise, being additive, was not propagated through the numerical model and does not have any effect on the true salinities. Measurement locations were spread at equal intervals throughout the estuary.

When the dynamic-stochastic model was applied to the data set, it was found to perform well. Shown in Figure 2 are typical results from one of 5 measurement locations (m=5). It can be seen that the estimated concentration (filter estimate) closely tracks the actual concentration (true state). However, when the deterministic numerical model (numerical model only) is applied to the same situation, the agreement is considerably worse. The lack of a measurement updating capability in the numerical model means this model cannot track the true state. The pronounced sinusoidal correlation structure in the errors of the numerical model values are attributable to the strong

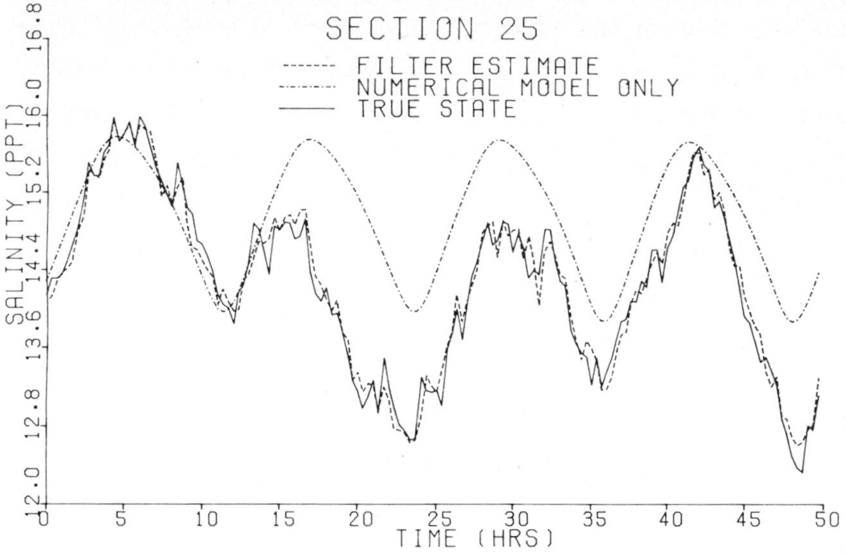

Fig. 2 Stochastic-dynamic model and deterministic numerical model estimates vs. the true state at a measurement location.

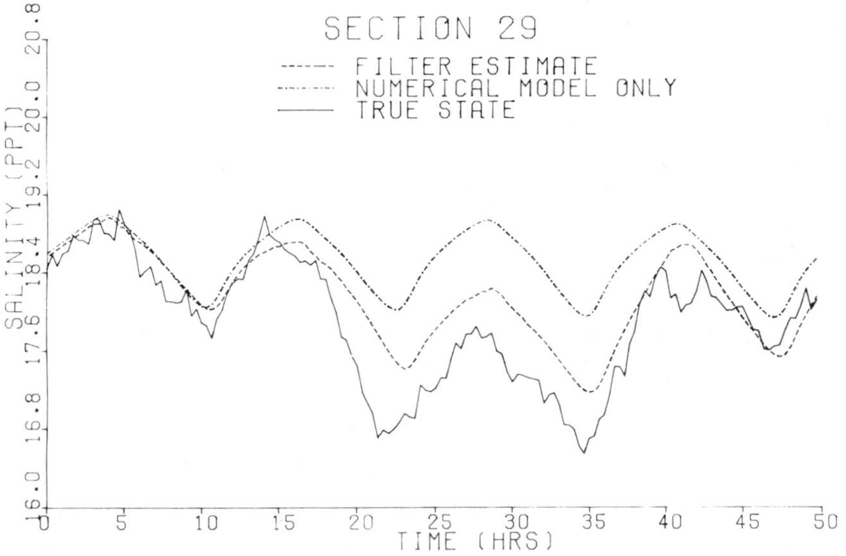

Fig. 3 Stochastic-dynamic model and deterministic numerical model estimates vs. the true state between measurement locations.

tidal forcing in the advective flux term $\partial(Qc)/\partial x$. Although the system noise input at each time step is uncorrelated, the noise is propagated through space and time, conditioned by the advection and dispersion terms in the mass balance equation. The advection term will tend to produce a harmonic correlation structure, whereas the dispersion term will tend to filter out high frequency and high wave number contributions to the correlation structure.

The accuracy of the dynamic-stochastic model estimates deteriorates with increasing distance from measurement locations. Data from grid point 29, situated half way between two sensors, is shown in Figure 3. The filter estimates do not follow the true state nearly as closely as at the measurement location as shown in Figure 2. Furthermore, the errors have a strong sinusoidal component. This is because system noise generated from 3 grid points on either side of the location in question is being advected and dispersed through it. The performance of the dynamic-stochastic model is still superior to that of the deterministic model because of the measurement updates carried out at grid points 25 and 33. As one moves farther from the location of a measurement update the filter performance degrades because of the cumulative effect of random system noise input.

In order to determine the level of uncertainty associated with the state estimates, it is necessary to examine the covariance of the state vector. Shown in Figure 4 is the longitudinal distribution of the mean square error (MSE) of the state estimates and the filter variance as estimated by the dynamic-stochastic model for a test case in which "observations" are available from a single sensor situated at the midpoint of the estuary. Both the mean square error and estimated variances have been averaged over 4 tidal cycles (150 time steps). The distance is relative to the location of the grid point number 1 in Figure 1. It can be seen from Figure 4 that at most locations the mean square error is larger than the variance estimated by the dynamic-stochastic model. The largest values for the estimated variance and

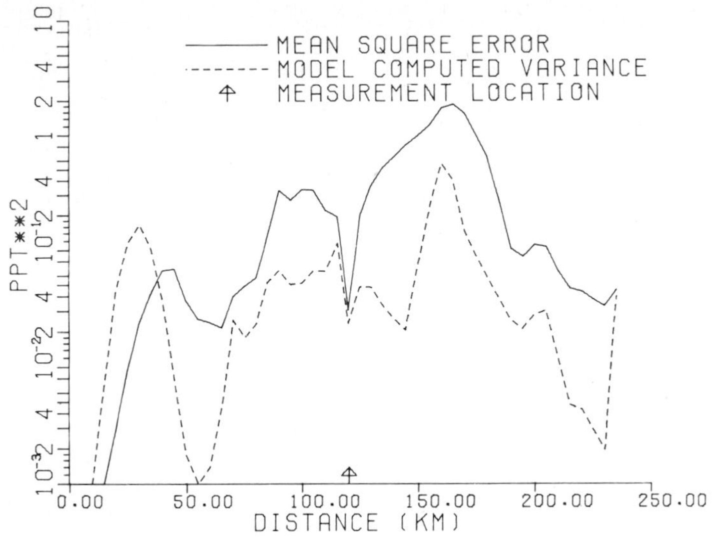

Fig. 4 Longitudinal distribution of the mean square error and model-computed variance with one measurement sensor.

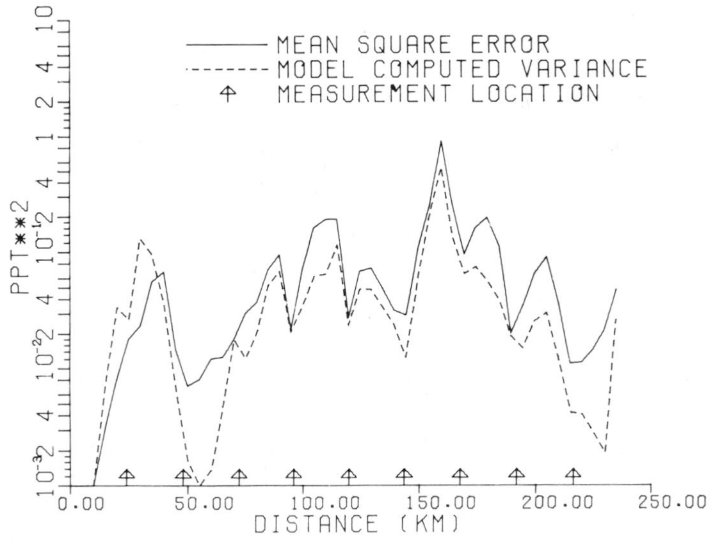

Fig. 5 Longitudinal distribution of the mean square error and model-computed variance with 9 measurement sensors.

MSE occur at constrictions in the channel while the minimum values occur at embayments and at the single measurement location. Constrictions tend to amplify the advective flux and thus the noise field. This tendency was reinforced in the creation of the data set by the insertion of system noise into the process such that the system noise variance is proportional to the variance in concentration attributable to tidal fluctuations. The reverse process occurs at embayments. At the measurement location, information is available from an observed time series to improve the estimate.

The dynamic-stochastic model estimates the probability distribution of the concentration field. Thus, if the probability distribution is Gaussian, not only the concentration but its variance must be estimated. It can be seen from Figure 5 that when data are available from 9 sensors distributed throughout the estuary, the MSE and variance estimated by the dynamic-stochastic model are in much closer agreement throughout the estuary than in the one-sensor case of Figure 4. Thus, the estimated variance more closely approximates the MSE over distance as the number of sensors is increased.

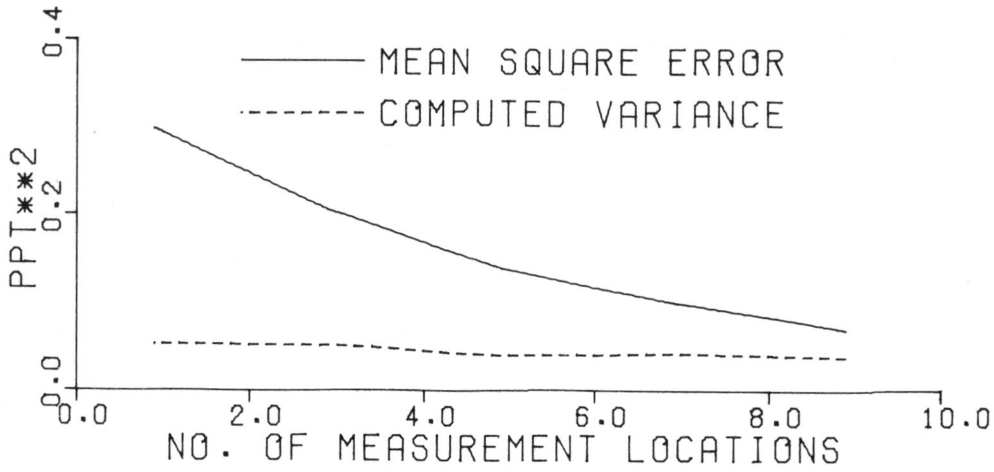

Figure 6. Mean square error and model-computed variance averaged over the estuary as a function of the number of measurement sensors.

The overall effect of spatial sampling density on the uncertainty of the state estimates is summarized in Figure 6. In this plot the MSE and estimated variance have been averaged over all the grid points in space as well as over 4 tidal cycles in time. While the model estimated variance is relatively invariant with the number of measurement sensors, the MSE decreases approximately as 1/m. The MSE approaches the estimated variance asymptotically, but for the MSE and estimated variance to be approximately equal, there must be more than 9 measurement locations ($m > 9$).

If there is prior knowledge of the system and measurement noise characteristics, a simulation such as that carried out in this study can provide a useful tool for experimental design. For example, Figures 3 and 4 suggest that placing sensors at grid points 9, 20 and 37 would reduce the uncertainty of concentration estimates considerably since these grid points are situated in regions of maximum variance. Furthermore, from a plot such as Figure 6, the number of sensors required to achieve a given level of accuracy can be obtained.

5. CONCLUSIONS

A model has been developed that combines a numerical solution to the one-dimensional advection-dispersion equation with a stochastic filter. The major portion of the variation of concentration over time and space in open channels can be described by the deterministic numerical model. By constructing a stochastic filter around the numerical model it is possible to compensate for errors incurred during the modelling process. A numerical model alone cannot be constrained to track the true system through the use of observed data. However, time series models such as autoregressive moving average series (Box and Jenkins, 1976) or simple Kalman filters (e.g., Chiu and Isu, 1977), into which observations are directly incorporated, are black box approaches that are unrelated to physics. The dynamic-stochastic model proposed in this paper retains the best features of the conventional deterministic and stochastic approaches.

The model equations are posed in state space form. The usual approach of applying a Kalman filter algorithm to obtain estimates of the concentration vector and covariance matrix could lead to filter instability due to covariance matrices becoming nonpositive definite. To circumvent this problem, the covariance matrix is factored into unit upper triangular and diagonal matrices. The time and measurement updates are performed on these factor matrices. This formulation ensures that the covariance matrices will remain positive definite and that the filtering algorithm will remain stable.

The dynamic-stochastic model has been tested using simulated salinity observations from an Arctic estuary. It was found that the model estimates of concentration closely track the true state in the vicinity of observations. The accuracy of the estimates deteriorates with increasing distance from measurement locations, but the accuracy is still superior to that of values produced by applying a deterministic numerical model to the same data set. As the number of measurement locations increases, the mean square error of the dynamic-stochastic model estimates approaches the computed filter variance. The mean square error decreases approximately as the reciprocal of the number of sensors.

If the system and measurement noise statistics are known, a simulation of the stochastic advection-dispersion process together with the dynamic-stochastic model can be used to select the number and locations of measurement sensors to be deployed in field programs.

REFERENCES

Bankoff, S.G. and Hanzevak, E.L., 1975. The adaptive-filtering transport model for prediction and control of pollutant concentration in an urban airshed. Atmos. Environ. 9:793-808.

Bierman, G.J., 1977. Factorization Methods for Discrete Sequential Estimation. Academic Press, New York, 241 pp.

Box, G.E.P. and Jenkins, G.M., 1976. Time Series Analysis: Forecasting and Control. Holden-Day, San Francisco, 575 pp.

Budgell, W.P., 1976. Tidal Propagation in Chesterfield Inlet, N.W.T. Manuscript Report Series No. 3, Ocean and Aquatic Sciences, Central Region, Environment Canada, Burlington, 99 pp.

Budgell, W.P., 1981., A Stochastic-Deterministic Model for Estimating Tides in Branched Estuaries. Manuscript Report Series No. 10, Ocean Science and Surveys, Fisheries and Oceans Canada, Burlington, 189 pp.

Chiu, C.L. (Editor), 1978. Applications of Kalman Filter Theory to Hydrology, Hydraulics and Water Resources. University of Pittsburg, Pittsburg, 783 pp.

Chiu, C.L. and Isu, E.O., 1977. Application of Kalman filter in modelling daily stream temperature. In: Proceedings of the Seventeenth Congress of the International Association for Hydraulic Research. I.A.H.R., Baden-Baden, Vol. 3, pp. 463-470.

Crank, J. and Nicholson, P., 1947. A practical method for numerical integration of solutions of partial differential equations of heat conduction type. Proc. Cambridge Philos. Soc. 43:50-67.

DeGuida, R.N., Connor, J.J. and Pearce, R.R., 1977. Application of estimation theory to design of sampling programs for verification of coastal dispersion predictions. In: Gray, W.G., Pinder, G.F. and Brebbia, C.A. (Editors), Finite Elements in Water Resources. Pentech, London, pp. 4.303-4.334.

Desalu, A.A., Gould, L.A. and Schweppe, F.C., 1974. Dynamic estimation of air pollution. IEEE Trans. Autom. Contr. 19:904-910.

Fischer, H.B., 1973. Longitudinal disperison and mixing in open-channel flow. Ann. Rev. Fluid Mech. 5:59-79.

Fischer, H.B., 1976. Mixing and dispersion in estuaries. Ann. Rev. Fluid Mech. 8:107-133.

Fronza, G., Spirito, A. and Tonielli, A. 1979. Real-time forecast of air pollution episodes in the Venetian region. Part 2: the Kalman predictor. Appl. Math. Model. 3:409-415.

Godin, G., 1972. The Analysis of Tides. University of Toronto Press, Toronto, 264 pp.

Hann, R.W. and Young, P.J., 1972. Mathematical models of water quality parameters for rivers and estuaries. Report TR-45, Texas Water Resources Institute, Texas A & M University, College Station, Texas.

Harleman, D.R.F., 1971. One-dimensional models. In: Ward, G.H. and Epsey, W.H. (Editors), Estuarine Modelling: An Assessment. Tracor Inc., Austin, pp. 34-89.

Hinwood, J.B. and Wallis, I.G., 1975. Classification of models of tidal waters. J. Hydraul. Div. Am. Soc. Civ. Engrs. 101:1315-1331.

Jazwinski, A.H., 1970. Stochastic Processes and Filtering Theory. Academic Press, New York, 376 pp.

Kalman, R.E., 1960. A new approach to linear filter and prediction problems. J. bas. Engng. 82:35-45.

Kalman, R.E. and Bucy, R.S., 1961. New results in linear filtering and prediction theory. J. bas. Engng. 83:95-108.

Koda, M. and Seinfeld, J.H., 1978. Estimation of urban air pollution. Automatica. 14:583-595.

Lam, D.C.L., 1977. Comparison of finite-element and finite-difference methods for nearshore advection-diffusion transport models. In: Gray, W.G., Pinder, G.F. and Brebbia, C.A. (Editors), Finite Elements in Water Resources. Pentech, London, pp. 1.115-1.129.

Roache, P.J., 1972. Computational Fluid Dynamics. Hermosa Publishers, Albuquerque, 446 pp.

Roff, J.C., Pett, R.J., Rogers, G.F. and Budgell, W.P., 1980. A study of plankton ecology in Chesterfield Inlet, Northwest Territories: an Arctic estuary. In: Kennedy, V.S. (Editor), Estuarine Perspectives. Academic Press, New York, pp. 185-197.

Stone, H.L. and Brian, P.L.T., 1963. Numerical solution of convective transport problems. Amer. Inst. Chem. Engrg. J. 9:681-688.

Thatcher, M.L. and Harleman, D.R.F., 1972. A mathematical model for the prediction of unsteady salinity intrusion in estuaries. Ralph M. Parsons Laboratory for Water Resources and Hydrodynamics, Report No. 144. Massachusetts Institute of Technology, Cambridge, Massachusetts, 232 pp.

LIST OF SYMBOLS

A	Cross-sectional area
$\underline{A}$	N×N coefficient matrix
$\underline{B}$	N×N coefficient matrix
c	Concentration
$\underline{c}$	Vector of length N specifying concentrations at grid points in the discretized channel
$\underline{D}$	N×N diagonal matrix in covariance matrix factorization
E	Longitudinal dispersion coefficient
$\underline{G}$	Nx2 matrix specifying the nature and location of boundary conditions
$\underline{h}$	Measurement control vector of length N
$\underline{H}$	mxN observation control matrix specifying the model grid points at which concentrations have been observed Kalman gain vector of length N
$\underline{k}$	Kalman gain vector of length N
$\underline{K}$	Nxm Kalman gain matrix
L	Length of the open channel

m	Number of measurement locations (sensors) in the channel
n	Time step index
N	Number of grid points in the discretized channel
$\underline{P}$	NxN covariance matrix
Q	Volume flow rate
$\underline{Q}$	NxN system noise covariance matrix
$\underline{R}$	Measurement noise covariance matrix
t	time
$\underline{u}$	vector of length 2 containing specified boundary conditions
$\underline{U}$	N×N unit upper triangular matrix in covariance matrix factorization
$\underline{w}$	Vector of length N specifying system noise input at each of the model grid points
x	Distance
$\underline{z}$	Vector of length m containing observations from m sensors
Δt	Time increment
$\underline{\zeta}$	Vector of length m containing measurement noise corresponding to $\underline{z}$
$\underline{\Phi}$	N×N state transition matrix
$\underline{\Omega}$	N×2 input control matrix
$\sim$	Filtered estimate, e.g. expected value of the variable at time $n\Delta t$ conditioned on information up to $n\Delta t$
$\hat{}$	One step ahead prediction, e.g. expected value at time $(n+1)\Delta t$ conditioned on information up to $n\Delta t$

THE MEAN AND VARIANCE OF WATER CURRENTS INDUCED BY IRREGULAR SURFACE WAVES

B. DE JONG and A.W. HEEMINK

Twente University of Technology, Enschede, and Data Processing Division of Rijkswaterstaat, The Netherlands

ABSTRACT

Irregular surface waves generate a net mean velocity of the fluid and the material in it while due to the random fluctuations of this velocity about its mean value there will be a dispersion of the material. Expressions are derived for the mean value and the variance of these current velocities for a one- and two-dimensional irregular wave field. Numerical results are given for a one-dimensional wave field. It appears that the familiar directional wave spectrum for a two-dimensional wave field is insufficient to derive usefull results.

INTRODUCTION

Tides and winds are found to be major sources of generation of residual currents as observed in seas and estuaries (Alfrink & Vreugdenhil, 1981). These currents generate not only a net transport of the center of gravity of substances suspended or dissolved in it but also effect a dispersion of this material relative to this center of gravity. In the present paper we study only the current due to irregular wind waves which we assume to have a velocity equal to the Lagrangian velocity generated by these waves. Expressions for the Lagrangian velocity generated by a harmonic surface wave are well-known and can be derived in the way as indicated for example by Phillips 1977. On the basis of these expressions we derive the Lagrangian velocity for irregular waves by conceiving the

Reprinted from *Time Series Methods in Hydrosciences*, by A.H. El-Shaarawi and S.R. Esterby (Editors)

irregular waves as a simultaneous amplitude, wave number and frequency varied harmonic wave with a locally and instantaneously defined frequency, phase, wave number and amplitude which are all random variables. In this concept the statistical properties of the irregular wave field are defined by the joint density of these quantities. By assuming the wave elevations normally distributed this joint density can be derived as indicated for example by Rice, 1954 and Cramer & Leadbetter, 1967. However, it will be seen that for a calculation of the spectral moments which appear as parameters in this density we need an expression for the joint wave number-frequency spectrum while for sea waves in general only the frequency spectrum is available. By using the dispersion relation which is only valid for constant atmospheric pressure an approximate expression is derived for the frequency-wave number spectrum. In the first part of the paper expressions are derived for the mean and variance of the residual current for a one-dimensional wave field. In the second part expressions are derived for the two-dimensional case. It is shown in the final part of the paper that these expressions lead to acceptable results for the one-dimensional wave field. However, it appears that this approximation is insufficient for two-dimensional irregular waves

THE MEAN AND VARIANCE OF THE RESIDUAL CURRENT GENERATED BY ONE-DIMENSIONAL IRREGULAR SURFACE WAVES

We assume the x- and y-axis of a Cartesian coordinate system in the still water surface and the z-axis positive in upward direction. A two-dimensional normally distributed random field $\zeta(x,t)$ of surface elevations due to one-dimensional irregular surface waves propagating in the x-direction can be represented by

$$\zeta(x,t) = \sum_{m=0}^{\infty} \sum_{n=-\infty}^{\infty} c_{mn} \cos(k_m x + \omega_n y + \phi_{mn}) \tag{1}$$

with $k_o = 0 < k_1 < k_2 \ldots, \ldots\ldots < \omega_{-1} < \omega_o = 0 < \omega_1 < \omega_2 < \ldots\ldots,$

$k_m - k_{m-1} = dk$, $\omega_m - \omega_{m-1} = d\omega$ and $c^2_{mn} = 2S(\omega_n, k_m)dkd\omega$ where $S(\omega,k)$ is the wave number-frequency spectral density function of the random field ζ defined on the range $k \geq 0$ and $-\infty < \omega < \infty$. The phase angles ϕ_{mn} are mutually independent stochastic variables, homogeneously distributed over $[0,2\pi]$. Eq. (1) is a discretization of a stochastic integral giving the spectral representation of a stationary random field, (see e.g. Wong, 1971). In our case the random field is real and normally distributed which enables us to start from a less general shape of the spectral representation. The envelope function $R(x,t)$ and phase function $\Theta(x,t)$ are defined by

$$\zeta(x,t) = R(x,t) \cos \Theta(x,t) \qquad \hat{\zeta}(x,t) = R(x,t) \sin \Theta(x,t) \tag{2}$$

where $\hat{\zeta}$ is the Hilbert transform of ζ given by

$$\hat{\zeta}(x,t) = \sum_{m=0}^{\infty} \sum_{n=-\infty}^{\infty} c_{mn} \sin(k_m x + \omega_n t + \phi_{mn}) \tag{3}$$

which has simular properties as ζ. For points close to a fixed point (x_o,t_o) we may write as a first approximation $\zeta(x,t) = R(x,t) \cos\{\Theta'(x-x_o) + \dot{\Theta}(t-t_o) + \Theta\}$ where $\dot{\Theta} = \partial\Theta(x_o,t_o)/\partial t$, $\Theta' = \partial\Theta(x_o,t_o)/\partial x$ and $\Theta = \Theta(x_o,t_o)$. So, approximately, the irregular waves behave locally and instantaneously as a regular wave propagating in the positive x-direction with amplitude R, phase Θ, wave number Θ' and frequency $-\dot{\Theta}$.We further assume that the irregular waves generate locally and instantaneously a Lagrangian velocity as if there is a regular progressive wave on the surface with above-mentioned values for the amplitude, phase, frequency and wave number. It can be readily derived by using methods as indicated by e.g. Phillips that a harmonic wave $\zeta = R \cos(kx \cos\chi + ky \sin\chi - \omega t + \Theta)$ which has an angle χ with the positive x-axis generates the following Langrangian velocity components in the x-, y- and z-direction respectively

$$u(\underline{x},t,\omega,R,k_x,k_y,\Theta) = \frac{k_x \omega R \cosh k(z+d)}{k \sinh kd} \cos\psi +$$

$$\frac{k_x^2 R^2 \omega \cosh^2 k(z+d)}{\sinh^2 kd} \sin^2\psi + \frac{\omega R^2 k_x \sinh^2 k(z+d)}{\sinh^2 kd} \cos^2\psi$$

(4)

$$v(\underline{x},t,\omega,R,k_x,k_y,\Theta) = \frac{k_y \omega R \cosh k(z+d)}{k \sinh kd} \cos\psi +$$

$$\frac{k_y R^2 \omega \cosh^2 k(z+d)}{\sinh^2 kd} \sin^2\psi + \frac{\omega R^2 k_y \sinh^2 k(z+d)}{\sinh^2 kd} \cos^2\psi$$

$$w(\underline{x},t,\omega,R,k_x,k_y,\Theta) = \frac{\omega R \sinh k(z+d)}{\sinh kd} \sin\psi$$

in which $\psi = kx - \omega t + \Theta$ and where $\underline{x} = (x,y,z)$ and $k_x = k \cos \chi$ and $k_y = k \sin \chi$ are the wave number components in x- and y-direction. Consequently, the Lagrangian velocity components u_1 and w_1 generated locally and instantaneously by the one-dimensional irregular wave field are derived from (4) by setting $\omega = -\dot{\Theta}$, $k_x = \Theta'$ and $k_y = 0$. The mean value of the horizontal current is given by

$$E\{u_1\} = \int_0^\infty dR \int_{-\infty}^\infty d\dot{\Theta} \int_{-\infty}^\infty d\Theta' \int_0^{2\pi} d\Theta u(\underline{x},t,-\dot{\Theta},R,\Theta',\Theta)\, p(R,\Theta,\dot{\Theta},\Theta') \quad (5)$$

in which $p(R,\Theta,\dot{\Theta},\Theta')$ is the joint density of $R,\Theta,\dot{\Theta}$ and Θ'. $E\{.\}$ denotes a mathematical expectation. The variance of the horizontal current is determined from $\sigma_{u_1}^2 = E\{u_1^2\} - (E\{u_1\})^2$. The mean value and the variance of the vertical velocity component w_1 are calculated in a similar way. The evaluation of the joint probability density $p(R,\Theta,\dot{\Theta},\Theta')$ is done in the usual way as indicated e.g. by Rice. The joint density $p(\zeta,\hat{\zeta},\dot{\zeta},\dot{\hat{\zeta}},\zeta',\hat{\zeta}')$ of ζ, $\hat{\zeta}$ and their partial derivatives $\dot{\zeta},\dot{\hat{\zeta}},\zeta'$ and $\hat{\zeta}'$ with respect to time and space is determined first. This density is jointly normal. For notational concenience we set

$$z_1 = \zeta, z_2 = \dot{\hat{\zeta}},\ z_3 = \hat{\zeta}',\ z_4 = \hat{\zeta},\ z_5 = \dot{\zeta} \text{ and } z_6 = \zeta' \tag{6}$$

Then we find

$$p(z_1, z_2, \ldots, z_6) = (2\pi)^{-3}\ |\underline{M}|^{-\frac{1}{2}}\ e^{-\frac{1}{2}\underline{z}^T \underline{M}^{-1} \underline{z}} \tag{7}$$

in which $\underline{z}$ is a column vector with components $z_1, \ldots, z_6$ and $\underline{z}^T$ its transpose. $\underline{M}$ represents the covariance matrix given by

$$\underline{M} = \begin{bmatrix} E(z_1^2) & E(z_1 z_2) & \cdots\cdots & E(z_1 z_6) \\ E(z_2 z_1) & E(z_2^2) & \cdots\cdots & \\ \vdots & & & \\ E(z_6 z_1) & & \cdots\cdots & E(z_6^2) \end{bmatrix} \tag{8}$$

Substituting eqs. (1) and (3) in (6) and the resulting expressions for $z_1, \ldots, z_6$ in (8) yields

$$\underline{M} = \begin{bmatrix} b_{00} & b_{10} & b_{01} & 0 & 0 & 0 \\ b_{10} & b_{20} & b_{11} & 0 & 0 & 0 \\ b_{01} & b_{11} & b_{02} & 0 & 0 & 0 \\ 0 & 0 & 0 & b_{00} & -b_{10} & -b_{01} \\ 0 & 0 & 0 & -b_{10} & b_{20} & b_{11} \\ 0 & 0 & 0 & -b_{01} & b_{11} & b_{02} \end{bmatrix} \tag{9}$$

where

$$b_{ij} = \int_0^\infty dk_x \int_{-\infty}^{\infty} d\omega\ \omega^i k_x^j S(\omega, k_x) \tag{10}$$

defines the spectral moments of the spectral density function $S(\omega,k)$ of the random field $\zeta(x,t)$. Substituting (9) in (7) yields after some algebra

$$p(z_1, z_2, \ldots, z_6) = \frac{1}{(2\pi)^3 B} \exp\Big[- \frac{1}{2B^2} \{B_o(z_1^2 + z_4^2) +$$

$$+ 2B_1(z_1 z_2 - z_4 z_5) - 2B_2(z_1 z_3 - z_4 z_6) + B_{22}(z_2^2 + z_5^2) - 2B_3(z_2 z_3 + z_5 z_6) +$$

$$+ B_4^2(z_3^2 + z_6^2)\}\Big] \tag{11}$$

where

$$B_0 = (b_{20}b_{02}-b^2_{11})B,\ B_1 = (b_{11}b_{01}-b_{10}b_{02})B, B_2 = -(b_{01}b_{20}-b_{10}b_{11})B,$$

$$B_{22} = (b_{00}b_{02}-b^2_{01})B,\ B_3 = -(b_{00}b_{11}-b_{10}b_{01})B, B_4 = (b_{00}b_{20}-b^2_{10})B,$$

$$B = b_{00}b_{02}b_{20} + 2b_{10}b_{01}b_{11} - b_{00}b^2_{11} - b^2_{10}b_{02} - b^2_{01}b_{20} \qquad (12)$$

Substituting eq. (2) in (6) and the resulting expressions for $z_1, \ldots, z_6$ in eq. (11) yields after multiplying with the Jacobian $\partial(z_1, \ldots, z_6)/\partial(R,\dot{R},R',\Theta,\dot{\Theta},\Theta') = R^3$ and after performing a considerable amount of algebra an expression for the following joint density

$$p(R,\dot{R},R',\Theta,\dot{\Theta},\Theta') = \frac{R^3}{(2\pi)^3 B} \exp[-\frac{1}{2B^2}\{B_0R^2+2B_1R^2\dot{\Theta}-2B_2R^2\Theta'+$$

$$+ B_{22}(\dot{R}^2+R^2\dot{\Theta}^2) - 2B_3(\dot{R}R'+R^2\dot{\Theta}\Theta') + B_4(R'^2+R^2\Theta'^2)\}]$$

The probability density $p(R,\Theta,\dot{\Theta},\Theta')$ is found by integrating this expression with respect to $\dot{R}$ and R', both over $(-\infty,\infty)$. It is found then from the resulting expression that Θ is independent of the other variables and has a homogeneous distribution over the range $[0,2\pi]$. So we may write

$$p(R,\dot{\Theta},\Theta',\Theta) = p(\Theta)\ p(R,\dot{\Theta},\Theta') = \frac{1}{2\pi}\ p(R,\dot{\Theta},\Theta') \qquad (13)$$

where

$$p(R,\dot{\Theta},\Theta') = \frac{R^3}{2\pi(b_{00}B)^{\frac{1}{2}}} \exp[-\frac{R^2}{2B^2}\{B_0+2B_1\dot{\Theta}-2B_2\Theta'+B_{22}\dot{\Theta}^2-2B_3\dot{\Theta}\Theta'+$$

$$+B_4\Theta'^2\}] \qquad (14)$$

where as may be seen from (12) the coefficients B_0, B_1, etc. are defined by the spectral moments b_{ij} which are defined by (10). However for sea waves an expression for the spectral density $S(\omega,k)$ is in general not available and we only have the frequency spectrum $S(\omega)$. For that reason an approximate expression for $S(\omega,k)$ is derived by assuming that the frequency and the wave number are related by the dispersion relation

$$\omega^2 = gk(1 + \frac{Tk}{\rho g})^2 \tanh kd \qquad (15)$$

where g is the gravitational acceleration, T the surface tension, d the depth and ρ the density of the fluid. By denoting the positive solution for k of (15) by $k = D(\omega)$ the approximation for $S(\omega,k)$ can be written as

$$S(\omega,k) = S(\omega).\delta(k - D(\omega)) \tag{16}$$

where $\delta(.)$ is the Dirac function. A serious restriction of the validity of this approximation is that (15) is only valid for constant atmospheric pressure above the waves which is in general not the case in practical circumstances. After a considerable amount of algebra it may be shown that for $S(\omega,k)$ with a shape as in eq. (16) expression (14) reduces to

$$p(R,\dot{\Theta},\Theta') = p(R,\dot{\Theta})\ \delta(\Theta' - D(-\dot{\Theta})) \tag{17}$$

where

$$p(R,\dot{\Theta}) = \frac{R^2}{(2\pi b_{00}\hat{B})^{\frac{1}{2}}} \exp\left[-\frac{R^2}{2\hat{B}}\{b_{20} + 2b_{10}\dot{\Theta} + b_{00}\dot{\Theta}^2\}\right] \tag{18}$$

with $\hat{B} = b_{00}b_{20} - b_{10}$.

Expression (17) is consistent with the dispersion relation $k = D(\omega)$. In deriving (17) we should make use of the properties $B_2/B_4 \to 0$, $B_3/B_4 \to 0$ and $B/B_4^{\frac{1}{2}} \downarrow 0$ as $S(\omega,k) \to S(\omega).\ \delta(k - D(\omega))$.

The joint density $p(R,\Theta,\dot{\Theta},\Theta')$ may now be determined from eqs. (13), (17) and (18). Substituting this expression in (5) and performing some integrations we find for the mean velocity of the horizontal component of the residual current

$$E\{u_1\} = -\frac{3\hat{B}^2}{4b_{00}^{\frac{1}{2}}} \int_{-\infty}^{\infty} \frac{\dot{\Theta}k\ \cosh\ 2k(z+d)}{\sinh^2\ kd(b_{20} + 2b_{10}\dot{\Theta} + b_{00}\dot{\Theta}^2)^{5/2}}\, d\dot{\Theta} \tag{19}$$

The variance of the horizontal current may be determined from $\sigma_{u_1}^2 = E\{u_1^2\} - (E\{u_1\})^2$ where $E\{u_1^2\}$ may be determined in an analogous way as $E\{u_1\}$. We find finally

$$E\{u_1^2\} = \int_{-\infty}^{\infty} d\dot{\Theta}\ \left[\frac{3\hat{B}^2}{4b_{00}^{\frac{1}{2}}} \cdot \frac{\dot{\Theta}^2\ \cosh^2 k(z+d)}{(b_{20} + 2b_{10}\dot{\Theta} + b_{00}\dot{\Theta}^2)^{5/2}.\sinh^2 kd} + \right. \tag{20}$$

$$\left. + \frac{15\hat{B}^3}{16b_{00}^{\frac{1}{2}}} \cdot \frac{k^2\dot{\Theta}^2\{3\cosh^4 k(z+d) + 3\sinh^4 k(z+d) + 2\cosh^2 k(z+d)\sinh^2 k(z+d)\}}{\sinh^4 kd(b_{20} + 2b_{10}\dot{\Theta} + b_{00}\dot{\Theta}^2)^{7/2}}\right]$$

In a similar way we may determine the mean value of the vertical velocity component. We find, as expected:

$$E\{w_1\} = 0 \tag{21}$$

The variance of the vertical velocity becomes

$$\sigma_{w_1}^2 = E\{w_1^2\} = \frac{3\hat{B}^2}{4b_{00}^{\frac{1}{2}}} \int_{-\infty}^{\infty} d\dot{\Theta} \frac{\dot{\Theta}^2 \sinh^2 k(z+d)}{\sinh^2 kd.(b_{20} + 2b_{10}\dot{\Theta} + b_{00}\dot{\Theta}^2)^{5/2}} \tag{22}$$

When, as is in general the case for sea waves, the spectrum has only non-zero values for $\omega > 0$ for waves propagating in the positive x-direction the spectral moments appearing in (19), (20) and (22) are defined by

$$b_{io} = (-1)^i \int_{-\infty}^{\infty} \omega^i S(\omega)\, d\omega \tag{23}$$

THE MEAN AND VARIANCE OF THE RESIDUAL CURRENT GENERATED BY TWO-DIMENSIONAL IRREGULAR SURFACE WAVES.

A three-dimensional normally distributed random field $\zeta(x,y,t)$ representing two-dimensional irregular surface waves propagating in the x-y plane can be given by

$$\zeta(x,y,t) = \sum_{l=0}^{\infty} \sum_{m=-\infty}^{\infty} \sum_{n=-\infty}^{\infty} c_{lmn} \cos(k_{x_l} x + k_{y_m} y + \omega_n t + \phi_{lmn}) \tag{24}$$

Which is again a discretization of a stochastic integral giving the spectral representation of a stationary random field. In this expression we have

$$k_{x_o} = 0 < k_{x_1} < k_{x_2} < \dots ; \dots k_{y_{-2}} < k_{y_{-1}} < k_{y_o} = 0 < k_{y_1} < k_{y_2} < \dots$$

$$\dots \omega_{-2} < \omega_{-1} < \omega_o = 0 < \omega_1 < \omega_2 < \dots \quad d\omega = \omega_{n+1} - \omega_n, \; dk_y = k_{y_{n+1}} - k_{y_n},$$

$dk_x = k_{x_{n+1}} - k_{x_n}$ and $c_{lmn}^2 = 2\, S(\omega_n, k_{x_l}, k_{y_m})\, dk_x dk_y d\omega$. The phase angles are mutually independent stochastic variables, homogeneously distributed over $[0,2\pi]$. The envelope function $R(x,y,t)$ and the phase function $\Theta(x,y,t)$ are defined by

$$\zeta(x,y,t) = R(x,y,t)\cos\Theta(x,y,t) \quad \hat{\zeta}(x,y,t) = R(x,y,t)\sin\Theta(x,y,t) \tag{25}$$

where $\hat{\zeta}$ is the Hilbert transform of ζ which is obtained from (24) by replacing the cosine by sine. The two-dimensional irregular wave field is locally and instantaneously approximated by the regular wave $R\cos(\Theta'_x x + \Theta'_y y + \dot{\Theta}t + \Theta)$ with frequency $-\dot{\Theta}$ and wave numbers $\Theta'_x = \partial\Theta/\partial x$ and $\Theta'_y = \partial\Theta/\partial y$ in the x- and y-direction. This wave generates locally and instantaneously a residual current (u_2,v_2,w_2) with $u_2 = u(\underline{x},t,-\dot{\Theta},R,\Theta'_x,\Theta'_y,\Theta)$, $v_2 = v(\ldots)$ and $w_2 = w(\ldots)$ where (u,v,w) is defined in (4). The first and second moments of these velocity components may be determined when the joint density $p(R,\Theta,\dot{\Theta},\Theta'_x,\Theta'_y)$ is given. This density may be determined in a similar way as the density $p(R,\Theta,\dot{\Theta},\Theta')$ for the one-dimensional wave field, by starting from the joint normal distribution $p(\zeta,\hat{\zeta},\dot{\zeta},\dot{\hat{\zeta}},\zeta'_x,\hat{\zeta}'_x,\zeta'_y,\hat{\zeta}'_y)$. The moments of the spectral density $S(\omega,k_x,k_y)$ appear as parameters in this density. However, since for sea waves in general only the directional spectrum is given a suitable approximation $S(\omega,k_x,k_y)$ has to be devised which gives also an approximate expression for $p(R,\Theta,\dot{\Theta},\Theta'_x,\Theta'_y)$. The directional spectrum is sometimes given in the form

$$S(\omega,\chi) = \frac{2}{\pi}S(\omega)\cos^2\chi, \quad -\infty < \omega < \infty \text{ and } -\frac{\pi}{2} < \chi < \frac{\pi}{2} \tag{26}$$

where χ is the angle with the main direction of the wave field which is in our case the x-axis. It is noted that $S(\omega,\chi)$ gives only the local time behaviour of the two-dimensional surface waves in the various directions and is only a function of the frequency ω and the ratio k_x/k_y of the wave number components of the harmonic components of the surface waves. No information is given with respect to the wave numbers k of the harmonic components. To meet this lack of information we relate the wave number to the frequency by using the dispersion relation. Consequently, we set

$$S(\omega,k_x,k_y) = S(\omega,\chi) . \delta(k - D(\omega)) \qquad (27)$$

Where we assumed a change of variables $(k_x, k_y) \rightarrow (\chi,k)$ by means of the relations $tg\ \chi = k_y/k_x$ and $k = (k_x^2 + k_y^2)^{\frac{1}{2}}$. In the same way as for the one-dimensional waves it can be shown that the density $p(r,\Theta,\dot{\Theta},\Theta'_x,\Theta'_y)$ can be written in a shape which is consistent with (27):

$$p(R,\Theta,\dot{\Theta},\Theta'_x,\Theta'_y) = p(R,\Theta,\dot{\Theta},\Theta'_x).\delta((\Theta'^2_x + \Theta'^2_y)^{\frac{1}{2}} - D(-\dot{\Theta})) \qquad (28)$$

It remains to determine the joint density $p(R,\Theta,\dot{\Theta},\Theta'_x)$ which refers only to the spatial behaviour of $\zeta(x,y,t)$ in the x-direction. Consequently, this propability density is related to the random field $\zeta(x,t)$ which is obtained from (24) dy deleting the y-dependent term:

$$\zeta(x,t) = \sum_{m=-\infty}^{\infty} \sum_{n=-\infty}^{\infty} c_{mn} \cos(k_n \cos\chi_m . x + \omega_n t + \phi_{mn}) \qquad (29)$$

in which we substituted $k = k \cos \chi$. The coefficients c_{mn} are given by $c^2_{mn} = 2S(\omega_n,\chi_m)d\omega d\chi$ where $d\chi = \chi_{m+1} - \chi_m$. The density $p(R,\Theta,\dot{\Theta},\Theta'_x)$ is determined in a similar way as the density $p(R,\Theta,\dot{\Theta},\Theta')$ for the one-dimensional wave field. We find

$$p(R,\Theta,\dot{\Theta},\Theta'_x) = p(\Theta)\ p(R,\dot{\Theta},\Theta'_x) = \frac{1}{2\pi}\ p(R,\dot{\Theta},\Theta'_x) \qquad (30)$$

with

$$p(R,\dot{\Theta},\Theta'_x) = \frac{R^3}{2\pi(b_{00}B)^{\frac{1}{2}}} .$$

$$\exp\ [-\frac{R^2}{2B^2}\{B_0+2B_1\dot{\Theta}-2B_2\Theta'_x+B_{22}\dot{\Theta}^2-2B_3\dot{\Theta}\Theta'_x+B_4\Theta'^2_x\}] \qquad (31)$$

in which the coefficients are as defined in eq. (12). However, for the spectral moments b_{ij} we should take

$$b_{ij} = \int_{-\frac{\pi}{2}}^{\frac{\pi}{2}} d\chi \int_{-\infty}^{\infty} d\omega\, \omega^{i} (k \cos\chi)^{j}\, S(\omega,\chi) \tag{32}$$

By substituting (26) we obtain

$$b_{ij} = (-1)^{i} \frac{2}{\pi} \int_{-\pi/2}^{\pi/2} d\chi \int_{-\infty}^{\infty} d\omega\, \omega^{i} (k \cos\chi)^{j}\, S(\omega) \cos^{2}\chi \tag{33}$$

where we assumed again that positive frequencies correspond with waves in the positive x-direction. The mean velocity component in the x-direction is determined from

$$E\{u_2\} = \int_0^{2\pi} d\Theta \int_{-\infty}^{\infty} d\dot{\Theta} \int_0^{\infty} dR \int_{-\infty}^{\infty} d\Theta'_x \int_{-\infty}^{\infty} d\Theta'_y\, u(\underline{x},t,-\dot{\Theta},R,\Theta'_x,\Theta'_y,\Theta) .$$

$$p(R,\Theta,\dot{\Theta},\Theta'_x,\Theta'_y) \tag{34}$$

in which we take $k = D(-\dot{\Theta})$. Substituting (4) and (30) yields after some algebra

$$E\{u_2\} = -\frac{3B_4}{4b_{00}^{\frac{1}{2}}B^2} \int_{-\infty}^{\infty} d\dot{\Theta}\, \frac{\dot{\Theta}(B_2+B_3\dot{\Theta})\cosh 2k(z+d)}{\sinh^2 kd\{b_{00}\dot{\Theta}^2 + 2b_{10}\dot{\Theta}+b_{20}\}^{5/2}} \tag{35}$$

The variance is calculated from $\sigma^2_{u_2} = E\{u^2_2\} - (E\{u_2\})^2$ where $E\{u^2_2\}$ may be determined in a similar way as $E\{u_2\}$. We find finally

$$E\{u^2_2\} = \frac{1}{b_{00}^{\frac{1}{2}}} \int_{-\infty}^{\infty} d\dot{\Theta} \left[\frac{\dot{\Theta}^2\cosh^2 k(z+d)}{4k^2\sinh^2 kd}\left\{\frac{3(B_2+b_3\dot{\Theta})^2}{B^2(b_{00}\dot{\Theta}^2+2b_{10}\dot{\Theta}+b_{20})^{5/2}} + \right.\right.$$

$$\left. + \frac{B}{(b_{00}\dot{\Theta}^2+2b_{10}\dot{\Theta}+b_{20})^{3/2}}\right\} +$$

$$+ \frac{3B_4\dot{\Theta}^2\{3\cosh^4 k(z+d)+3\sinh^4 k(z+d)+2\cosh^2 k(z+d)\sinh^2 k(z+d)\}}{16\sinh^4 kd} \tag{36}$$

$$\left. .\left\{\frac{5(B_2+B_3\dot{\Theta})^2}{B^3(b_{00}\dot{\Theta}^2+2b_{10}\dot{\Theta}+b_{20})^{7/2}} + \frac{1}{(b_{00}\dot{\Theta}^2+2b_{10}\dot{\Theta}+b_{20})^{5/2}}\right\}\right]$$

It is easy to verify that $E\{v_2\} = 0$ and $E\{w_2\} = 0$. It is further easily seen from eqs. (4) that between the horizontal velocity u_1 of the one-dimensional wave field and the horizontal velocity components u_2 and v_2 of the two-dimensional wave field the relation $u_1^2 = u_2^2 + v_2^2$ exists. The variance $\sigma_{v_2}^2$ of the transverse velocity v_2 is therefore given by $\sigma_{v_2}^2 = E\{u_1^2\} - E\{u_2^2\}$. Finally, we may derive that the variances of the vertical velocity components are identical for the one- and two-dimensional wave field.

NUMERICAL RESULTS AND CONCLUSIONS.

In figures 1 numerical results are given for the mean and standard deviations of the horizontal transport velocity for a one-dimensional wave field for several water depths and wind speeds using the Pierson-Moskowitz wave spectrum:

$$S(\omega) = 0.0081\, g^2/\omega^5\, e^{-0.74(g/V\omega)^4} \quad \omega > 0, \; S(\omega) = 0 \quad \omega < 0$$

where V is the wind speed at a height of 19.5 m. above the sea surface. In figures 2 the standard deviation of the vertical transport velocity is given for various wind speeds and water depths. In performing these calculations we did not bother about the fact whether or not such a spectrum can be realized on a sea of finite depth, the more as only the first two moments of the spectrum are needed in these calculations. It is noted that the variations of the transport velocity about its means value are in the range of 5 to 15 percent of the wind velocity. This indicates that irregular waves have a high dispersive effect. It is seen from eqs. (19), (20) and (22) that the wave spectrum influences the mean value and the variances of the tranport velocities by the values of the spectral moments b_{oo}, b_{1o} and b_{2o}. The integrands in the integral expressions for these moments converge to zero at least as ω^{-3} as $\omega \to \infty$. For the frequency range which gives the major contribution to these integrals the surface

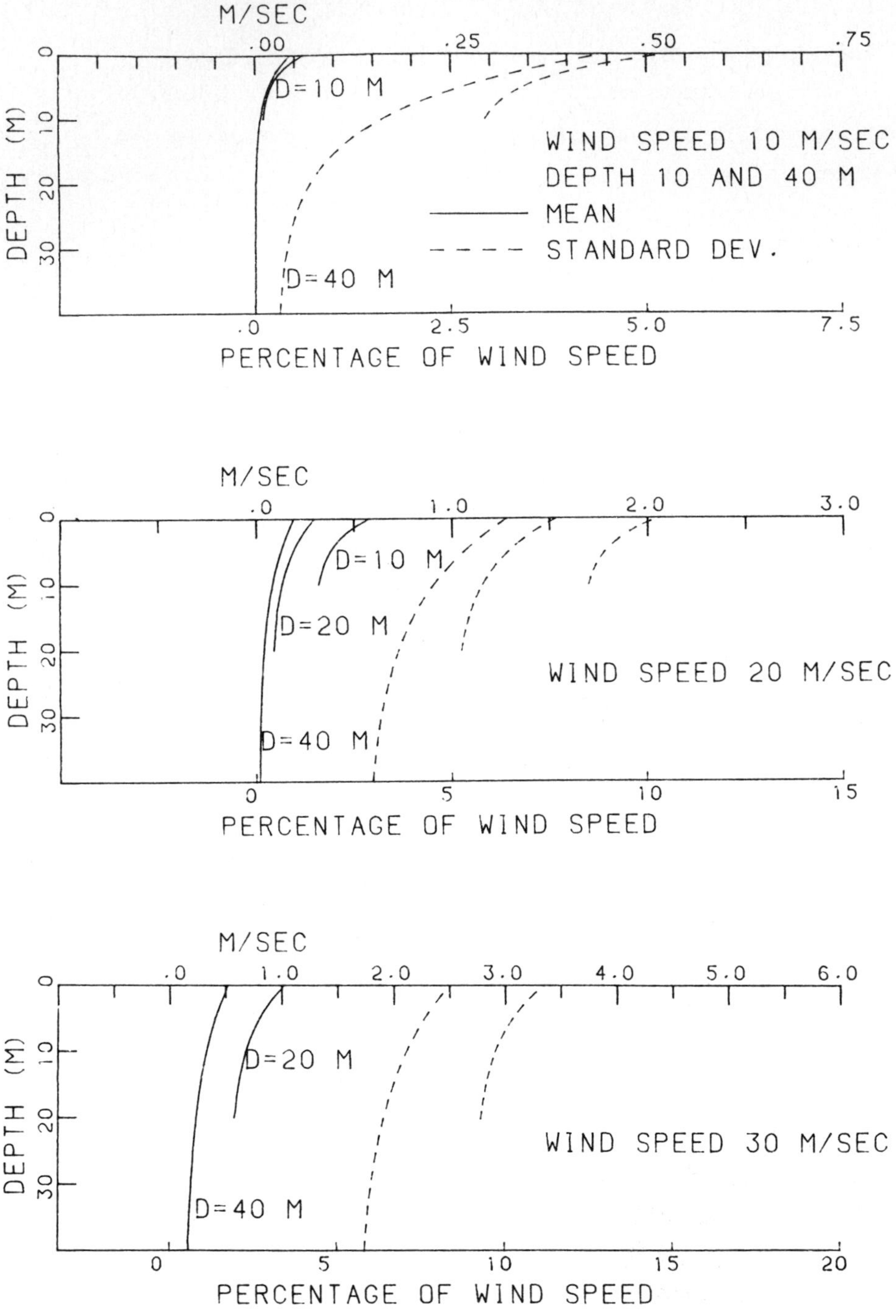

Fig. 1. Mean and standard deviation of the horizontal residual current for a one-dimensional irregular field.

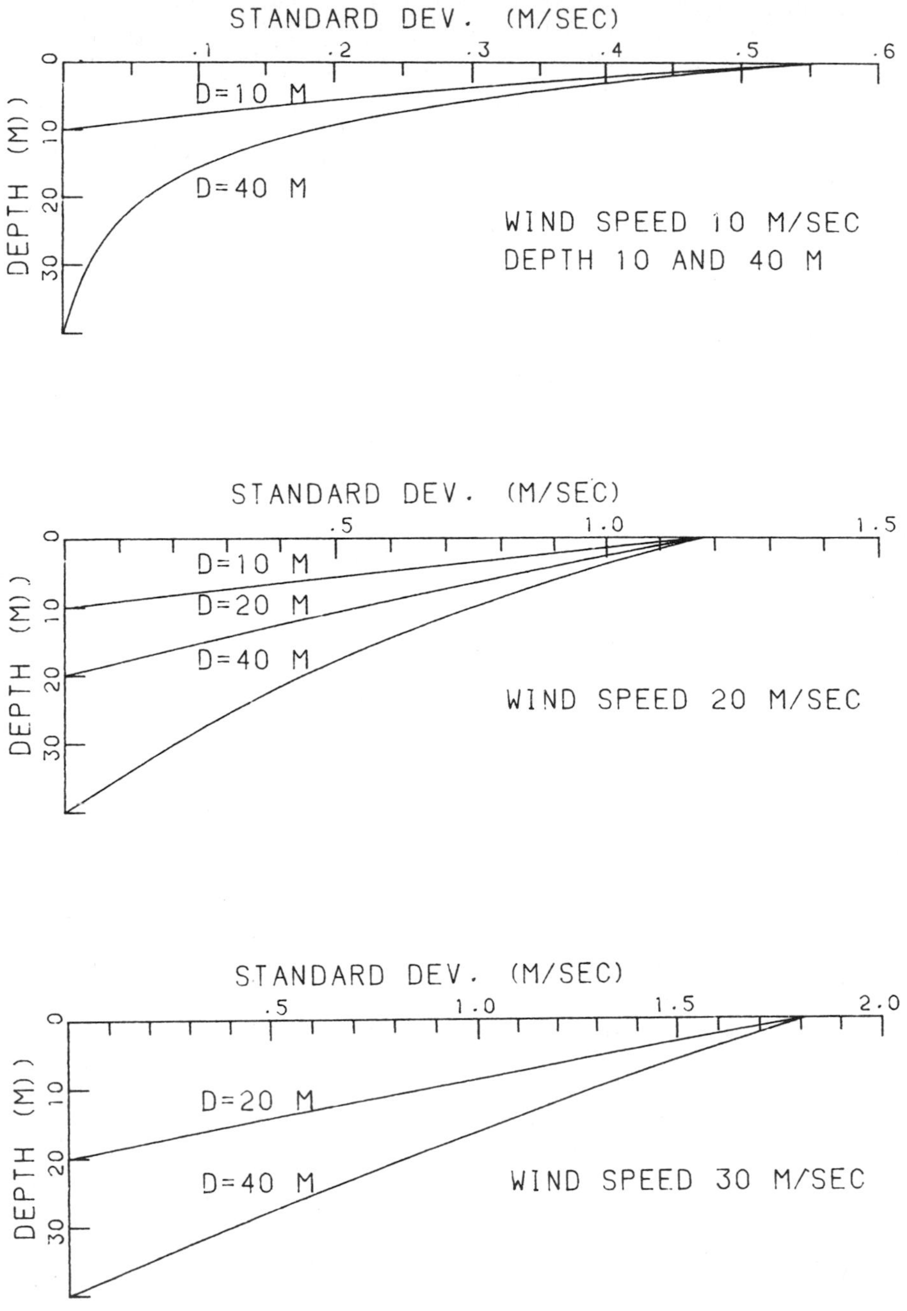

Fig. 2. Standard deviation of vertical transport velocity.

tension term in the dispersion relation is negligible as compared to the gravity term. Consequently, for the one-dimensional irregular waves only the gravity waves give a significant contribution to the mean and variances of the transport velocities. It is noted that the approximation (16) for the spectrum $s(\omega,k)$ is only valid for constant atmospheric pressure above the waves. More accurate expressions for $S(\omega,k)$ should give better results for the mean and variance of the transport velocity. With respect to the two-dimensional irregular wave field it is seen from eqs. (12), (35) and (36) that the mean and variance of the transport velocity in the x-direction are in addition to b_{00}, b_{10} and b_{20} also determined by the spectral moments b_{01}, b_{11} and b_{02} which are given by

$$b_{i1} = \frac{8}{3\pi}\int_0^\infty d\omega\, \omega^i k\, S(\omega),\ i = 0,1 \qquad b_{02} = \frac{3}{4}\int_0^\infty d\omega\, k^2\, S(\omega)$$

Since according to the dispersion relation we have for gravity waves $k \sim \omega^2$ as $\omega \to \infty$ the integral expression b_{02} diverges. Using the dispersion realtion (15) which includes the surfaces tension we have $k \sim \omega$ as $\omega \to \infty$. Then b_{02} has a finite value. Our conclusion may be that the approximation (27) for the two-dimensional waves has as a consequence that the ripples which are waves with a lenght less than around 0.1 meter and which cover less than one percent of the total energy play a decisive part. This does not agree with our physical intuition. To eliminate the influence of the higher frequencies we collected in table 1 some results which are obtained by restricting the interval of integration for the spectral moments to the range [0,1.5]. For a velocity of 20m/s this range covers 99.2% of the wave energy. More reliable results may be obtained by prescribing a more accurate expression for the frequence-wave number spectrum $S(\omega,k_x,k_y)$.

TABLE 1.

Mean and variance of transport velocity at the surface for a Pierson-Moskowitz wave spectrum. The integration in the spectral moment expression has been restricted to the frequency range

$0 \leq \omega \leq 1.5$. Wind speed in 20m/s.

Depth in meters	1-dim. wave field		2-dim. wave field	
	$E\{u_1\}$	$\sigma^2_{u_1}$	$E\{u_2\}$	$\sigma^2_{u_2}$
10	.684	5.576	.594	4.242
20	.343	2.937	.300	2.356
40	.217	1.876	.195	1.649

REFERENCES

Alfrink, B.J. and Vreugdenhil, C.B., 1981. Residual Currents, Analysis of Mechanisms and Model Types. Delft Hydraulic Laboratory, report R1469-II.

Cramer, H. and Leadbetter, M.R., 1967. Stationary and Related Stochastic Processes. John Wiley & Sons, New York.

Phillips, O.M., 1977. The dynamics of the Upper Ocean. Cambridge University Press, Cambridge.

Rice, S.O. Mathematical Analysis of Random Noise. In: Wax, N. (Editor), 1954. Selected Papers on Noise and Stochastic Processes. Dover, New York.

Wong, E., 1971. Stochastic Processes in Information and Dynamic Systems. Mc Graw-Hill, New York.

GENERATION OF WEEKLY STREAMFLOW DATA FOR THE RIVER DANUBE-RIVER MAIN-SYSTEM

EXPERIENCES WITH AN AUTOREGRESSIVE MULTIVARIATE MULTILAG MODEL

L.A. SIEGERSTETTER AND W. WAHLIß

Technical University of Munich

ABSTRACT

In search for an optimal operating strategy for a complex hydrological system simulation runs required hundreds of years of synthetic weekly streamflow data. A multivariate multilag model was used for the generation of both time and space correlated hydrological series for eight gauging stations. As the model assumes the input to be normally distributed standardized variables extensive data transformation had to be performed. Stepwise regression is applied to keep the number of parameters defining the multiple autoregressive process as low as possible.

1. INTRODUCTION

Statistical simulation of river flows is considered a powerful tool in the design and operation of water resources projects. Quite often historical records are too short to serve as a secure basis for the analysis of the behaviour of the system in question. Thus, long synthetic flow series have to be generated which truly reproduce the characteristics of the original data. Furthermore, for complex systems cross-correlated series for

Reprinted from *Time Series Methods in Hydrosciences*, by A.H. El-Shaarawi and S.R. Esterby (Editors)

various locations may have to be considered simultaneously. As a consequence a multivariate model is required describing not only the internal structures of the respective series but also all spacial interactions.

A well known multivariate approach was introduced by G.K. Young and W.C. Pisano when they extended a first order autoregressive process (AR(1)-process) suggested by N.C. Matalas. However, flow series usually prove to possess significant autocorrelation coeffients for time lags higher than lag-1 and should not be treated as AR(1)-processes to fully use all available information. In the following a higher order autoregressive model and its application as a planning tool is discussed.

2. THE MODEL

A multivariate autoregressive model that theoretically allows for any p time lags is given by the matrix equation

$$X^{(i)} = A1\ X^{(i-1)} + A2\ X^{(i-2)} + \ldots + Ap\ X^{(i-p)} + B \cdot E^{(i)} \quad (1)$$

where $X^{(i)}, X^{(i-1)}, \ldots, X^{(i-p)}$: (n)-vectors containing flow measurements for time intervals i,...,i-p for all n gauges

$A1, A2, \ldots, Ap, B$: (n n)-matrices with model parameters

$E^{(i)}$: (n)-vector representing white noise

p : order of the autoregressive process

Persistence of the runoff is modelled by assuming that any flow realization is influenced by p predecessors observed at the same and all other gauges. An independent stochastic component $B \cdot E^{(i)}$ added to the autoregressive part of eq.(1)

accounts for all such variance that cannot be derived from the history of the runoff process.

2.1 Estimation of Model Parameters

In comparison with an univariate model of the same order the number of model parameters in eq. (1) is increased by n^2. Practical application, however, demands that the runoff structure be reproduced in the most economical way, i.e. with the lowest possible number of coefficients. Stepwise linear regression was therefore used in the estimation of the matrices A1, ... , Ap. Matrix equation (1) is split into separate row requations each being interpreted as a linear regression of the form

$$
\begin{aligned}
x_j^{(i)} &= a1_{j,1}\, x_1^{(i-1)} + a1_{j,2}\, x_2^{(i-1)} + \ldots + a1_{j,n}\, x_n^{(i-1)} + \\
&+ a2_{j,1}\, x_1^{(i-2)} + a2_{j,2}\, x_2^{(i-2)} + \ldots + a2_{j,n}\, x_n^{(i-2)} + \ldots \\
&\ldots + ap_{j,1}\, x_1^{(i-p)} + ap_{j,2}\, x_2^{(i-p)} + \ldots + ap_{j,n}\, x_n^{(i-p)} = \\
&= \sum_{k=1}^{n} a1_{j,k}\, x_k^{(i-1)} + \sum_{k=1}^{n} a2_{j,k}\, x_k^{(i-2)} + \ldots + \sum_{k=1}^{n} ap_{j,k}\, x_k^{(i-p)}
\end{aligned}
\tag{2}
$$

where

$ap_{j,k}$:	regression coefficient for the pth predecessor at gauge k
$j=1, \ldots, n$:	row number in eq. (1)
$i=p+1, \ldots, m$:	index
m:	total number of observations
$k=1, \ldots, n$:	gauge number

Stepwise regression searches eq. (2) for the optimal combination of independent variables in the description of the response variable. Finally, a dependence is found that

(i) contains significant regression coefficients only

(ii) minimizes the variance of deviations from observed to computed values for the response variable.

Remaining (significant) regression coefficients are inserted in matrices A1, ... , Ap of eq. (1) while all other matrix elements are assumed to be zero. In estimating the stochastic matrix B the difference between the observed values for the response variable and the respective computed values is calculated for each time interval as

$$D^{(i)} = X_o^{(i)} - [A1 \cdot X^{(i-1)} + \dots + Ap \cdot X^{(i-p)}] = B \cdot E^{(i)} \qquad (3)$$

Provided that the vectors $E^{(i)}$ consist of (0,1)-normally distributed random numbers the expected value of $B \cdot B^T$ can be derived from D as

$$D \cdot D^T = B \cdot E \cdot (B \cdot E)^T = B \cdot E \cdot E^T \cdot B^T = B \cdot I \cdot B^T = B \cdot B^T \qquad (4)$$

with I: identity matrix

B^T, E^T: transpose matrices of B and E

and $E = [E^{(1)}, E^{(2)}, \dots, E^{(i)}, \dots, E^{(m-p)}]$

$D = [D^{(1)}, D^{(2)}, \dots, D^{(m-p)}]$

B can then easily be found using an iterative procedure described by G.K. Young and W.C. Pisano.

2.2 Data Transformation

In general the observed flow series cannot be used in the calibration of parameters without statistical manipulation as the suggested model requires stationary and normally distributed input data. A double data transformation approximately fulfils these requirements:

(i) To achieve normality the following simple mathematical functions were used

$$nx = x^{1/2} \tag{5}$$

$$nx = x^{1/3} \tag{6}$$

$$nx = \log(x) \tag{7}$$

A more complex function transforms the distributions Pearson-Type III into a normal distribution

$$nx = \frac{6}{cs} \{ [\frac{cs}{2} K + 1]^{1/3} - 1 \} + \frac{cs}{6} \tag{8}$$

where cs: coefficient of skewness

K: standardized parameter of Pearson-Type III distribution

(ii) The second transformation elimates cyclical components by a seasonal standardization

$$rx = (nx - \overline{nx}_z)/s_z$$

nx: normalized runoff values

rx: cyclically standardized runoff values (residuals)

$\overline{nx}_z, s_z$: mean and standard deviation of nx in time interval z

z: cyclical index (z = 1, 2, ... , 52 for weeks)

The residuals rx can now be used as input data for the model. Consequently, the algorithm will produce synthetic residuals which obviously have to undergo the inverse transformations to (i) and (ii) to represent the desired artificial runoff series.

3. APPLICATION OF THE MODEL

The model was tested for a complex hydrological and water resources system (Fig. 1). Water from the Danube watershed will be transfered to the River Main watershed

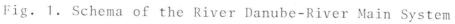

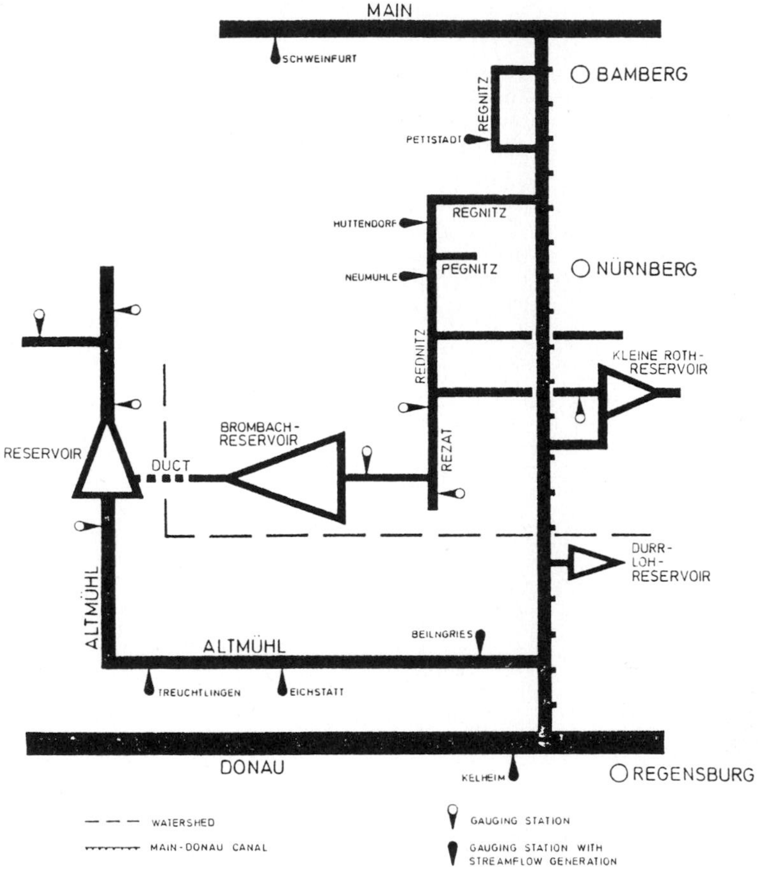

Fig. 1. Schema of the River Danube-River Main System

to improve its water balance. The system includes the Main-Danube-Canal, three reservoirs and various natural or artificial connecting ducts. To obtain optimal strategies for the long term operation of the system a computer routine has been developed on the basis of a weekly simulation. As input data numerous 100-year synthetic series for 8 water gauges in the respective areas were required.

Following an extensive data analysis and pilot runs an autoregressive third order process was considered suitable for the generation of the synthetic series. Stepwise regression reduced the number of significant elements in the matrices A1, A2 and A3 to appr. 40 percent of the maximum possible value with a minimum of 14 percent for A3. It should be noted that A3 was basically a diagonal matrix which indicates that spacial dependencies were neglectable at lag-3 and higher in this specific case.

4. RESULTS

4.1 Transformations

The generating algorithm choses the most suitable transformation function out of (5) - (8) with optimality being defined as

(i) the lowest possible skewness is observed in the transformed data

(ii) the null hypothesis of a normal distribution is not rejected at the 5 percent level.

When using just one transformation function for a complete runoff record it was found that while some of the weekly series fulfilled the above criteria others were significantly rejected. Hence, a special transformation was selected for each week at each station with extremely satisfying results as only 4 out of a total of 416 weekly distributions did not meet requirements (i) and (ii) simultaneoulsy. In such cases condition (i) was considered to to be sufficient.

Whereas normality of the residuals could be achieved in good approximation for most weeks unrealistically high synthetic runoff values were in some few cases observed with log and Pearson-Type III. The result was an over-estimation of the skewness of the artificial data. The explanation may be purely mathematical as the inverse transformations contain e^x and x^3. In general, a more sophisticated estimation (maximum likelihood) of the population parameters should be used if extreme values are of major importance. In addition, the skewness criterion may be responsible for the production of too high runoff values as it is very sensitive to extreme values in the observed data. After an effort to cope with a single large observation the inverse transformation can lead to a relatively high proportion of large synthetic values.

In the final version of the algorithm the log-transformation (7) was used in 64 percent of all cases. Second came Pearson-Type III (8) with approximately 30 percent whereas functions (5) and (6) were selected in about 2 percent and 4 percent of the weeks, respectively.

4.2 Statistical Parameters

The statistical parameters mean, standard deviation and skewness were computed for both the generated residuals and the synthetic runoff. Constant comparison of these parameters clearly showed the effects of the chosen transformation and the quality of the model. Figures 2 and 3 present the results of a typical 100-year production run for the gauge 'Hüttendorf' with respect to means and standard deviations. Figure 4 compares the results for five 100-years series with one 500-year series. It should be noted that 100-year series still show fairly large sampling deviations which have to be considered in any application.

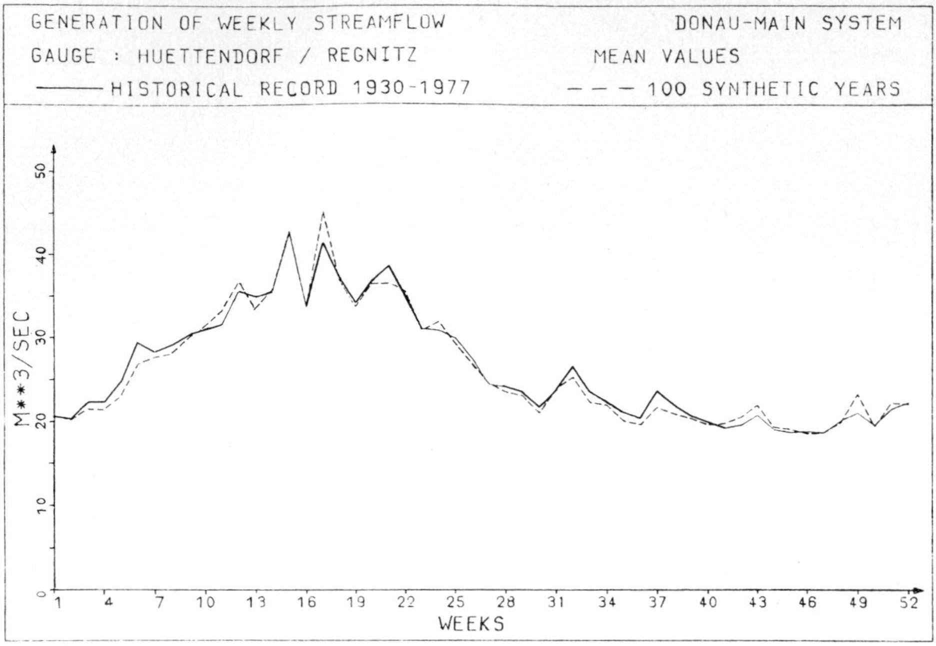

Fig. 2. Weekly means at gauge Hüttendorf (100-year simulation)

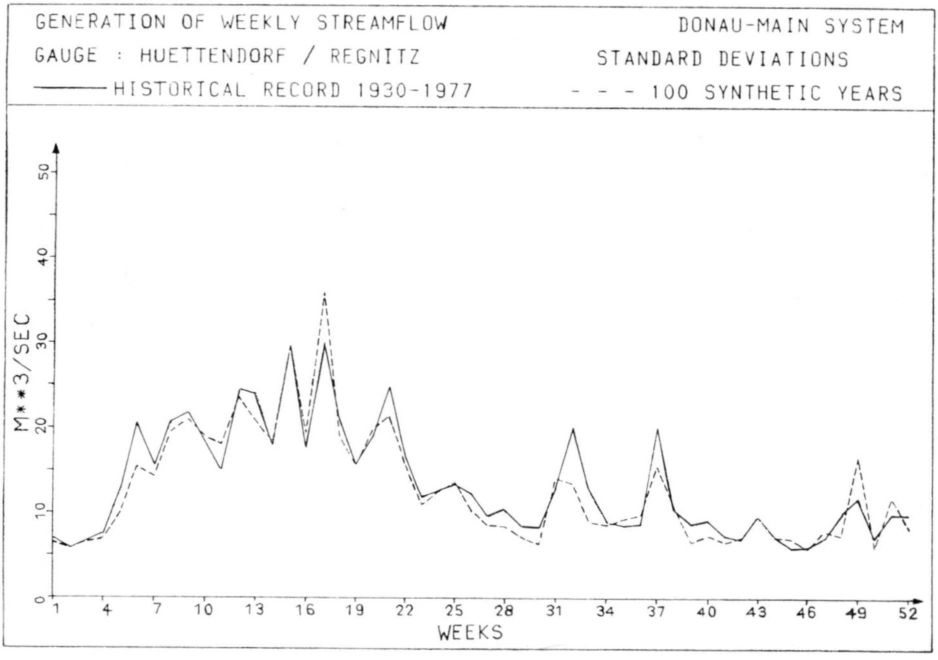

Fig. 3. Weekly standard deviations at gauge Hüttendorf (100-year simulation)

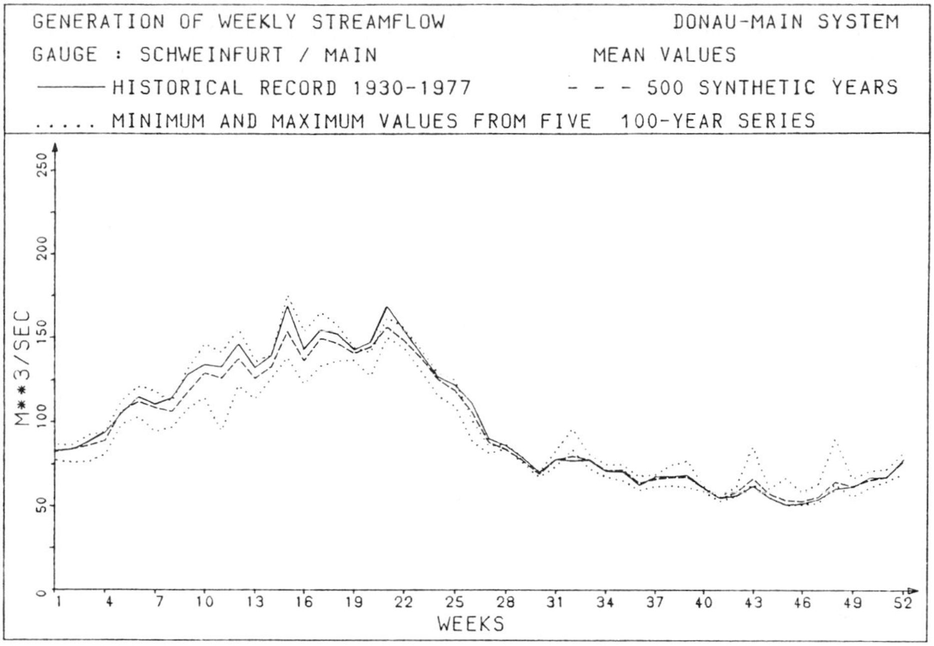

Fig. 4. Weekly means at gauge Schweinfurt (500-year simulation and extreme values out of five 100-year simulations)

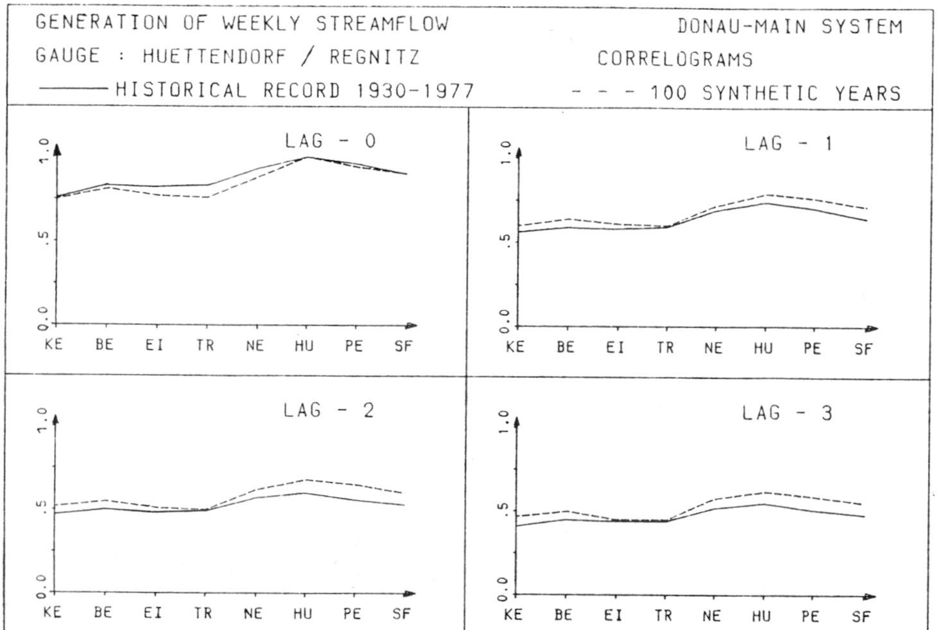

Fig. 5. Auto- and crosscotrelation coefficients at gauge Hüttendorf for time lags 0 to 3 (100-year simulation)

4.3 Correlation Structure

In Figure 5 the reproduction of the original correlation structure of the hydrological system is presented for time lags 0 to 3. There is some indication that correlation is overestimated for higher time lags. However, this can only be observed for the synthetic runoff, i.e. after inverse transformation, whereas residuals reflect the correlation structure precisely and without systematic deviations. How transformations do affect correlation is still to be investigated.

4.3 Negative Runoff

When sampling from a symmetrical distribution function with range $-\infty < x < +\infty$ negative values are likely to occur for long series with a probability depending on mean and standard deviation. The proportion of negative runoff produced for the Danube-Main-System was around 0,05 percent. Without significantly affecting the statistical properties of the series the values were equalled nought.

In multivariate simulation negative values of intermediate runoff can be observed, especially if runoff sequences possess high coefficients of variation together with almost identical means as in the case of the gauges along the River Altmühl. The number of negative values generated totalled approximately 4 percent. It should be mentioned that negative intermediate runoff was found in the Altmühl observations, too. At all other gauges no such irregularities occurred.

5. CONCLUSION

A multivariate regression model for the generation of both time and space correlated synthetic runoff sequences is described. Time dependence can be reproduced for any number of time lags required. In an application weekly runoff series were generated.

For the necessary transformation of the observed data into normal distribution various functions were tested with log (x) and Pearson-Type III proving to be most suitable. The results with respect to the reproduction of statistical parameters as means, standard deviations, auto- and crosscorrelation coefficients are shown for some typical sample sequences. It should be mentioned that the model has also been used for the generation of monthly series as well as sequences of decades with satisfying results.

REFERENCES

Fiering, M.B.: Multivariate Technique for Synthetic Hydrology. Journal of the Hydraulics Division ASCE 90 (1964) No. HY 5, pp. 43 - 60.

Kindler, J., W. Zuberek: On some Multi-Site Multi-Season Streamflow Generation Models. International Institute for Applied System Analysis, RM-76-76, Laxenburg 1967.

Matalas, N-C.: Mathematical Assessment of Synthetic Hydrology. Water Resources Research 3 (1967) No. 4, pp. 937 - 945.

O'Connell, P.E.: Multivariate Synthetic Hydrology: A Correction. Journal of the Hydraulics Division ASCE 99 (1973) No. HY 12, pp. 2393 - 2396.

Pegram, G.G.S., W. James: Multilag Multivariate Autoregressive Model for the Generation of Operational Hydrology. Water Resources Research 8 (1972) No. 4, pp. 1074 - 1076.

Schneider, K., R. Harboe: Anwendung des Young-Pisano-Modells zur Erzeugung gleichzeitiger künstlicher Abflußreihen für verschiedene Stellen in einem Einzugsgebiet. Wasserwirtschaft 69 (1979) No. 7/8, pp. 219 - 225.

Schramm, M.: Zur mathematischen Darstellung und Simulation des natürlichen Durchflußprozesses. Acta Hydrophysica 19 (1975) No. 2/3, pp. 77 - 191.

Young, G.K.: Discussion of "Mathematical Assessment of Synthetic Hydrology" by N.C. Matalas. Water Resources Research 4 (1968) No. 3, pp. 681 - 682.

Young, G.K., W.C. Pisano: Operational Hydrology Using Residuals. Journal of the Hydraulics Division ASCE 94 (1968) No. HY 4, pp. 909 - 923.

PROBABILISTIC CHARACTERIZATION OF POINT AND MEAN AREAL RAINFALLS

VAN-THANH-VAN NGUYEN AND JEAN ROUSSELLE

Universite du Quebec a Chicoutimi, Chicoutimi, Quebec and Ecole Polytechnique de Montreal, Montreal, Quebec, Canada

INTRODUCTION

Information on rainfall distributions over time and space are both important in various types of hydrologic studies concerning the determination of runoff characteristics. The objective of the study to be reported was to consider, from a probabilistic perpestive, two of the rainfall characteristics essential for the planning and design of urban drainage systems: the temporal pattern and the areal correction of storms.

First, the probabilistic characterization of temporal storm patterns was investigated wherein a storm was defined as an uninterrupted sequence of consecutive hourly rainfalls. A stochastic model was developed to determine the probability distributions of rainfall accumulated at the end of each time unit within a total storm duration. Secondly, a theoretical methodology is proposed to establish a relationship between the rainfall at a fixed point and the associated mean rainfall over a geographically fixed area.

TEMPORAL STORM PATTERN

Consider an interval of time which consists of n hours. With the definition of a storm stated in the previous section, the probabilistic characterization of a temporal storm pattern can be achieved by finding the probability distribution function of accumulated rainfall amounts at the end of each hour within the n-hour storm duration.

Reprinted from *Time Series Methods in Hydrosciences*, by A.H. El-Shaarawi and S.R. Esterby (Editors)

Let M_n be defined as the number of consecutive rainy hours starting from the first hour of the n-hour period. By this definition the random variable M_n can assume the values 0, 1, 2,...n. Let ε_ν denote the hourly rainfall depth in the ν-th rainy hour. The accumulated amount of rainfall S(n) during M_n consecutive rainy hours can be defined as

$$S(n) = \sum_{\nu=0}^{M_n} \varepsilon_\nu \; . \tag{1}$$

Then the distribution function $F_n(x)$ of S(n) can be written as follows (Nguyen and Rousselle, 1981a):

$$F_n(x) = P\{S(n) \leq x\} = \sum_{k=0}^{n} P\{X_K \leq x,\ M_n = k\} \tag{2}$$

in which $P\{\cdot\}$ denotes the probability, $X_o = 0$ and $X_K = \sum_{\nu=0}^{k} \varepsilon_\nu$ for k = 0, 1, 2,...,n.

In the following, for hourly rainfall process, we assume that:

i) the hourly rainfall depths $\varepsilon_1, \varepsilon_2, \ldots \varepsilon_n$ are independent, identically distributed random variables with $P\{\varepsilon_\nu \leq x\} = 1 - e^{-\lambda x}$ for every ν = 1, 2,...n;

ii) $\varepsilon_1, \varepsilon_2, \ldots \varepsilon_n$ are independent of M_n; and

iii) the sequences of rainy hours can be represented by first and second-order Markov chains.

Under these assumptions, we can write (Nguyen and Rousselle, 1981a):

$$F_n(x) = P\{M_n=0\} + \sum_{k=1}^{n} P\{X_K \leq k\}\ P\{M_n=k\} \tag{3}$$

or

$$F_n^*(x) = \frac{1}{1 - P\{M_n=0\}} \cdot \sum_{k=1}^{n} \left(\frac{\lambda^k}{\Gamma(k)} \int_0^x u^{k-1}\, e^{-\lambda u}\, du \right) P\{M_n=k\} \tag{4}$$

where $\Gamma(k) = (k-1)!$.

An illustrative application of the theoretical model to an actual record of hourly rainfall has been made, in which the Dorval Airport hourly rainfall data for 1943-1974 have been used (Nguyen and Rousselle, 1981a). For purposes of illustration, Figure 1 shows comparisons between the empirical and theoretical distributions for the periods starting from the second hour of the day and consisting of n hours, n = 2, 10 by using the hourly rainfall record for July. It was found that the bigger the value of n the larger the discrepancies. A more detailed discussion of the agreement between observed and theoretical results can be found in Nguyen and Rousselle (1981a).

In summary, the stochastic model proposed is more general and more flexible than those used in previous investigations. Using the methodology reported here, a temporal storm pattern can be characterized in terms of the time of storm occurence, the total storm duration, the total storm depth, and the probability of occurence of cumulative rainfalls within the storm.

AREAL CORRECTION FACTOR

In the planning and design of urban drainage systems, engineers have been thwarted in attempting to relate an expected mean rainfall over a jurisdiction to a historical record at a first-order weather station. This is because research has been confined to spatial properties of moving storms rather than to catchments and reference raingages fixed in space. That is, attention has been focused on either a moving area coupled with a moving reference point or a fixed area and a moving reference point. With the moving reference point being the peak portion of the storm rainfall at each particular time interval, it follows that areal mean rainfall so evaluated will always be an attenuation of the reference point rainfall.

Reported in this paper are the findings from the works by Nguyen et al. (1981b) where mean rainfalls for a fixed area and the associated rainfall for a fixed point in that area have been analyzed from a probabilistic perspective.

The approach adopted consists of assuming an exponential distribution for point rainfall depth at each raingage, and from this deriving theoretically the distribution function for mean areal

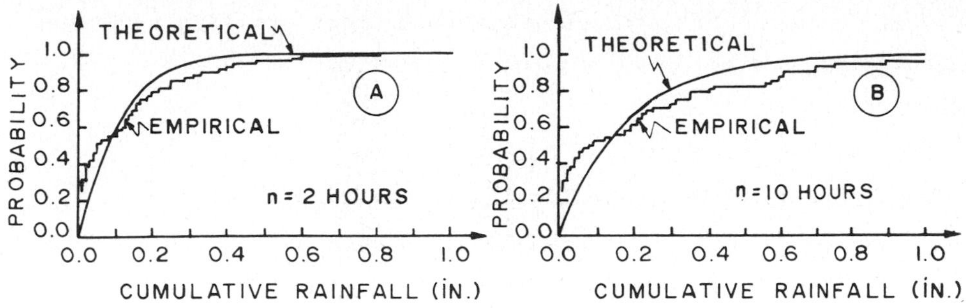

Fig. 1 Comparison between theoretical and empirical distributions of accumulated rainfalls in the n-hour period. (A) n = 2 hours, (B) n = 10 hours.

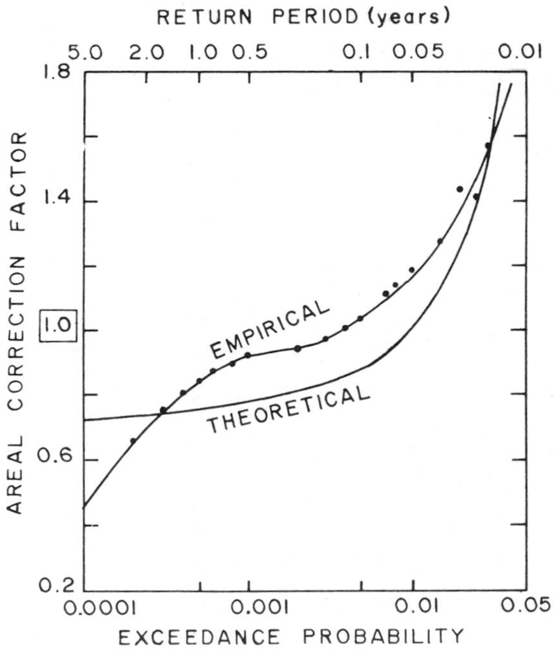

Fig. 2 Theoretical and empirical probability distributions of areal correction factor.

rainfall values. The areal correction factors are then computed with the same level of exceedance probability for point and areal means rainfalls.

An illustrative numerical application has been made for Montreal Island, with an area of 466 km^2 (Nguyen et al., 1981b). The gage at Dorval Airport was chosen as the reference gage because it is the only first-order weather station in the immediate region of Montreal. The Thiessen polygon method was used to compute the mean rainfall over Montreal Island by using the hourly rainfall data from the 5 recording rain gages appropriate to the Thiessen method. The longest concurrent record for these 5 gages was for the period September 1, 1969 through October 31, 1974; but, for the moment, only hourly rainfalls in the summer season (June through September) have been considered.

Figure 2 shows the comparison between the empirical and the theoretical distributions for areal correction factor. A more complete discussion of the results has been detailed by Nguyen et al. (1981b). From these results, it was found that the areal correction factor is not constant for all return periods and its values may not be less than one as was the necessary outcome under the premises of previous investigations.

REFERENCES

Nguyen, V.T.V. and Rousselle, J., 1981a. A Stochastic Model for the Time Distribution of Hourly Rainfall Depth. Water Resour. Res., Vol. 17, No. 2, pp. 399-409.

Nguyen, V.T.V., Rousselle, J. and McPherson, M.B., 1981b. Evaluation of Areal Versus Point Rainfall with Sparse Data, Can. Jour. Civ. Eng., Vol. 8, No. 2, pp. 173-178.

A RAINFALL-RUNOFF MODEL FOR DAILY FLOW SYNTHESIS

M. Mimikou, Civil Engineering Department, Athens Technical University and Public Power Corporation, Athens, Greece
A. Ramachandra Rao, School of Civil Engineering, Purdue University, West Lafayette, IN 47907, U.S.A.

ABSTRACT

A dynamic stochastic model for synthesis of daily flows and flood hydrographs at a certain site in a river reach is presented in this paper. The inputs to the model are the inflows at an upstream point of the reach and the rainfall on the basin between the upstream and downstream points. The model consists of three parts: an inflow-outflow transfer function model, a Kalman filter for the lateral flow contribution of the drainage basin and a second order autoregressive model for the noise component. The model for the lateral flow contribution to the outflow was necessary to correct the outflows given by the inflow-outflow transfer function of the system and it enables proper accounting of the lateral flow into the reach. The model is tested by using observed data. Daily flows and flood hydrographs at the downstream location simulated by using the model have been shown to agree with the historic data with remarkable accuracy. Besides, the synthesized data preserve the historical mean, skewness, kurtosis, lag-one autocorrelation and inflow-outflow cross-correlation coefficients at the 90% confidence level and the variance at the 95% confidence level. The model is found to be very efficient in extending daily flow records, estimating missing daily data and flow hydrographs, and for simulating daily reservoir inflows in real time, etc. The model is especially useful for synthesizing flows from basins of irregular hydrological characteristics and for which only limited data are available.

INTRODUCTION

Estimation of daily flows is important for efficient management of water resource systems. Estimation of daily flows at a location where

Reprinted from *Time Series Methods in Hydrosciences*, by A.H. El-Shaarawi and S.R. Esterby (Editors)

a stable rating curve can be established is a routine task (Nemec (1972)). However, better methods are needed to estimate daily flows at sites where flows are not gaged or at gaged sites where field measurements are not available because of various reasons such as equipment failure. The present paper deals with the development and testing a stochastic model for estimation of daily flows and flood hydrographs. The model was developed to generate data which are used for designing a dam in Greece and for use in operating the reservoir once it is built.

The model considered herein has three components. The first component is an input-output transfer function model. The residuals from this model are shown to be correlated with precipitation thereby indicating the presence of lateral inflow contribution in the residual sequence. The lateral inflow is estimated from the residual sequence by using Kalman filter techniques and this constitutes the second component of the model. The residual sequence from the second component of the model is also usually correlated. This residual sequence can be easily modeled by a low order Autoregressive Model which constitutes the third component of the model. The residuals from the AR model can be shown to be white and without periodicities.

The model is tested by using the daily precipitation and flow data from the Aoos river between Vovoussa (inlet) and Konitsa (outlet) measuring stations in Northern Greece (Fig. 1). The model is calibrated by using daily inflows and outflows of the reach and precipitation measured during 1971-75. The model performance is verified by using the data measured during 1975-77 which are not used for calibrating the model. The accuracy of the model to estimate daily flows and to estimate flood hydrographs are tested and the performance of the model is shown to be satisfactory.

The importance of estimation of lateral inflows, the extraction of hydrologically meaningful information from residual sequences and the limitations of simple transfer function models are discussed in the paper. Models of the type developed in this paper are compared to conceptual models (Crawford and Linsley (1966)) with special reference to watersheds in which the physical characteristics vary considerably and

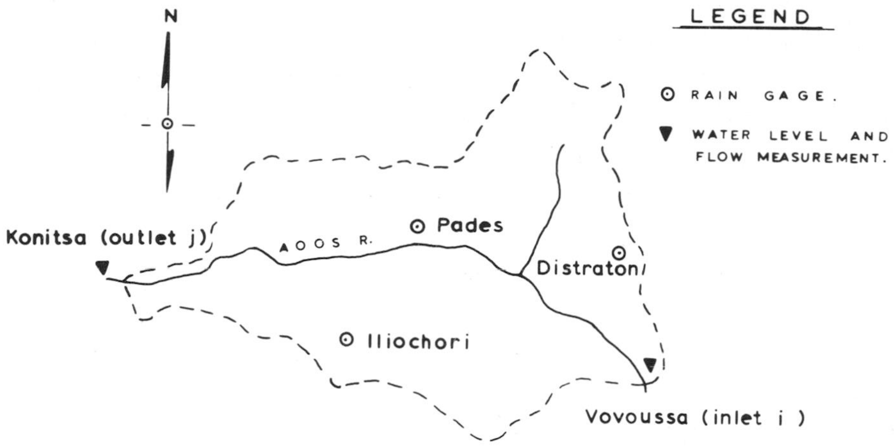

Fig. 1 Aoos river basin between Vovoussa (i) and Konitsa (j) stations.

where limited hydrological data are available.

The paper is organized as follows. The structure of the model is discussed in the second section. The data used in the study are briefly discussed in the third section. The details of model calibration are given in the fourth section. In the fifth section results of model verification are given. A set of conclusions are given in the last section.

MODEL DESCRIPTION

The inflow-outflow transfer function model.

The watershed between inlet (i) and outlet (j) is represented as a black box system. The inputs to the system are the inflows $q_{i,t}$ and the basin rainfall P_t. The outflows $q_{j,t}$ are the outputs from the system. Assuming a linear relationship between the output at time t and one of the inputs $q_{i,t}$ at times t-b, t-1-b,...,t-m-b,..., their relation can be written as:

$$q_{j,t} = U_o q_{i,t-b} + U_1 q_{i,t-1-b} + \ldots + U_m q_{1,t-m-b} + \ldots + S_L \quad (1)$$

where, U_o, $U_1,\ldots$, $U_m,\ldots$, are the weights of the impulsive response function of the system, b is the delay time between input and output, and S_t is a residual series. Using the notation of Box and Jenkins (1970), equ.(1) can be rewritten as:

$$q_{j,t} = (U_o + U_1B + \ldots + U_mB^m + \ldots)\, q_{i,t-b} + S_t = U(B)B^b q_{i,t} + S_t \qquad (2)$$

where B is the backwards shift operator and U(B) is the transfer function of the system. This transfer function may be written as a product of two polynomials of B of orders r and s by the following expressions:

$$U(B)=\delta^{-1}(B)\omega(B)\ldots \; (3) \quad \delta(B)=1-\delta_1B-\ldots-\delta_rB^r\ldots \; (4) \quad \omega(B)=\omega_o-\omega_1B-\ldots-\omega_sB^s \; (5)$$

where, $\delta_1,\delta_2,\ldots,\delta_r$, and $\omega_o,\omega_1,\ldots,\omega_s$, are two sets of parameters. Therefore, the final form of the inflow-outflow transfer function model is given by:

$$q_{j,t} = (1-\delta_1B-\ldots-\delta_rB^r)^{-1}(\omega_o-\omega_1B-\ldots-\omega_sB^s)B^b q_{i,t}+S_t. \qquad (6)$$

For constant inflow values, when the system progressively reaches an equilibrium, the outflow-inflow ratio, the gain g of the system, is given by equ. (6a) as:

$$g = (1 - \delta_1 - \ldots - \delta_r)^{-1}\,(\omega_o - \omega_1 - \ldots - \omega_s) \qquad (6a)$$

The parameters of the model in equ. (6) is estimated in the present study by using the cross-correlation coefficients between input and output sequences. The nonstationarity of the daily flows series is removed in order to estimate the parameters by differencing the time series. It should be noted that by differencing the inflow and outflow series the parameters of the model of equ. (6) do not change. The transfer function or the impulse response function was estimated by using equ. (7):

$$\begin{bmatrix} \gamma_{ij}(0) \\ \gamma_{ij}(1) \\ \vdots \\ \gamma_{ij}(m) \end{bmatrix} = \begin{bmatrix} \gamma_{ii}(0) & \cdots & \gamma_{ii}(m) \\ \gamma_{ii}(1) & \cdots & \gamma_{ii}(m-1) \\ \vdots & \cdots & \vdots \\ \gamma_{ii}(m) & \cdots & \gamma_{ii}(0) \end{bmatrix} \times \begin{bmatrix} U_o \\ U_1 \\ \vdots \\ U_m \end{bmatrix} \qquad (7)$$

where, $\gamma_{ij}(m)$ is the inflow-outflow cross-covariance function, $\gamma_{ii}(m)$ is the inflow autocovariance function and lag m is selected so that U_m approaches zero. The parameters δ_i and ω_j of the model in equ. (6) are estimated by equating the coefficients of B in equ. (3) as follows:

$$\begin{aligned} U_j &= 0 , && J < b \\ U_j &= \delta_1 U_{j-1}+\ldots+\delta_r U_{j-r}+\omega_o , && J = b \\ U_j &= \delta_1 U_{j-1}+\ldots+\delta_r U_{j-r}-\omega_{j-b} , && J = b+1,\ b+2,\ \ldots,\ b+s \\ U_j &= \delta_1 U_{j-1}+\ldots+\delta_r U_{j-r} , && J > b + s \end{aligned} \qquad (8)$$

It is apparent that in order to solve equ. (8), the order (r,s,b) of the model equ. (6) must be estimated. The order r is here estimated by using a general characteristic of the inflow-outflow process discussed below and the adequacy of this estimate is further confirmed by data analysis. The orders s and b, on the other hand, which are specific characteristics of the hydrological system under consideration are estimated by trial and error during the calibration of the model.

The order r is assumed to be equal to the order of the differential equation which can be used to represent the inflow-outflow relationship in the reach between (i) and (j). When the system is undisturbed, the inflow-outflow transfer function is simple and linear, given by:

$$q_{j,t} = g q_{i,t} \qquad (9)$$

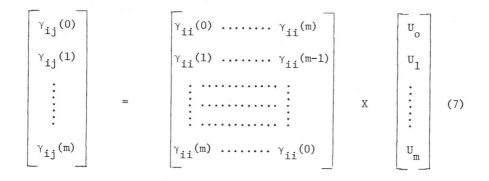

with g the steady state gain of the system. For unsteady flows, the difference ($gq_{i,t-b}-q_{j,t}$) where b the time lag, is usually different from zero. This difference ($gq_{i,t-b}-q_{j,t}$) obeys the differential equation:

$$\frac{dq_{j,t}}{dt} = \frac{1}{T}(gq_{i,t-b} - q_{j,t}), \tag{10}$$

where T is a time constant of the system (Box and Jenkins (1970)). Defining $dq_{j,t}/dt = Dq_{j,t}$ equ. (10) may be rewritten as equ. (11), which is a first

$$(1+TD)q_{j,t} = gq_{i,t-b}, \tag{11}$$

order differential equation representing the inflow-outflow relationship in a reach. Therefore, the order r is assumed equal to one.

So far, the other input to the system, namely the rainfall over the basin, which produces a lateral flow contribution to the outflow, has not been considered in the transfer function model. In relatively small basins, such as the one under consideration, the rainfall over the basin is related to inflows because of the uniformity of storm occurrences. In such cases, the inflow-outflow transfer function model takes into account a part of the lateral flow contribution, which depends on the significance of the rainfall-inflow relationship. In any case, it is difficult to estimate the lateral flow contribution accurately, mainly because it is strongly affected by local character of storms. The significance of lateral flow estimates can be checked by using the residuals S_t of the model in equ. (6). If the lateral flows are a predominant component of outflows, then these residuals would be significantly cross-correlated to the rainfall over the basin. Otherwise, the transfer function would account for the inflow-outflow relationships and hence the cross-correlations between the inflows and residuals would be small. If the residuals are strongly cross-correlated to the rainfall then the lateral flows must be modeled separately and used with the transfer function model. Besides, the transfer function gain of the system given by equ. (6a), which may be used as an indication of the mean lateral flow contribution from the basin (Whitehead et al. (1979)), must be also corrected.

The lateral inflow model.

The residual series S_t of the model in equ. (6) is a combination of lateral inflows and noise and is given by equ. (12), where $\hat{q}_{j,t}$ is the

$$S_t = q_{j,t} - \hat{q}_{j,t} \tag{12}$$

estimated and $q_{j,t}$ the observed outflow. Thus the lateral inflows LS_t must be filtered from the noise in S_t. This can be accomplished by a Kalman filter with time varying parameters (Schwartz and Shaw (1975)). The model used for lateral inflow estimation is given in equ. (13), for which the mean daily basin rainfall P_t is the input and a_t and b_t

$$LS_t = a_t LS_{t-1} + b_t P_t, \quad |a_t| < 1, \tag{13}$$

parameters of the filter. Based on the following two conditions, Schwartz and Shaw (1975) give the expressions for the estimation of the parameters of the filter:

(1) Input series P_t is a stationary stochastic process.

(2) Sample series S_t can be modeled as a first order autoregressive process driven by a zero-mean white noise ω_t.

$$S_t = \alpha S_{t-1} + \omega_t \tag{13a}$$

Thus the noise variance σ_ω^2 is related to the variance σ_s^2 of the sample series through the formula: $\sigma_\omega^2 = \sigma_s^2\,(1-\alpha^2)$, where α is the first order autocorrelation coefficient of the sample series. The relation between sample series S_t and the input series P_t is obscured by an additive zero-mean observation white noise n_t, as shown in equ. (14).

$$P_t = S_t + n_t \tag{14}$$

The noise variance σ_n^2 is related to the variances σ_p^2 and σ_s^2 of series P_t and S_t, by the expression: $\sigma_n^2 = \sigma_p^2 - \sigma_s^2$.

Under these assumptions, the expressions for the parameters a_t and b_t in equ. (13) are as follows.

$b_t=(A+\alpha^2 b_{t-1}/1+A\alpha^2 b_{t-1})$ (15) $\quad a_t=\alpha(1-b_t), \; |a_t| < 1, \; |\alpha| \leqq 1,$ (16)

where α is already defined and A is the ratio of the variance of the white-noise ω_t to the variance of the observation noise σ_n^2.

$$A = \sigma_\omega^2/\sigma_n^2 = \sigma_s^2(1-\alpha^2)/\sigma_p^2 - \sigma_s^2 \tag{17}$$

The estimated b_t values from equ. (15) has very limited variability, since it rapidly converges to a steady value. Equation (15) is valid only when equ. (14) is realized, i.e., when nothing else besides noise n_t interferes between the input series P_t and the signal S_t. This situation can be only realized when the basin is saturated, when all components such as infiltration, interception, overland flow, etc. reach their maximum values. Another observation is that as b_t is smaller than its steady value, equ. (15) is an increasing function on the (b_t, b_{t-1}) plane. Since b_t can be physically interpreted as the daily runoff coefficient of the basin (runoff-rainfall ratio), it is strongly related to rainfall. Experience has shown that the runoff coefficient of the basin increases (up to a steady limiting value), when rainfall increases or when it remains constant. Therefore, equ. (15) is more useful for periods of heavy (capable to saturate the basin) and increasing or constant rainfall. For the other periods of the year some modifications of equ. (15) are necessary.

Replacing LS_{t-1} in equ. (13) by its expression from the same equation at the previous step and again doing the same for LS_{t-2}, etc, one can get the following expression for the Kalman filter equ. (18) where B is

$$LS_t = a_t t_{t-1} \cdots a_{t-k} LS_{t-k-1} + \left[\sum_{n=0}^{k-1} (a_t a_{t-1} \cdots a_{t-n} B^{n+1}) + 1\right] b_t P_t, \tag{18}$$

the backwards shift operator. The first term of equ. (18) is very small and can be omitted. Therefore, the filter for the lateral flow contribution becomes equ. (19) where k is the memory of the rainfall-lateral inflow process.

$$LS_t = \left[\sum_{n=0}^{k-1} (a_t a_{t-1} \cdots a_{t-n} B^{n+1}) + 1\right] b_t P_t, \tag{19}$$

The input-output model for daily flow synthesis.

Adding the signal LS_t to the right hand side of equ. (6), improved estimates are obtained for the daily outflows $\hat{q}_{j,t}$. The new residuals R_t from the historical values $q_{j,t}$, called now second stage residuals of the model, must be uncorrelated to the rainfall series P_t. The eventual autocorrelation structure of this residual series is due to the presence of a noise component R_t, which can be removed by the following ARMA (p,q) model, equ. (20) (Box and Jenkins (1970), Kashyap and Rao (1976))

$$R_t = \phi^{-1}(B)\theta(B)\varepsilon_t, \tag{20}$$

where $\phi(B)$, $\theta(B)$ are polynomials of B of order p and q correspondingly, and ε_t is a zero-mean white noise series representing the final third stage residuals of the model. Thus, finally, the input-output model for daily flows synthesis takes the form:

$$q_{j,t} = \delta^{-1}(B)\omega(B)B^b q_{i,t} + \left[\sum_{n=0}^{k-1} (a_t a_{t-1} \cdots a_{t-n} B^{n+1}) + 1\right] b_t P_t + \phi^{-1}(B)\theta(B)\varepsilon_t \tag{21}$$

DATA USED IN THE STUDY

The model is tested by using the data of a reach of Aoos river in Northern Greece, between the inlet station (i) at Vovoussa and the outlet station (j) at Konitsa. The area of the drainage basin is 461 km^2 and the reach is 34 km long. Daily flow data are available at the Vovoussa and Konitsa stations from 1965 and 1971 respectively. Daily rainfall measured at Pades, Distraton and Iliochori stations in the basin are also available. However, only the station at Pades is continually working since 1965, and the data from the other two stations are discontinuous. Furthermore, the Pades station can be considered as a representative station, because it is located at the center of the basin and at an altitude close to the mean basin altitude. Consequently, the daily rainfall at Pades station, after multiplication by a point-to-surface rainfall coefficient of 0.91 is assumed to give the mean daily basin rainfall. Rainfall values were converted to m^3/sec and used. Six years of data (1971-77) of daily flows and rainfall were used in the study. The first four years (1971-75) of data were used for the calibration of the model and the last two years (1975-77) of data were used for verification.

CALIBRATION OF THE MODEL

Transfer function model calibration.

The auto and cross-correlation coefficients of the inflow, outflow and their differenced series indicated a weakly nonstationary behavior. Daily series are usually seasonal with yearly and within-the-year cyclicity. In this particular case, the autocorrelation structure of the daily flows indicated only weakly periodic behavior. For this reason only a nonseasonal first order differencing operation was initially tried. The efficiency of this differencing alone to significantly reduce the periodic component of the series is checked by the auto and cross-correlograms of the differenced series. The correlograms of the first differenced series exhibited nonperiodic stationary behavior, and fluctuated inside the 97.5% confidence limits (Jenkins and Watts (1969)) after the first few lags. Besides, the first and second order autocorrelation coefficients of the inflow and outflow differenced series indicated the nonperiodic behavior of the autocorrelation coefficients. Therefore, the nonseasonal first order differencing was considered adequate to render the flow series stationary.

The auto and cross-covariance functions of the differenced series are used to estimate the weights of the impulse response function of the system equ. (7). The maximum lag m in equ. (7) was found by sensitivity analysis. It was found that the maximum value of m was about 10 and the impulse response function ordinates U_{11}, U_{12}, ..., etc, were practically zero. The impulse response values indicated that the first two values of the impulse response function are by far more important than the rest.

The cross-correlogram of the differenced inflow-outflow processes which had the highest value at zero lag, suggested that the delay parameter b is zero. The memory of the inflow-outflow process was significant upto the third lag. In other words, the parameter s was equal to three. It has been already shown that the order r of the inflow-outflow transfer process can be assumed to be equal to one. This assumption was also verified by using the estimated impulse response function, which satisfies the difference equation: $U_j-\delta_1 U_{j-1}=0$, with $\delta_1=0.36$, for $j>3$. Besides, examining the difference equation: $U_j-\delta U_{j-1}-\delta_2 U_{j-2}=0$ for $j>3$, for a second order model, it was found that, for $j = 4$, and $j = 5$, this equation gives a very small value of 0.006 for δ_2, which is insignificant compared to δ_1.

These results were used to fix the values of r, s and b. By using $r = 1$, $s = 3$, $b = 0$, equ. (8) were solved to estimate the parameters δ_i and ω_j. Values of these estimates with their standard errors in parentheses are: $\delta_1 = 0.36$ (0.03), $\omega_o = 1.81$ (0.07), $\omega_1 = 0.54$ (0.04), $\omega_2 = 0.12$ (0.02), $\omega_3 = 0.06$ (0.01). The transfer function gain of the system is thus $g = (1-0.36)^{-1}$ $(1.81-0.54-0.12-0.06) = 1.70$. These initial estimates may be refined by using optimization techniques (Box and Jenkins (1970)). But if the third stage residual series of the model are a white noise series, these estimates would be very close to the optimal estimates.

Estimation of the parameters of the lateral inflow model.

Applying the transfer function model to the data of the calibration period, one can obtain the first stage residual series S_t of equ. (12). The cross-correlogram of residuals S_t and basin rainfall series exhibited strong dependence of S_t on rainfall, which is partly due to the presence of the lateral flow contribution in the residuals. The cross-correlation between S_t and inflows on the other hand, indicated very weak dependence of the S_t series on the inflows. The correlogram between first stage residuals and rainfall had the maximum value at lag zero rapidly decayed after the third lag, which implied that the memory parameter k of the model of equ. (19) was equal to three.

The parameters a_t and b_t of the model for lateral inflow are estimated from equations (15) and (16), after verifying that the two previously described (Schwartz and Shaw (1975)) conditions are met. The first condition, about the stationarity and the stochasticity of the input series is met, because the series of daily rainfall at Pades can be considered to be a stationary series. This was verified by examining the correlogram of the rainfall series which fluctuated inside the 97.5% confidence limits after the first few lags. The second condition is also met, as the correlogram of the first stage residual series S_t had an exponential form and resembles the correlogram of a first order Markov process. Therefore, equations (15) and (6) can be used to estimate the parameters a_t and b_t. The parameters a and A are estimated to be equal to 0.36 and 0.10 respectively.

Equation (15) can be written as equ. (22) after defining $L=1/1+A+\alpha^2 b_{t-1}$, and is valid for time periods of increasing or constant rainfall,

$$b_t = 1 - L \tag{22}$$

greater than or equal to a value which saturates the basin, as explained earlier. For the rest of the periods, equ. (15) and more specifically its slope on the (b_t, b_{t-1}) plane must be modified. The modifications are made in accordance to the changes which have been observed in the relationship between the historical daily rainfall and the runoff coefficient (which is the physical analog of b_t) of the Aoos river basin. Several such modifications were used in the present study and are discussed below.

The slope of equ. (15), using the transformation L, is given by equ. (23).

$$db_t/db_{t-1} = \alpha^2 L^2 \tag{23}$$

The following modifications are made in estimating b_t. (1) For rainfall values less than a critical value, below which no direct surface runoff is produced, b_t is set equal to b_{t-1}. This critical value was found to be approximately equal to 0.5 mm/day for the present data. (2) Above this critical value and for increasing or constant rainfall up to a value which saturates the basin, the increasing rate of the runoff coefficient of the basin and therefore, b_t is estimated to be equal to the slope given by equ. (23) to become $\alpha(\alpha \leq 1)$ times smaller; namely it becomes milder and equal to:

$$(db_t/db_{t-1}) = \alpha^3 L^2 \tag{24}$$

Integrating equ. (23), one gets the expression for b_t in equ. (25),

$$b_t = \alpha(1-L) \quad (25)$$

The upper zone in moderately covered basins, with steep slopes and with reasonably uniform rainfall throughout the year, is approximately saturated for daily rainfall values equal to: $0.06(4+\bar{P}_t/8)$, with $\bar{P}_t$ the mean yearly basin rainfall in inch (25.4 mm), (Crawford and Linsley (1966)) equal here to 900 mm. Therefore, the rainfall value for the saturation of the basin is approximately equal to 12.0 mm/day. (3) For decreasing rainfall, the runoff coefficient and therefore b_t is estimated to decrease also and the slope of equ. (23) is changing direction; the expression found to fit the recession limb of b_t well is the following:

$$(db_t/db_{t-1}) = 1-m(\alpha^2L^2)^{\mu} \quad (26)$$

For $P_t \geq 12.0$ mm/day, it is estimated that b_t is slowly decreasing, according to a rate given by equ. (26) with $m = 1$, $\mu = 1$, (mild slope), and the expression for b_t becomes approximately:

$$b_t = b_{t-1}(1-\alpha^2L^2) \quad (27)$$

For $P_t < 12.0$ mm/day, it is estimated that b_t is rapidly decreasing, according to a rate given by equ. (26) with $m = 1$, $\mu = 1/2$, (steep slope), and the expression for b_t becomes approximately:

$$b_t = b_{t-1}(1 - \alpha L) \quad (28)$$

Starting with an arbitrary value for b_1 and using the parameters a and A the time varying parameter b_t and hence a_t are estimated. The efficiency of the calibrated filter of equ. (19) to extract the lateral inflows LS_t from the series S_t is tested by investigating the cross-correlogram of the second stage residual series R_t of the model with the rainfall series. The cross-correlogram of R_t and P_t indicated that these two series are uncorrelated.

The mean lateral inflow signal value for the calibration period was found to be equal to 3.70 m^3/sec, which accounts for 40% of the mean inflow value. In other words, the inflow-outflow transfer function gain g found equal to 1.70, must be corrected to: $g' = 1.70 + 0.40 = 2.10$. The intermediate basin contributes 110% of the mean flow at Konitsa station and not 70%, as the gain g is indicating.

The noise model.

The autocorrelation structure of the residual series R_t, resembled the autocorrelation structure of a second order autoregressive process. This means that the autocorrelated signal R_t, inherent in the residuals, can be modeled as in equ. (29).

$$R_t = \phi^{-1}(B)\varepsilon_t = (1 - \phi_1 B - \phi_2 B^2)^{-1} \varepsilon_t \tag{29}$$

The estimated parameters ϕ_1, ϕ_2 and σ_ε, the standard deviation of noise, are: $\phi_1 = \phi_2 = 0.25$ and $\sigma_\varepsilon = 7.5$ m^3/sec.

Diagnostic checking of the model.

The white noise series ε_t of equ. (29) may be estimated by using equ. (21). Without b and $\theta(B)$, equ. (21) gives the following expression for the residuals $\hat{\varepsilon}_t$:

$$\hat{\varepsilon}_t = \phi(B)[q_{j,t} - \delta^{-1}(B)\omega(B)q_{i,t} - LS_t] \tag{30}$$

Applying (30) to the data of the calibration period, series $\hat{\varepsilon}_t$ is obtained and then used for diagnostic checking of the model and of the efficiency of its parameters estimates. If the model is correct and its parameters have been efficiently estimated, this residual series must be a zero-mean white noise series. In addition it has to satisfy the Darbin-Watson statistic d, given by:

$$d = \sum_{i=2}^{N} (\hat{\varepsilon}_t - \hat{\varepsilon}_{t-1}) \ / \ \sum_{i=1}^{N} \hat{\varepsilon}_t^2 , \tag{31}$$

which was found (Kendall (1973)) to be equal to zero for autocorrelated sequences and close to two for random sequences with N values.

Indeed, $\hat{\varepsilon}_t$ was found to be a zero-mean white noise series, with variance equal to 7.5 m^3/sec and with statistic d equal to 1.96. The empirical probability density function of the series had a high peak and was approximately symmetrical and bounded. One can assume that its high kurtosis in comparison with a normal distribution may have resulted from over or under-removal of harmonics in periodic components of the daily flow series (Yevjevich (1976)). Based only on the skewness test for normality discussed by Hipel et al (1977), the skewness coefficient was found to be not significantly different from zero at 97.5% confidence level. Therefore, series ε_t is assumed to be normal, which can be generated by the following mechanism, where t_i are standard normal variates from N(0,1). The final model takes the form:

$$q_{j,t} = (1-0.36B)^{-1}(1.81-0.54B-0.12B^2-0.06B^3)q_{i,t} +$$
$$[(a_tB+a_ta_{t-1}B^2+a_ta_{t-1}a_{t-2}B^3)+1]b_tP_t + (1-0.25B-0.25B^2)^{-1}\sigma_\varepsilon t_i \quad (33)$$

Synthesis of daily flows in the calibration period.

The efficiency of the model of (33) in synthesizing daily flows in the calibration period was checked by the following criteria:

(1) The mean accuracy in simulation, which is given by:

$$M.A. = \frac{1}{N}\sum_{t=1}^{N}\left|\frac{q_{j,t} - \hat{q}_{j,t}}{q_{j,t}}\right| \quad (34)$$

where $q_{j,t}$ and $\hat{q}_{j,t}$ are respectively the historical and simulated flows and N the sample size. This value was found to be equal to 14%. The application of the model in (33) without the lateral inflows and the noise model, in other words only the transfer function model equ. (6), had an M.A. value of 27%. Addition of the other models thus significantly improves the accuracy of the transfer function model.

(2) The ability of the model to preserve some historical statistical characteristics. The statistical characteristics of the historical and estimated data were also evaluated for different years. The null hypothesis (Benjamin and Cornell (1970)), that each estimated characteristic is not significantly different from the historical one can be accepted at the 90% confidence level for the mean, skewness, kurtosis and first order autocorrelation and inflow-outflow cross-correlation coefficients and at the 95% confidence level for the variance.

MODEL VERIFICATION AND APPLICATIONS

Synthesis of daily data which were not used for calibration.

The model (33) was used to synthesize daily flows which were not used to calibrate it. The synthetic outflows estimated by using the model were compared to the observed outflows. The efficiency of the model is again checked by the two previously described criteria. The simulated and the historical series were in good agreement. The mean accuracy of the model was good and found to be equal to 17%. The application of only the transfer function equ. (6) gave an accuracy of 30%. Thus once again the addition of the two other models to the transfer function model signficantly improves the accuracy of daily flow estimation. Finally, the model preserves the historical statistical characteristics of the series at the 95% confidence level for the variance and at the 90% level for the rest of the characteristics.

Verification of the model for daily flood routing.

Six separate flood hydrographs, recorded at the inlet of the basin during the year 1976-77, were routed in order to estimate the corresponding outflow hydrographs at the outlet station. Historical outflow data were not used in the computations and the synthetic hydrographs were compared with the observed hydrographs. The mean accuracy in routing is again measured by (34), where N is the duration of each hydrograph in days. Another measure of accuracy was also used in this phase of the study. It is the "peak accuracy", defined as:

$$P.A. = \left| \frac{q^p_{j,t} - \hat{q}^p_{j,t}}{q^p_{j,t}} \right| , \qquad (35)$$

where $q^p_{j,t}$ and $\hat{q}^p_{j,t}$ are the historical and synthetic peak flows respectively. The mean and peak accuracies estimated from (34) and (35) are given for each of the six hydrographs. It was again apparent that there is a significant improvement in the accuracy of the model with the addition of the lateral inflow and noise models.

CONCLUSIONS

The following conclusions are arrived at from this study.

(1) The input-output stochastic model developed in this study can be efficiently applied for daily flows and daily flood hydrograph synethesis. The model preserves the historical yearly mean, skewness, kurtosis and

first order autocorrelation and inflow-outflow cross-correlation. Coefficients at the 90% confidence level and the variance at the 95% level.

(2) The addition of the lateral inflow and the noise models to the inflow-outflow transfer function model significantly improves its accuracy in daily flow estimation. The addition of the lateral inflow model is necessary whenever the lateral inflow component is identified in the residuals of the inflow-outflow transfer function model.

(3) The model connected to an automatic network with a small computer can be operationally used on-line for real time daily flows estimation, which is especially useful at sites with reservoirs in operation.

(4) The model, because of its limited data need, is especially useful for estimating runoff from watersheds with highly variable physical characteristics (roughness, rating curves, etc.) and where limited data are available.

REFERENCES

Benjamin, T.R. and A.C. Cornell, 1970. Probability, Statistics and Decision for Civil Engineers, McGraw-Hill Co., New York, 684 pp.

Box, G.P. and G.M. Jenkins, 1970. Time Series Analysis-Forecasting and Control, Holden-Day Co., San Francisco, 553 pp.

Crawford, N.H. and R.K. Linsley, 1966. Digital Simulation in Hydrology: Stanford Watershed Model IV, Tech. Rept. No. 39, Stanford University, California, 210 pp.

Eagleson, S.P., 1970. Dynamic Hydrology, McGraw-Hill Co., New York, 462 pp.

Hipel, K.W., A.I. McLeod and W.C. Lennox, 1977. Advances in Box-Jenkins Modeling, 1-Model Construction, Water Resour. Res., 13(3), 567-576.

Jenkins, G.M. and D.G. Watts, 1969. Spectral Analysis and Its Applications, Holden-Day Co., San Francisco, 525 pp.

Kashyap, R.L. and A.R. Rao, 1976. "Dynamic Stochastic Models from Empirical Data", Academic Press, New York, New York.

Kendall, M.G., 1973. Time Series, Griffin, London, 330 pp.

Nemec, J., 1972. Engineering Hydrology, McGraw-Hill Co., England, 316 pp.

Schwartz, M. and L. Shaw, 1975. Signal Processing, Discrete Spectral Analysis, Detection and Estimation, McGraw-Hill Co., New York, 396 pp.

Whitehead, P., G. Hornberger and R. Black, 1979. Effects of Parameter Uncertainty in a Flow Routing Model, Hydrol. Sc. Bull., 24/4, 445-463.

Yevjevich, V., 1976. Structure of Natural Hydrologic Time Processes. In: H.W. Shen (Editor), Stochastic Approaches to Water Resources, Vol. I:2. 1-2.59.

ANALYSIS OF FLOOD SERIES BY STOCHASTIC MODELS

P. VERSACE, M. FIORENTINO AND F. ROSSI

Dip. Difesa del Suolo, Universita della Calabria, and Ist. Idraulica e Costruzioni Idrauliche, Universita di Napoli, Italy

ABSTRACT

Flood analysis for regions, like Southern Italy, where the annual flood series exhibits outliers (and, then, high skewness), associated with disastrous storms, requires building suitable stochastic models. In such cases, the usual simple model (Model A), which assumes the largest annual flood to be the maximum of a Poissonian number of indepen dent random variables with common exponential distribution function, proves to be inadequate. Better models can be built by replacing the hy potheses on which Model A is based with others, phenomenologically closer to reality, namely, that the number of exceedances in a year is still a non-homogeneous Poisson process, but the exceedance values are not identically distributed random variables. Of the two models considered, i.e., a time-dependent distribution for the exceedance magnitude (Model B) and a mixed exponential distribution (Model C), the latter is found to give a better statistical fit. There is also better phenomenological support for Model C in that disastrous storms occur more rarely but with much larger intensities than others, and they are accordingly better modelled as belonging to different populations.

1 INTRODUCTION

The analysis of floods has been the object of investigations by many authors. Among the approaches followed, two distinct ones, respectively empirical and theoretical, may be identified. The former consists in guessing which theoretical distribution best fits the observed frequency distribution of the largest annual flood peak. Following this approach, for example, the log Pearson Type-3 distribution has been recommended in the USA (U.S.W.R.C., 1977). While this particular choice has met with much adverse criticism (Landwehr et al., 1978), more generally the empirical approach is objected to, in principle, on several grounds. Thus, it makes no use

Reprinted from *Time Series Methods in Hydrosciences*, by A.H. El-Shaarawi and S.R. Esterby (Editors)

of the partial flood series, which retains more information than is the case with the annual flood series (Todorovic,1978). Goodness-of-fit tests, used to compare the performance of different distributions, yield largely inconclusive results even with the longer records (N.E.R.C., 1975). Furthermore, the approach takes no account of physical aspects of the phenomena investigated. Finally, with the distributions most commonly used, one is unable to account for the high observed variance of the skewness or for the presence of outliers, as are sometimes the case in the data observed (Rossi and Versace, 1981). By contrast, the theoretical approach endeavours to construct a model, based on phenomenological considerations. The data are then used merely to verify the model and, possibly, to suggest which if any modifications are needed. In recent years this approach has undergone much development and it would seem to offer the best basis for the analysis and prediction of floods.

2 MATHEMATICAL MODELS

Let us consider the stochastic process described by the stream-flow hydrograph $\{Q(t);\ t\geq 0\}$ and let us select a base level q_0. The sequence of the hydrograph peaks above q_0 (referred to as the process of exceedances) is a marked point process (Snyder, 1975) characterized by:

- a sequence $\tau_1, \tau_2, \ldots, \tau_i, \ldots$, where τ_i is the instant of time when the i-th exceedance occurs;
- a sequence $Z_1, Z_2, \ldots, Z_i, \ldots$, where $Z_i = Q(\tau_i) - q_0$ is the magnitude of the exceedance at time τ_i.

Both occurrence times and exceedance values are random variables. The process is further characterized by the random variable K_t, the number of exceedances within a fixed interval $[0, t]$, which can assume, for every $t\geq 0$, the integer values $k = 0, 1, 2, \ldots$:

$$K_t = \max \{i;\ \tau_i \leq t\}; \tag{1}$$

so $\{K_t;\ t\geq 0\}$ is a counting process.

Let X'_t denote the magnitude of the largest exceedance within $[0, t]$, i.e.,

$$X'_t = \max_{\tau_i \leq t} Z_i = \max_{0\leq i\leq K_t} Z_i; \tag{2}$$

so X'_t is the maximun among a random number of random variables.

Accordingly the distribution of X'_t will depend on both the counting process $\{K_t;\ t\geq 0\}$ and the distribution of $\{Z_i\}$.

The process $\{K_t;\ t\geq 0\}$ is usually assumed to be a non-homogeneous

Poisson counting process (Zelenhasic, 1970; Todorovic and Zelenhasic, 1970; Dauty, 1972; North, 1980; Rossi and Versace, 1981) with

$$P_{K_t}(k) = P[K_t = k] = \frac{(\Lambda_t)^k \exp(-\Lambda_t)}{k!}, \quad k = 0, 1, 2, \ldots \tag{3}$$

where

$$\Lambda_t = E[K_t] \tag{4}$$

is the parameter function of the Poisson process. The derivative $\lambda(t)$ of Λ_t is the intensity function of the process, i.e.,

$$\Lambda_t = \int_0^t \lambda(u)du. \tag{5}$$

For a high enough base level q_0, the variables Z_i may be assumed to be mutually independent. Many authors (Zelenhasic, 1970; Todorovic and Zelenhasic, 1970; Dauty, 1972) introduce the further assumption that the Z_i's are identically distributed random variables, their common distribution being of the exponential type:

$$F_Z(z) = P[Z \leq z] = 1 - e^{-\beta z} \tag{6}$$

where

$$E[Z] = 1/\beta. \tag{7}$$

In this case the distribution of X'_t, the largest exceedance within $[0, t]$, is

$$F_{X'_t}(x') = P[X'_t \leq x'] = \exp(-\Lambda_t e^{-\beta x'}) \tag{8}$$

If the interval $[0, t]$ is a year and we assume $\Lambda_t = \exp[\alpha(\varepsilon - q_0)]$ and $\beta = \alpha$, it follows from (8) that

$$F_X(x) = P[X \leq x] = \exp[-e^{-\alpha(x-\varepsilon)}] \tag{9}$$

where X denotes the largest annual flood. Equation (9) is the well-known Gumbel's distribution (model A) with parameters α and ε.

In many cases there is a good agreement between Gumbel's distribution and observed annual flood series, indicating that the assumptions introduced in the derivation above are basically correct. There are cases, however, when, using Gumbel's distribution, the observed and fitted distributions of the largest annual floods exhibit appreciable discrepancy, and the need for more refined models arises. One may proceed in this direction, by removing the strongest of the above hypotheses, namely, that the Z_i are identically distributed random variables.

As it has been remarked by many authors (Todorovic and Rousselle, 1971; Rousselle, 1972; North, 1980), the distribution of Z_i is actually dependent on τ_i. This time dependence may be allowed for by retaining an exponential distribution and then assuming its parameter β to be time dependent, i.e.,

$$F_Z(z) = P[Z \leq z] = 1 - e^{-\beta(t)z}. \tag{10}$$

On this assumption, the distribution of the largest exceedance X_t' within $[0, t]$ will be given by the expression

$$F_{X_t'}(x') = P[X_t' \leq x'] = \exp\left\{-\int_0^t e^{-\beta(u)x'} \lambda(u)du\right\} \tag{11}$$

This distribution will be referred to as Model B.

Another model deserving consideration is obtained by assuming that Z_i arises as the mixture of two components, both exponentially distributed. Its distribution is accordingly written:

$$F_Z(z) = pF_{Z_1}(z) + (1-p)F_{Z_2}(z) = p(1-e^{-\beta_1 z}) + (1-p)(1-e^{-\beta_2 z}) \tag{12}$$

Z_1 and Z_2 being the component random variables and p the proportion of Z_1 in the mixture.

The underlying assumption of this model allows for the existence of two distinct types of precipitation, as is the case in some regions like Southern Italy (Penta et al., 1980).

If the numbers of exceedances of the two components in a year, K_1 and K_2 follow Poisson processes of parameters Λ_1 and Λ_2 respectively, we have

$$F_{X'}(x') = \exp(-\Lambda_1 e^{-\beta_1 x'} - \Lambda_2 e^{-\beta_2 x'}) \tag{13}$$

where X' is the largest exceedance in a year, and

$$\frac{\Lambda_1}{\Lambda_1 + \Lambda_2} = p \tag{14}$$

As is readily shown, the distribution of the largest annual flood may be written:

$$F_X(x) = \exp\left[-e^{-\alpha_1(x-\varepsilon_2)} - e^{-\alpha_2(x-\varepsilon_2)}\right] = F_{X_1}(x)\, F_{X_2}(x) \tag{15}$$

i.e., as the product of two Gumbel's distributions of parameters α_1,

ε_1 and α_2, ε_2 respectively, i.e., the largest-annual-flood distributions of the individual components.

This third model shall be referred to as Model C.

3 APPLICATIONS

The above three models were applied to analysing several series of largest annual flood peaks in Southern Italy. As a typical example, an account is here given of such analysis for the daily flows at the Amato River, at Marino station (Calabria), for which a 36-year record is available. Compared with Gumbel's distribution, the annual flood series exhibits an outlier, the largest and next largest observed values being $x_{(n)} = 185\ m^3sec^{-1}$ and $x_{(n-1)} = 81\ m^3sec^{-1}$ respectively. As a result, the observed skewness coefficient ($\hat{\gamma}_1 = 2.80$) is much too high for a Gumbel distribution with $n = 36$, for which the expected value and standard deviation of the sample skewness coefficient are $E[\hat{\gamma}_1] = 0.88$ and $\sigma[\hat{\gamma}_1] = 0.54$ respectively (Matalas et al., 1975).

To investigate the validity of the hypotheses on which Model A is based, let us consider the partial duration series. The number of independent exceedances occured was taken to equal 74. Observed and theoretical distribution functions of the number of exceedances in a year are shown in Fig. 1 ($\hat{\Lambda} = \bar{k} = 2.06$). The good agreement between

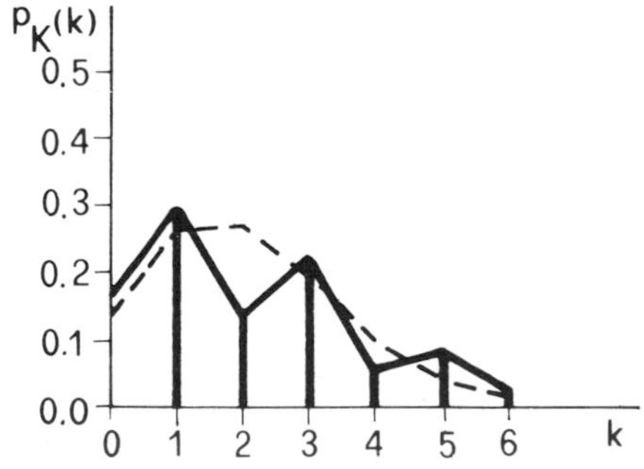

Fig. 1. Amato River at Marino. Observed (solid line) and theoretical Poisson (broken line) distribution functions of the number of exceedances in a year ($\hat{\Lambda} = \bar{k} = 2.06$).

the distributions lends support to the hypothesis that the process $\{K_t, t \geq 0\}$ is a Poisson counting process. The conclusion is also warranted by the value of the test statistic R, equalling the ratio of the observed variance to the observed mean ($R = 1.30$ against the critical value at the 5% level, $R = 1.42$).

Observed and theoretical (exponential) distribution functions for the magnitude of the exceedances are shown in Fig. 2. The estimates

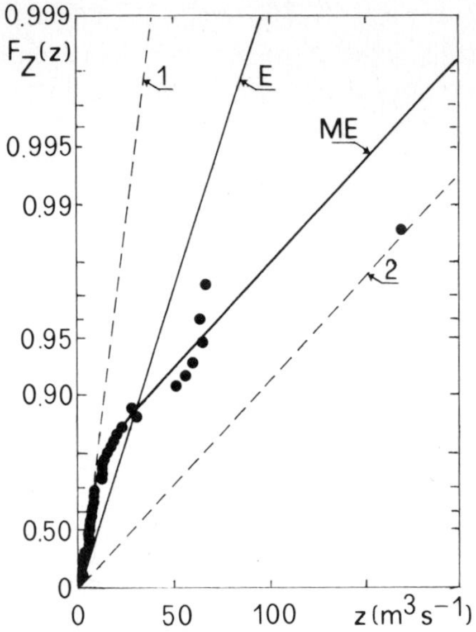

Fig. 2. Amato River at Marino. Observed (black points) distribution function of exceedance values. Exponential (Curve E) and mixed exponential (Curve ME) theoretical distribution functions. Distribution functions for the individual components (Curves 1 and 2) of the ME model.

for the parameters q_0 and β in (6) were obtained by the best linear unbiased estimators (Sarhan, 1954). The theoretical distribution (curve E) is a poor fit to the observed data, particularly at the largest values which are significantly underestimated.

In Fig. 3 the annual flood series also indicates a poor fit by Model A. Furthermore, were Model A applicable, the observed largest value $x_{(n)}$ = 185 m^3sec^{-1} would correspond to a cumulative exceedance probability close to unity both for the maximum annual flood distribution $F_X(x)$ and the maximum-in-36-years flood distribution $F_n(x)$.

Let us now pass to consider alternative models, starting from Model B which pays tribute to the underlying time dependence of the exceedance magnitude. In Fig. 4a are shown, superimposed, the exceedances of the 36-year record. There are indications supporting the assumption of a piecewise constant $\beta(t)$ in (11) (see, e.g., Todorovic an Rousselle,

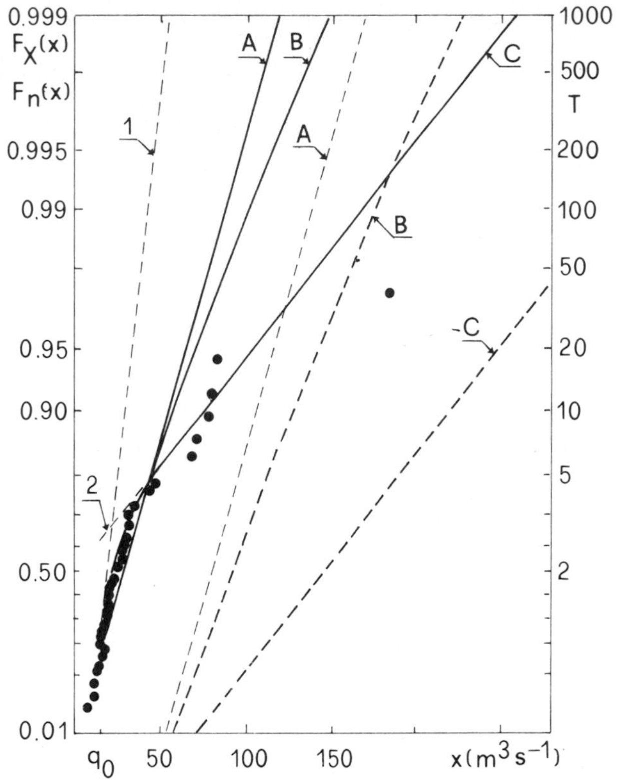

Fig. 3. Amato River at Marino. Observed distribution function of annual flood series (black points). Theoretical distribution functions of maximum annual flow (solid curves) and of the maximum-in-36-year flow (broken curves) for Models A, B, C. Distribution functions of the maximum annual flow for the components of Models C (curves 1 and 2)

1971) and the costancy intervals may here be identified with one-month periods. Monthly values of the exceedance mean magnitude and mean number in a year may be read in Figs. 4a and 4b respectively. With such a piecewise constant $\beta(t)$ in (11) the distribution is easily evaluated.

Consider finally Model C. The parameters p, β_1 and β_2 in (12) were estimated by the maximum-likelihood method (Hasselblad 1969). The distribution function of the individual components and resulting mixed-exponential (ME) distribution function thus obtained are shown in Fig. 2. It is seen that the ME distribution fits the observed data much better than is the case of the plain exponential distribution.

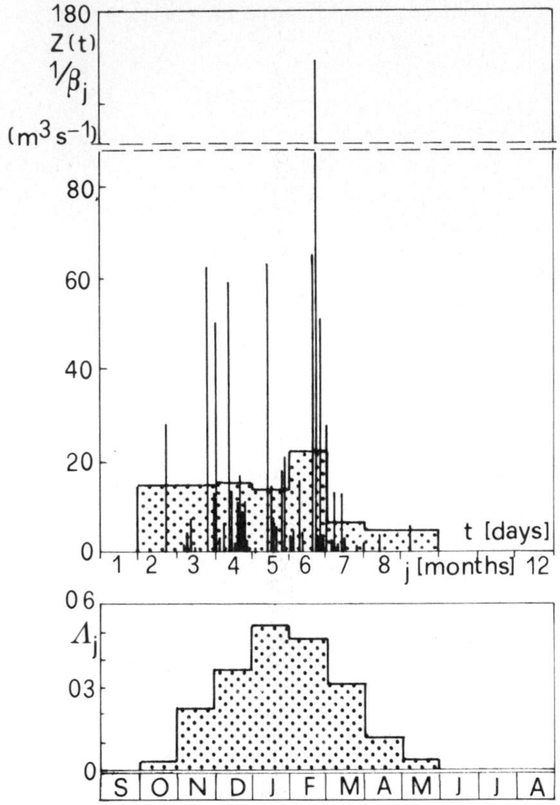

Fig. 4. Amato River at Marino. (a) Observed values of exceedances and their monthly means; (b) observed monthly means of number of exceedances in a year.

As the expected value of the number of exceedances $\Lambda = \Lambda_1 + \Lambda_2$ is known, all parameters in (15) may be determined.

Figure 3 also shows the distribution functions of both the maximum annual flow and the maximum-in-36-year flow when Model B or Model C holds.

Of all models considered, the latter (Model C) shows the best fitting of the observed data and, in particular, it would seem to account for the largest observed values.

4 CONCLUSIONS

The analysis of flood data for the Amato River and other rivers of Southern Italy suggests the following conclusions:

1) The flood peaks exceeding a given base level may be treated as

a marked point process.
2) The number of exceedances K_t within a fixed interval of time $[0, t]$ is a non-homogeneous Poisson counting process.
3) In some cases the exceedance values Z_i above a base level q_0 do not lend themselves to be modelled as independent random variables with common exponential distribution. More refined models, i.e., a time-dependent distribution for the Z_i (Model B) or a mixed exponential distribution (Model C) prove to be more correct.
4) In many areas of Southern Italy the annual flood series exhibit statistical outliers (and, accordingly, high values of skewness), associated with disastrous storms. Model C, which accounts for them by modelling the flood population as the mixture of two distinct populations, is in keeping with the fact that disastrous storms occur more rarely but with larger intensity than others. By contrast, there is little phenomenological evidence in support of Model B, for disastrous storms may occur at any time during the year. The superiority of Model C is confirmed by the better fit it provides to the data, in spite of the fact that it has fewer parameters than is the case of Model B.

ACKNOWLEDGEMENTS This work was supported by CNR "Progetto Finalizzato Conservazione del Suolo" sottoprogetto Dinamica Fluviale - Pubbl. n. 154.

REFERENCES

Dauty, J., 1972. Méthodes des processus stochastiques pour la determination de lois de probabilité des crues. Atti del Convegno Internazionale Piene: loro previsione e difesa del suolo. Roma, 11 pp.
Hasselblad, V., 1969. Estimation of Finite Mixtures of Distributions from the Exponential Family. J. Amer. Statist. Assoc., 64: 1459-71.
Landweher, J., Matalas, N.C. and Wallis, J.R., 1978. Some Comparisons of Flood Statistics in Real and Log Space. Water Resour. Res., 14: 902-920.
Matalas, N.C., Slack, J.R. and Wallis, J.R., 1975. Regional Skew in Search of a Parent. Water Resour. Res., 11: 815-826.
Natural Environment Research Council, 1975. Flood Studies Report. NERC Publications. London.
North, M., 1980. Time-Dependent Stochastic Model of Floods. Proceedings Am. Soc. Civ. Eng., 106: 649-665.
Penta, A., Rossi, F., Silvagni, G., Veltri, M. and Versace, P., 1980. Un modello stocastico per l'analisi delle massime piogge giornaliere in presenza di grandi nubifragi. Atti del XVII Convegno di Idraulica e Costruzioni Idrauliche. Palermo, 17 pp.

Rossi, F. and Versace, P., 1981. Criteri e metodi per l'analisi statistica delle piene. Pubblicazione Programma Finalizzato Conservazione Suolo. Roma, 61 pp.

Rousselle, J., 1972. On Some Problems of Flood Analysis. Ph.D. Thesis, Colorado State University. Fort Collins, 226 pp.

Sarhan, A.E., 1954. Estimation of the mean and standard deviation by order statistics. Ann. Math. Statist., 25: 317-328.

Snyder, D.L., 1975. Random Point Processes. John Wiley and Sons. New York, 485 pp.

Todorovic, P., 1976. Stochastic Models of Floods. Water Resour. Res., 14: 345-356.

Todorovic, P. and Rousselle, J., 1971. Some Problems of Flood Analysis. Water Resour. Res., 7: 1144-1150.

Todorovic, P. and Zelenhasic, E., 1970. A Stochastic Model for Flood Analysis. Water Resour. Res., 6: 1641-1648.

U.S. Water Resources Council, 1977. Guidelines for Determining Flood Flow Frequency. Hydrologic Committee, Bull. 17 A. Washington.

Zelenhasic, E., 1970. Theoretical Probability Distribution for Flood Peaks. Hydrology Paper 42. Colorado State University, Fort Collins.

A MODEL FOR SIMULATING DRY AND WET PERIODS OF ANNUAL FLOW SERIES

M. BAYAZIT

Department of Hydraulics and Water Power, Technical University, Istanbul, Turkey

ABSTRACT

A two-stage model has been developed with the purpose of simulating periods of flows of various magnitudes. Observed annual flows of a river are arranged into n subsets in view of their positions with respect to the suitably chosen truncation levels. Elements of the transition matrix between the states are determined from the observations. In the first stage of the simulation states of flows are generated by a Markovian process which preserves the transition matrix. In the second stage actual values of flows are produced by means of a first-order autoregressive model. Two-state and three-state versions of the model are described. Two-state model simulates dry periods at a certain truncation level whereas three-state model preserves the positive and negative run-lengths of observed flows which may have different values. Thus these two models may account for extreme droughts and differential persistence, respectively. Applications of the model to the simulation of annual flows of a river exhibiting differential persistence are presented.

INTRODUCTION

Hydrologic data are prerequisites for all engineering studies aimed at developing water resources. Decisions to be made in the planning and operation of a water-resource system depend to a great extent on the available hydrologic information. Hydrologic variables, being of random character, can only be expressed in

Reprinted from *Time Series Methods in Hydrosciences*, by A.H. El-Shaarawi and S.R. Esterby (Editors)

terms of their statistical properties. Samples of sufficient size are required in order to estimate the statistical parameters of the population with an acceptable precision. Stochastic dependence increases the required size of sample for a given degree of accuracy. Streamflows, which are the most important inputs of hydrologic studies, usually exhibit considerable sequential dependence. On the other hand, series of recorded streamflows are generally too short. This situation has led to the development of synthetic hydrology which attempts at generating synthetic series of flows based on a mathematical model of the stochastic process.

Synthetic flow series are mostly used in reservoir operation studies where it is expected that the information contained in the observations will be used more efficiently and the risks corresponding to various decisions can be estimated, especially by simulating the extreme dry and wet periods that might not be contained in the observed data. Therefore it is essential that the generated series represent these periods adequately.

Serious difficulties are encountered in the modelling of hydrologic processes. The choice of the model type and the estimation of its parameters are rendered difficult due to the limited time-span of the available records. In order to minimize the errors arising from this situation it has been recommended to use simple models that have as few parameters as possible. As no model can be expected to represent all aspects of the flow process which depends on the complex physical characteristics of the river basin, it should be attempted to select a model which can reproduce the properties of the flows related to the problem in hand. The model most frequently used to generate annual streamflows is the first-order linear autoregressive model:

$$x_k = \mu + \rho\,(x_{k-1} - \mu) + \sigma(1-\rho^2)^{\frac{1}{2}}\,\varepsilon_k \tag{1}$$

where x_k and x_{k-1} are the flows of years k and k-1, respectively.

The model has three parameters: mean (μ), standard deviation (σ) and lag-one autocorrelation coefficient (ρ) of annual flows. ε_k is the standard normal variate. It was pointed out (Askew, et al., 1971; Bayazit, 1974) that dry periods generated by this model were not so severe as those recorded in some rivers. This is a serious deficiency of the model since reservoir operation is very sensitive to periods of extreme drought. Higher order autoregressive models or more general ARIMA type of models have too many parameters and do not still guarantee to preserve the characteristics of extreme flows.

In this study a two-stage model is developed with the purpose of simulating periods of flows of various magnitudes correctly. In the first stage the model generates flow states (such as dry, normal, wet). In the next stage actual flows belonging to these states are generated by means of a modified first-order autoregressive process that can preserve mean, standard deviation and lag-one autocorrelation coefficient of the flows. The advantage of the model is that it can preserve the distribution of lengths of dry and wet periods as well.

PROPOSED MODEL

Consider a stationary stochastic process consisting of normally distributed variables x_k which can be regarded as the flow of year k. An appropriate transformation (such as logarithmic) should be applied first if x_k are not distributed normally. Let the flows be divided into n classes such that the probability of a flow being in class interval i is q_i:

$$P[x_{i-1} < x_k < x_i] = q_i \qquad (2)$$

where, obviously, $\sum_{i=1}^{n} q_i = 1$.

The transition matrix of the n-state Markov process can be defined as $P_{ij} = [a_{ij}]$ where a_{ij} is the probability of a flow in class i to be followed by a flow in class j:

$$a_{ij} = P[x_{j-1} < x_k < x_j \mid x_{i-1} < x_{k-1} < x_i] \tag{3}$$

Transition probabilities satisfy the following equations:

$$\sum_{i=1}^{n} a_{ij} \; q_i = q_j \qquad j = 1,2,\ldots,n \tag{4}$$

$$\sum_{j=1}^{n} a_{ij} = 1 \qquad i = 1,2,\ldots,n \tag{5}$$

a_{ij} values can be estimated from the recorded data by counting the numbers of observed transitions between the states.

Having decided the number of classes n and their probabilities q_i and determined the elements of the transition matrix, synthetic flows can be generated by the following two-stage scheme.

Stage I.

An initial value $x_{1,i}$ is chosen and a sequence of flow states of desired length is generated by means of a random number generator simulating the n-state transition matrix P_{ij}. At the end of this stage states of synthetic flows have been determined but not their actual values.

Stage II.

Once it is determined that x_k belongs to state j, its value can be computed as:

$$x_{k,j} = \mu + \sigma \; \varepsilon_{k,j} \tag{6}$$

where $\varepsilon_{k,j}$ has a truncated normal distribution such that:

$$\frac{x_{j-1}-\mu}{\sigma} < \varepsilon_{k,j} < \frac{x_j-\mu}{\sigma} \qquad (7)$$

x_{j-1} and x_j are the limits of the class interval j.

The sequence of flows generated in this way preserves the population mean μ, standard deviation σ, and transition matrix P_{ij}. It has a built-in autocorrelation coefficient ρ which can be computed as follows:

$$\rho = \frac{\left[\sum_{i=1}^{n} \sum_{j=1}^{n} P(i,j)\ E(x_{k-1,i}\ x_{k,j})\right] - \mu^2}{\sigma^2} \qquad (8)$$

where P(i,j) is the probability of the flow of the year k-1 to be in class i and the flow of the next year to be in class j, which is equal to:

$$P(i,j) = q_i\ a_{ij} \qquad (9)$$

Expected value of the product of $x_{k-1,i}$ and $x_{k,j}$ equals:

$$E(x_{k-1,i}\ x_{k,j}) = \mu_i\ \mu_j \qquad (10)$$

where μ_i and μ_j are means of the flows in classes i and j, respectively. Substituting these into eq.8:

$$\rho = \frac{\left[\sum_{i=1}^{n} \sum_{j=1}^{n} q_i\ a_{ij}\ \mu_i\ \mu_j\right] - \mu^2}{\sigma^2} \qquad (11)$$

ρ value computed as above will usually be lower than the observed autocorrelation coefficient of the process since correlations between the successive flows are not considered fully in this scheme.

In order to preserve the observed autocorrelation coefficient, following first-order autoregressive model should be used:

$$x_{k,j} = \mu + \rho' (x_{k-1,i} - \mu) + \sigma' (1-\rho'^2)^{\frac{1}{2}} \eta_{i,j} \qquad (12)$$

where $x_{k-1,i}$ is the flow of the year k-1 which is in class i, and $x_{k,j}$ is the flow of the year k which is in class j. $\eta_{i,j}$ is a random variate drawn from the standard normal distribution with the condition that $x_{k,j}$ computed by eq.(12) takes indeed a value belonging to class j. This can be accomplished by means of a random number generator which produces standard normal variates but then rejects those which do not satisfy the condition that $x_{k,j}$ computed by eq.(12) is in class j.

The standard deviation σ' and lag-one autocorrelation coefficient ρ' of the generating scheme given by eq.(12) can be expressed in terms of σ and ρ of the population of annual flows. It can be shown (see Appendix) that σ and ρ will be preserved when σ' and ρ' are chosen such as to satisfy the following equations:

$$(1-\rho'^2) \left(\left(\frac{\sigma'}{\sigma}\right)^2 - 1\right) + 2 \rho' \frac{\sigma'}{\sigma} D = 0 \qquad (13)$$

$$\rho' - \rho + \frac{\sigma'}{\sigma} D = 0 \qquad (14)$$

where:

$$D = \sum_{i=1}^{n} \sum_{j=1}^{n} q_i \; a_{ij} \; d_{ij} \qquad (15)$$

d_{ij} in eq.(15) are defined as follows:

$$d_{ij} = (1-\rho'^2)^{\frac{1}{2}} \; E_{ij} \left(\frac{x_{k-1,i} - \mu}{\sigma} \; \eta_{i,j}\right) \qquad (16)$$

where E_{ij} denotes the expected value of the variable in brackets for the subset of flows in class i followed by those in class j. For a certain value of n, d_{ij} are functions of ρ' and q_i ($i = 1, 2,\ldots,n$), and can be determined experimentally as will be described later on.

Obviously the number of states n to be used in the model should be small in order to be able to estimate a_{ij} values from the observed data with a sufficient accuracy. Below, two-state and three-state versions of the model are going to be discussed.

TWO-STATE MODEL

The simplest case of the model developed in the previous section is the two-state model where one of the states corresponds to dry periods below a certain truncation level x_1 (with probability q_1) and the other to wet (or normal) periods above that level (with probability $q_2 = 1-q_1$). A model of this kind was introduced by Jackson (1975 a) with the purpose of preserving the observed persistence of droughts. Her model, however, differs from that given by eq.(12) in that actual flow values are generated by the following scheme:

$$x_{k,j} = \mu_j + \frac{\sigma_j}{\sigma_i} \rho'(x_{k-1,1}-\mu_i) + \sigma_i(1-\rho'^2)^{\frac{1}{2}} \varepsilon_k \qquad (17)$$

where σ_i and σ_j are standard deviations of the flows in classes i and j, respectively. ε_k is the standard normal variate. The trouble with this model is that eq.(17) does not guarantee that the value of x_k generated by this scheme will belong to state j indeed as prescribed by the transition matrix. It should be replaced by eq.(12) where ρ' and σ' are to be computed from eqs.(13) and (14). This model will preserve μ,σ and ρ as well as P_{ij}. Expected values of the negative and positive run-lengths at the truncation level x_1 are related to the transition probabilities a_{11} and a_{22} by the following equations (Bayazit and Sen, 1979):

$$a_{11} = 1-(1/E(N_n)), \quad a_{22} = 1-(1/E(N_p)) \tag{18}$$

Therefore $E(N_n)$, mean length of dry periods, and $E(N_p)$, mean length of wet periods, will also be preserved by this generating scheme.

$d_{i,j}(i,j = 1,2)$ values defined by eq.(16) have been determined as functions of ρ' by the data generation method for two cases:

$q_1 = 0.4$, $q_2 = 0.6$ (Fig. 1) and $q_1 = q_2 = 0.5$ (Fig. 2).

THREE-STATE MODEL

Let the flows be divided into three class intervals, such as low flows below x_1(with probability q), normal flows between x_1 and x_2 (with probability q_2), and high flows above x_2 (with probability $q_3 = 1-q_1-q_2$). This process can be represented by a three-state model where states correspond to dry, normal and wet periods. The transition matrix have 9 elements, only 4 of which are independent as they have to satisfy the relations expressed by eq.(4) and (5). a_{11} and a_{33} represent the probabilities of transitions from dry to dry and wet to wet states, respectively, which are related to $E(N_n)$ at the level x_1 and $E(N_p)$ at the level x_2 as follows:

$$a_{11} = 1-(1/E(N_n)), \quad a_{33} = 1-(1/E(N_p)) \tag{19}$$

It can be concluded that this model will preserve the expected values of negative and positive run-lengths at chosen truncation levels, and hence it can be used to simulate flow series with differential persistence. Jackson (1975 b) showed that some annual flow records exhibited differential persistence, i.e. the low flows were more persistent than high flows, and she proposed a birth-death model to simulate such sequences.

In order to generate such flow sequences using the present model, elements of the transition matrix $P_{ij} = [a_{ij}]$ $(i,j = 1,2,3)$ are computed from the data, and successive flow states are first generated.

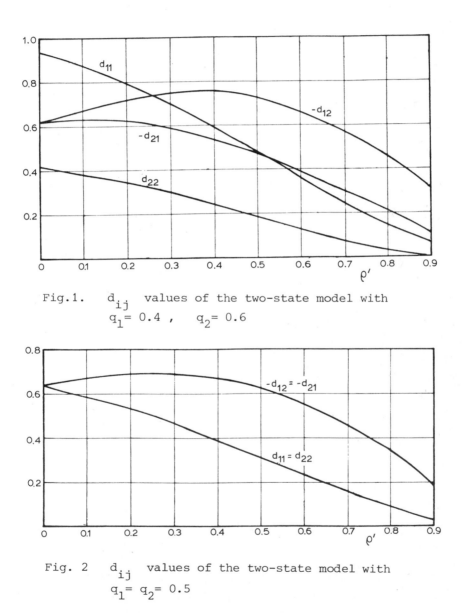

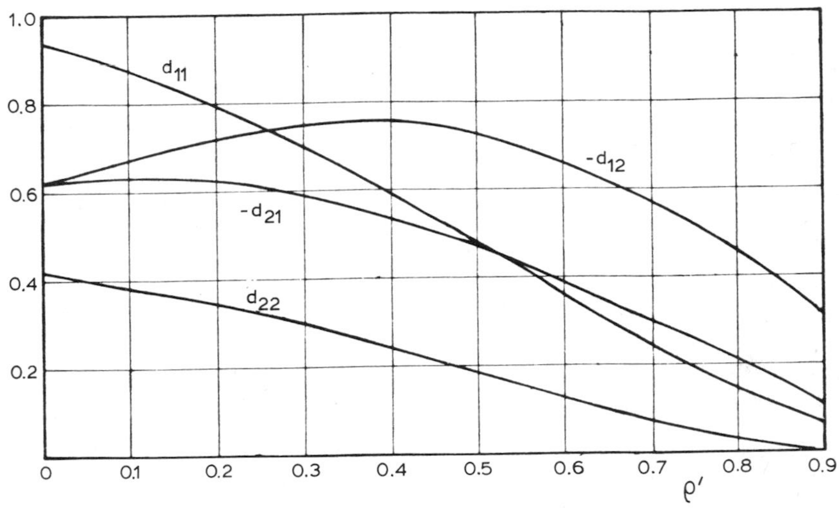

Fig.1. d_{ij} values of the two-state model with $q_1 = 0.4$, $q_2 = 0.6$

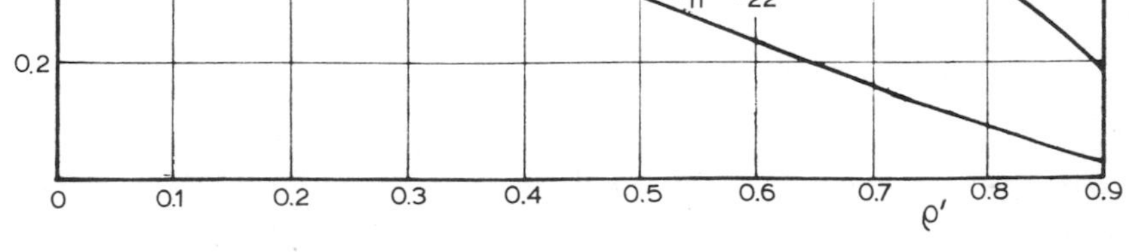

Fig. 2 d_{ij} values of the two-state model with $q_1 = q_2 = 0.5$

Then actual flow values are computed by eq.(12) where σ' and ρ' are to be determined from eqs.(13) and (14) with a trial-and-error procedure such as to preserve the observed standard deviation and lag-one autocorrelation coefficient of annual flows.

d_{ij} (i,j = 1,2,3) values of the three-state model have been determined by the data generation method for the following cases:

$q_1 = q_3 = 0.3$, $q_2 = 0.4$ (Fig.3), and $q_1 = q_3 = 0.4$, $q_2 = 0.2$ (Fig.4).

APPLICATIONS

The proposed model has been applied to annual flows of St. Mary's river in Canada. Observed flows of this river in the periods 1860-1964 were published by Unesco (1971). These flows are normally distributed ($\chi^2 = 3.63$ for 7 degrees of freedom) with $\mu = 2103\ m^3/s$, $\sigma = 326\ m^3/s$ and $\rho = 0.57$.

Two-state model

The truncation level for the two-state model was chosen as $x_1 = 2021\ m^3/s$ which corresponds to a probability of exceedence of $q_2 = 0.6$. Thus flows below 2021 m^3/s are in class 1 (dry flows with probability $q_1 = 0.4$), and flows above 2021 m^3/s are in class 2 (normal flows with $q_2 = 0.6$). Elements of the transition matrix were determined by counting the number of transitions between the states in the recorded series with the following results:

$$a_{11} = 0.63,\quad a_{12} = 0.37,\quad a_{21} = 0.24,\quad a_{22} = 0.76$$

ρ' and σ' values were determined from eqs.(13) and (14) with the aid of Fig. 1. In this case above equations are satisfied approximately when $\rho' = \rho$ and $\sigma' = \sigma$. 5000 years long synthetic flow trace generated using eq.(12) has the following statistics:

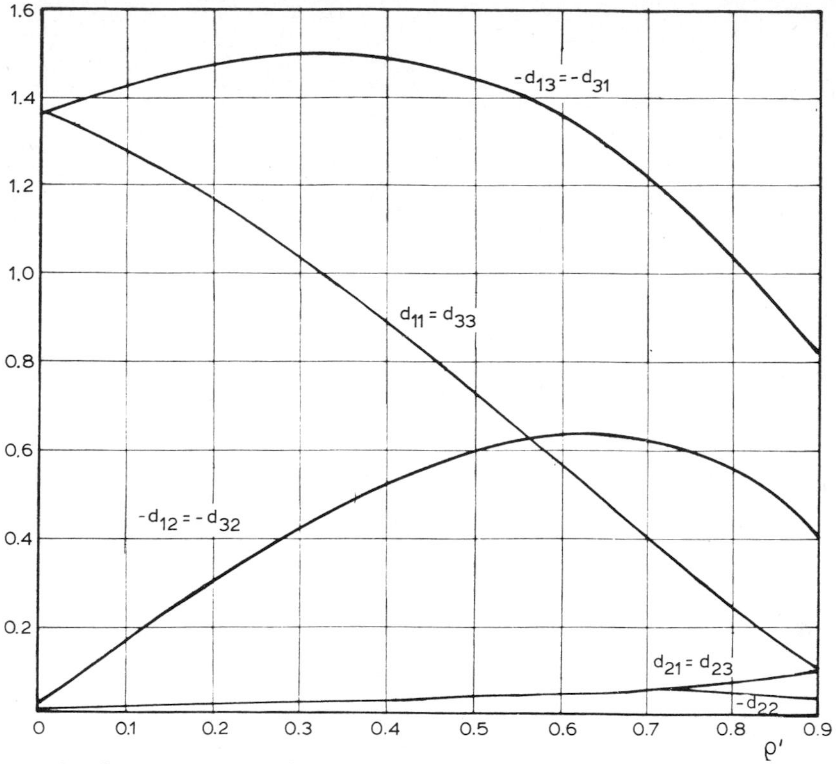

Fig.3. d_{ij} values of the three-state model with $q_1 = q_3 = 0.3$, $q_2 = 0.4$

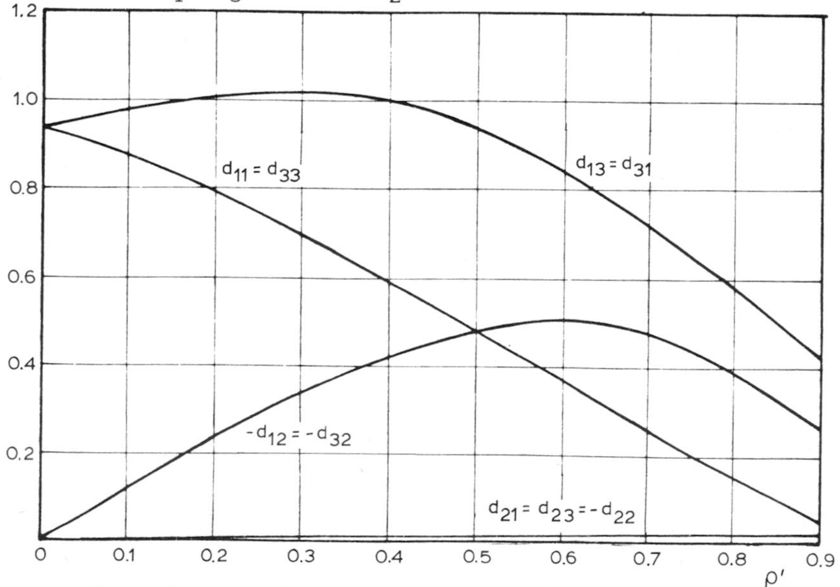

Fig.4. d_{ij} values of the three-state model with $q_1 = q_3 = 0.4$, $q_2 = 0.2$

$\mu = 2102,\quad \sigma = 290,\quad \rho = 0.52$

$a_{11} = 0.62,\quad a_{12} = 0.38,\quad a_{21} = 0.23,\quad a_{22} = 0.77$

$E(N_n) = 2.74,\quad E(N_p) = 4.34$

For comparison, statistics of the synthetic flow series of equal length generated by the simple first-order autoregressive model are given below:

$\mu = 2106,\quad \sigma = 319,\quad \rho = 0.69$

$a_{11} = 0.68,\quad a_{12} = 0.32,\quad a_{21} = 0.21,\quad a_{22} = 0.79$

$E(N_n) = 3.37,\quad E(N_p) = 5.55$

Mean negative and positive run-lengths of the observed flow series are 2.73 and 4.28, respectively, which agree favorably with those generated by the present model.

Three-state model

As mentioned before, the three-state model can be used to generate streamflow traces with differential persistence. In applying this model to the annual flows of St. Mary's river, the truncation levels were chosen such that $q_1 = q_3 = 0.3$, i.e. the lower 30% of the annual flows belonged to dry years and the upper 30% to wet years. Flows exhibited strong differential persistence as was evidenced by the fact that the observed mean negative run-length at this level was 2.73 years whereas the observed positive run-length was 1.87 years, although the probabilities were equal ($q_1 = q_3 = 0.3$). Transition probabilities from dry-to-dry and wet-to-wet states can be computed by eq.(19) as:

$a_{11} = 1-(1/2.73) = 0.64,\quad a_{33} = 1-(1/1.87) = 0.47$

Elements of the three-state transition matrix were computed from the observed data with the following results:

$$P_{ij} = \begin{vmatrix} 0.64 & 0.27 & 0.09 \\ 0.13 & 0.54 & 0.33 \\ 0.18 & 0.35 & 0.47 \end{vmatrix}$$

It should be remarked that a_{11} and a_{33} values computed in this way are the same as those obtained through the use of eq.(19).

σ' and ρ' of the model are computed from eqs.(13) and (14) using as Fig. 3 $\sigma' = 365$ and $\rho' = 0.67$. Synthetic flows of 5000 years generated by eq.(12) with the above values of σ' and ρ' have the following statistics:

$\mu = 2108$, $\sigma = 327$, $\rho = 0.49$

$E(N_n) = 2.95$, $E(N_p) = 1.85$

$$P_{ij} = \begin{vmatrix} 0.65 & 0.26 & 0.09 \\ 0.13 & 0.53 & 0.34 \\ 0.16 & 0.36 & 0.48 \end{vmatrix}$$

First-order autoregressive model with $\rho = 0.57$ generated a series of equal length with the statistics:

$\mu = 2106$, $\sigma = 319$, $\rho = 0.69$

$E(N_n) = 2.66$, $E(N_p) = 2.90$

$$P_{ij} = \begin{vmatrix} 0.60 & 0.35 & 0.05 \\ 0.26 & 0.52 & 0.22 \\ 0.04 & 0.32 & 0.64 \end{vmatrix}$$

It is seen clearly that the simple first-order autoregressive model cannot simulate the differential persistence in this case whereas the three-state model can.

CONCLUSIONS

It has been shown that a two-stage Markov model can be employed successfully to generate synthetic traces of annual flows which preserve the mean, variance, lag-one autocorrelation coefficient of the process as well as the transition matrix between the states of flows. The two-state version of the model generates sequences with the desired mean length of droughts. The three-state version can be used in modeling differential persistence.

APPENDIX

Expressions for the variance and lag-one autocorrelation coefficient of the variable x_k generated by the scheme of eq.(12) can be derived as follows:

Variance. Squaring both sides of eq.(12):

$$(x_{k,j}-\mu)^2 = \rho'^2(x_{k-1,i}-\mu)^2 + 2\rho'\sigma'(1-\rho'^2)^{\frac{1}{2}}(x_{k-1,i}-\mu)\eta_{i,j} + \sigma'^2(1-\rho'^2)\eta_{i,j}^2 \qquad \text{(A.1)}$$

Expected values of the terms in eq.(A.1) can be computed as follows:

$$E[(x_{k,j}-\mu)^2] = E[(x_{k-1,i}-\mu)^2] = \sigma^2 \qquad \text{(A.2)}$$

$$E[(1-\rho'^2)^{\frac{1}{2}}(x_{k-1,i}-\mu)\,\eta_{i,j}] = \sigma(1-\rho'^2)^{\frac{1}{2}}E[\frac{x_{k-1,i}-\mu}{\sigma}\eta_{i,j}]$$

$$= \sigma\sum_{i=1}^{n}\sum_{j=1}^{n} q_i\, a_{ij}\, d_{ij} = \sigma D \qquad \text{(A.3)}$$

where D and d_{ij} were defined by eqs.(15) and (16).

$$E(\eta_{i,j}^2) = 1 \qquad \text{(A.4)}$$

Substituting these into eq.(A.1):

$$\sigma^2 = \rho'^2\sigma^2 + 2\rho'\sigma'\,\sigma D + \sigma'^2(1-\rho'^2) \qquad \text{(A.5)}$$

Dividing by σ^2 and rearranging:

$$(1-\rho'^2)\left(\left(\frac{\sigma'}{\sigma}\right)^2-1\right) + 2\rho'\frac{\sigma'}{\sigma}D = 0 \qquad \text{(A.6)}$$

Autocorrelation coefficient. Multiplying both sides of eq.(12) by $x_{k-1,i}$:

$$x_{k-1,i}\, x_{k,j} = x_{k-1,i}\, \mu + \rho'\, x_{k-1,i}\,(x_{k-1,i}-\mu) + \sigma'(1-\rho'^2)^{\frac{1}{2}}\, x_{k-1,i}\, \eta_{i,j} \qquad (A.7)$$

Expected values of the terms in this equation can be computed as follows:

$$E(x_{k-1,i}\mu) = \mu^2 \qquad (A.8)$$

$$E[\rho'\, x_{k-1,i}(x_{k-1,i}-\mu)] = \rho'\, \sigma^2 \qquad (A.9)$$

$$E[\sigma'(1-\rho'^2)^{\frac{1}{2}}\, x_{k-1,i}\, \eta_{i,j}]$$

$$= E[\sigma'\, \sigma(1-\rho'^2)^{\frac{1}{2}}\, \frac{x_{k-1,i}-\mu}{\sigma}\, \eta_{i,j} + \mu\, \sigma'(1-\rho'^2)^{\frac{1}{2}}\, \eta_{i,j}]$$

$$= \sigma'\sigma \sum_{i=1}^{n} \sum_{j=1}^{n} q_i\, a_{ij}\, d_{ij} = \sigma'\sigma D \qquad (A.10)$$

Substituting these into eq.(A.7):

$$E(x_{k-1,i}\, x_{k,j}) = \mu^2 + \rho'\sigma^2 + \sigma'\sigma D \qquad (A.11)$$

$$\rho = \frac{E(x_{k-1,i}\, x_{k,j})-\mu^2}{\sigma^2} = \rho' + \frac{\sigma'}{\sigma} D \qquad (A.12)$$

$$\rho'-\rho + \frac{\sigma'}{\sigma} D = 0 \qquad (A.13)$$

ACKNOWLEDGMENT

The author is greatful to Mrs. Beyhan Oguz for her assistance in computer programming for this study.

REFERENCES

Askew, A.J., Yeh, W.W.-G, and Hall, W.A. (1971): "A Comparative Study of Critical Drought Simulation", Water Resources Research, Vol.7, No.1, p.p.52-62.
Bayazit, M. (1974): "Statistical Analysis of Dry Periods in Turkish Rivers", Bulletin of the Technical University of Istanbul, Vol.27, No.2, pp.24-35.
Bayazit, M., and Sen, Z. (1979): "Dry Period Statistics of Monthly Flow Models", Modeling Hydrologic Processes, ed. by H.J. Morel-Seytoux et.al., Water Resources Publications, Littleton, Colorado.
Jackson, B.B. (1975 a): "Markov Mixture Models for Drought Lengths", Water Resources Research, Vol.11, No. 1, pp.64-74.
Jackson, B.B. (1975 b): "Birth-Death Models for Differential Persistence", Water Resources Research, Vol.11, No.1, pp.75-95.
Unesco (1971): Discharge of Selected Rivers of the World, Vol.II, Paris.

A COMBINED SNOWMELT AND RAINFALL RUNOFF

KAZUMASA MIZUMURA

Kanazawa Institute of Technology, Ishikawa, Japan

1. INTRODUCTION

The area faced to the sea of Japan are known as the region with heavy snowfall in Japan. The main cause is the monsoon blowing from the high pressure developed over the Siberia to the low pressure over the Pacific Ocean. The monsoon becomes contained much moisture during passing over the sea of Japan and makes heavy snowfall when it rises along the high mountaneous zone in the Honshu Island. The snowmelt becomes very important water resources such as electricity, rice growing, and drinking water. The snowmelt runoff is storaged in reservoirs and the accurate prediction of that is necessary for water level controls in reservoirs. During snowmelt period the rainfall and snowmelt runoffs are the dominant source of streamflow. The study area used for the rainfall-runoff process and the snow accumulation-snowmelt runoff process is the Sai river watershed located in the southeast of Kanazawa city in Japan. The watershed area is 56.1 km^2 and the meteolorogical data are observed at the measuring station. The elevation of this station is almost 300 m above the sea and many mountains from 1000 m to 1500 m exist within the watershed.

2. TANKS MODEL SYSTEM

In this study four tanks model (Sugawara, 1978) was used for rainfall-runoff analysis. The reason is dependent on the watershed area.

An additional tank for snowfall is located on the upper position of the four tanks as illustrated in Fig.1. The snowfall is storaged in the upper tank and it melts when air temperature becomes higher than 0 c. The snowmelt runoff or/and the rainfall runoff are poured into the second tank. The most part of the snowmelt runoff and rainfall

Reprinted from *Time Series Methods in Hydrosciences*, by A.H. El-Shaarawi and S.R. Esterby (Editors)

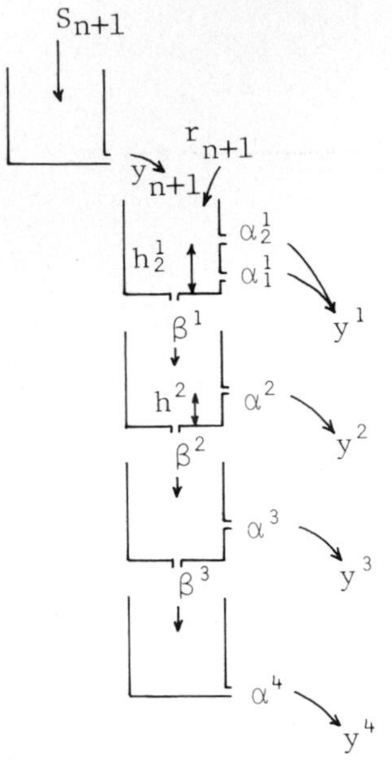

Fig. 1. Tanks model

runoff is discharged by the side outlets and the remains infiltrate into the third tank. Therefore, there is no interaction between rainfall and snowfall, that is, the rain does not melt the snow. This process is also reported by Sugawara (1978).

Let us define the rainfall and the snowmelt in the tank r_n and y_n^0. The snowmelt does not occur when air temperature T is less than 0 c or the snowfall accumulation h_n^0 in the first tank is zero. Accordingly, the snowmelt can be expressed by the following equations.

$$y_n^0 = \begin{cases} 0 & \text{if } h_n^0 = 0 \text{ or } T_n \leq 0 , \\ h_n^0 & \text{if } h_n^0 < mT_n \text{ and } T_n > 0 , \\ mT_n & \text{if } h_n^0 \geq mT_n \text{ and } T_n > 0. \end{cases} \tag{1}$$

in which $h_n^0 = \sum_{i=1}^{n-1} (\lambda S_i - y_i^0)$, m and λ = the parameters to be identified, and S_i = snowfall at i-step. As the snowfall data in this watershed the data at the measuring station are used, but in general the average snowfall in this watershed can be several times of that at the measuring station. So, λS_i is considered to be the average snowfall in this watershed. To simplify the model, the snowmelt can be assumed to be proportional to the air temperature T_n and it is expressed by mT_n.

Eq.(1) can be written as

$$y_n^0 = \min \left\{ \sum_{i=1}^{n-1} (\lambda S_i - y_i^0) , \quad mT_n \right\} \frac{T_n + |T_n|}{2T_n} \tag{2}$$

in which the sign min {A, B} means the selection of smaller one. If $x_n = y_n^0 + r_n$, the runoff from the second tank can be obtained as:

$$y_n = \begin{cases} 0 & \text{if } X_n^1 \le h_1^1, \\ \alpha_1^1 (X_n^1 - h_1^1) & \text{if } h_1^1 < X_n^1 \le h_2^1, \\ \alpha_2^1 (X_n^1 - h_2^1) + \alpha_1^1 (X_n^1 - h_1^1) & \text{if } h_2^1 < X_n^1. \end{cases} \quad (3)$$

$$z_n^1 = \beta^1 X_n^1 \quad (4)$$

$$X_{n+1}^1 = X_n^1 - y_n^1 - z_n^1 + x_{n+1} \quad (5)$$

in which z_n^1 = discharge from the tank bottom, X_n^1 = storage in the tank, n = time step, α_1^1, α_2^1, and β^1 = discharge coefficients, h_1^1 and h_2^1 = the elevations of outlets from the tank bottom, and the superscript 1 shows the second tank. The third tank is also formulated as

$$y_n^2 = \begin{cases} 0 & \text{if } X_n^2 \le h^2, \\ \alpha^2 (X_n^2 - h^2) & \text{if } X_n^2 > h^2. \end{cases} \quad (6)$$

$$z_n^2 = \beta^2 X_n^2 \quad (7)$$

$$X_{n+1}^2 = X_n^2 - y_n^2 - z_n^2 + z_{n+1}^1 \quad (8)$$

in which z_n^2 = discharge from the tank bottom, X_n^2 = storage in the tank, n = time step, α^2 and β^3 = discharge coefficients, and h^2 = the elevation of a outlet from the tank bottom. The calculations in the forth and fifth tanks are same as that in the third tank..The used data for this procedure are rainfall, snowfall, snowfall accumulation, runoff, and air temperature at 9 a.m. at the measuring station. The rainfall, snowfall, and runoff are daily averaged from 9 a.m. to 9 p.m. The temperature is also much influenced by the sea of Japan and the minimum of that in a year appears in February. And it remarkably increases during the snowmelt period from March to May. The higher temperatures than 20 c found in April are caused by the foehn phenomenon. Fig.2 shows rainfall and snowfall at the measuring station. This watershed belongs to the heavy snowfall district in Japan and the precipitation in January and in February is principally due to snowfall. The snowfall at the measuring station starts in the first part

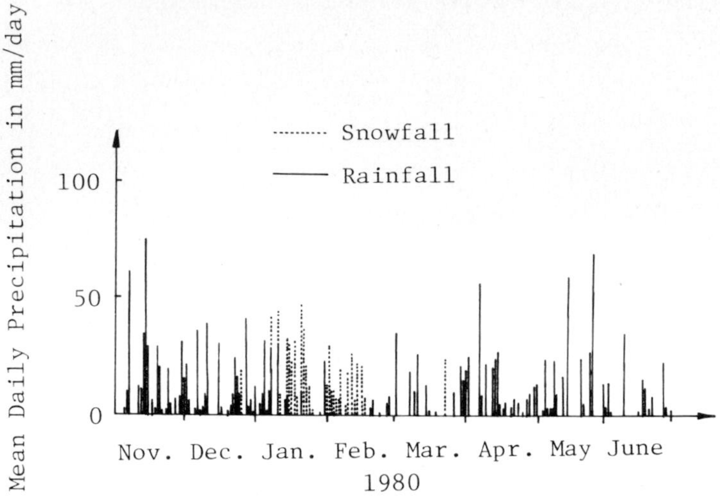

Fig. 2. Observed precipitation

of December and ends in the last part of March. But in mountains of this watershed it starts in the first part of November.

3. MAXIMUM A POSTERIORI ESTIMATION

The maximum a posteriori estimation is equivalent to an appropriate least-squares curve fit, using as weighting matrices the inverses of the plant- and measurement-noise covariances. We use for the optimum estimate the value of θ which maximizes $p_{\theta|z}(\Theta|\, Z)$. The discrete message and observation models are given by

$$\underline{X}(n+1) = \underline{\phi}(\underline{X}(n),\ n) + \underline{w}(n) \tag{9}$$

$$\underline{z}(n) = \underline{h}(\underline{X}(n),\ n) + \underline{v}(n) \tag{10}$$

in which $\underline{X}(n)$ = N-dimensional state vector, $\underline{\phi}(\underline{X}(n),\ n)$ = N-dimensional vector-valued function, $\underline{w}(n)$ = M-dimensional plant-noise vector, $\underline{z}(n)$ = R-dimensional observation vector, $\underline{h}(\underline{X}(n),\ n)$ = R-dimensional vector-valued function, and $\underline{v}(n)$ = R-dimensional observation noise vector.

For the discrete-estimation model, $\underline{w}(n)$ and $\underline{v}(n)$ are assumed to be independent zero-mean Gaussian white sequences such that

$$E\{\underline{w}(n)\ \underline{w}^T(j)\} = V_w(n)\ \delta(n-j) \tag{11}$$

$$E\{\underline{v}(n)\ \underline{v}^T(j)\} = V_v(n)\ \delta(n-j) \tag{12}$$

in which $\delta_k(n-j)$ is the Kronecker delta function, and $V_w(n)$ and $V_v(n)$ are non-negative definite MxM and RxR covariance matrices, respectively. The estimate is derived from maximizing the conditional probability function. The one-stage prediction is given by

$$\underline{\hat{X}}(n+1|n) = \underline{\phi}(\underline{\hat{X}}(n),\ n) \tag{13}$$

The priori error-variance algorithm is

$$V_{\underline{\tilde{X}}}(n+1|n) = \{\frac{\partial}{\partial\underline{\hat{X}}(n)}\ \underline{\phi}(\underline{\hat{X}}(n),\ n)\}\ V_{\underline{\tilde{X}}}(n)\ \{\frac{\partial}{\partial\underline{\hat{X}}(n)}\ \underline{\phi}^T(\underline{\hat{X}}(n),\ n)\} + V_w(n) \tag{14}$$

The error-variance algorothm is also given by

$$V_{\underline{\tilde{X}}}(n+1) = V_{\underline{\tilde{X}}}(n+1|n) - V_{\underline{\tilde{X}}}(n+1|n)\{\frac{\partial}{\partial\underline{\hat{X}}(n+1|n)}\ \underline{h}^T(\underline{\hat{X}}(n+1\ n),\ n+1)\}$$

$$\Bigg[\{\frac{\partial}{\partial\underline{\hat{X}}(n+1|n)}\ \underline{h}(\underline{\hat{X}}(n+1|n),\ n+1)\}\ V_{\underline{\tilde{X}}}(n+1|n)\{\frac{\partial}{\partial\underline{\hat{X}}(n+1|n)}\ \underline{h}^T(\underline{X}(n+1|n)\}$$

$$+ V_v(n+1)\Bigg]^{-1}\ \{\frac{\partial}{\partial\underline{\hat{X}}(n+1|n)}\ \underline{h}(\underline{\hat{X}}(n+1|n),\ n+1)\}\ V_{\underline{\tilde{X}}}(n+1|n) \tag{15}$$

The filter algorithm becomes

$$\underline{\hat{X}}(n+1) = \underline{\hat{X}}(n+1|n) + \underline{K}(n+1)\{\underline{z}(n+1) - \underline{h}(\underline{\hat{X}}(n+1|n),\ n+1)\} \tag{16}$$

in which

$$\underline{K}(n+1) = V_{\underline{\tilde{X}}}(n+1)\{\frac{\partial}{\partial\underline{\hat{X}}(n+1|n)}\underline{h}^T(\underline{\hat{X}}(n+1|n),\ n+1)\}$$

The flow chart of the maximum a posteriori estimation is given in the reference (Sage et al., 1971). To apply the maximum a posteriori estimation to the tanks model described in the previous section, let us define the state vector as follows:

$$\underline{X}(n) = (y_n^1, y_n^2, y_n^3, y_n^4, X_n^1, X_n^2, X_n^3, X_n^4, \alpha_1^1, \alpha_2^1, \alpha^2, \alpha^3, \alpha^4, \beta^1, \beta^2, \beta^3, h_1^1, h_2^1, h^2, h^3, m, \lambda)^T \tag{17}$$

The vector function $\underline{\phi}$ can be calculated from the relation in the tanks model and $\underline{h}(\underline{X}(n), n)$ becomes as:

$$\underline{h}(\underline{X}(n), n) = y_n^1 + y_n^2 + y_n^3 + y_n^4 \tag{18}$$

4. NUMERICAL EXAMPLE

The initial vector $\hat{\underline{X}}(0)$ and the initial covariance matrix $V_{\tilde{\underline{X}}}(0)$ must be given beforehand to apply the algorithm of the maximum a posteriori estimation. In the consideration of the watershed area the result of Sugawara (1978) was used for the initial values of the parameters of the tanks model. Further we assumed $m = 3$ and $\lambda = 2$ as the initial values of the parameters related with snowmelt. So the initial state vector was

$$\underline{X}(0) = (0, 0, 0, 0, 0, 0, 0, 0, 0.2, 0.2, 0.04, 0.01, 0.001, 0.2, 0.04, 0.01, 10, 20, 10, 10, 3, 2)^T \tag{19}$$

The determination of the initial matrix $V_{\tilde{\underline{X}}}(0)$ was made by the following way. For simplicity the orders of y_n^i and X_n^i for $i = 1$ to 4 were assumed as:

Order of $X_n^1 = 1$
Order of $y_n^1 = 1$
Order of X_n^2 = Order of $\beta^1 X_n^1$
Order of y_n^2 = Order of $\alpha^2 X_n^2$
Order of X_n^3 = Order of $\beta^2 X_n^2$

Order of y_n^3 = Order of $\alpha^3 X_n^3$
Order of X_n^4 = Order of $\beta^3 X_n^3$
Order of y_n^4 = Order of $\alpha^4 X_n^4$
And the order of the other parameters were assumed to be the initial values. For example, the (1, 10) component of the covariance matrix $V_{\tilde{X}}(0)$ was 0.2. The sign of each component of $V_{\tilde{X}}(0)$ was determined as follows. By considering that the increase of y_n^1 corresponds to the increase of α_2^1, the correlation was positive and so is the sign. If there is no correlation between them such as α_1^1 and α_2^1, the component is zero. So the (1, 10) component of $V_{\tilde{X}}(0)$ was 0.2. Next, we assumed that $V_w(n) = 0$ and $V_v(n) = 1$, because the variation range of runoff data was approximately from 1 to 100. And the results of the calculations with $V_v(n) = 1$ and $V_v(n) = 10$ were better in the following four predictions ($V_v(n) = 0.1, 1, 10, 100$). The numerical calculation started from the first of September.

5. PARAMETER IDENTIFICATION AND RUNOFF PREDICTION

The parameters to construct the tanks model are α_1^1, α_2^1, α^2, α^3, β^1, β^2, β^3, h_1^1, h_2^1, h^2, h^3, λ, and m. In the tanks model introduced by Sugawara (1978) these parameters are assumed to be constant. In this study these parameters are identified step by step by the algorithm of the maximum a posteriori estimation. The parameters such as the coefficients of the third, forth, and fifth tanks were almost constant.

For small discharge these parameters do not show remarkable changes, but for large discharge α_1^1 and α_2^1 increase and h_2^1, h_2^1, and β^1 decrease. And λ and m are correlated with snowfall and snowmelt, respectively, also increased. In Fig. 3 the observed snowfall accumulation at the measuring station was compared with the estimated average snowfall accumulation in this watershed. The former is much smaller than the latter, because of the elevation difference. There is still much snowfall in mountains even if there is no snowfall at the measuring station. In the estimation snowfall exists in the last of May and it explains the condition of remaining snow very well. Fig. 4 represents the observed and predicted runoff. The prediction was made by using

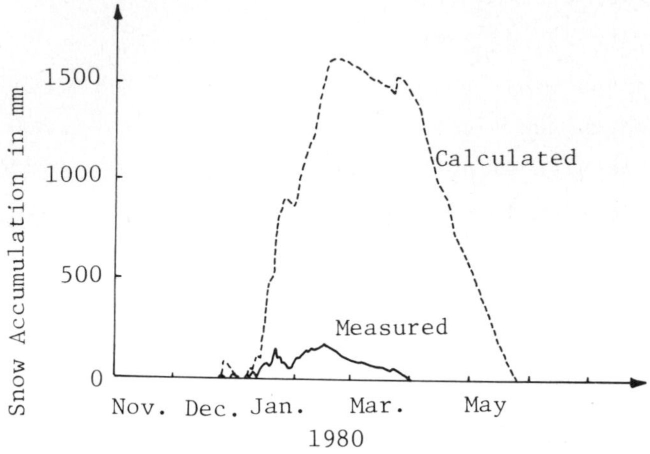

Fig. 3. Measured and calculated snow accumulations

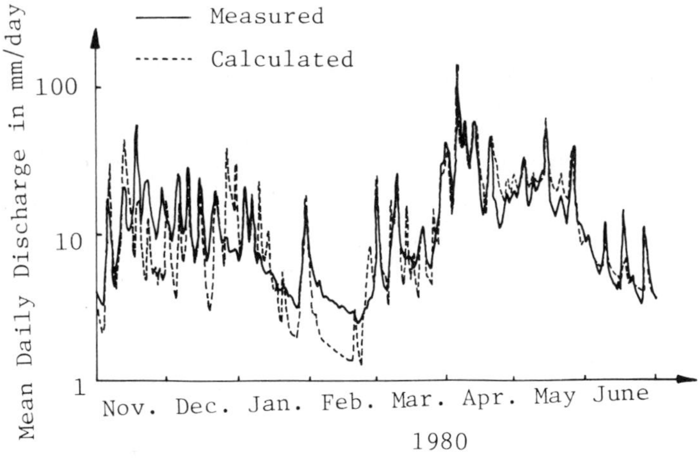

Fig. 4. Measured and estimated snowmelt hydrograph

the data in the previous day. The prediction from November to February exceeded the observed runoff around peaks and it was below the observed one around receding runoff. The prediction of runoff becomes in good agreement from March. It can be explained that the precipitation in mountaneous area becomes snow already in November and it melts gradually, but it still rains at the measuring station in the same season because of the elevation difference and it discharges immediately.

6. SUMMARY AND CONCLUSIONS

Through this study the following results were obtained.

(1) The parameters in the tanks model were identified by the algorithm of the maximum a posteriori estimation on each time step.

(2) By using the method described herein it become possible to predict snowfall in mountains during winter.

(3) It becomes possible that the prediction of mean daily runoff combined rainfall and snowfall runoff in the previous day by meteorological factors.

(4) Since the measuring station is located 300 m above the sea, it snows in mountains and it rains at the measuring station in the same time in November or December. The incorrect description of precipitation gives the error in runoff. Therefore, in future study an additional parameter will be considered during this period.

(5) The usage of air temperature at 9 a.m. results in error, for example, in the case of which the air temperature at 9 a.m. is less than 0 c and the maximum temperature in a day more than 0 c the runoff is zero in calculations but it has some value in data.

(6) In the application of the tanks model to predict runoff the model in this study is suggested by Sugawara (1978) and the parameters are constant. But since the parameters are estimated step by step, the problem on over parameterization occurs. This will be discussed in the future study.

REFERENCES

Sage, A.P. and Melsa, J.L., 1971. Estimation theory with applications to communications and control. McGraw-Hill Book Company, New York, pp. 441-457.

Sugawara, M., 1978. Runoff analyses. Kyoritsu Shuppan Book Company, Tokyo (in Japanese).

ANALYSIS OF CURRENT METER DATA FOR PREDICTING POLLUTANT DISPERSION

PHILIP J.W. ROBERTS

School of Civil Engineering, Georgia Institute of Technology

INTRODUCTION

Although current meter data are frequently collected during the design of major ocean outfalls, the data are rarely subject to extensive analyses to aid in the design of these systems. An exception to this occurred during oceanographic investigations for outfalls proposed for the City of San Francisco, California. In this case current meter data collected were subjected to fairly extensive analyses which aided considerably in the design and prediction of performance of the outfalls. The purpose of this paper is to present the results of some of these analyses.

STUDY SITE

The study site and proposed outfall design are shown in Figure 1 (Roberts, 1980). The proposed discharge site lies in the Pacific Ocean off San Francisco, just South of the Golden Gate Bridge. When completed, the outfall will discharge both domestic and industrial sewage, and during wet weather, a mixture of sewage and stormwater runoff. Note that the design has been modified from that discussed previously (Roberts, 1980) in that the old wet weather outfalls have been eliminated, and now all flows will discharge through the one long outfall.

The location of the moored current meters are also shown in Figure 1. Site 7 contains two meters, 7A nearer the surface, and 7B nearer the bottom. Of these sites, numbers 7, 8, and 12 were occupied continuously for one year, and the rest were occupied

Reprinted from *Time Series Methods in Hydrosciences*, by A.H. El-Shaarawi and S.R. Esterby (Editors)

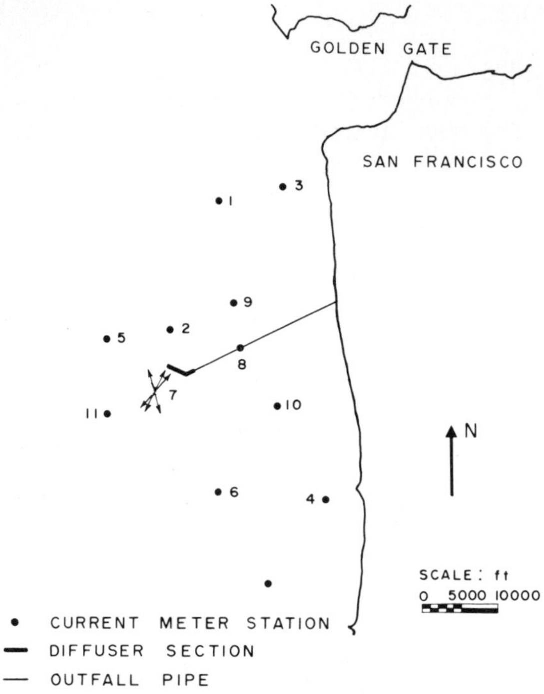

Figure 1. Study site. Vectors on Station 7 are the principal component directions.

intermittently for periods of one to two months during the year. The total period of investigation was February 1977 to January 1978. The current meters were Endeco type 105 set to record speed and direction averaged every half-hour; the duration of each data set was nominally one month. The analyses discussed below are for the period September 2 to September 30, 1977, when seven meters were operating.

CURRENT METER ANALYSES

The speed and direction were first converted to orthogonal speed components, one in a Northerly direction and one in an Easterly direction. As these directions do not have any inherent physical significance, the directions of the principal axes were computed. These are defined as the directions of the eigenvectors of the matrix formed by the covariances between the two speed component time series. These axes also maximize and minimize, respectively, the variance of the currents projected onto them. It was found that the directions of the first principal components all point towards the Golden Gate, with their variance decreasing with distance from the Gate. The first principal component is strongly tidal, the second less so, and the first principal component contains much more energy than the second. These preliminary findings are discussed in Roberts, 1980.

Oceanic motions occur over a very wide range of timescales, each of which has different effects on the fate of the discharged wastewater. To illustrate this, time series plots of the first and second principal components at Station 7A are shown in Figure 2, and a power spectral estimate of the first principal component in Figure 3. (The spectra was computed by an FFT algorithm after applying an approximation to the Parzen window to 1024 points of the raw data. No averaging of the resulting coefficients was employed, although the spectrum was smoothed by one pass of a recursive filter.) The spectrum shows the energy to be strongly peaked at the diurnal and semidiurnal tidal frequencies, with most energy at the semidiurnal frequency. Relatively little energy is contained in the higher frequencies. The low frequency shows increasing energy with decreasing frequency.

In order to better discuss the effects of the different timescales on the wastefield, the currents were divided into three frequency bands by the application of digital filters. The frequency bands are shown in Figure 3. To do this, the raw Northerly and Easterly speed components were both subjected in turn to: First, a low-pass filter with a cut-off frequency of 0.84 cpd; Second, a band pass filter with half-power cutoffs at 2.40 cpd and 0.84 cpd, and third, a high-pass filter with a half-power cutoff at 2.40 cpd. The filters used were of the linear

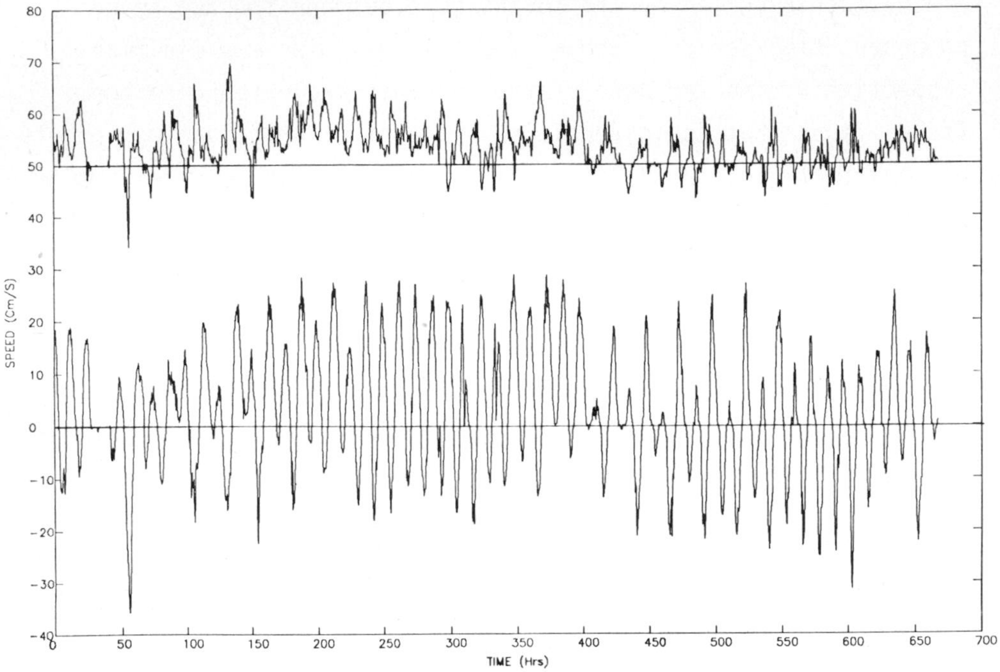

Figure 2. Principal components of currents at Station 7A. The second principal component (top) is displaced 50 cm/s above the first (bottom).

phase finite impulse response type using a Kaiser window. After filtering, each frequency band pair of time series were subjected to a principal component analysis. For Station 7A, the resulting directions of the principal axes of the low, tidal, and high frequency bands are shown in Figure 1, and the time series of the low frequency currents in Figure 4. An alternative analysis would be to compute the rotary spectra of the currents. Filtering is used here to preserve the time series, particularly of the low frequency content. The characteristics of each band are discussed separately below.

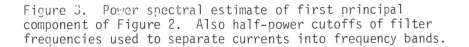

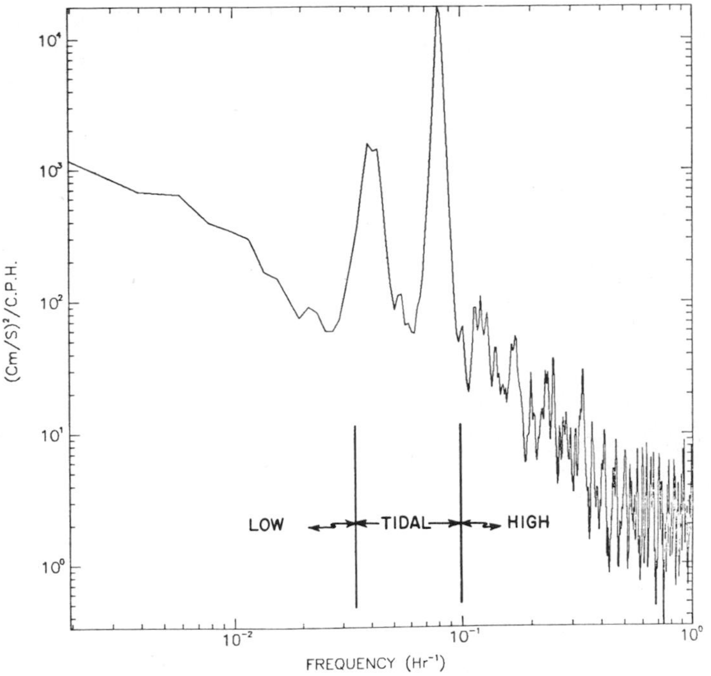

Figure 3. Power spectral estimate of first principal component of Figure 2. Also half-power cutoffs of filter frequencies used to separate currents into frequency bands.

The first few minutes following release of the sewage consists of the initial dilution phase. As this initial dilution is strongly increased by ambient currents, it would be expected that the most effect on the sewage for short times would result from the most energetic currents. These currents are contained in the tidal band. Because diffusers placed perpendicular to a current result in the highest initial dilution (Roberts, 1979), the diffuser was placed perpendicular to the first principal current direction (see Figure 1). The direction of the tidal principal components shown is an average over the diurnal and semidiurnal components. Although this would not generally be desirable, in this case the

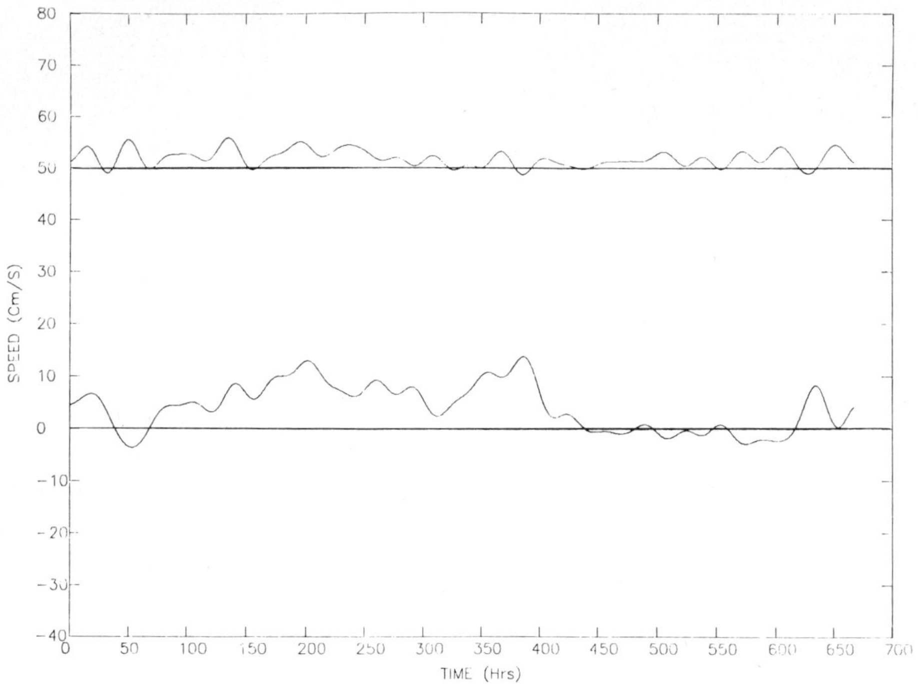

Figure 4. Low frequency band principal components at Station 7A. The second principal component (top) is displaced 50 cm/s above the second (bottom).

tidal currents are dominated by the Golden Gate and have similar directions. Further discussion of the effect of diffuser orientation and simulation of the effect of currents on dilution for this outfall are given in Roberts, 1980, and the mechanics of wastewater dispersion are discussed in Koh and Brooks, 1975.

For the first few hours following its establishment, the wastefield is advected by ocean currents and diffused by oceanic turbulence. This turbulence is contained in the high frequency band, and so this band affects the wastewater for these time scales. Although the directions of the first principal axis of the high frequency band is similar to that of the tide, i.e. pointing towards the Golden Gate, the direction is not particularly significant, as this band is almost isotropic. The tidal band also affects the wastefield on this time scale in two ways. First, the currents on this time scale responsible for advecting the wastefield are most probably tidal. Second, the excursion of the wastefield is also tidal.

For very long periods of days to months or even years, the buildup of pollutants in the vicinity of the discharge is governed by the low frequency currents. These are the currents which sweep by the site and flush the area. The direction of the low frequency band is towards the Northwest, and is unrelated to the tidal and higher frequency currents. Typical velocities (Figure 4) are less than 10 cm/s. Considerable variability exists in the currents, and it is apparent from Figure 4 that two distinct periods exist. Up until about 400 hours a consistent, well-defined flow to the Northwest exists. From about 400 hours to 620 hours, however, the current becomes very slow, with an ill-defined direction.

Because of the great importance of these low frequency currents, they were investigated in more detail. To do this a lowpass filter with a half-power cutoff of 0.60 cpd was applied to the Northerly and Easterly current components of all meters operating during this period. These meters were numbers 5, 6, 7A, 7B, 8, 11, and 12. A multiple principal component analysis was then applied to the resulting 14 time series. (The multiple principal component analysis is described in detail by Koh, 1977, and also in Kundu, et al, 1976, where it is known as an Empirical Orghogonal Function, or EOF analysis.) Briefly, the principal components are formed by the product of the eigenvectors of the covariance matrix of the original time series. The principal components are arranged from one to fourteen in order of decreasing variance.

The results of the multiple principal component analysis are shown in Figures 5 and 6. Figure 5 shows the first five principal components, and Figure 6 represents the direction and magnitude of the contribution of the first principal component to the total

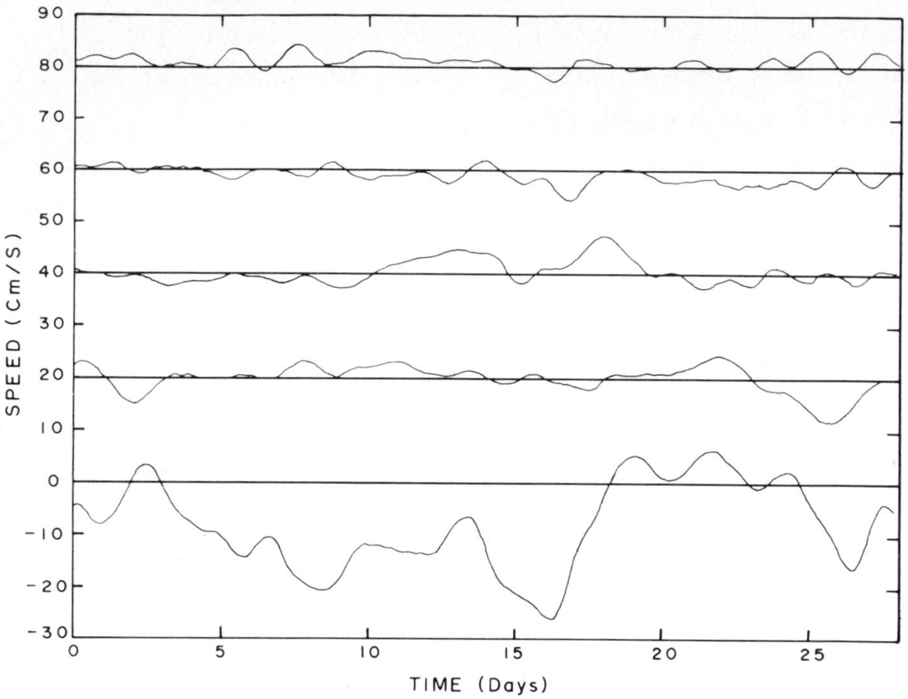

Figure 5. First five principal components of low frequency currents. Arranged in decreasing order of variance from bottom to top.

low frequency current at each station. (The actual contribution of the first principal component is found by multiplying the vector of Figure 6 by the first principal component of Figure 5). Note that because the first principal component is predominantly negative, the low frequency currents flow in a direction opposite to the vectors shown in Figure 6. The first principal component in this case accounted for 76% of the total variance, and the first four principal components accounted for 91% of the total variance.

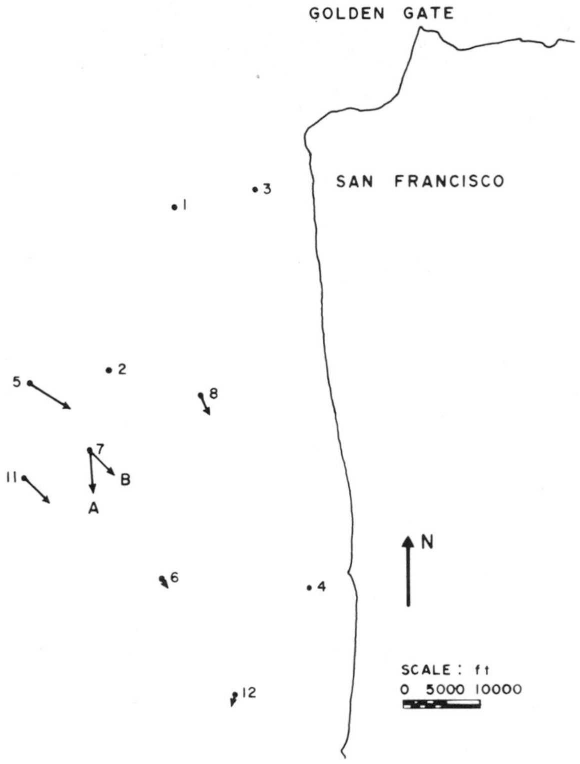

Figure 6. Magnitude and direction of contributions of first principal component of low frequency currents to total current.

The first principal component shows the characteristics of the currents at Station 7A (Figure 4) to be common to all sites. That is, there is a period of rapid flushing towards the Northwest, which ends at about 400 hours (approximately day 17), followed by a period of sluggish circulation from day 17 to day 25. Investigation of six months of record showed the predominant current pattern over this period to be towards the Northwest. This current pattern will therefore be termed the "normal" pattern. Many periods existed, in some cases lasting for several weeks, however, when the current was very small. These periods will be termed "abnormal."

The multiple principal component analysis has also been applied to the normal and abnormal periods separately. It was found that for the normal period, the directions and magnitudes of the first principal component were very similar to those shown in Figure 6 with a generally decreasing magnitude to the Southeast. For the abnormal period, however, the directions and magnitudes of the first principal component varied considerably from site to site, with no well-defined flow field. The picture that emerges is that the currents can be divided into two periods. First, a well-defined flow to the Northwest exists in which the principal component analysis would be very useful in describing the flow field. The currents are highly correlated in space and time during this period. Second, a period in which the currents are slow and no well-defined circulation pattern exists. The currents are not well correlated for this period, and the multiple principal component approximations would be poor.

These abnormal periods are of considerable importance to the environmental impact of the outfall. During these periods flushing of pollutants from the area is very slow, and as they can extend for several weeks a significant build-up of contaminants could occur. Simulations of wastefield transport for these periods showed that during the normal period, the wastefield moved steadily to the Northwest, while moving back and forth with the tide along an axis pointing approximately to the Golden Gate. During the abnormal period the wastefield still moved back and forth with the tide, but was not transported out of the immediate discharge site.

SUMMARY AND CONCLUSIONS

An analysis of currents measured near a proposed ocean outfall is presented in order to examine the effects of the currents on the design and transport of the discharged sewage. The currents were discussed in terms of their low, tidal, and high frequency content. It was shown that the tidal currents affect the wastewater on time scales of minutes to hours after release from the diffuser. The high frequency content affects the wastewater on time scales of hours, primarily through turbulent diffusion. The low frequency currents affect the long-term buildup of pollutants.

Two distinct periods of low frequency current behavior existed. One had fairly rapid currents to the Northwest, during which rapid flushing of the discharge site existed. These currents would be well approximated by the first component of a multiple principal component analysis. During the other periods, which can last for several weeks, the currents were slow and poorly defined. Poor flushing would be expected for these periods, and the multiple principal component analysis may not be useful for these times.

REFERENCES

Koh, R.C.Y., 1977, "Analysis of multiple time series by principal components with application to ocean currents off San Francisco," Tech. Memo. 77-4, W.M. Keck Laboratory of Hydraulics and Water Resources, California Institute of Technology.

Koh, R.C.Y., and Brooks, N.H., 1975, "Fluid Mechanics of Wastewater Disposal in the Ocean," Ann. Rev. of Fluid Mechanics, Vol. 7: 187-211.

Kundu, P.K. et al., 1976, "Modal decomposition of the velocity field near the Oregon Coast," J. Physical Oceanography, 5: 683-704.

Roberts, P.J.W., 1979, "Line Plume and Ocean Outfall Dispersion," J. Hydraulics Division, ASCE, 105 (HY4): 313-331.

Roberts, P.J.W., 1980, "Ocean Outfall Dilution: Effects of Currents," J. Hydraulics Division, ASCE, 106 (HY5): 769-782.

SHOULD WE SEARCH FOR PERIODICITIES IN ANNUAL RUNOFF AGAIN?

ANDERS WILLEN
Vassdragsregulantenes Forening, P.O. Box 145, N-1371 Asker, Norway

ABSTRACT

This paper deals with annual series of runoff and precipitation. The search for hidden periodicities in annual hydroclimatological series that took place earlier, has been replaced by a "stochastic modeling" approach during the last 15 years. According to this school, the cosine components one may possibly find in annual series are a result (almost) entirely of chance, and hence should not be used in any modeling or forecasting procedures. While recognising the importance of stochastic elements in hydroclimatological phenomena, the conclusion of this paper is that the use of simple harmonic components could be justified in some cases, if used by care. This conclusion is supported by analysis of some series from and in the neighbourhood of Norway and a discussion of arguments for and against the use of cosine components in stochastic modeling of hydroclimatological phenomena. In the analysis the split-sample technique was applied, where each series was divided into two or three parts. For most periods the phases and amplitudes were different, a result that was expected. However, a period of about 19 years was found in both parts of a Norwegian runoff record of 108 years, and both the amplitudes and phases were in accord. Moreover, similar results were found for some other series.

INTRODUCTION

The search for periodicities in annual runoff was more or less

Reprinted from *Time Series Methods in Hydrosciences*, by A.H. El-Shaarawi and S.R. Esterby (Editors)

abandonned some 15 years ago. The new idea, introduced by scientists interested in stochastic processes, was that the periodicities that may exist in historical series are a result (almost) entirely of chance. Instead of looking for periodicities, one should regard runoff as a stochastic process of one kind or another. Different models have been suggested, from the very simple stationary white noise model to complicated models like for example Fractional Gaussian Noise. The use of simple cosine-components as part of a stochastic model was (almost) never applied, however. There are several arguments against the use of such components, for example:

a) The extracted harmonic components are not statistically significant.
b) If the split-sample technique is applied, the period lengths, phases and amplitudes will be quite different in different subsamples of a record. This argument then would make cosine components quite useless in forecasting procedures.
c) Even if a cosine component really would exist in the "real" process that produces runoff at a site, this component only explains very little of the total variation in the annual values.

The last argument is not very strong. It is certainly true, that it is usually not possible to make any deterministic predictions of next year runoff by using cosine components, at least no predictions with confidence. However, similar arguments could be raised against essentially all models in use today.

Turning to argument a), the lack of statistical significance one usually obtains may very well be a result of the short record(s) at hand. Owing to this it is usually difficult to obtain significance for any kind of model or coefficient, and often even the simple stationary white noise model seems reasonable according to standard tests of significance. However, there are several facts that indicates that the simple stationary white noise model is too simple. And if the task is to estimate for example the probabilities of combinations of dry years, the computed risks often will be too low, if the small (but probably real) departures from the stationary white noise model are not taken into account.

The author considers point b) to be the strongest argument against the use of cosine components, and the anlysis below is devoted partly to this argument.

ANALYSIS METHODS

In this study only simple methods are used. A number of period lengths has been applied to the series. For each period length and series a least square procedure has been applied according to:

$$crit(p) = \min_{A_o, A, B} D = \sum_{k=1}^{T} \left(A_o + A \cdot \sin \frac{2\pi k}{p} + B \cdot \cos \frac{2\pi k}{p} - Q_k\right)^2 \qquad (1)$$

Q_k	runoff year number k
k	year number
T	length of the series (number of years)
p	length of period ("wavelength")
A_o, A, B	parameters

It should be noticed that this is not quite the same as conventional Fourier-analysis, because in that method the series is decomposed into a number of harmonic components, where all period lengths could be written as p=T/N (N integer, N=1,2,...,T/2. Here we have more freedom in the choice of p, we could for example have T=50 and p=20.

For fixed values of p eq. (1) is a linear least square problem. A number of values of p were tried, and as a result the criterion (crit) was obtained as a function of p for each series.

In the output of the computer programe we also obtained the harmonic components in the form

$$QP_k = A_o + C \cdot \cos\left[\frac{2\pi}{p}(k - ph)\right] \qquad (2)$$

QP_k value of the cosine component for year k-1900

C amplitude of the cosine component

ph phase. Here the reference year for phases is 1900. This means that the harmonic component obtains its maximum at years 1900+m ph, m integer

One possibility in the search for periodicities would be to simply choose the value of p that gives the best fit, restricting p to "reasonable" values. Argument b) above strongly opposes this method however. Instead the split-sample technique was applied, where the records were divided into two or three parts. Then the analysis method was applied to the different parts as well as to the whole record. This method requires long records, and hence strongly restricts the choice of stations to use.

RESULTS AND CHOICE OF STATIONS

Unfortunately only very few stations with long enough records of good quality exists in Norway. For Elverum (147 km north of Oslo) at river Glomma we have a runoff record starting in 1872 and at Ås near Oslo the precipitation record starts in 1874, both these stations are still in operation. There exists some precipitation records starting in 1890:s, and very few starting earlier than that.

The quality of the longer precipitation series can be questionned however. A precipitation series is very much influenced by very local phenomena (as for example change in vegetation in the immediate vicinity of the gauge) and by changes in shielding procedures. As the river profiles in Norway usually are stable and as runoff is an integrator over the catchment area, river runoff is probably a better indicator of long-term variations of water availability than precipitation.

Starting with the streamflow record from Elverum (1872-1979) it was split up into two parts, 1872-1922 and 1923-1979. Applying different values of p in eq. (1), the result was (as expected) that no single value of p gave a harmonic component that explained very much of the total variance in runoff. Anyhow, crit (p) as a function of p was obtained, and in fig. 1 is shown $\sqrt{R^2}$ as function of p (R^2=(total variance - explained variance)/total variance)) for the whole record. It should be noticed that the study and figure is not very detailed for low values of p, where local maxima and minima may

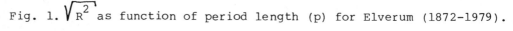

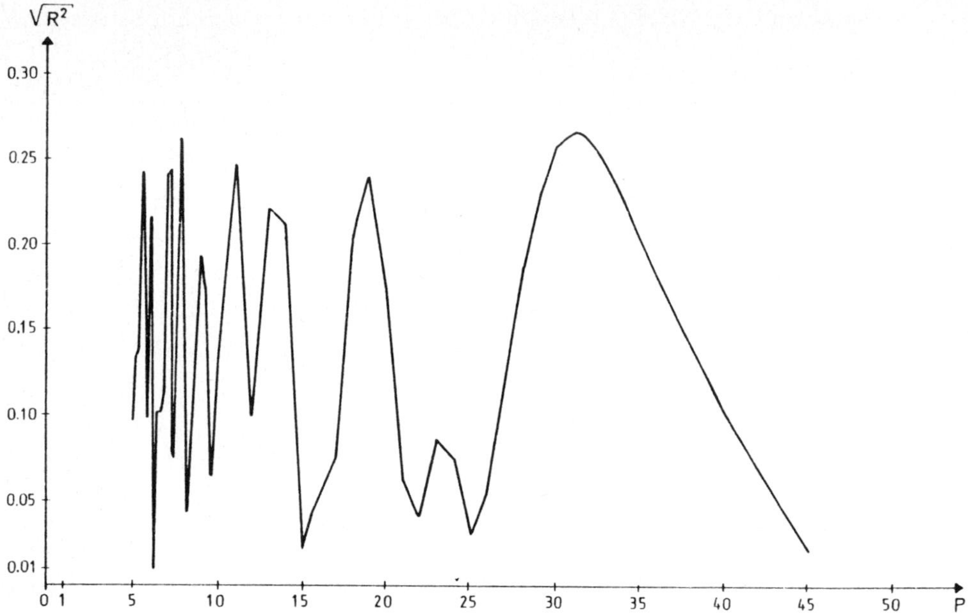

Fig. 1. $\sqrt{R^2}$ as function of period length (p) for Elverum (1872-1979).

exist not shown in the figure. Anyhow, if cosine terms are to be used in a stochastic model, the author thinks they should be used in modeling long-term phenomena, i.e. phenomena which may be the result of large-scale climatic variations. Short-term phenomena (with "periodicities" of about a few years) are easily considered a combined result of local effects and chance, and hence are easily modeled by low order autoregressive terms.

Ignoring period lengths shorter than 10 years, the figure shows peaks for periods of about 11, 13, 20 and 30 years. These peaks certainly are not significant according to standard statistical tests, and indeed do not explain very much of the total variance.

Moreover, the amplitude and/or phases for cosine-terms corresponding to these periods were different in the two parts of the record, except for p about 20 years. Considering a period length of 31 years for example, the amplitude is 8.86 in the first part and 21.32 in the second part of the record (unit: m^3/s).

The best integer value of p near 20 was 19, and for this value of p the following table was obtained.

TABLE 1.

Results for cosine term with period length 19 years for Elverum (runoff). Unit: m^3/s.

Series	1872-1922	1923-1979	1872-1979
Stand.dev.	43.47	42.96	43.08
A_o	243.20	248.33	246.40
C	13.75	15.81	14.58
ph	9.79	8.13	8.81
R^2	0.0524	0.0689	0.0576

Peaks near a period length of 20 years also were obtained for the two parts of the series, the best integer values were 20 and 19 years for 1872-1922 and 1923-1979 respectively.

Looking at table 1, one notices that for a cosine-term with a period length of 19 years, the amplitudes and phases are (approximately) consistent in the two parts of the record. This consistence made the author inclined to accept a cosine term with this wavelength as a "reality", although indeed this component explains very little of the variance. In figure 2 is shown the Elverum record and the cosine term with p=19 years fitted to the whole record.

A second-order autoregressive (AR-2) model was also tried. The order of the AR-model was chosen after applying the split sample technique, the highest order that gave consistence in the parameters between the two parts of the record was accepted. The AR-2 model explained slightly more of the variance than the periodic component with p=19 years (R^2=0.0728 for the whole record) for years starting at 1:st of January. However, when changing the starting date for the year, the parameters of the AR-2 model changed and sometimes an R^2-value as low as 0.0360 was obtained. For the cosine-term with p=19 years, the parameters remained essentially constant however.

Turning to the precipitation series from Ås (1874-1979) it was divided into two parts, 1874-1924 and 1925-1979, and analyzed in the same way as Elverum.

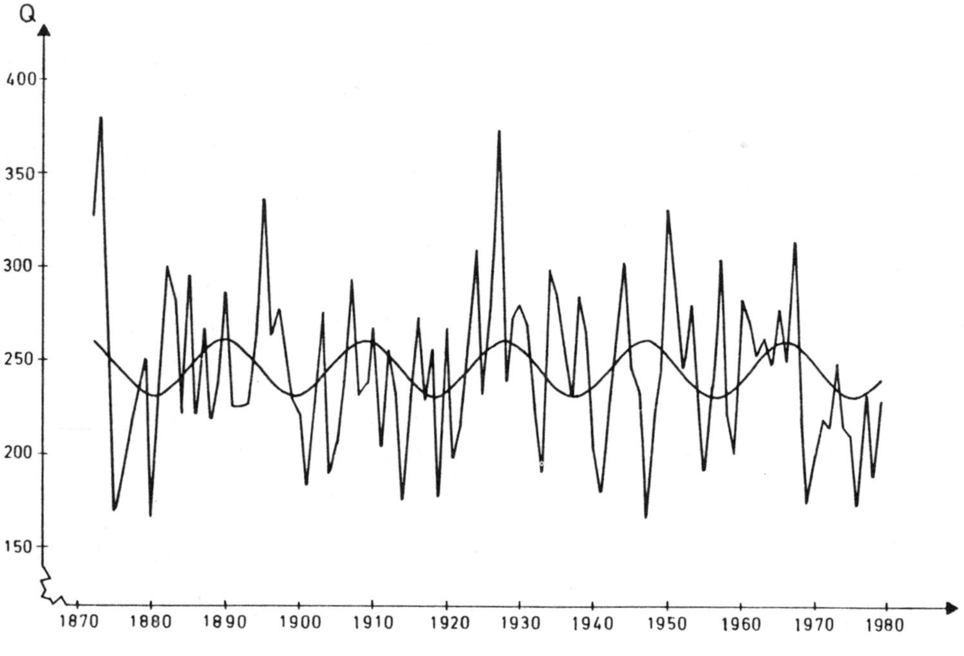

Fig. 2. The Elverum runoff series (1872-1979) and the fitted cosine term with p=19 years. Unit: m^3/s.

Also this record had a peak in R^2 for p near 20 years for both parts of the record as well as for the whole record. For p=19 years, table 2 was obtained.

Obviously the amplitudes were not quite in accord, but the phases were nearly coincident and also near the phases obtained for Elverum.

TABLE 2.

Results for cosine term with period length 19 years for Ås (precipitation). Unit: mm.

Series	1874-1924	1925-1979	1874-1979
Stand.dev.	141.32	147.15	144.37
A_o	754.48	779.75	768.06
C	32.85	52.53	44.25
ph	9.42	8.47	8.86
R^2	0.0528	0.0668	0.0468

For this precipitation series better agreement were obtained for p=14 years, with amplitudes 48.26, 36.69 and 41.20 and phases 9.73, 8.78 and 9.35 respectively. One precipitation series from Trøndelag in Norway (Lien in Selbu 1896-1978) also showed internal agreement in phases (5.90, 4.90, 4.36) for the series (1896-1936, 1937-1978 and 1896-1978) for a period of 19 years, but the amplitudes differed even more than for Ås. For the Lien series shielding procedures has not been homogeneous however, and also the series is somewhat shorter.

At last the runoff record from Göta Älv river (1807-1964) obtained from Unesco (1971) was analyzed. The station is located in the west of Sweden some 330 km south of Oslo. Table 3 shows the results obtained for a cosine-term with a wavelength of 19 years.

TABLE 3.

Results for cosine term with period length 19 years for Göta Älv (runoff). Unit: m^3/s.

Series	1807-1871	1872-1964	1872-1922	1923-1964
Stand.dev.	105.47	97.32	96.22	99.51
A_o	542.26	529.10	532.48	523.61
C	5.32	28.68	30.29	28.39
ph	5.30	11.66	12.24	10.81
R^2	0.0013	0.0443	0.0527	0.0389

Obviously the results for the first part of the series (1807-1871) differed from the results for the rest of the record. But for the three other analyzed parts of the record (the first of which is simply the combination of the last two records) the phases and amplitudes were approximately consistent. And these three time periods cover nearly the same years as the Norwegian series discussed above.

Returning to the Elverum record, eq. (2) with p=19 years was applied, and the residuals were analyzed. These did not appear quite like white noise, and simple low-order Ar-models were applied to the residuals. Again, the choice of order of the AR-model was determined by the criterium, that the highest possible order that gave consistence in parameter estimates in the two subsamples was accepted. The result was a second-order model, i.e. the same order as that obtained when analyzing the streamflow directly. The residuals from this combined model appeared rather white. The suggested stochastic model for Elverum is thus:

$$Q_k = QP_k + AA(1)\cdot(Q_{k-1} - QP_{k-1}) + AA(2)\cdot(Q_{k-2} - QP_{k-2}) + EPS_k \qquad (3)$$

QP_k	value of harmonic component year k
AA (1), AA (2)	autoregressive coefficients
EPS_k	random number (with zero mean) corresponding to year k
SEPS	standard deviation of EPS

In this preliminary study, we simply used the normal distribution for EPS. Indeed, the choice of distribution should be devoted a study, but this is beyond the scope of this paper. The parameters are shown in table 4.

TABLE 4.

Parameters obtained for combined model for Elverum. Unit: m^3/s.

Series	1872-1922	1923-1979	1872-1979
A_o	243.20	248.33	246.33
C	13.75	15.81	14.58
ph	9.79	8.13	8.81
AA(1)	0.2010	0.2433	0.2271
AA(2)	-0.1356	-0.1696	-0.1585
SEPS	41.32	40.03	40.53
R^2	0.0966	0.1320	0.1150

DISCUSSION

Do not this search for cosine-terms in annual series means a big risk, that the cosine-terms which is applied here are the results of a quite different process? And do not the model applied mean that we fix the period lengths (in generated sequences) to 19 years, while these lengths should vary? There are many critical questions to raise against the model and analysis applied, some of them were stated already in the introduction.

Turning to the first question (the risk that we choose a model to fit what is in reality a result of a quite different process), this risk exists whatever model one may use.

Considering the length of "periods" in generated sequences, it is not fixed to 19 years. Owing to the rather large variance of the residuals (EPS), generated sequences will contain "periods" with different lengths. The most serious problem concerns the stability of the model parameters when estimated from different parts of a record. How to interprete the results of the split-sample technique applied to the series? As stated above, the author has more confidence in long records of runoff than in long records of precipitation. Thus, the difference in amplitudes for even the long precipitation record from Ås should be given lower weight than the coincidence for the Elverum series, and for the Göta Älv record as regards the years 1872-1922 and 1922-1964. And these two time periods cover about the same years as do the Elverum series.

But what about the small amplitude for the 1807-1871 part of the Göta Älv record, is this not a confirmation that we should not use cosine terms? If we really are interested in (theoretical) long term stationary conditions, then certainly this argument is a strong one. But in practise, the main interest often is on the next 10 to 20 years rather than on the next 150 years. Then we may hope, that not only the phases but also the amplitudes will remain stable for the time span of main interest.

The fact that the amplitudes and phases etc., may be the result of a process of quite different kind, does not prevent us from using them. Our hope is, that the real process is in a state of "near stationarity", i.e. in a state where the real states of the real process are changing very slowly. Thus we may hope, that the parameters chosen can be considered stable for at least some years ahead. But of course we cannot consider the parameters as stable as would be required if we were to use the model for deterministic predictions. As stated above, similar arguments apply to most other models.

Another matter of discussion is how many parameters to use in a model. One school favours the use of as few parameters and as simple models as possible, only the very significant parameters should be accepted, the other parameters might be a result of chance. This idea has its merits when discussing scientific reseach. Suppose a scientist has found that the 50-year lag autocorrelation for temperature usually is negative. Then, if he wants to claim this as the discovery of a new climatic theory, indeed we should require strong evidence before accepting the theory. But if the interest is to estimate risks of combinations of dry years during some years ahead, then these estimates may be seriously biased (probably too low) if rejecting all parameters not passing the conventional statistical significance tests. At least parameters, whose estimates appear stable under the split-sample technique should be taken into account, unless this leads to excessive computing costs.

Considering the period length of 19 years, Hibler III and Johnsen (1979) analyzed very long records (1244-1971) from Greenland ice cores, and they claimed to have found a 20-yr cycle.

REFERENCES

Hibler III, W.D. and Johnsen, S.J., 1979. The 20-yr cycle in Greenland ice core records. Nature, 280: 481-483.

UNESCO, 1971. Discharge of Selected Rivers of the Wold. Vol II. Unesco, Paris.

STEP AHEAD STREAMFLOW FORECASTING USING PATTERN ANALYSIS

M.G. GOEBEL and T.E. UNNY

Department of Civil Engineering, University of Waterloo,
Waterloo, Ontario, Canada

SYNOPSIS

Persistence in streamflow time series is usually specified through correlation coefficients of order one and larger. These coefficients are calculated using the whole time series record. However, persistence effects are bound to be variable in different parts of the record. Consideration of small collections of successive data as distinct entities effectively incorporates persistence in analysis of data. This is the basic method involved in pattern analysis.

The historic time series is segmented into a series of representational objects which define local shapes within the time series. These objects are in turn defined by a set of pattern vectors. Cluster analysis is used to specify time indexed regions or classes in n-dimensional pattern space according to which future patterns may be classified.

The current streamflow observations are used for step ahead forecasting based on the concept that a complimentary datum is required to complete a pattern.

INTRODUCTION

Streamflow data recorded at monthly intervals is characterized by the presence of seasonality caused by reoccurring geophysical events within a year.Thus one observes groups of high or low flows according to wet or dry periods. Interdependence of successive flow data in monthly streamflow time series is termed persistence and can be explained by watershed storages as well as by long term meteorological

Reprinted from *Time Series Methods in Hydrosciences*, by A.H. El-Shaarawi and S.R. Esterby (Editors)

events.

Persistence in data is usually specified through correlation coefficients of order one and larger. However, because historical data is only a single realization of a stochastic sequence, correlation coefficients are bound to be variable with increases in the length of record with time. Furthermore, correlation coefficients are derived by scanning the data from beginning to end; thus if at all they represent the persistence effect, it could be nothing more than an averaged effect across the whole time series. On the other hand, in physical situations dealing with real world data, the persistence effect is bound to vary change in different par of the record. An effective way of incorporating persistence in data analysis is through consideration of small collections of data as distinct entities in the analysis. Such an approach is taken through the use of pattern analysis.

Pattern analysis is a well developed discipline by itself and it has been applied in diverse fields such as satellite imagery, medical diagnosis and cybernetics; however, it's application in hydrology is recent. Panu (1978) has developed a model to synthesize streamflow records using pattern analysis. Within the context of analysis of time dependent hydrologic data Panu, Unny, MacInnes and Wong (1981) give a well documented review.

In simple terms, a pattern is a shape representation of a physical object. In this application a rising limb, the peak or the falling limb of a streamflow time series (a streamflow time waveform or STW) are all considered separate physical objects. Measurements on these objects define the shape or pattern of the objects. In practice the actual streamflow data is simply arranged into a set of n-dimensional pattern vectors with each pattern element being numerically equal to the field measurement.

Selecting an appropriate dimensionality for the pattern vectors is based on the degree of interdependence between successive streamflow observations. For monthly data it would be difficult to justify any significant dependence beyond a two month period; hence, three dimensional pattern vectors are adequate. Overlapping the objects such that each pattern vector contains only one new datum allows all possible shapes to be represented. Pattern analysis is carried out on these pattern vectors and gives interpretable information regarding the nature of the streamflow time series. Pattern vector formation is illustrated in Figure 1. Feature extraction, a process that maps a pattern $\underline{X}_n$ onto a feature vector $\underline{Y}_m$, is commonly used to reduce dimensionality and enhance separability. This process was not required, here therefore feature vectors $\underline{Y}_m$ are synonomous with pattern vectors.

Step ahead forecasting is carried out by considering the pattern found from current and most recent observations as being incomplete in the sense that the future observation is required to provide the data to complete the pattern vector. The information gathered through pattern analysis of the historical record is used to predict the missing datum. Dixon (1979) used pattern analysis to estimate data in data sets with missing observations. The methodology presented herein differs from Dixon's approach in that the step ahead forecast is made by finding the expected value of the future streamflow given the possible classification of the incomplete pattern.

Data from the Black River watershed (gauging station 02EC002, near Washago, Ontario) was used to simulate step ahead forecasting. For this river 30 years of historical record was used to perform cluster analysis. An additional 28 years of historical data was then used to make step ahead forecasts which were then compared to actual observations. Further forecasts were always based upon the actual observations rather than forecasted values.

CLUSTER ANALYSIS

A pattern class can be considered to be a region in n-dimensional pattern space containing the set of all possible patterns which are sufficiently alike according to some specified measure. Any mathematical description of a class is a statistical summary of

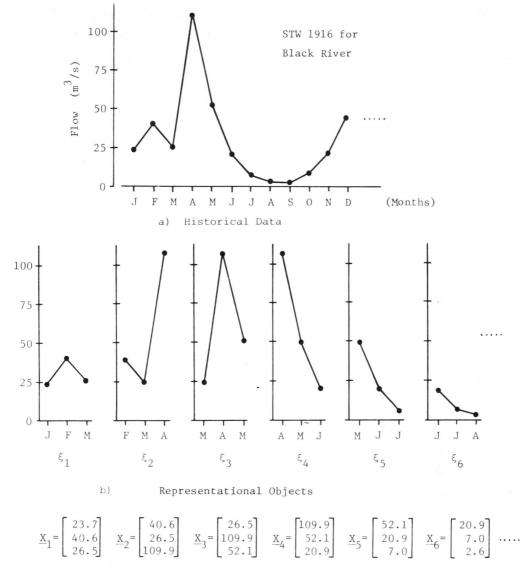

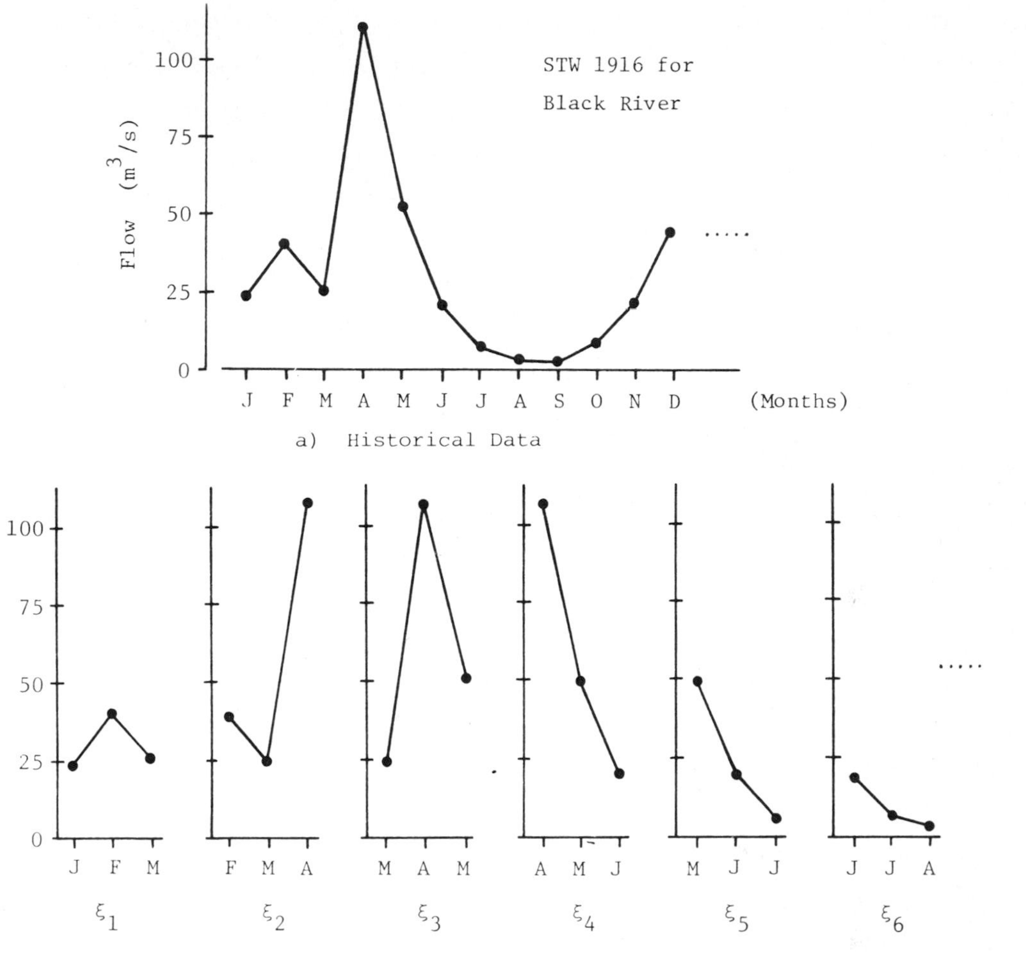

b) Representational Objects

$$\underline{X}_1 = \begin{bmatrix} 23.7 \\ 40.6 \\ 26.5 \end{bmatrix} \quad \underline{X}_2 = \begin{bmatrix} 40.6 \\ 26.5 \\ 109.9 \end{bmatrix} \quad \underline{X}_3 = \begin{bmatrix} 26.5 \\ 109.9 \\ 52.1 \end{bmatrix} \quad \underline{X}_4 = \begin{bmatrix} 109.9 \\ 52.1 \\ 20.9 \end{bmatrix} \quad \underline{X}_5 = \begin{bmatrix} 52.1 \\ 20.9 \\ 7.0 \end{bmatrix} \quad \underline{X}_6 = \begin{bmatrix} 20.9 \\ 7.0 \\ 2.6 \end{bmatrix} \quad \ldots..$$

c) Pattern Vectors

Figure 1 Pattern Vector Formation From Streamflow Data For Black River its member patterns. In many instances of pattern analysis, including the one in this application the true classes are unknown. All information concerning class structure must be estimated from a training set of unlabelled sample patterns through a procedure referred to as clustering or unsupervised learning.

Various clustering algorithms have been developed, the simplest and better known of which are the K-means algorithm (Tou and Gonzales, 1974) and the Isodata algorithm (Ball and Hall, 1965). In the clustering algorithm developed for monthly streamflow patterns it was assumed that there are K = 12 distinct pattern classes. All the patterns from one time position are initially assumed to belong to one group, G^k; that is they form an initial cluster. The cluster center is a reference vector, $\underline{m}^k$, calculated as follows:

$$\underline{m}^k = \frac{1}{N^k} \sum_{j=1}^{N^k} \underline{Y}_j \tag{1}$$

where N^k is the number of patterns (feature vectors $\underline{Y}_j$) in group G^k. These initial clusters are characterized by a rather large variation. This variation is described by a covariance matrix C^k as follows:

$$C^k = \frac{1}{N^k} \sum_{j=1}^{N^k} (\underline{Y}_j - \underline{m}^k)(\underline{Y}_j - \underline{m}^k)^T \tag{2}$$

Another way of examining the variation within a group is to calculate the distance from each pattern to its reference vector. Using a Euclidean distance measure:

$$d_E(\underline{Y}_j, \underline{m}^k) = \left[\sum_{i=1}^{n} (y_i - m_i^k)^2 \right]^{\frac{1}{2}} \tag{3}$$

The mean intra group distance is:

$$\bar{d}_E^k = \frac{1}{N^k} \sum_{j=1}^{N^k} [d_E(\underline{Y}_j, \underline{m}^k)] \tag{4}$$

Another useful distance is the inter group distance or the distance between reference vector $\underline{m}^k$ and any other reference vector $\underline{m}^j$:

$$d_E(\underline{m}^k, \underline{m}^j) = [\sum_{i=1}^{n} (m_i^k - m_i^j)^2]^{\frac{1}{2}} \tag{5}$$

A boundary between two groups can be visualized as a distance half-way between any two cluster centres. The hyperplane describing such a boundary is a decision surface according to which a particular pattern is said to belong to one group or the other. For example:

$$\underline{Y} \in G^k$$

$$\text{if} \quad d_E(\underline{Y}, \underline{m}^k) < \text{all} \quad d_E(\underline{Y}, \underline{m}^j) \quad k \neq j \tag{6}$$

Clustering proceeds as follows:

1. Determine initial cluster centres and boundaries as noted above.
2. Using Equation (3), determine for every pattern the distance to its own cluster centre and the distance to the centre of the two clusters that are situated time contiguous on both sides (i.e., neighbouring clusters).
3. Reassign all the patterns to any of the three time contiguous clusters according to the criterion of Equation (6). The centres of these clusters can be considered to be the centres of the pattern classes of which the clusters form a sub-group. The procedure ensures that a proper trajectory progressing in time is formed by connecting the centres of the classes. The clusters so formed are time-dependent clusters and not global clusters.
4. For the clusters obtained in Step 3 above, new cluster centres and boundaries are calculated and the procedure repeated from Step 2 to Step 3 with the objective of reducing the variance of the groups, and thus the intra cluster distances are also reduced.

Further, the inter class distances are enhanced providing for increased separability of the classes.

This clustering algorithm stops when the last complete cycle of iteration was carried out and it was found that there was no further change in the membership of any of the clusters. Typically, about fourteen iterations were required to cluster the patterns. The final clusters represent sample estimates of the pattern classes (ω^k). These classes are characterized by a narrow multivariate probability distribution having a mean and some variance about the mean. The regions in m-dimensional feature space representing each of the twelve classes also have definite boundaries according to which any new unlabelled pattern can be properly classified. Two typical clusters for Black River are shown in Figure 2.

The ordered progression from one class centre to the next is called a mean annual cycle (See Figure 3). The outer segment of the curve having large values of y_1, y_2 or y_3 or combinations of all three, represent the regions in pattern space that are equivalent to peak flows. The two loops represent streamflows that have a large annual peak flow (as occurs in the spring run off) and a smaller peak occurring annually in the fall. A larger distance between two adjacent reference vectors on the annual cycle signifies a greater variability of the member patterns within the class. Cluster boundaries bisect the distance between adjacent reference vectors. Clusters may be relatively close together in terms of distance in the 3-dimensional feature space. However, if the reference vectors are on different loops, they are far apart in time (i.e., they are not time contiguous).

STEP AHEAD FORECASTING

Forecasting is based on all of the information gathered through the use of cluster analysis. The most recent pattern is considered to be incomplete in that it has two known features, y_1 and y_2, but the third feature, y_3, representing the step ahead flow, is unknown.

Figure 2 Typical Clusters Represented by Objects from the STW for Black River. (Not all objects are shown)

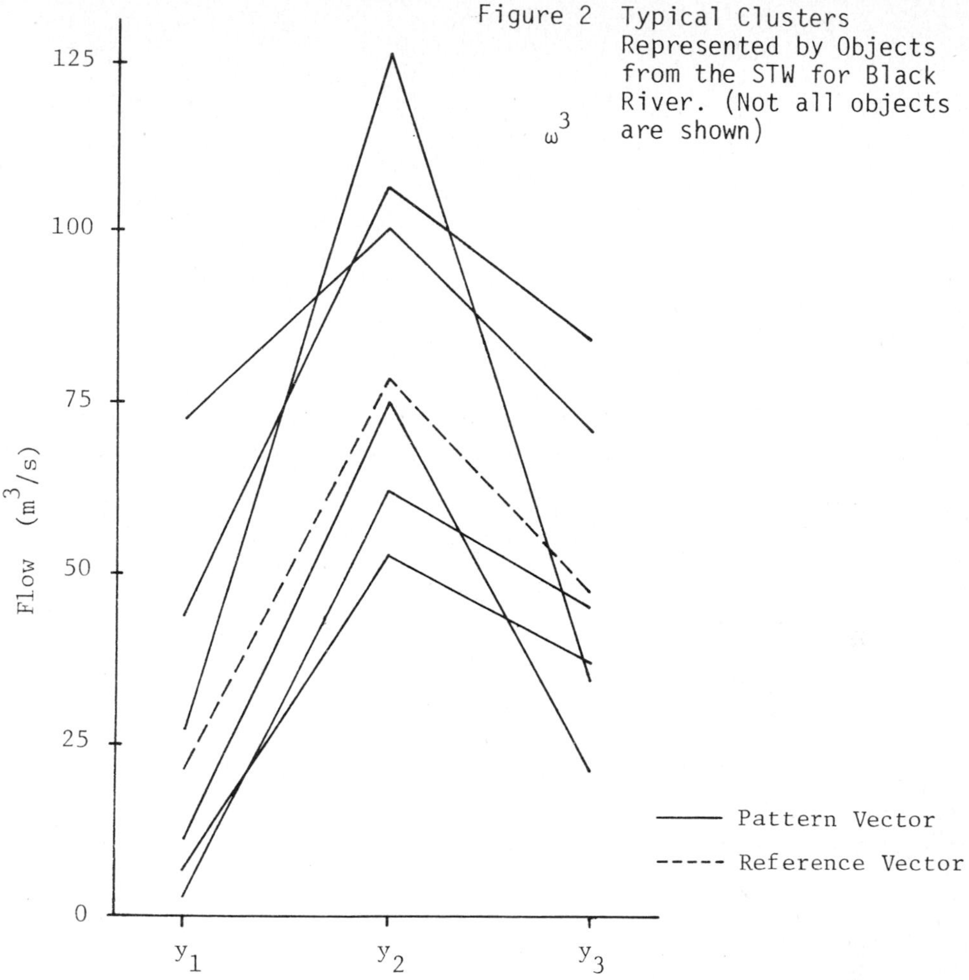

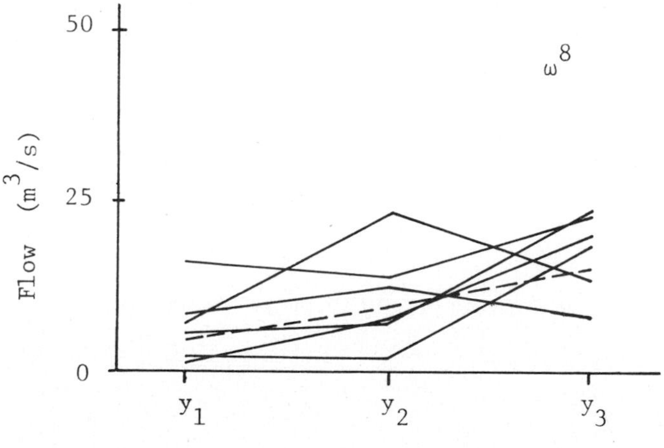

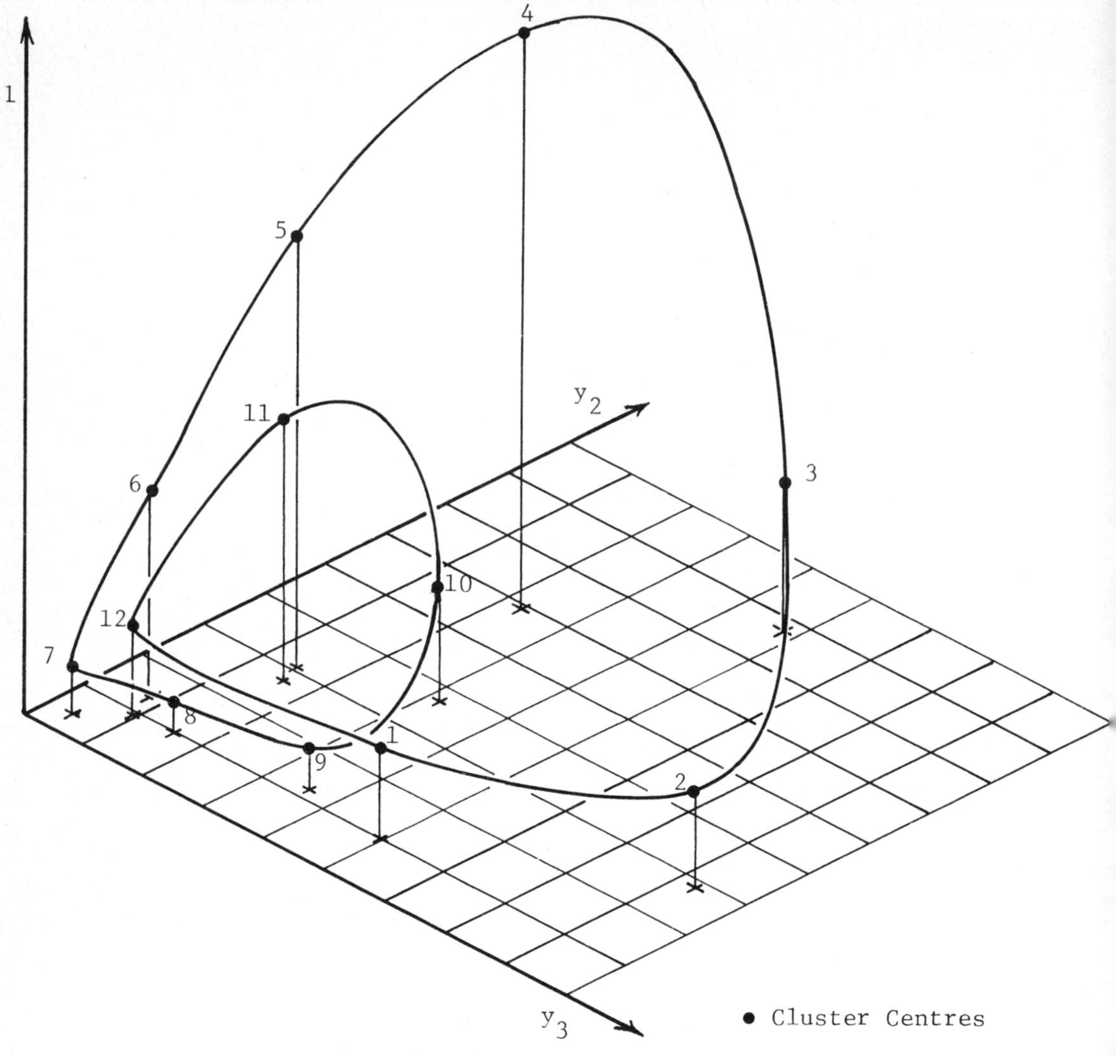

Figure 3. Mean Annual Cycle For the Black River represented in the three dimensional feature space.

A step ahead forecast is made by estimating the value of y_3.

The initial step is to "classify" the incomplete pattern. Because the third element of the pattern vector is missing, proper classification is not always possible. However, classes to which the pattern is likely to belong can be determined with a finite probability. The classes to which an unlabelled pattern may belong are called candidate classes. These candidate classes are determined using the first two elements of the incomplete pattern.

Forecasting is carried out based on the probability of the occurrence of y_3 within a candidate class as well as the probability of occurrence of the candidate class within the pattern space. Thus the forecast is the expected value of y_3.

It was observed in the cluster analysis that patterns tended to cluster in groups whose member patterns were derived from times of origin that are the same or neighbouring months of the year. The time of origin of a pattern is therefore an a priori information for determination of candidate classes. Two criteria must be satisfied for a class to be a candidate class:

1. For the features y_1 and y_2 from an incomplete pattern, a finite value for y_3 must exist within the class boundaries.
2. The probability of the class must be non-zero given the months of origin of the incomplete pattern class to be a candidate class.

Pattern classification is based on the a posteriori probability of the class which can be found using Bayes Theorem. Written in terms of the application:

$$P(\omega^k|y_1,y_2) = \frac{p(y_1,y_2|\omega^k)\ P(\omega^k)}{p(y_1,y_2)} \tag{7}$$

where $p(y_1,y_2|\omega^k)$ is the class-conditional probability density and is estimated from the marginal probability as follows:

$$p(y_1,y_2|\omega^k) = \frac{1}{2\pi \det \begin{bmatrix} \sigma_{11} & \sigma_{12} \\ \sigma_{21} & \sigma_{22} \end{bmatrix}_k^{\frac{1}{2}}} e^{-\frac{1}{2}\begin{pmatrix} y_1-m_1^k \\ y_2-m_2^k \end{pmatrix}^T \begin{bmatrix} \sigma_{11} & \sigma_{12} \\ \sigma_{21} & \sigma_{22} \end{bmatrix}_k^{-1} \begin{pmatrix} y_1-m_1^k \\ y_2-m_2^k \end{pmatrix}} \qquad (8)$$

$\begin{bmatrix} \sigma_{11} & \sigma_{12} \\ \sigma_{21} & \sigma_{22} \end{bmatrix}_k$ is the partial covariance matrix of class ω^k which is obtained from the first two rows and columns of the 3x3 matrix C^k.
m_1^k, m_2^k are the first two features of the reference vector for class ω^k.
$P(\omega^k)$ is the probability for the pattern as determined by the number of patterns from one time of origin in the class

$$p(y_1,y_2) = \sum_k^K p(y_1,y_2|\omega^k)\, P(\omega^k) \qquad (9)$$

The a posteriori probabilities for all $k \leqslant 12$ candidate classes are such that

$$\sum_k^K P(\omega^k|y_1,y_2) = 1 \qquad (10)$$

It must be noted here that all probabilities are based on calculations using data from the cluster analysis. Should a slightly different set of patterns be used for clustering then it is quite likely that the a posteriori probabilities will change given the same incomplete pattern. Such a case would occur as new streamflow measurements continuously update the historical record. This situation is common to any model that uses a finite and, in most cases, limited amount of samples.

The pattern classes are assumed to have a multivariate normal distribution of sample patterns. The missing feature y_3 of an incomplete pattern is found by calculating the expected value of y_3. For any y_1 and y_2 from a particular candidate class there is a probability associated with any y_3. This conditional probability is:

$$p(y_3|y_1,y_2)^k = \frac{p(y_1,y_2,y_3)^k}{p(y_1,y_2)^k} \tag{11}$$

where $$p(y_1,y_2,y_3)^k = \frac{1}{2\pi^{3/2}\,\det[C^k]^{\frac{1}{2}}}\, e^{-\frac{1}{2}(\underline{Y}-\underline{m}^k)^T[C^k]^{-1}(\underline{Y}-\underline{m}^k)}$$

and $$p(y_1,y_2)^k = p(y_1,y_2|\omega^k)$$

as in Equation (8).

Not all values of y_3 are feasible. Feasible values of y_3 are determined by finding points on the boundary of the class in m-dimensional feature space along the line given by y_1 and y_2. These values for y_3 range from, say, A to B where A is on the boundary between class ω^k and ω^{k+1} and B is on the boundary between ω^k and ω^{k-1}.

The expected value of y_3 given the candidate class is:

$$\exp(y_3|\omega^k) = \frac{\int_A^B y_3\, p(y_3|y_1,y_2)^k\, dy_3}{\int_A^B p(y_3|y_1,y_2)^k\, dy_3} \tag{12}$$

The variance of y_3 can also be calculated by simply including an additional term in the expression above. In all cases the estimate of the most probable value of y_3 is weighted according to the class a posteriori probabilities. The expected value of y_3 is:

$$\exp(y_3) = \frac{\sum_{k=1}^{K} \exp(y_3|\omega^k) \;\; P(\omega^k|y_1,y_2)}{\sum_{k=1}^{K} P(\omega^k|y_1,y_2)} \tag{13}$$

STEP AHEAD FORECASTING RESULTS

The methodology described above was applied to an additional 28 years of historical data to produce a set of step ahead forecasts based on actual observations. This made it possible to compare the step ahead forecasts with the streamflows that actually occurred. It is clear that there are always some forecast errors. These errors being the difference between predicted and observed streamflows are assumed to be independent random variables. The mean overall error for the entire series of forecasts is given as follows:

$$\bar{\varepsilon} = \frac{1}{n} \sum_{i=1}^{n} \hat{\varepsilon}_i \tag{14}$$

where n is the number of step ahead forecasts.

The standard deviation is calculated as:

$$s = [\sum_{i=1}^{n} (\hat{\varepsilon}_i - \bar{\varepsilon})^2 / (n-1)]^{1/2} \tag{15}$$

The two sided Student's t Test can be used to test whether the overall error is significantly different from zero. The t test is as follows:

$$\left| \frac{\bar{\varepsilon}\sqrt{n}}{s} \right| > t_{n-1;0.025} \tag{16}$$

The forecast errors were also evaluated on a monthly basis using the method given above. A 5% significance level was used for the critical value of t. Results for the Black River are given in the following table:

Summary of Forecast Errors by Months for Black River

Month	$\bar{\varepsilon}$	s	$\frac{\lvert\bar{\varepsilon}\sqrt{n}\rvert}{s}$	$t_{n-1;0.025}$
January	-0.07	11.27	0.032	2.052
February	-1.93	6.44	1.583	2.052
March	-0.33	20.46	0.086	2.052
April	3.54	26.76	0.699	2.052
May	8.03	13.49	3.150	2.052
June	3.05	7.88	2.048	2.052
July	-0.36	6.46	0.292	2.052
August	2.17	3.03	3.790	2.052
September	0.89	3.17	1.493	2.052
October	-1.77	11.47	0.816	2.052
November	-3.81	10.83	1.860	2.052
December	-3.37	12.95	1.379	2.052
Overall	0.50	13.2	0.70	1.960

Figure 4 shows the first two years of historical data used for the simulation of forecasting. Superimposed on the time series are the step ahead forecasts including the 2 standard deviation limits as calculated by the forecasting methodology.

CONCLUDING REMARKS

A methodology is given for step ahead streamflow forecasting using pattern analysis which accounts for varying persistence effects within the time series. Clustering the historic data provides the information that characterizes the streamflows. The actual forecasting procedure uses this information to complete a pattern vector on a probabilistic basis. This approach gives satisfactory results for the case study presented here.

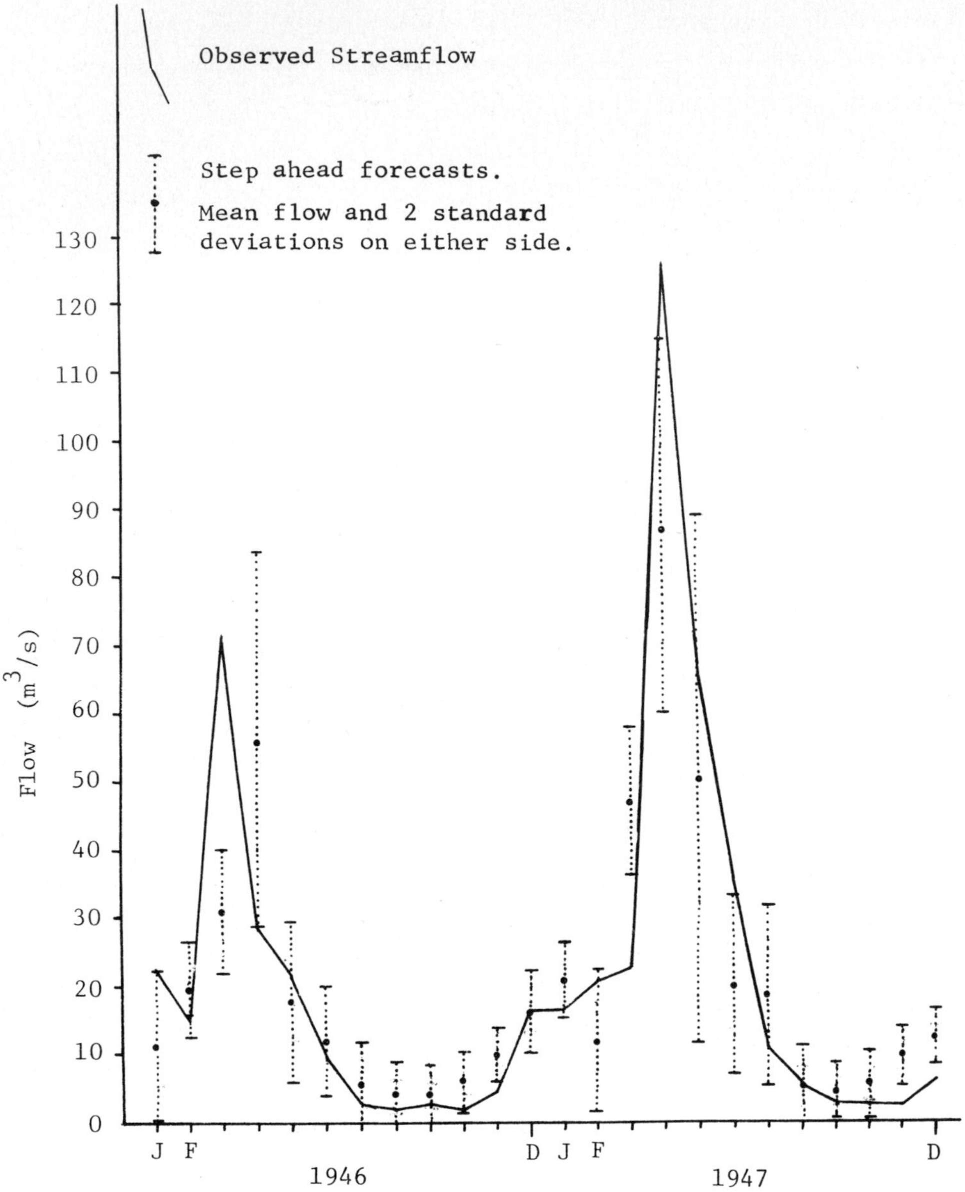

Figure 4. Step Ahead Forecasted Streamflows Superimposed on the Observed Time Series For the Black River

Further developments in this use of pattern analysis will be to extend the methodology to weekly and daily streamflow time series. Pattern vectors of higher dimensionality would then be needed. In addition pattern vectors can be used to incorporate other pertinent information in the analysis. This would include such data as snow accumulation, temperatures, the amount of water stored in the watershed, or weather forecasts from independent sources. Data from tributaries or other rivers in the region could be included as well.

ACKNOWLEDGEMENTS

The senior author gratefully acknowledges the support provided during the research period by grants from the National Science and Engineering Research Council of Canada.

REFERENCES

Ball, G.H., and Hall, D.J., "Isodata, An Iterative Method of Multivariate Analysis and Pattern Classification", Proceedings of the IFIPS Congress, 1965.

Dixon, J.K., "Pattern Recognition with Partly Missing Data", IEEE Trans. on Systems, Man and Cybernetics, Vol. SMC-9, No. 10, October 1979.

Panu, U.S., "Stochastic Synthesis of Monthly Streamflows Based on Pattern Recognition", Doctoral Dissertation, Department Of Civil Engineering, University of Waterloo, Canada, May, 1978.

Tou, J.T., and Gonzalez, R.C., "Pattern Recognition Principles", Addison-Wesley Publishing Company, Massachusetts, 1974, rev. 1977.

Unny, T.E., Panu, U.S., MacInnes, C.D., and Wong, A.K.C., "Pattern Analysis and Synthesis of Time-Dependent Hydrologic Data", Advances in Hydroscience, Academic Press, New York, Vol. 12, p.p. 195-297, 1981.

WALSH SOLUTIONS IN HYDROSCIENCE

ZEKAI SEN

Civil Engineering Faculty, Technical University of Istanbul, Turkey

ABSTRACT

Orthogonal Walsh functions have potential application possibilities in hydrosystem problems such as the statistical description, simulation, real time prediction, porous media description, differential equation solution etc. This paper presents, first of all, the fundamentals of these square-wave functions which assume alternatively only +1 and -1 values over a specified time interval during the operation of hydrosystem concerned. The most attractive properties of them are that they are piecewise linea linear, orthonormal and their multiplications require simple mathematical operations.

Four different ways of Walsh function applications in the hydroscience are described herein. These are; (1) the analysis of stochastic-periodic time series with a minimum afford of computations and great accuracy; (2) statistical description of porous media geometry and its comparison with the Fourier functions; (3) an adaptive Walsh predictor coupled with the Kalman technique capable of making real time predictions

Visiting Professor, King Abdulaziz University, Faculty of Earth Sciences, P.O. Box 1744 - Jeddah, Kingdom of Saudi Arabia.

Reprinted from *Time Series Methods in Hydrosciences*, by A.H. El-Shaarawi and S.R. Esterby (Editors, and its application for a monthly sequence observed in Turkey; and (4) Walsh functions as an alternative to the finite element method which are extensively employed in the solutions of various differential equations related to ground water flow.

INTRODUCTION

The Walsh functions are a set of two valued orthogonal functions that promise potential use in various hydroscience problems. The original definition of Walsh functions has been presented by Walsh (1923) using a set of recursive relations. In general, the Walsh functions form an ordered set of square-waves taking two specified values +1 and -1 only. Two arguments, namely, time period, t, and an ordering number, n, are required for their definition, hence n-th order Walsh function is denoted by WAL(n,t). The discrete version of these functions has been given in a compact form by Brown(1977) as,

$$WAL(0,n) = 1 \qquad \text{for } n = 1,2,\ldots,N.$$

$$WAL(1,n) = \begin{cases} 1 & \text{for } n = 1,2,\ldots,N/2 \\ -1 & \text{for } n = (N/2)+1,(N/2)+2,\ldots,N \end{cases} \tag{1}$$

$$WAL(m,n) = WAL([m/2],2n) \cdot WAL(m - 2[m/2],n)$$

where N depicts the total number of orthogonal functions in a complete set; and $[\cdot]$ is the integer part of the argument. An alternative definition of the Walsh functions has been proposed by Paley(1932) as a product of another orthogonal set called Rademacher (1922) functions. However, Paley's definition is different from Walsh's only in the ordering. Paley definition of Walsh functions has been summarized by Sen(1981). On the basis of Eq.(1) the resulting complete set of Walsh functions for N=8 is shown in Figure 1.

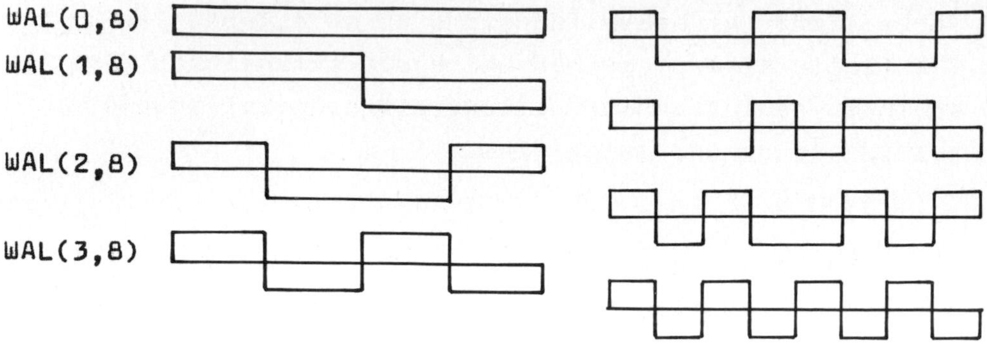

Figure 1. Walsh functions for N = 8.

A necessary requirement for a successful application of Walsh functions is that $N=2^q$ where q is any convenient positive integer power. The orthogonality property of Walsh functions is given by the following equation

$$\sum_{n=0}^{N} WAL(i,n).WAL(j,n) = \begin{cases} 0 & \text{for} \quad (i \neq j) \\ 1 & \text{for} \quad (i = j) \end{cases} \tag{2}$$

As is obvious from Figure 1 the Walsh functions are symmetric about their mid time points. Hence, this property enables a symmetry relationship as,

$$WAL(i,n) = WAL(n,i) \tag{3}$$

The practical importance of this is that the Walsh transforms and their inverses represent the same mathematical operation.

TIME SERIES REPRESENTATION

An effective use of Walsh functions occurs in representing a time series by the superposition of members of a set of orthogonal simple functions. In order to represent a time series X_i (i = 1,2, . . ,n) completely by the Walsh functions, it is necessary that the number of data points in the time series be equal to the minimum sequency order, N. Hence in general, Walsh representation is ,

$$X_i = \sum_{j=0}^{N} C_j \cdot WAL(j,i) \qquad (4)$$

where C_j are the coefficients to be chosen such that the mean-square approximation error is minimum; that is

$$M.S.E. = \sum_{i=0}^{N} \left[X_i - \sum_{j=0}^{N} C_j \cdot WAL(j,i) \right]^2 \qquad (5)$$

as a result of which one can obtain

$$C_j = \frac{1}{N} \sum_{j=0}^{N} X_j \cdot WAL(j,i) \qquad (6)$$

The first Walsh coefficient, C_0, is, in fact, equal to the mean value of the time series concerned, since WAL (0,i) = 1 for all i values. Şen(1981) has represented monthly flow time series periodic part by a complete set of Walsh functions with a maximum sequency number N = 16; Hence, the periodic stochastic process turns out to have the following mathematical form,

$$X_i = \sum_{j=0}^{N} C_j \cdot WAL(j,k) + \varepsilon_i \qquad (7)$$

where ε_i is the stochastic component and $k = i - 15[i/15]$; here $[\cdot]$ is the integer part of the argument. The relevant Walsh coefficient can be calculated according to Eq.(6). The application of Walsh decomposition to Colombia river (USA) monthly flows yields the coefficients shown in Figure 2. The seperation of the periodic component according to Eq.(7) gives the stochastic part. Fitting of the first order Markov process to this part results in a satisfactory solution. In time series analysis, the Walsh functions are capable of depicting the periodic component with minimum effort of computation and great accuracy.

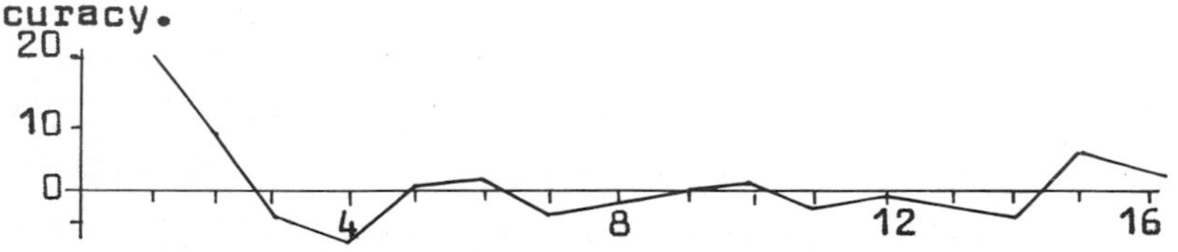

Figure 2. Walsh coefficients for Columbia River.

POROUS MEDIA DESCRIPTION

In order to represent statistically the porous medium it must be replaced by a convenient mathematical abstraction. To this end, the porous medium will be represented by a characteristic function, $f_i(s)$, which is defined as a random sequence of +1's and -1's. Herein, i denotes the i-th realization out of an ensemble and s the arc length of any point on the line from an arbitrarily chosen origin. In such a representation, the occurences of +1's and -1's imply grain and void spaces, respectively. Such a function is referred to as the sample characteristic function.

Fara and Scheideggar(1961) made an attempt to characterize a given porous medium from a photomicrographically read off data. In their study, the spectral analysis in terms of harmonic functions and of other orthogonal functions together with a spectral analysis of a specially constructed function of the porous medium have been considered as possible descriptors of the medium geometry. However, these methods cannot account for discontinuities in the porous media sample function due to Gibbs phenomenon. In order to alleviate this situation the use of Walsh functions are capable to digest effectively the existing discontinuities in the sample function.

To illustrate the Walsh function application, the porous medium is assumed to have a sample characteristic function given in Figure 3a. The Walsh as well as the Fourier analysis are then applied to this sample characteristic function which yields mathematically obtained counterparts as in Figure 3b, c and d. Figure 3b shows the transformation and reconstruction of the sample characteristic function by Walsh functions of the sequency order 32.

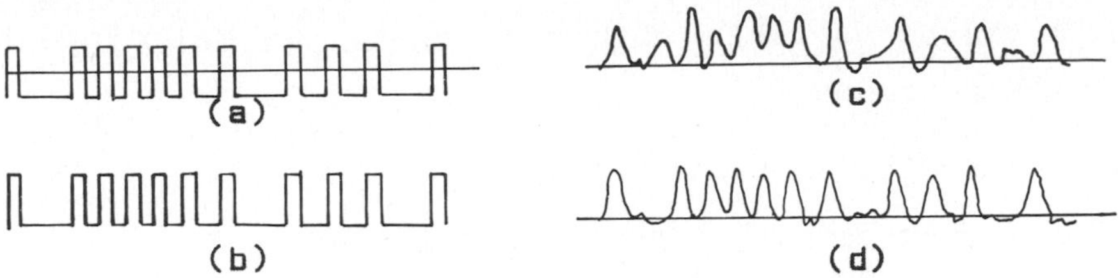

Figure 3a. An illustrative sample characteristic function.
b. Walsh transform with 32 terms.
c. Fourier transform with 18 terms.
d. Fourier transform with 44 terms.

The first ten Walsh coefficients are presented in Figure 4. On the other hand, Figures 3c and d show the Fourier approximation to the sample characteristic function with 24 and 32 harmonics, respectively. Comparison of Figures 3a, b, c and d yield how effective are the Walsh functions in the porous medium description. Virtually, all of the characteristic function is represented with its discontinuities totally by the Walsh series. However, Fourier approach gives a general pattern similar to the original characteristic function but lacks in depicting the discontinuities i.e. corners.

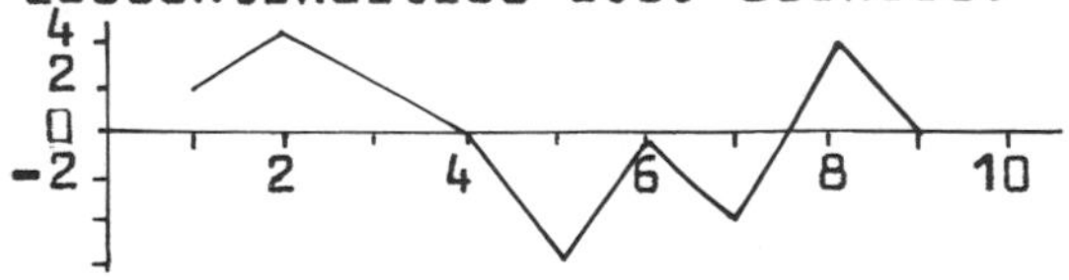

Figure 4. Walsh coefficients of sample function.

REAL-TIME PREDICTION

Real-time prediction of any periodic data requires construction of a recursive model (Şen, 1980). In order to produce such a recursive model of a periodic stochastic process the difference $X_i - X_{i-1}$ is performed by considering Eq.(7) leading to,

$$X_i = X_{i-1} + \sum_{k=1}^{15} C_k \cdot [WAL(k.i) - WAL(k,i-1)] + e_i \qquad (8)$$

where i=2,3, . . ,16n ; n being the number of years and e_i is a random variable with zero mean. By defining the Walsh difference as,

$$WAD(k,i) = WAL(k,i) - WAL(k,i-1)$$

Eq.(8) can then be rewritten succinctly as,

$$X_i = X_{i-1} + \sum_{k=1}^{15} C_k \cdot WAD(k,i) + e_i \qquad (9)$$

Herein, the coefficients, C_k, are unknown and need to be estimated from the available monthly data. However, it is assumed that the coefficients are time independent i.e. they do not change with time. Hence, Eq.(10) can be written in matrix notation as :

$$\begin{bmatrix} X \\ C_0 \\ C_1 \\ \cdot \\ \cdot \\ C_{15} \end{bmatrix}_i = \begin{bmatrix} 1 & 0 & w_1 & w_2 & \cdot & \cdot & w_{15} \\ 0 & 1 & 0 & 0 & \cdot & \cdot & 0 \\ 0 & 0 & 1 & 0 & \cdot & \cdot & 0 \\ \cdot & \cdot & \cdot & \cdot & \cdot & \cdot & \cdot \\ \cdot & \cdot & \cdot & \cdot & \cdot & \cdot & \cdot \\ 0 & 0 & 0 & 0 & 0 & 0 & 1 \end{bmatrix} \begin{bmatrix} X \\ C_0 \\ C_1 \\ \cdot \\ \cdot \\ C_{15} \end{bmatrix}_{i-1} + \begin{bmatrix} e \\ 0 \\ 0 \\ \cdot \\ \cdot \\ 0 \end{bmatrix}_i \qquad (10)$$

where w_i's are short versions of WAD(k,i)'s. On the other hand, in an implicit matrix notation Eq.(10) becomes,

$$Y_i = \Phi_{i,i-1} \cdot Y_{i-1} + W_i \qquad (11)$$

where Y_i is a (17x1) vector of state variables. The transition matrix $\Phi_{i,i-1}$ from state (i-1) to state i has a dimension of (17x17) and W_i is (17x1) vector of independent errors including 16 elements which are all equal to zero.

Kalman filter can be applied to the system equation given in Eq.(11) provided that a suitable measurement equation is supplied, (Kalman, 1960). At the time instant i there is only one measured state variable that is the monthly flow value. Hence, the measurement equation which renders the state variables into measurements can be as,

$$Z_i = H_i \cdot Y_i + V_i \qquad (12)$$

where H_i is the measurement dynamics vector with its first element equal to 1 others being all zero. As the accuracy of the measurement increases the error contribution, V_i, diminishes. Herein, the measurements are assumed to be perfect which is the case when $V_i = 0$. With Eqs.(11) and (12) at hand, the Kalman filter application is straightforward (see Gelb,1974). The state estimate, $Y_{i/i-1}$, and error covariance, $P_{i/i-1}$, extrapolations are,

$$Y_{i/i-1} = \Phi_{i,i-1} \cdot Y_{i-1/i-1} \qquad (13)$$

$$P_{i/i-1} = \Phi_{i,i-1} \cdot P_{i-1/i-1} \cdot \Phi^T_{i,i-1} + Q_i \qquad (14)$$

respectively. The Kalman gain matrix, K_i, is

$$K_i = P_{i/i-1} \cdot H_i^T \; H_i \cdot P_{i/i-1} \cdot H_i^T + R_i \;^{-1} \qquad (15)$$

Finally, the state estimation and error covariance updates after the measurements turns out to be,

$$Y_{i/i} = Y_{i/i-1} + K_i \; Z_i - H_i \cdot Y_{i/i-1} \qquad (16)$$

respectively.

Application of the method is presented for the Seyhan river in the southern Turkey. The successive execution of Eqs.(13)-(17) on a digital computer require initial values $Y_{0/0}$, $P_{0/0}$ and Q. The diagonal elements of the covariance matrix are all taken as 1000's and off diagonal elements are equal to 100. All of the initial state vector elements are adopted as zeros. The system noise variance is selected as Q=1000. However, the measurement noise is taken as zero i.e. the measurements are assumed perfect. With these initial values the filtering equations (Eqs. 13-17) are executed one by one and finally the Walsh coefficients are obtained as in Table 1.

TABLE 1.
Seyhan river Walsh coefficients.

Sequency	Coefficient
1	-3.49
2	0.56
3	-0.60
4	-4.03
5	-0.60
6	-3.60
7	-4.12
8	-8.27
9	-4.12
10	-2.22
11	-5.66
12	-9.13
13	-5.66
14	-5.47
15	-3.90
16	4.85

Figure 5 represents true and filtered monthly runoff values of the Seyhan river flows for the first five years.

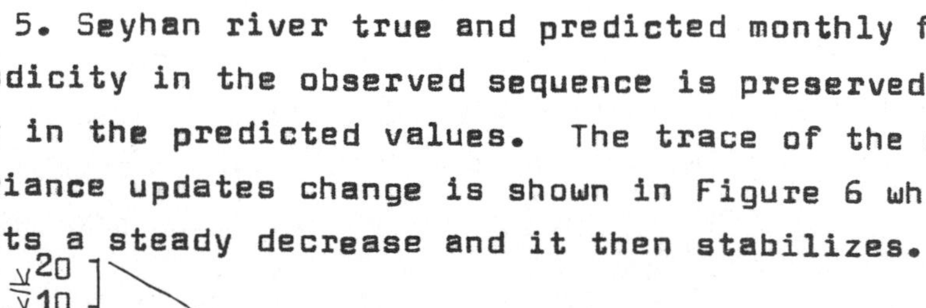

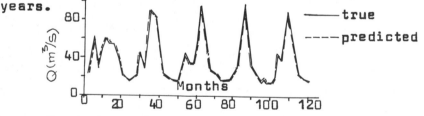

Figure 5. Seyhan river true and predicted monthly flows.

Periodicity in the observed sequence is preserved similarly in the predicted values. The trace of the error covariance updates change is shown in Figure 6 which exibits a steady decrease and it then stabilizes.

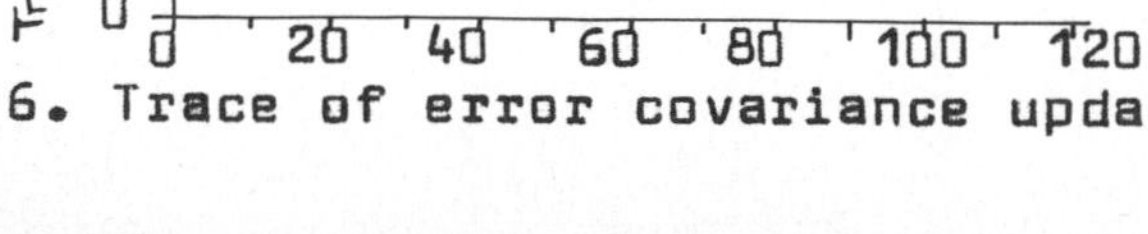

Figure 6. Trace of error covariance update matrix.

DIFFERENTIAL EQUATION SOLUTION

Another very potential application area of the Walsh functions in the differential equation solution as an alternative to the classical finite element technique. Applications to this end have already been undertaken in systems engineering by Chen and Hsiao(1975) ; Paraskevopoulos and Bounas(1978) and Shih and Han(1978). Since, the basic Walsh functions are piecewise constant at either +1 or -1 their integration yields simple piecewise linear functions which are in fact triangles. The integrations are shown in Figure 7 for sequency order of 2^3. The Walsh integrations can be expressed in terms of basic Walsh functions. After performing the necessary analytical evaluations one can obtain for $N=2^3$ the following equations :

$$
\begin{aligned}
\int WAL(0,t)dt &= (1/2)WAL(0,t)-(1/4)WAL(1,t)-(1/8)WAL(2,t) \\
&\qquad -(1/16)WAL(4,t) \\
\int WAL(1,t)dt &= (1/4)WAL(0,t)-(1/8)WAL(3,t)-(1/16)WAL(5,t) \\
\int WAL(2,t)dt &= (1/8)WAL(0,t)-(1/16)WAL(6,t) \\
\int WAL(3,t)dt &= (1/8)WAL(1,t)-(1/16)WAL(7,t) \\
\int WAL(4,t)dt &= (1/16)WAL(0,t) \\
\int WAL(5,t)dt &= (1/16)WAL(1,t) \\
\int WAL(6,t)dt &= (1/16)WAL(2,t) \\
\int WAL(7,t)dt &= (1/16)WAL(3,t)
\end{aligned}
\tag{18}
$$

or in matrix notation succinctly,

$$\int W_8 dt = T_{(8\times8)} \cdot W_8 \tag{19}$$

where W_8 is (8x1) vector of the basic Walsh functions and $T_{(8\times8)}$ is the transition matrix. The general form of this matrix is given by Chen and Hsiao(1975) as,

$$T_{(N\times N)} = \left[\begin{array}{c:c:c:c} \ddots\ 1/2\ \ddots & -(2/N)\cdot I_{N/8} & & \\ \hdashline (2/N)\cdot I_{N/8} & 0_{N/8} & -(1/N)\cdot I_{N/4} & \\ \hdashline (1/N)\cdot I_{N/4} & & 0_{N/4} & -(1/2N)\cdot I_{N/2} \\ \hdashline (1/2N)\cdot I_{N/2} & & & 0_{N/2} \end{array}\right]$$

where $I_{N/4}$ and $0_{N/2}$ are the identity and zero matrices, respectively. It is evident from the above calculations that if any mathematical expression is written in terms of Walsh functions then through the aforementioned transition matrix its integration can be achieved by the Walsh functions. Hence, the integration procedure becomes the problem of finding the relevant Walsh coefficients. Let us now consider a simple illustration as in the following example,

$$dx/dt = 2x \tag{21}$$

with the boundary conditions (x=1 at t=0). If the derivative is expanded into Walsh set with $N=2^2$, then

$$dx/dt = C_0 WAL(0,t)+C_1 WAL(1,t)+C_2 WAL(2,t)+C_3 WAL(3,t)$$

taking the integration leads to,

$$x = C_0\ WAL(0,t)dt+C_1\ WAL(1,t)dt+C_2\ WAL(2,t)+C_3\ WAL(3,t)dt + x_0$$

or briefly,

$$x = C_4\cdot T_{(4\times 4)}\cdot W_4 + 1$$

where $C_4^T=[C_0\ \ C_1\ \ C_2\ \ C_3]$ is the coefficients vector. Explicitly, the substitution of the necessary Walsh function into Eq.(21) yields,

$$C_4\cdot W_8 = -4C_4\cdot T_{(4\times 4)}\cdot W_4 + [-4\ \ 0\ \ 0\ \ 0\ \ W_4]$$

The only unknown is the coefficients vector which can be

found as,

$$C_4 = [I_4 + 4T_{(4\times4)}]^{-1}[-4\ 0\ 0\ 0] = -(1/81)[80\ 64\ 40\ 32]$$

Exact and approximate Walsh solutions are shown in Figure 8. It is obvious that increase in the Walsh sequency will result in more refined approximations. Theoretically, infinity of sequency number corresponds with the exact solution.

Unsteady one-dimensional flow in an unconfined aquifer without recharge is given by the following partial differential equation,

$$\frac{\delta s(x,t)}{\delta x^2} = \frac{S}{T}\frac{\delta s(x,t)}{\delta t} \qquad (22)$$

where s(x,t), S and T are the drawdown, storativity and transmissivity of the aquifer, respectively ; x and t are the space and time variables. There is no general solution for this differential equation and only for specific boundary conditions its solution is possible. Herein, the Walsh solution is presented as a brief summary. Let us integrate Eq.(22) twice with respect to x and once with respect to t with boundary conditions s(x,0)=0 and s(0,t) = Δt. After some algebraic manipulations it leads to,

$$\int_0^t s(x,t)dt + a\int_0^x\int_0^x s(x,t)dxdx - \Delta t = 0 \qquad (23)$$

where a = S/T and assumed to be known. Two variable function, s(x,t) can be expanded into a double Walsh series as,

$$s(x,t) = \sum_{j=1}^{\infty}\sum_{i=1}^{\infty} s_{ij}\cdot \varphi_i(t)\cdot \varphi'_j(x) \qquad (24)$$

where $\varphi_i(t)$ and $\varphi'_j(x)$ are the Walsh functions with respect to variables t and x, respectively ; s_{ij}'s are the double Walsh coefficients which are given by,

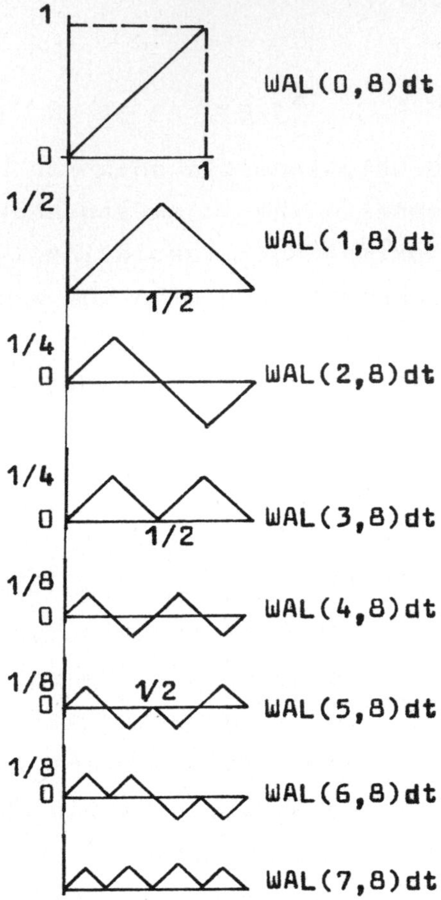

Figure. 7. Walsh functions integrals for N =8.

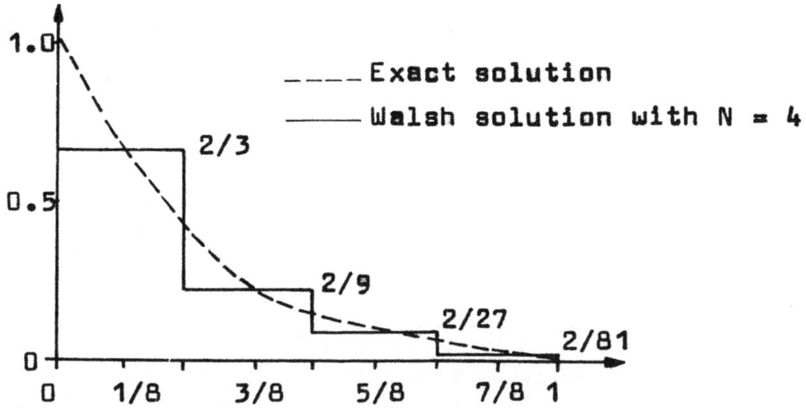

Figure 8. Walsh series solution to differential equation (after Chen and Hsiao, 1975)

$$s_{ij} = \int_0^1 \int_0^1 s(x,t) \cdot \varphi_i(t) \cdot \varphi'_j(x) \cdot dx \cdot dt$$

However, an approximation of s(x,t) will then be,

$$s(x,t) \sum_{j=0}^{M-1} \sum_{i=0}^{N-1} s_{ij} \cdot \varphi_i(t) \cdot \varphi'_j(x)$$

or in matrix notation

$$s(x,t) = \Phi'^{T}_{M}(x) \cdot S_{MN} \cdot \Phi_N(t) \qquad (25)$$

where superscript T denotes the matrix transposition and S_{MN} is (MxN) matrix of drawdown variable.

Integration of Eq.(25) α times with respect to time and β times with respect to the space has been derived by Paraskevapoulos and Bounas as,

$$\int_0^x \cdots \int_0^x \int_0^t \cdots \int_0^t \Phi'^{T}_{M}(x) \cdot S_{MN} \cdot \Phi_N(t) \underbrace{dt \ldots dt}_{\alpha \text{ times}} \underbrace{dx \ldots dx}_{\beta \text{ times}} \qquad (26)$$

or succinctly this integration is given in its implicit matrix notation as,

$$\Phi^T_N(t) [T^T_{(N\times N)}]^{\alpha} \cdot S^T_{MN} [T_{(M\times M)}]^{\beta} \Phi'_M(x) \qquad (27)$$

Under the light of these explainations Eq.(23) can be written in Walsh expansion form as,

$$S_{MN}[T_{(N\times N)}]^2 + a[T^T_{(N\times N)}]^2 S_{MN} \cdot T_{(N\times N)} - \Delta t = 0$$

Hence, through the matrix algebra the unknown, S_{MN}, can be solved through a similar procedure presented by Paraskevapoulos and Bounas.

CONCLUSIONS

The Walsh function expansions have been applied to various problems encountered within the context of hydroscience. They decompose successfully a given time series into linear and simple components. Their mathematical manipulations are based on simple addition and/or subtraction. A complete set of Walsh functions is the most convenient transformation for representing the porous media sample characteristic function. The periodic-stochastic data can be effectively assessed by an adaptive Walsh procedure. This procedure does not give only the parameter estimations at each time instant but also simultaneously decompose series into periodic and stochastic parts. The Walsh functions can be effectively applied in the solution of ordinary or partial differential equations.

REFERENCES

Brown, R.D. 1977. A recursive algorithm for sequency-ordered fast Walsh transforms. IEEE Trans. on Comp., C-26(8):879-882.

Chen, C.F., and Hsiao, C.H., 1975. Walsh series analysis in optimal control. Vol. 21, No. 6: 881-897.

Fara, H.D., and Scheidegger, A.E., 1961. Statistical geometry of porous media. Jour. of Geophys. Res., Vol.66, No. 10: 3279-3284.

Gelb, A., 1974. Applied optimal estimation. M.I.T. Press Cambridge, Mass.: 379 pp.

Kalman, R.E., 1960. A new approach to linear filtering and prediction theory. Trans. ASME, Ser. D., Jour. Basic Engrg., Vol. 83: 35-45.

Paraskevopoulos, P.N., and Bounas, A.C., 1978. Distributed parameter system identification via Walsh functions. Int. Jour. Systems Sci., Vol. 9, No. 1: 75-83.

Paley, R.E.A.C., 1932. A remarkable series of orthogonal functions. Proc. London Math. Soc., Vol. 34: 241-279.

Shih, Y., and Han, J., 1978. Double Walsh series Solution of first-order partial differential equations. Int. Jour. Systems Sci., Vol. 9, No. 5: 569-578.

Sen, Z., 1980. Adaptive Fourier analysis of periodic-stochastic hydrologic sequences. Jour. Hydrol., Vol. 46: 239-249.

Sen, Z., 1981. Walsh analysis of monthly flow volumes. Int. Symp. on Rainfall-Runoff Modeling, Mississippi.

Walsh, J.L., 1923. A closed set of orthogonal functions. Amer. Jour. Math., Vol. 45: 5-24.

MODELLING THE ERROR IN FLOOD DISCHARGE MEASUREMENTS

KENNETH W. POTTER AND JOHN F. WALKER

Department of Civil and Environmental Engineering, University of Wisconsin-Madison

ABSTRACT

The measurement of peak discharge is typically composed of three distinct processes. Low magnitude floods are determined with an established rating curve. Intermediate magnitude floods are inferred by extrapolating the established rating curve. High magnitude floods are usually determined through a field survey. The measurement error characteristics for each process are different, a phenomenon termed discontinuous measurement error (dme). Monte Carlo experiments with an error model that approximates the peak measurement process reveal a bias in the measured coefficients of variation, skewness, and kurtosis. This bias is significant and has important implications with regard to·flood frequency analysis.

INTRODUCTION

As the statistical methods available to the hydrologist become more sophisticated, it is essential that closer attention be paid to the operational procedures by which hydrologic data are collected. Failure to do so can lead to serious misinterpretation of statistical results. A notable example of such a failure is the Hurst phenomenon. Clearly in many instances high Hurst coefficients are merely artifacts of consistent measurement error, such as results from the relocation of a precipitation gage. We believe that similar artifacts may result from the way in which flood discharge records are constructed.

For low magnitude floods, peak stages are recorded at the gage and the corresponding discharges are estimated from a rating curve established by current meter measurements. For high magnitude floods, peak stages are usually inferred from high-water marks and

Reprinted from *Time Series Methods in Hydrosciences*, by A.H. El-Shaarawi and S.R. Esterby (Editors)

the discharges are estimated by rating curve extension or by an indirect means, such as the slope-area method. Clearly the variance of the discharge estimates is much higher for high magnitude floods. We have termed this phenomenon "discontinuous measurement error" (dme), and have shown that it causes the coefficients of variation, skewness, and kurtosis of the measured flood distribution to be much higher than the corresponding coefficients of the parent flood distribution (Potter and Walker, 1981).

There are two limitations to our initial study of dme. First, we documented biases in population coefficients of variation, skewness, and kurtosis, rather than in small-sample coefficients. The latter, of course, are also subject to bias due to small-sample boundedness (Wallis et al., 1974). Second, our model of dme was based on the assumption that errors in estimates of high-magnitude floods are homescedastic and independent. Such a model might be appropriate if the slope-area method were independently applied to all high magnitude floods. This is not, of course, the case. For many high magnitude floods the discharges are estimated by simple rating-curve extension. This results in errors which are both correlated and heteroscedastic. Furthermore, if the extended rating curve is adjusted to be consistent with available slope-area estimates, the error variance also decreases with time. The problem may be further complicated by temporal changes in the measurement procedures and in the hydraulic conditions at the gage. In this paper we develop a model which more realistically mimics this complex measurement process. We then use this model in a limited exploration of the small-sample effects of dme.

NEW MODEL OF DME

Our new model is based on a three-tiered measurement process, as depicted in Figure 1. For stages below a certain value (S_1), the rating curve established by current meter measurements applies. Typically S_1 would be bankful stage. Between S_1 and S_2, the rating curve is linearly extended in log-space. (In Figure 1 the sloping

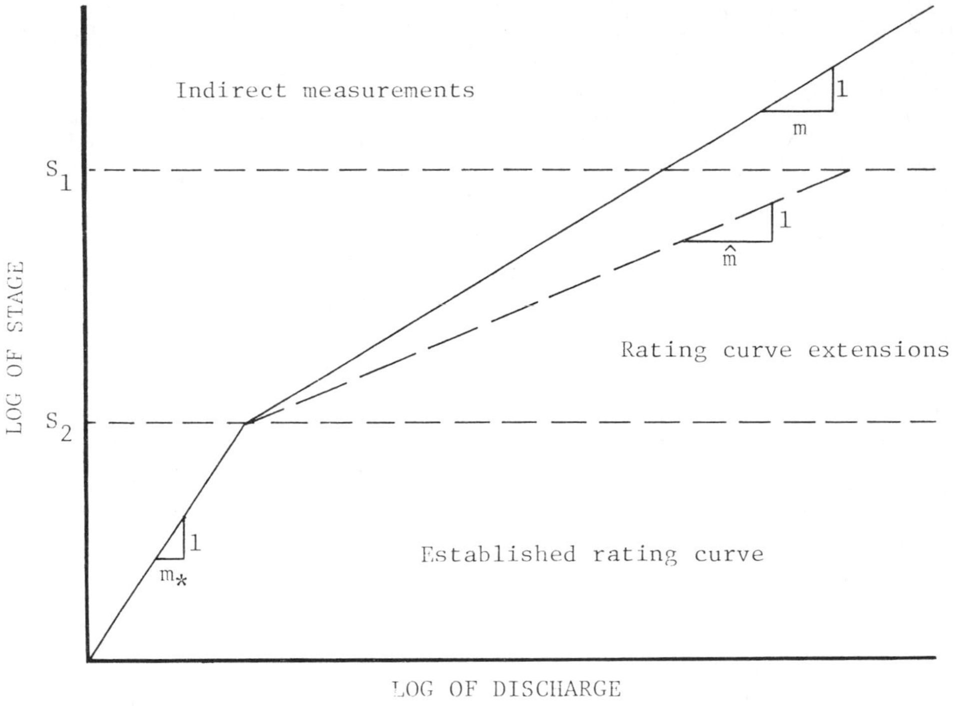

Fig. 1. A three-tiered error model.

dashed line represents the extended rating curve; the sloping solid line represents the actual, but unknown rating curve.) For stages beyond S_2 indirect measurement is assumed. Our model of this three-tiered process is based on the following assumptions:

1. Stage measurements are made without error.
2. The lower rating curve is linear in log-space, is known without error, and is unchanging in time.
3. The true (but unknown) rating curve extension is linear in log-space and is unchanging in time.
4. The initial estimate of the inverse of the slope of the rating curve extension is a random variable with a three-parameter lognormal distribution.

5. The estimated rating curve extension is continuously updated to comply with indirect discharge measurements.
6. The errors in indirect discharge measurements are independent and homescedastic.

Below S_1 and above S_2, the new model is the same as the previous one, except that the variance of the low-discharge error is assumed to be zero (see Potter and Walker, 1981, for details of the simple model). This is done to give the extended rating curve a fixed point from which to begin. Because rating-curve errors are generally much smaller than the errors associated with rating-curve extensions or indirect measurements, it is believed that this simplification is reasonable.

In the intermediate region between S_1 and S_2, the rating curve is linearly extended in log-space. This can be represented by

$$\log_e Q_m = \log_e q_1 + \hat{m}(\log_e S - \log_e S_1) \qquad (1)$$

where Q_m is the estimated discharge with stage S, q_1 is the discharge associated with S_1, and $1/\hat{m}$ is the assumed slope of the extended rating curve. Equation (1) can be rewritten as

$$Q_m = q_1(Q_a/q_1)^{\hat{m}/m} \qquad (2)$$

where $1/m$ is the true (but unknown) slope of the extended rating curve. Letting $X = \hat{m}/m$, we assume that X is a random variable having a 3-parameter lognormal distribution with a shift parameter equal to m_*/m, where $1/m_*$ is the slope of the lower rating curve. Thus $\log_e(X - m_*/m)$ is normally distributed. We also assume that $E[X] = 1$ and $V[X] = \sigma_X^2$.

The assumption of unit mean insures that estimates of discharge based on rating-curve extensions are unbiased. Furthermore, the assumed error structure insures that $\hat{m}$, the inverse of the estimated slope of the extended rating curve, is always between m_* and infinity. The lower bound is consistent with the usual flattening of the rating curve beyond bankful stage.

In actual field situations, rating curve extensions are adjusted as additional information is obtained by indirect-discharge measurements. This facet of the measurement process is incorporated in our model by adjusting $\hat{m}$ whenever an indirect measurement is made. The adjustment is made by simple linear regression in log-space on all available indirect measurements. Thus the error in $\hat{m}$ decreases (on average) with each additional indirect measurement.

The new model of dme can be summarized by

$$Q_m = \begin{cases} Q_a & \text{for } 0 \leq S < S_1 \\ q_1(Q_a/q_1)^{\hat{m}/m} & \text{for } S_1 \leq S \leq S_2 \\ tQ_a & \text{for } S_2 < S \leq \infty \end{cases} \tag{3}$$

Note that $\hat{m}$ and t are random variables. The variable t represents the multiplicative error associated with indirect measurement, with $E[t] = 1$ and $V[t] = \sigma^2$. As indirect measurements are made, the estimate of $\hat{m}$ is modified.

EFFECTS OF DME ON SMALL SAMPLES

Monte Carlo experiments were conducted to explore the effects of dme on small-sample statistics.

Our new model of dme involves seven parameters:

S_1 - stage of first error discontinuity;

q_1 - discharge corresponding to S_1;

m - inverse of slope of lower rating curve;

m_* - inverse of slope of true extended rating curve;

σ_x^2 - variance of $\hat{m}/m$, where $1/\hat{m}$ is the estimated slope of the extended rating curve;

S_2 - stage of second error discontinuity;

σ^2 - variance of multiplicative error associated with indirect measurements.

In order to reduce the parameter space, S_2 was fixed at the stage at which the variance of a rating curve extension just equalled the variance of an indirect measurement. This is a reasonable constraint, since use of the rating curve extension beyond this point would, on average, yield larger errors than use of indirect measurements. Based on this constraint,

$$S_2 = S_1 \exp\{[\log_e(1 + \sigma^2)]^{1/2} / m\sigma_x\} \qquad (4)$$

In addition to the model parameters, which have been reduced to six, it is necessary to specify the mean and coefficient of variation of the parent flood distribution, μ_a and σ_a/μ_a, and the length of the sample, n. As in the case of our initial model of dme, the parent flood distribution is assumed to be lognormal.

Monte Carlo runs with numerous parameter combinations indicated that the small sample coefficients of variation, skewness, and kurtosis of the measured flood discharges are unaffected by the choice of S_1, m, m_*, and μ_a. Therefore the relevant parameter set is q_1, σ_x^2, σ, σ_a/μ_a, and n.

Figure 2 illustrates selected results of the Monte Carlo experiments. These results are expressed in terms of the relative bias in the small-sample coefficients of variation, skewness, and kurtosis, where relative bias is defined as the ratio of the average small-sample estimate of the population coefficient to the population value. Average small-sample estimates are based on 1000 realizations. Also shown are the relative biases which result when there is sampling without error, due to the boundedness of the sample coefficients. For the case illustrated, the lower discharge threshold (q_1) is the 2-year event; the standard deviation of $\hat{m}/m$ (σ_x) is 1.0; the standard deviation of the indirect measurement error (σ) is 0.2; and the parent flood population coefficient of variation (σ_a/μ_a) is 0.4. In each plot, there are two relative bias curves - one for the dme error model and one for the case of no measurement error. In the latter case, with increasing n the relative bias curve converges from below to the zero-bias case, represented by the dashed line. This, of course, reflects the easing of the effects of small-sample boundedness with increasing sample size.

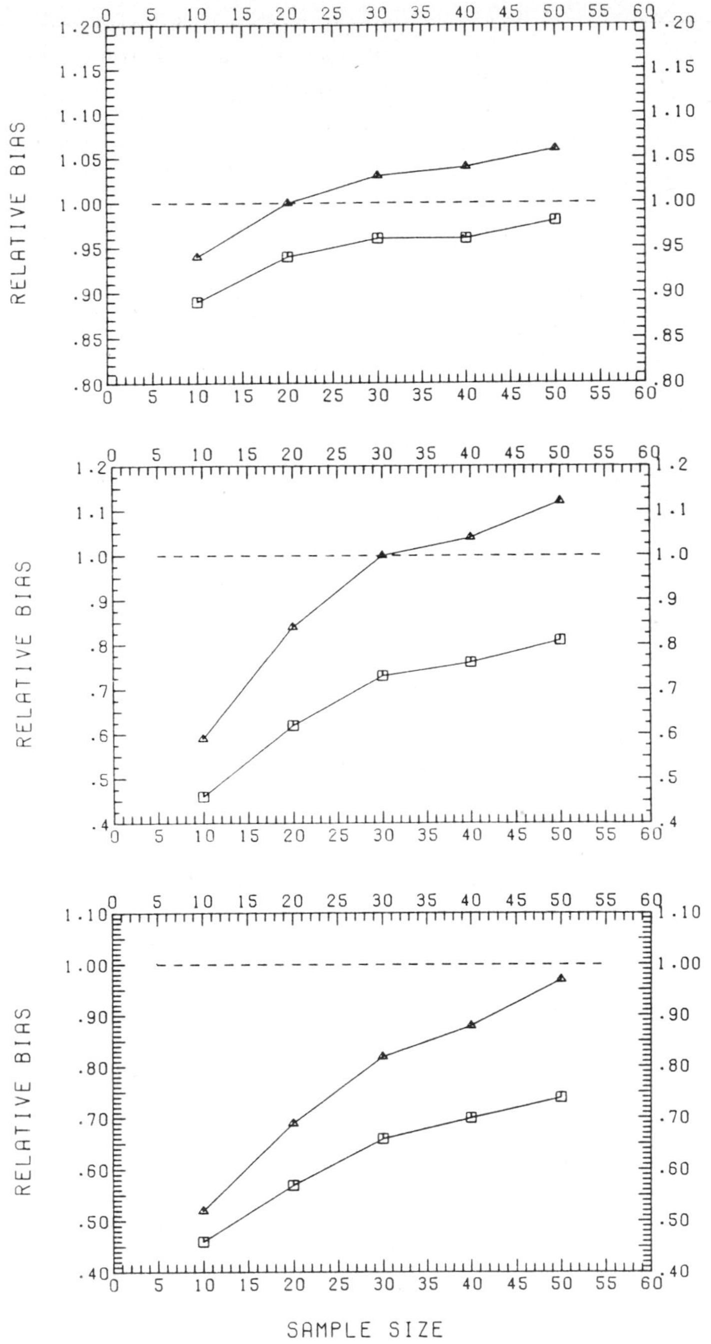

Fig. 2. Coefficients of variation, skewness, and kurtosis, from top to bottom, for $\sigma_a/\mu_a = 0.4$. The square and triangle symbols represent no error and the dme error model, respectively.

Figure 2 illustrates that for parent flood distributions with average coefficients of variation (σ_a/μ_a = 0.4), the small-sample effects of dme are very striking. For all three coefficients the relative biases increase with sample size. In two cases the effects of small-sample boundedness are offset for low sample sizes. Thus for n greater than 30, dme leads to expected coefficients of variation and skewness greater than the population values. In all cases the relative bias for the dme error model is considerably higher than the zero-error case.

By varying the model parameters, it was discovered that the small-sample biases due to dme diminish as the parent population coefficient of variation increases. Thus for smaller coefficients of variation, the biases are even more dramatic than the biases shown in Figure 2. Furthermore, the coefficient of skewness proved to be the most sensitive to the effects of dme. Clearly fitting techniques relying on the coefficient of skewness are highly suspect!

CONCLUSIONS AND RECOMMENDATIONS

(1) It is clear from our results that dme has an important effect on the coefficients of variation, skewness, and kurtosis of measured flood discharges, particularly when the coefficient of variation of the parent flood distribution is low. This effect is to offset downward bias due to small-sample boundedness. Furthermore the bias due to dme <u>increases</u> with sample size, unlike the bias due to most other sources.

(2) Immediate attention should be focused on estimating the variance of errors in rating-curve extensions and indirect measurements, in order to determine the magnitude of the problem caused by dme.

(3) If, as expected, dme proves to be an important problem in flood-frequency estimation, ways must be developed to deal with it. One obvious way is to abandon the coefficient of skewness.

(4) The new model of dme presented in this paper seems to be at least conceptually adequate for the case of rivers with stable sections at their gages. It is clearly not adequate for the unstable case.

ACKNOWLEDGEMENTS

Funding for this research was provided by the Graduate School and the Engineering Experiment Station of the University of Wisconsin. We would also like to thank Ruth Wyss for her usual splendid job of typing.

REFERENCES

Potter, K.W. and Walker, John F., 1981. A model of discontinuous measurement error and its effects on the probability distribution of flood discharge measurements, Water Resour. Res., 17(5), 1505-1509.

Wallis, J.R., Matalas, N.C. and Slack, J.R., 1974. Just a moment!, Water Resour. Res., 10(2), 211-219.

INFORMATION THEORETICAL CHARACTERISTICS OF SOME STATISTICAL MODELS IN THE HYDROSCIENCES

W.F. CASELTON

The University of British Columbia, Vancouver, B.C., Canada

1 INTRODUCTION

Within the hydrosciences, there is a growing need to model complex environmental phenomena which are both spatial and random in nature. Statistical models capable of accommodating the large number of variables involved are, in themselves, highly complex. A broad understanding of the characteristics of such models, the influence of their underlying assumptions, the implications of not meeting these assumptions in practice, and their potential performance, are all desirable to the practitioner but difficult to achieve. Theoretical investigations into the characteristics of any particular multivariate model are often involved and the results difficult to generalize.

The author originally encountered questions concerning the performance characteristics of multivariate spatial models in connection with the design of monitoring networks. There is a distinct risk of producing a network design which is more a reflection of the type of spatial model adopted in the design method rather than an optimal collector of information for the region served. Caselton and Husain [1980] have shown that the need to adopt any form of spatial model can be avoided when the network performance objective is specified in information theoretic terms and information transmission is maximized.

Information theoretic approaches to establishing estimation performance bounds have been presented by Weidemann and Stear [1969] for the case of parameter estimation and by Tomita et al. [1976] in connection with the Kalman filter. The latter showing that the optimal filter is also the optimal information

Reprinted from *Time Series Methods in Hydrosciences*, by A.H. El-Shaarawi and S.R. Esterby (Editors)

transmitter. Both of these papers also describe aspects of the general case of estimation in information terms and these are summarized here. This summary shows the relationship between the entropies of the principal variables involved in estimation and the information transmissions between these variables. Some specific types of models which arise in the hydrosciences are then considered and include: a simple form of spatial estimation; models involving serial dependency; and models involving Gaussian errors.

2 A MEASURE OF INFORMATION

The idea of quantitatively measuring information has considerable appeal in many scientific and engineering situations. Many concepts of information measures have been proposed but the three most often encountered are Fisher's, Shannon's, and the one which arises in statistical decision analysis. A useful comparative review of these three measures in a hydrosciences context has been provided by Dyhr-Nielsen [1972]. Only Shannon's measure will be discussed here.

Shannon's information is tied directly to the concept of message uncertainty prior to receiving a transmitted signal and after receipt of this signal. The quantitative measure of uncertainty used is the function

$$H(X) = -\sum_i P(x_i)\log P(x_i)$$

where x_i is, for example, a discrete message or outcome c random source X. The quantity H(X) is referred to as th entropy of X. Upon receipt of a signal y_j, from the sign source Y, the probabilities concerning the message are am probabilities and uncertainty is now reflected by the condi entropy

$$H(X|Y) = -\sum_i \sum_j P(x_i,y_j) \log P(x_i|y_j)$$

The amount of information transmitted about X by Y is I(X;Y) and the reduction in uncertainty attributable to the signal. This is given by

$$I(X,Y) = H(X) - H(X|Y) \qquad (2.3)$$

or equivalently

$$I(X;Y) = H(X) + H(Y) - H(X,Y) \qquad (2.4)$$

when the joint entropy is defined as

$$H(X,Y) = -\sum_i \sum_j p(x_i,y_j) \log P(x_i,y_j) \qquad (2.5)$$

These definitions extend naturally to the case where X and Y are random vectors so that, if the message vector is $X_1,X_2,...,X_m$ and the signal vector is $Y_1,Y_2,...Y_n$ (where n is not necessarily equal to m), then the information transmitted is

$$I(X_1,X_2,...X_m;Y_1,Y_2,...,Y_n) = H(X_1,X_2,...,X_m) + H(Y_1,Y_2,...,Y_n) - H(X_1,X_2,...,X_m;Y_1,Y_2,...,Y_n) \qquad (2.6)$$

3 INFORMATION AND ESTIMATION

3.1 Estimation Error

The process of estimation will be described in a simple context of an m dimensional state vector X representing the true value of the quantity to be estimated. An n dimensional measurement vector Z will represent measurements which are in some way related to X and upon which an m dimensional estimate $\hat{X}$ of X will be based. The estimator used to produce the estimate $\hat{X}$ is represented by F(Z). No restrictions as to the form of F( ) will be imposed other than being unbiased.

In information terms X is the message which is being transmitted first by the intermediate signal Z which is further processed to produce an output signal $\hat{X}$. The m dimensional estimation error vector ε is defined by

$$\varepsilon = X - \hat{X} \quad (3.1)$$

or

$$\varepsilon = X - F(Z) \quad (3.2)$$

Because of the dependency of the estimate on the measurement it follows that

$$p(X,Z) = p(X-F(Z),Z) = p(\varepsilon,Z) \quad (3.3)$$

$$p(\varepsilon,\hat{X}) = p(X-\hat{X},\hat{X}) = p(X,\hat{X}) \quad (3.4)$$

so that, from the definition of entropy

$$H(X,Z) = H(\varepsilon,Z) \quad (3.5)$$

$$H(\varepsilon,\hat{X}) = H(X,\hat{X}) \quad (3.6)$$

An informational quantity of particular interest is $I(X,\hat{X})$, the information transmitted by the estimate about the true state. From Equation 2.5

$$I(X;\hat{X}) = H(X) + H(\hat{X}) - H(X,\hat{X}) \quad (3.7)$$

Adding and subtracting $H(\varepsilon)$ from the R.H.S. of equation 3.7 and substituting equivalent joint entropies from Equation 3.6 yields

$$I(X;\hat{X}) = H(X) - H(\varepsilon) + H(\hat{X}) + H(\varepsilon) - H(\varepsilon,\hat{X})$$

$$= H(X) - H(\varepsilon) + I(\varepsilon;\hat{X}) \quad (3.8)$$

Equation 3.8 expresses the information transmitted by estimation in terms of the estimate and its error only. An alternative expression involving the measurement Z is obtained by adding and subtracting $H(Z)$ from the RHS of Equation 3.8 which yields

$$I(X;\hat{X}) = H(X) + H(Z) - H(\varepsilon) - H(Z) + I(\varepsilon;\hat{X})$$

$$= I(X;Z) + H(X,Z) - I(\varepsilon;Z) - H(\varepsilon;Z) + I(\varepsilon;\hat{X})$$

$$I(X;\hat{X}) = I(X;Z) - [I(\varepsilon;Z) - I(\varepsilon;\hat{X})] \quad (3.9)$$

Finally combining equation 3.6 and 3.9 yields an expression similar to Equation 3.8. From 3.9 $I(X;Z) = I(X;\hat{X}) + I(\varepsilon;Z) - I(\varepsilon;\hat{X})$

$$= H(X) + H(\hat{X}) - H(X,\hat{X}) + H(\varepsilon) + H(Z) - H(\varepsilon,Z)$$

$$- H(\varepsilon) - H(\hat{X}) + H(\varepsilon,\hat{X})$$

$$I(X;Z) = H(X) - H(\varepsilon) + I(\varepsilon;Z) \qquad (3.10)$$

These expressions will be used in the following section to establish bounds for information transmission and error entropy.

3.2 PERFORMANCE BOUNDS

3.2.1 Information Transmission

At each step in a feed forward estimation process information can only be preserved or lost, it cannot be increased. In general

$$I(X;Y) \geqslant I(X;G(Y))$$

where $G(\ )$ is any function of Y (Gallager [1968]). Thus the information transmitted by the estimate cannot exceed the information transmitted by the measurement, i.e.

$$I(X;Z) \geqslant I(X;\hat{X}) \qquad (3.11)$$

Hence the upper bound of information transmission by the estimate is established by the information, or lack of information, in the measurement. An overall upper bound to both $I(X;Z)$ and $I(X;\hat{X})$ is the entropy of the message $H(X)$ and this is only achieved when all uncertainty concerning X is resolved by Z or by $\hat{X}$.

3.2.2 Error Entropy

Information transmission is always non-negative so that in Equations 3.8 and 3.10

$$I(\varepsilon;Z) \geqslant 0 \quad \text{and} \quad I(\varepsilon;\hat{X}) \geqslant 0$$

and equations 3.8 and 3.10 can be written as inequalities

$$H(\varepsilon) \geqslant H(X) - I(X;\hat{X}) \qquad (3.12)$$

$$H(\varepsilon) \geqslant H(X) - I(X;Z) \qquad (3.13)$$

Because

$I(X;Z) \geqslant I(X;X)$

then equation 3.12 will generally yield a higher value for the lower bound of error entropy. Equation 3.13 has the advantage, however, of being able to predict the lower bound of estimation error entropy without the need to define the optimal estimator or even restrict the form of the estimator. This equation was proposed by Weidemann and Stear [1969] for purposes of performance prediction in parameter estimation. The lower bounds of error entropy are both achieved when the error is independent of both measurement and estimate. Under this (commonly assumed) condition, the information transmission $I(X;\hat{X})$ is maximized and equations 3.8 and 3.10 become

$$I(X;Z) = I(X;\hat{X}) = H(X) - H(\varepsilon) \tag{3.14}$$

This indicates that the estimator preserves all information contained in the measurement. In effect, the error arises from an additive noise source in the measurement process.

4 SPATIAL ESTIMATION

An elementary case of spatial estimation involves the estimation of events at q discrete locations in a region on the basis of measurements made at just a few of these locations, say n (where $n \ll q$). X and $\hat{X}$ are q dimensional vectors representing actual and estimated concurrent events at the q locations, while X^n will represent error free measurements made at the n locations of a monitoring network. It is important that these n locations be a subset of the q location.

An informational quantity of interest is the information transmitted by the network measurements concerning X. From Equation 2.4

$$I(X;X^n) = H(X) + H(X^n) - H(X,X^n)$$

But as X^n is a subvector of X then the joint entropy term $H(X,X^n)$ reduces to $H(X)$ so that

$$I(X;X^n) = H(X^n) \tag{4.1}$$

Substituting X^n for the measurement Z in Equation 3.9 yields the information transmitted by the estimate about X

$$I(X;X) = I(X;X^n) - [I(\varepsilon;X^n) - I(\varepsilon;\hat{X})]$$

Under conditions where the estimate preserves the information contained in X^n then $I(\varepsilon;X_i^n) = I(\varepsilon;\hat{X})$, or when the error is independent of both the measurement and the estimate, then $I(X;\hat{X})$ achieves an upper bound which is given by

$$I(X;\hat{X}) = I(X;X^n) = H(X^n) \tag{4.2}$$

The error entropy is given by Equation 3.8

$$H(\varepsilon) = H(X) - I(X;X^n) + I(\varepsilon;X^n) \tag{4.3}$$

and achieves its minimum value when the error is independent of the measurement. Thus the lower bound for error entropy is

$$H(\varepsilon) = H(X) - H(X^n) \tag{4.4}$$

Equations 4.1, 4.2, 4.3, and 4.4 suggest the desirability of maximizing the information transmitted by the network measurements, and under certain conditions, their joint entropy, when designing monitoring networks. This approach forms the basis of the network design method suggested by Caselton and Husain [1980]. The implications of this approach to network design when the (spatial) properties of X are described by a multivariate normal model are discussed by Caselton and Zidek [1981].

5 MODELS INVOLVING SERIAL DEPENDENCY

Serial processes, where an event at time t is influenced to some degree by prior events, often arise in the hydrosciences. A measure of the extent of this influence is the information which is transmitted by prior events concerning the next event. X_t

will represent a vector describing events at time t and X_{t-1}, X_{t-2}, etc. represent events at earlier times. The informational quantity of interest $I(X_t;X_{t-1},X_{t-2},\ldots,X_{t-r})$ for prior events to time $t-r$ is given by Equation 2.4

$$I(X_t;X_{t-1},X_{t-2},\ldots,X_{t-r}) = H(X_t) + H(X_{t-1},X_{t-2},\ldots,X_{t-r}) - H(X_t,X_{t-1},X_{t-2},\ldots,X_{t-r}) = H(X_t) - \Delta H_r \tag{5.1}$$

where ΔH_r is the increase in joint entropy when X_t is added to the array $\{X_{t-1},X_{t-2},\ldots,X_{t-r}\}$. In the case where the process is stationary, an upper bound for ΔH_r can be established when X_t is a "free extension" of the process, i.e. when

$$P(X_t|X_{t-1},X_{t-2},\ldots,X_{t-r}) = P(X_t|X_{t-1},X_{t-2},\ldots,X_{t-r},X_{t-(r+1)})$$

If ΔH_{r+1} is defined by

$$\Delta H_{r+1} = H(X_{t-1},X_{t-2},\ldots,X_{t-(r+1)}) - H(X_{t-1},X_{t-2},\ldots,X_{t-r})$$

then from Hartmanis [1959]

$$\Delta H_r = \Delta H_{r+1} \tag{5.2}$$

and when the extension is not "free" then

$$\Delta H_r > \Delta H_{r+1} \tag{5.3}$$

Thus, if r is large enough to ensure that Equation 5.2 holds for both r and all larger values of r, then the maximum amount of information transmitted by the past process to the next event is given by Equation 5.1. If an estimate of the next event is to be based upon the values of r past events then the lower bound of error entropy can be obtained from Equation 3.13. without specifying the optimal estimation model.

$$H(\varepsilon) \geq H(X_t) - I(X_t;X_{t-1},X_{t-2},\ldots,X_{t-r})$$

$$\geq H(X_t) - H(X_t) - H(X_{t-1},X_{t-2},\ldots,X_{t-r})$$

$$+ H(X_t,X_{t-1},X_{t-2},\ldots,X_{t-r} \qquad (5.4)$$

$$H(\varepsilon) \geq \Delta H_r \qquad (5.5)$$

The lower bound of error entropy is given, in this case, by the increase in the joint entropy caused by adding the next event.

6 INFORMATION TRANSMISSION FOR SOME SPECIFIC ESTIMATION MODELS

The previous sections have shown how certain bounds on information transmission and error entropy can be established without the estimation model or error characteristics being specified. In this final section, the information transmission characteristics of some commonly specified types of estimation model are mentioned.

A requirement of many estimation models is that the error be Gaussian and unbiased. The relationship between the joint entropy of an m dimensional error vector and the error covariance matrix R_ε for Gaussian errors is given by

$$H(\varepsilon) = \frac{1}{2}\log_e\{(2\pi e)^m \det(R_\varepsilon)\} \qquad (6.1)$$

where det( ) denotes the determinant.

Equation 6.1 confirms that minimizing the determinant of the error covariance matrix is synonymous with minimizing error entropy. Thus, the general case of regression, where errors are required to be unbiased and uncorrelated and are commonly assumed to be Gaussian, also conforms to the minimization of error joint entropy and the maximization of information transmission $I(X;\hat{X})$.

The Kalman filter is representative of a more complex type of dynamic model which has received attention in the hydrosciences.

A property of the optimal linear filter is that the optimal estimate and its error are uncorrelated (Gelb [1974]). Since the process $\{\hat{X}_k, \varepsilon_k\}$ is jointly Gaussian then the value of $I(\hat{X}_k;\varepsilon_k)$ is zero and Equation 3.8 reduces to

$$I(X_k;\hat{X}_k) = H(X_k) - H(\varepsilon_k) \tag{6.2}$$

Tomita et al. [1976] show that the optimal linear filter minimizes the determinant of the error covariance matrix and therefore, for Gaussian error, minimizes the error entropy. Equation 6.2 confirms that the information transmission between the state and its estimate is also maximized.

REFERENCES

Caselton, W.F. and Husain, T., "Hydrologic Networks: Information Transmission", Journal of the Water Resources Planning and Management Division, ASCE, Vol. 106, No. WR2, July 1980, pp. 503-520.
Caselton, W.F. and Zidek, J.V., "The Use of a Proper Local Utility in the Design of Monitoring Networks", Technical Report No. 81-11. Institute of Applied Mathematics and Statistics, UBC, June 1981.
Dyhr-Nielsen, M. "Loss of Information By Discretizing Hydrologic Series", Hydrology Papers, Colorado State University, Oct. 72.
Gallager, R.G. "Information Theory and Reliable Communication", John Wiley & Sons, 1968.
Gelb, A. "Applied Optimal Estimation", M.I.T. Press 1974.
Hartmanis, J. "Application of Some Basic Inequalities for Entropy", Information and Control, Vol. 2, 1959, pp.199-213.
Raiffa, H. "Decision Analysis", Addison Wesley Publishing Company 1970.
Shannon, C.E. "A Mathematical Theory of Communications", Bell Systems Technical Journal, Vol. 27, 1948, p.379-423, 623-656.
Tomita, Y., Ohmatsu, S., Soeda, T. "An Application of Information Theory to Estimation Problems", Information and Control, Vol. 32, 1976, pp.101-111.
Weidemann, H.L., Stear, E.B. "Entropy analysis of parameter estimation", Information and Control, Vol. 14, 1969, pp.493-506.

VALIDATION OF SYNTHETIC STREAMFLOW MODELS[a]

DENNIS P. LETTENMAIER[b] AND STEPHEN J. BURGES[c]

INTRODUCTION

In reviewing the last two decades of work in stochastic hydrology, one finds an overwhelming predominance of effort in model development, as opposed to parameter estimation (calibration) and verification. Recent work, however, (Klemes, et al., 1981; Burges and Lettenmaier, 1981) has demonstrated that the practical implications of differences in model form may not be nearly as great as suggested by much of the work on model development. Therefore, increased emphasis on model validation, which we consider to include calibration and verification, appears in order.

The common approach to implementation of stochastic (or synthetic) hydrologic models is to estimate a parameter vector from a set of historic records, then to proceed as if the estimated parameters were the population values. For instance, a parameter vector $\underline{\hat{\theta}}^T = [\hat{\theta}_1, \ldots, \hat{\theta}_m]$ is estimated from a historic data matrix $X^T = [\underline{X}_i^T, \ldots, \underline{X}_n^T]$, with $\underline{X}_i = (X_{i1}, \ldots, X_{is})$, where n is the number of years of historic record and s is the number of sites. Although the record length, n, is implied to be equal for all sites, this need not be the case. Further, the observations X_{ij}

---

[a]Presented at International Conference on Time Series Methods in Hydrosciences, Canada Centre for Inland Waters, Burlington, Ontario, October 6-8, 1981.

[b]Research Associate Professor, Department of Civil Engineering, University of Washington, Seattle, WA 98195.

[c]Professor, Department of Civil Engineering, University of Washington, Seattle, WA 98195.

Reprinted from *Time Series Methods in Hydrosciences*, by A.H. El-Shaarawi and S.R. Esterby (Editors)

may consist of p seasonal flow volumes, i.e., $X_{ij} = \{X_{ij1}, \ldots, X_{ijp}\}$ with $\sum_{k=1}^{n} X_{ijk} = X_{ij}$.

The time-honored approach to model validation, once $\hat{\underline{\theta}}$ has been estimated, is to generated multiple sequences of synthetic data, $\overline{Y}_i(\ell)$, i = 1, ...,T; where T is the number of sequences generated, and ℓ is the length of the synthetic sequences (not necessarily equal to n, the historic record length). From the multiple synthetic data sequences, low order moments, $M_{ijk}^{(r)}$ are computed. For validation of multisite sequences, cross moments, usually of the second order, may also be computed. The averages of these estimated moments over the T synthetic sequences are then compared to the (single) moments estimated from the historic record. If the replication is, in some sense, satisfactory, the model is considered to be valid. Although quantitative measures of validation and model quality are rarely made, the estimated standard error of the synthetic moments provides a possible measure for evaluating the quality of replication.

At least three difficulties are encountered in such moment-oriented validations. First, the historic record lengths are usually so short that it is not practical to reserve any of the record for validation, i.e., the historic moments and the parameter set $\hat{\underline{\theta}}$ from which the synthetic sequences are derived, (and synthetic moments estimated) are the same historic record. Thus the historic and synthetic moments are not independent, and the degree of replication which should be expected is open to question. Second, estimates of moments beyond the first order are biased as a result of the persistence structure inherent in hydrologic time series and, in some cases, small sample effects. Additional biases may be introduced by transformations invoked to allow synthesis of raw sequences in the normal (gaussian) domain, an intermediate step in many models. Third, the multiplicity of moments in a multisite, multiseason model, coupled with short historic records from which model parameters are estimated, make it doubtful that any properly parameterized stochastic model can, or should, preserve all moments. Clearly, if enough parameters are included in the model (leaving a small number of degrees of freedom) the synthetic sequences become increasingly similar to the historic, until unltimately the synthetic sequences are identical to the historic record. This, of course, subverts the purpose of synthetic hydrology, which is to provide a representation of alternative, equally likely scenarios which are statistically representative of the past. When suitable model identification procedures (such as the Akaike (1974) information criterion) are invoked the resulting model may be unable to preserve some, or even many, moments. This is particularly true when the model estimation procedure is not moment-based (e.g., maximum likelihood). Although to some extent such infidelity may represent a shortcoming of the model used, in a more general sense it simply reflects limitations imposed by short data records.

ALTERNATIVE VALIDATION MEASURES

Sample moments, of course, are not the only possible validation measures. Others, which we term statistical, include autocorrelations, as well as other measures of model persistence structure, such as the Hurst coefficient. The basic computational approach is the same as that outlined above; averages of the estimates over many synthetic traces are compared with historic estimates.

All statistical measures, moments and others, have a common shortcoming: they do not reflect the ultimate use of the synthetically generated sequences. Although specifics vary, most applications of synthetic hydrology address problems either of estimating the required initial or incremental size of a storage facility, or of estimating performance characteristics (e.g., reliability) of an existing (or proposed) system. Thus, validation measures which reflect compatibility of synthetic and historic sequences with respect to system performance or sizing may be more relevant to acceptance or rejection of a model than statistical measures.

The two performance-related validation measures used here are the zero failure storage capacity, computed using the sequent peak algorithm, and the critical extraction rate, or maximum uninterrupted withdrawal rate for a fixed storage size. In both cases, simple mass curve techniques are used. Although such techniques are crude in that they assume each streamflow sequence to be deterministic, the estimates provided are helpful in estimating baseline storage requirements and/or extraction rates. The sequent peak algorithm, in determining the storage necessary to meet a prescribed demand, is applicable to system expansion problems. It has been found to be quite sensitive to long term persistence, or low frequency effects (Burges and Lettenmaier, 1977). The critical extraction rate is more representative of performance of an existing system. Other performance measures, discussed by Palmer and Lettenmaier (1981) might also be used. However, as demonstrated by Klemes, et al. (1981), many measures of system performance which allow limited supply shortfalls, e.g., number, severity, or length of shortfalls for a fixed demand and reservoir size, are relatively insensitive to model form. Thus, while the "no fail" algorithms used here may not be physically or economically realistic for many systems, they are desirable as validation tools in that they are relatively sensitive to model form and parameter magnitudes.

A final validation measure considered here which is, in a sense, statistical, is the crossing distribution of the normalized (deseasonalized) flow sequences. The crossing distribution is the frequency with which a given level, expressed as standard deviations from the mean, partitions sequential normalized (seasonal) flow levels. Normalized flows are computed by subtracting the seasonal means from each observation and dividing by the seasonal standard deviation. For example, a crossing of normalized level K is said to occur between times t and t + 1 if $F_t^* < K$ and $F_{t+1}^* > K$ or $F_t^* > K$ and $F_{t+1}^* < K$, where F_t^* is the normalized flow at time t. Since F_t^* is normalized, the levels K are the number of standard deviations from the mean, and the region of interest is typically $-3 \lesssim K \lesssim 3$; for $|K| \gtrsim 3$ few crossings occur. A single estimated crossing distribution is estimated for each flow sequence, synthetic or historic; for synthetic flows the average over the estimated crossing distributions is computed.

The significance of the crossing distribution is that it represents the combined effect of the marginal distribution (variability at a given time) and persistence. For instance, changes in skewness alter the crossing distribution by changing the number of low level relative to high level crossings, while increasing the persistence decreases the number of crossings at all levels. Persistence effects are perhaps of greater interest because they are related to drought frequency.

IMPLEMENTATION

A computer program was developed to provide graphical displays of the statistical and performance measures discussed for single and multiple site analysis. For single sites, coefficients of variation and skew coefficients are estimated on a seasonal and annual basis. Seasonal lag one correlation coefficients are computed for adjacent seasons, and annual lag one correlation coefficients and Hurst coefficients are estimated. For the latter, Hurst's K estimator is used, primarily because it requires much less computer time than the GH estimator evaluated by Wallis and Matalas (1970). Sequent peak storage (Fiering, 1967) and critical extraction rate are computed for the entire (seasonal) historic and synthetic records, where annual demand is a fixed quantity apportioned equally to each season. Annual demand level for the sequent peak storage determination, and storage capacity for the critical extraction determination are specified a priori. Experience has shown that the most useful results are produced when sequent peak demand is relatively high, and critical extraction storage is low, so that the hypothetical systems are relatively highly stressed. Finally, crossing distributions are computed for the individual sites as described in the previous section.

For multiple site validation, many of the same measures used for single sites are applied to a hypothetical aggregate record formed by adding the predicted flows at selected site pairs. Two options are provided; the first takes the simple average of the flows at the two sites, while the second computes an aggregate flow equal to one half of the sum of the flow at the first site and the weighted flow at the second site, where the weight is the ratio of the annual mean at the first site to the annual mean at the second site. The latter option has the advantage that differences in mean flow do not allow either station to dominate, assuring that multiple site effects are represented in the aggregate flow.

The statistical indicators computed for station pairs are the coefficients of variation and skew coefficients for seasonal and annual aggregate flow, lag one correlation coefficients and estimated Hurst coefficient of aggregate annual flows, and seasonal and annual lag zero cross correlations. Sequent peak storage, critical extraction rate, and crossing distributions are also estimated for the aggregate flows. At the individual sites, coefficients of variation and skew coefficients are estimates for seasonal and annual flows, in addition to seasonal lag one correlations and annual Hurst coefficients and lag one correlation coefficients. Sequent peak storage, critical extraction rate, and crossing distributions are also estimated at the individual sites.

The results of the analyses are plotted as empirical cumulative distribution functions on a normal probability scale. The normal probability scale has no particular significance other than its wide familiarity and its ability (in common with other probability scales) to transform extreme events for ease of graphical comparison. No implication is made that the distributions estimated are, or should be, normal. The greatest advantage in presenting full distributions of validation measures, rather than summary measures (such as mean and variance) is that the analyst is made immediately aware of how far the historic estimates lie from the predominance of synthetic estimates. For instance, in extreme cases it may be that all of the synthetic estimates lie above or below the historic, which would be a rare occurrence if the synthetic model were in fact representative of the historic process. Comparison of means and variances of synthetic and historic measures does not provide nearly as clear an indication of model performance, as should become clear from the results presented in the following section.

APPLICATION

The validation procedure discussed above was applied to three two-site models of the Cedar River, and the North Fork of the Snoqualmie River, Washington. These models were estimated from 48 years of coincident record at the two stations. The raw flow record consisted of recorded monthly volumes and estimates provided by the U.S. Army Corps of Engineers during periods when observations were not made. The monthly flow volumes were aggregated to three seasons per (water) year: October-February, March-June, and July-September. The runoff response in both basins is dominated by snow accumulation during the winter months, and melt during the spring and early summer. Therefore, the seasons were chosen to reflect periods of dominant snow accumulation, melt, and snow-free conditions. The three models considered for generation of annual flow volumes were A) multivariate lag one Markov (Clarke, 1973); B) multi-site ARMA (1,1) (Ledolter, 1978), a maximum likelihood approach, and C) multiple site ARMA (1,1) with modified maximum likelihood estimation (Lettenmaier, 1980). Each annual model was used to generate 100 sequences of length 48 years in normalized (zero mean, unit variance, normal marginal distribution) form. These normalized sequences were then disaggregated using Lane's (1979) multiple site disaggregation model in the normal domain, and subsequently transformed to synthetic flows using three parameter log normal transformations. Therefore, each synthetic model consists of an annual model coupled with a seasonal disaggregation model, where the latter is the same for all three annual models.

Although the purpose of this paper is to discuss a validation technique, and not to assess models *per se*, a brief comment regarding the annual models should be made. The lag one Markov model (Model A) is a short term persistence model which has a relatively small low frequency component, hence long periods of flow above or below the mean are not represented. For this reason, considerable work in stochastic hydrology has been directed towards development of models similar to the second two, which may be described as long term persistent (Models B and C). These models are capable of generating very long periods of deficit (drought) and excess flows. One of the difficulties with multiple site generation of flows with long term persistence is assuring that the flows at the individual sites are similar with those that would have been generated had the sites been represented by a univariate model. To accomplish this, Lettenmaier (1980) proposed a modified maximum likelihood estimation technique, incorporated in model C, which penalizes parameter estimates that give rise to autocorrelation structures much different from those represented by single site estimates of the Hurst coefficient.

RESULTS

The computer program developed to perform the model validation can provide output in the form of lineprinter plots on a computer terminal or hard copy, or as continuous plots on a graphics terminal or ink plotter. Given the large number of plots involved in multiple site applications, we have found the first options preferable. Due to space limitations here, only a small subset of the plots generated for each model can be shown. In practice, given the great number of plots generated in validation of a multisite, multiseason model, we find it is much quicker to review a group of lineprinter plots as hard copy, rather than individual plots on a computer terminal screen

Figures la-d show the empirical distributions of the coefficient of variation for the aggregate flows generated from Model A for seasons 1-3 and for annual flow volumes. In these figures, as in those that follow, the (single) historic estimate is plotted as a dashed line between cumulative probability levels 10 and 90 per cent. If the synthetic sequences were independent of the historic record, the empirical cumulative distribution from the synthetic flows would be

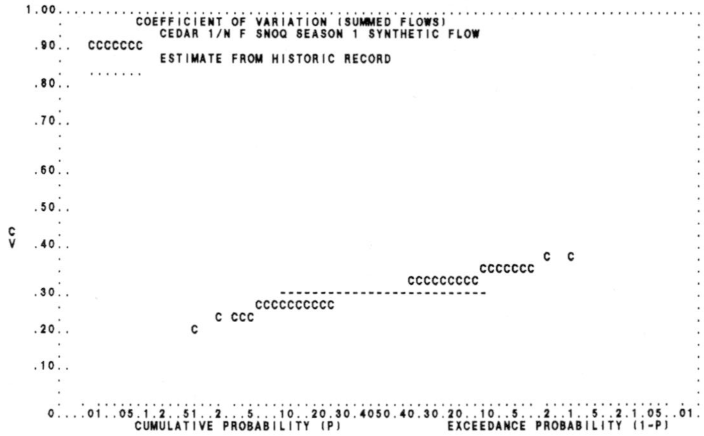

1a. Season 1

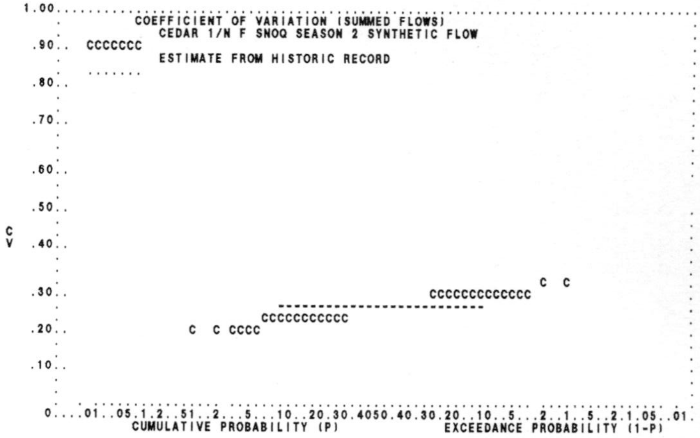

1b. Season 2

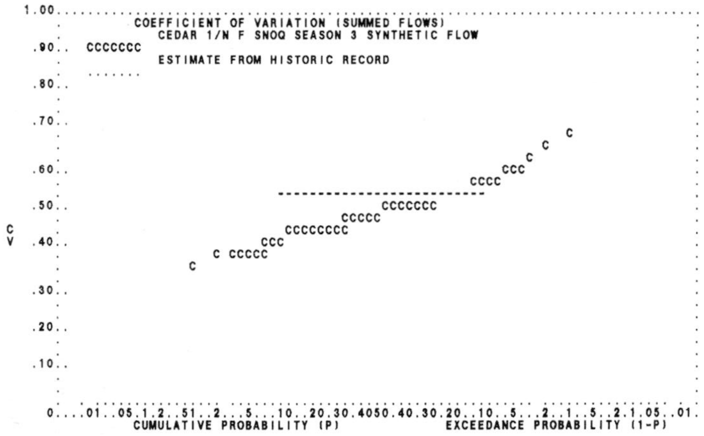

1c. Season 3

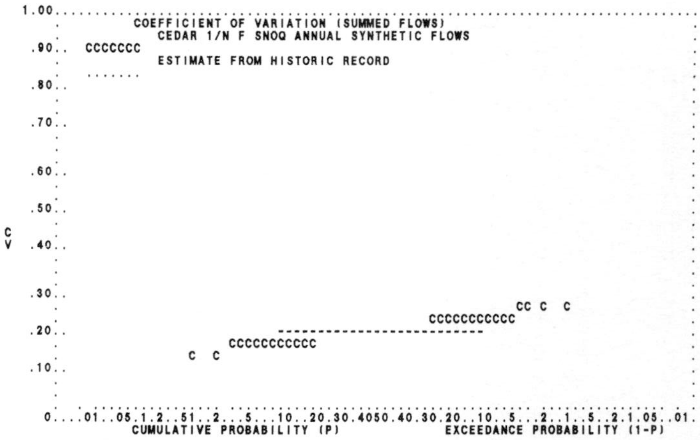

1d. Annual

Figure 1. Empirical cumulative distribution functions of coefficient of variation for aggregated flows, Model A, Seasons 1-3 and Annual.

expected to cross the dashed line with about 80 per cent confidence. Since historic and synthetic flow sequences are not independent, the true confidence region represented by this line should be considerably higher than 80 per cent.

Since the seasonal coefficients of variation are largely determined by the disaggregation model, and not the annual generator, similar results to those shown in Figures 1a-c were indicated for models B and C. As shown in Figure 1c, the season 3 coefficient of variation was slightly underestimated. This may be a result of the method used in the disaggregation model to conserve mass, which sometimes results in slight inconsistencies in the first and/or last season of the year. Figure 1d indicates that annual coefficients of variation from the synthetic sequences were similar to the historic value (median approximately equal to historic annual coefficient of variation). However, results for models B and C (not shown) indicated that the synthetic coefficients of variation were slightly lower than the historic. This may be the result of biasing of the estimator by the higher autocorrelations present in models B and C.

Figures 2a-d show results from Model A aggregate flows for seasons 1-3 and annual totals. As with the coefficient of variation, the results are quite similar for all models. The empirical distributions for synthetic flows generally reflect the downward biasing of the estimator of the skew coefficient (Wallis, et al., 1974). Although this bias could be corrected at the parameter estimation stage, correction of moments for bias may be counterproductive, as shown by Stedinger (1980); therefore no bias correction was attempted. As for the coefficient of variation, the results for season 3 are apparently anomolous. We believe that this is also related to the mass conservation adjustment made in the disaggregation model.

Figures 3a-d show the empirical distribution of K-estimators of the Hurst statistic, as well as the historic estimates for annual flow volumes at both sites, for all three models. The Markov model (Figures 3a and 3b) yields estimates that generally are more compatible (lower) than the ARMA models, although the modified ARMA

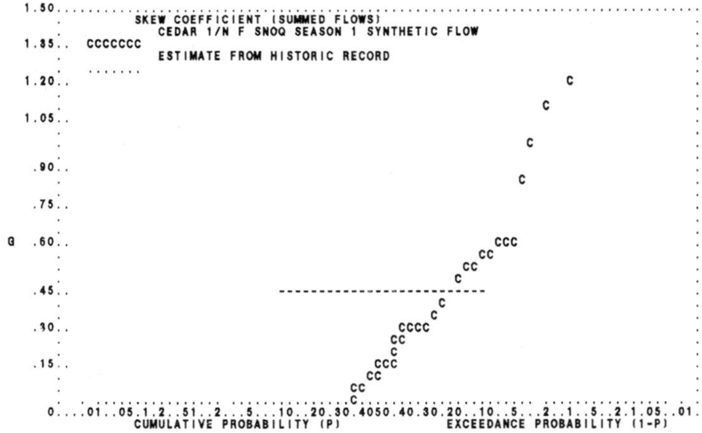

2a. Season 1

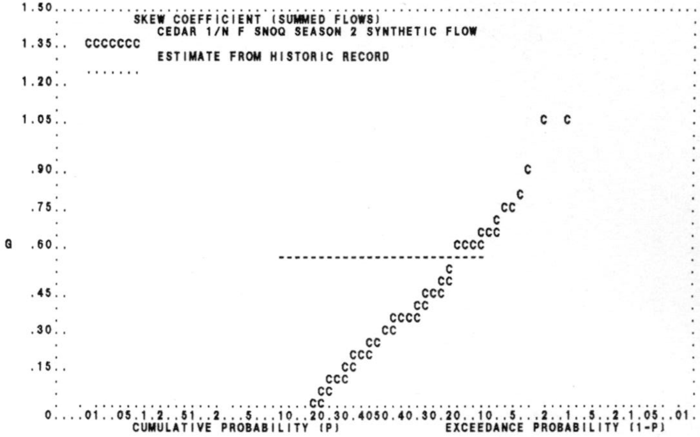

2b. Season 2

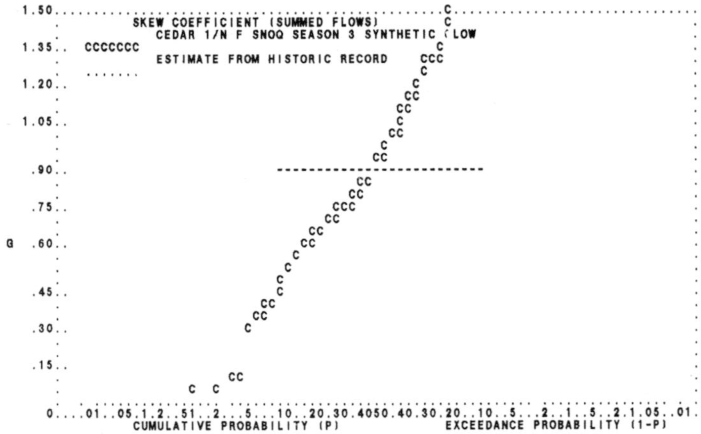

2c. Season 3

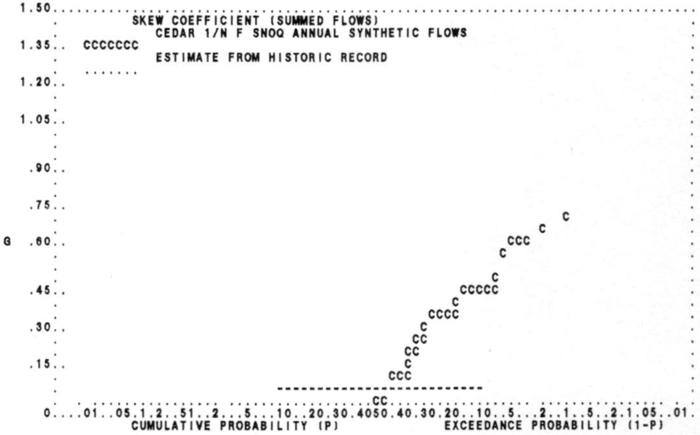

2d. Annual

Figure 2. Empirical cumulative distribution functions of skew coefficient for aggregated flows, Model A, Seasons 1-3 and Annual.

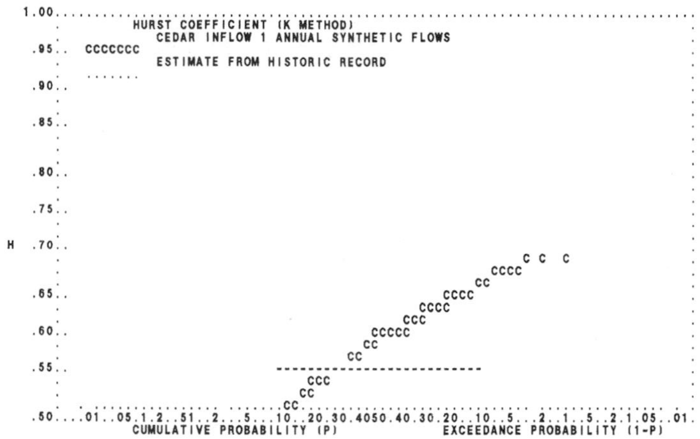

3a. Cedar River, Model A

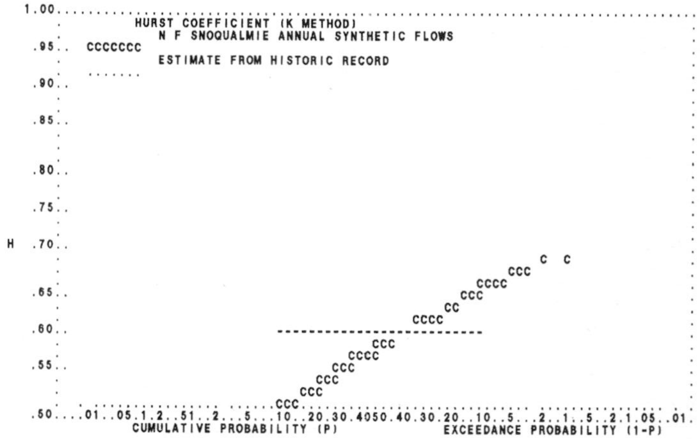

3b. North Fork Snoqualmie River, Model A

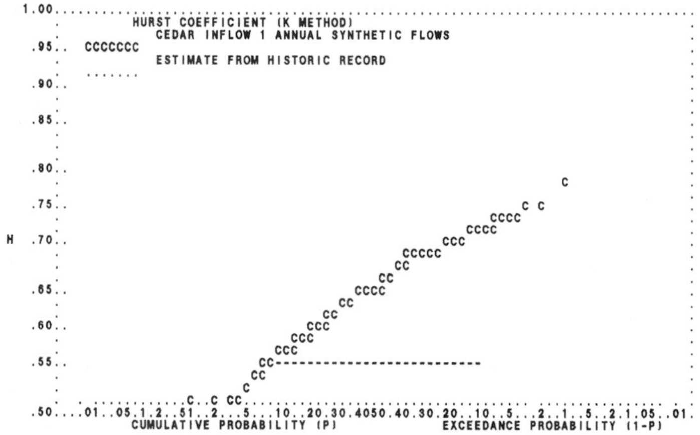

3c. Cedar River, Model B

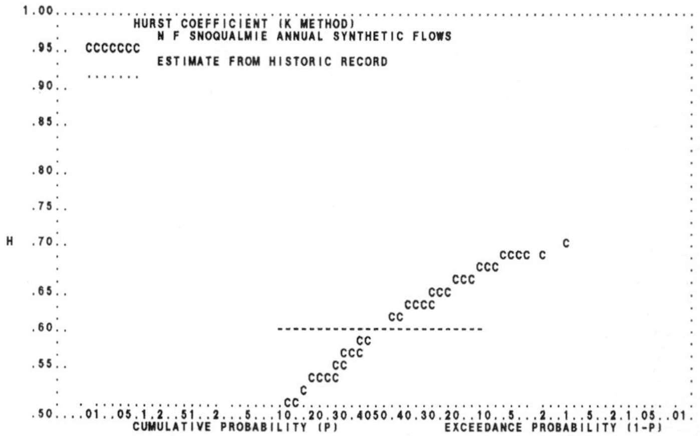

3d. North Fork Snoqualmie River, Model B

Figure 3. Empirical cumulative distribution functions of Hurst coefficient (K estimator), Cedar River and North Fork Snoqualmie River, Models A and B.

model (not shown) appeared preferable to model B (figures 3c and 3d) in this respect. Since the primary motivation behind long term persistence models is to preserve the so-called Hurst effect, represented by the Hurst coefficient, it is not surprising that Models B and C will generally have higher Hurst coefficients, since the estimator K is biased upwards for small ($\lesssim 0.7$) values of H. The results do emphasize, however, that bias in the estimator may itself be a sufficient explanation of the Hurst effect, since Model A, in expectation, has an H of 0.5, considerably less than the historic estimate at either site, even although the median synthetic K estimate is approximately equal to the historic value.

Figures 4a-d show empirical distributions of seasonal lag one correlations for Model A, seasons 2 and 3 at both sites. In these figures the correlation for the season indicated represents the lag one correlation with the subsequent season. Since the seasonal correlations are determined by the disaggregation model, results are essentially identical for all three annual models. Season 1 correlations (not shown) were near zero for both the historic estimate and the synthetic median at both sites. Although the low correlation between season 1 (winter) and season 2 (spring) may seem counterintuitive, it is the result of a mixture of effects. During winter seasons with normal or below normal temperatures, low runoff occurs as much of the precipitation is stored in the snowpack, and subsequently contributes to spring runoff suggesting a negative correlation. If precipitation is below normal and/or temperatures are above normal, on the other hand, a positive correlation with spring runoff is indicated. The net effect is to make the correlation between these two seasons approximately zero. Season 2 and season 3 correlations are positive, indicative of flow persistence which is usually observed in rain-affected watersheds. The only possible model inadequacy indicated by Figure 4 is that season 3 correlations for the second site are slightly overestimated. Generally, however, the synthetic and historic flows appear to be compatible with respec to seasonal correlations.

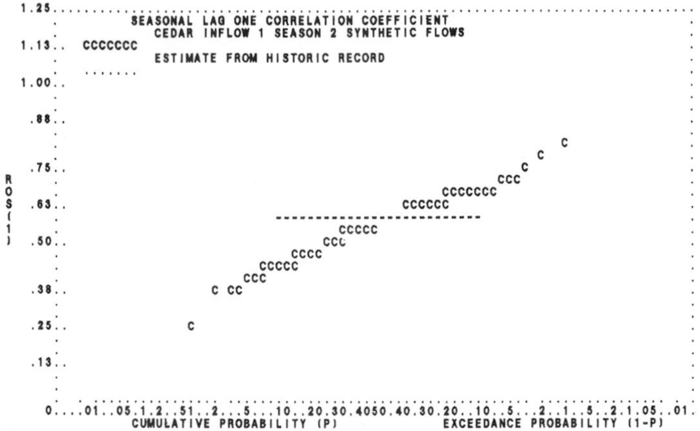

4a. Cedar River, Season 2

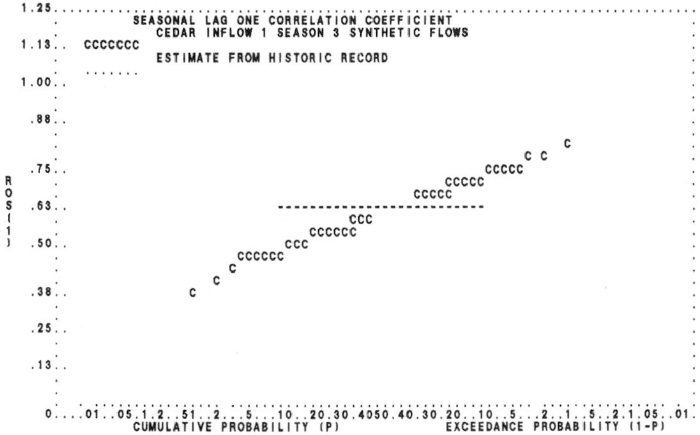

4b. Cedar River, Season 3

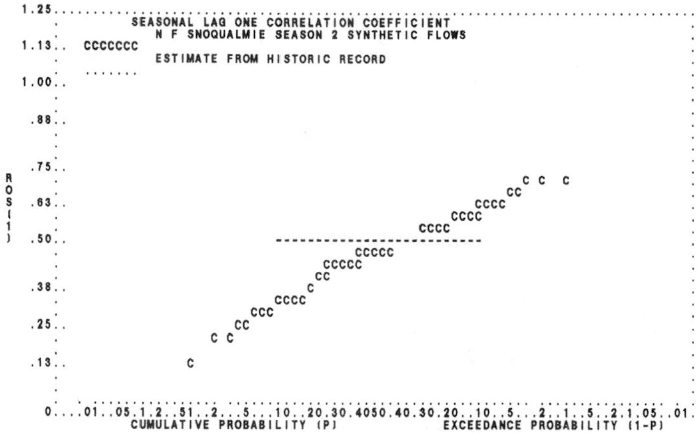

4c. North Fork Snoqualmie River, Season 2

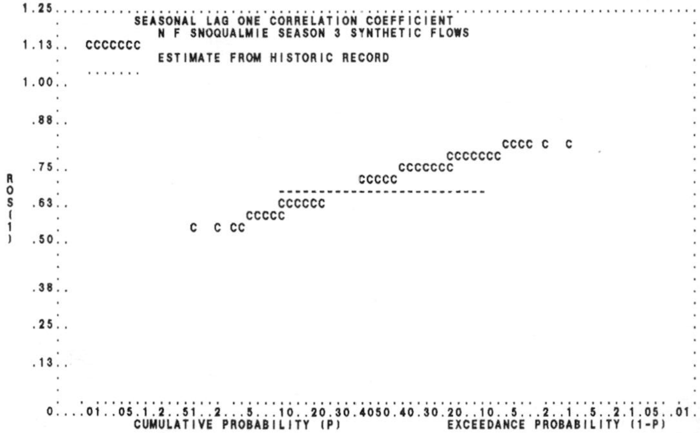

4d. North Fork Snoqualmie River, Season 3

Figure 4. Empirical cumulative distribution functions of seasonal lag one correlation coefficient for Model A, Seasons 2 and 3, Cedar River and North Fork Snoqualmie River.

Figures 5a-c show annual lag zero cross correlations for all three models. The results represent one of the most significant differences between models. The median Model A synthetic lag zero correlations are quite close to the historic estimate, while the Model B and Model C correlations are much less than the historic. This is directly attributable to the parameter estimation methods used by the three models: the annual cross correlation matrix is an explicit parameter set in the Markov model, while it is not in the ARMA models which use maximum likelihood estimators. Sensitivity analysis suggests that the multisite ARMA models achieve long term persistence at the individual sites by reducing cross correlations, therefore a tradeoff is indicated in these models between preservation of single site and cross site properties.

Figures 5d-f show annual lag one correlation coefficients for the aggregate flows. These figures do not provide a complete picture of the differences in correlation structure between models, for instance although lag one synthetic correlations for Models A and C are similar, the autocorrelation function for Model A (Figure 5d) decays much more rapidly than for Model C. These figures do indicate, however, that in terms of high frequency effects models A and C appear to be preferable to model B, which generally overestimates low lag correlations.

Figures 6a-f show aggregate flow sequent peak storage and critical extraction for the three models. As discussed earlier, selection of the demand level for the sequent peak algorithm, and storage capacity for the critical extraction computation determine sensitivity of these indicators to possible model inadequacies. It was determined that a demand level of 0.90 times mean annual flow was appropriate for the sequent peak computations, and a reservoir storage of 0.25 times the mean annual inflow for critical extraction determinations. The indicated model differences are similar to those suggested by the Hurst coefficient; storage requirements for Model A are slightly underestimated, while Models B and C appear to be more compatible with the historic record in requiring larger storage. Also of

5a. Annual Cross Correlation, Model A.

5b. Annual Cross Correlation, Model B.

5c. Annual Cross Correlation, Model C.

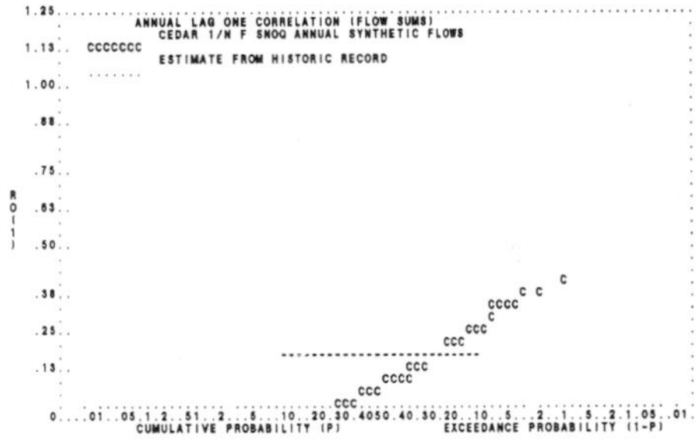

5d. Annual Lag One Correlation, Model A.

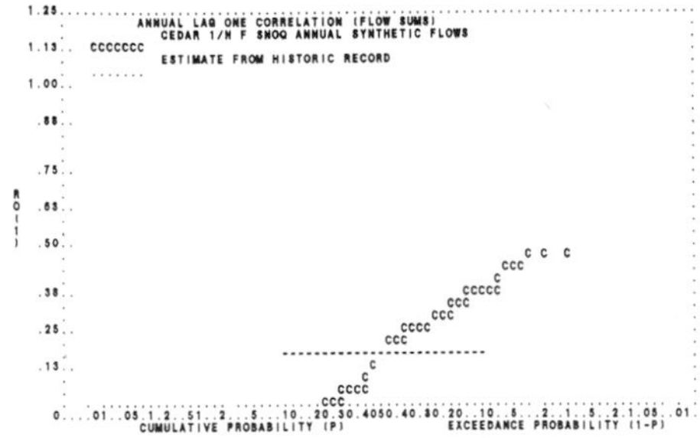

5e. Annual Lag One Correlation, Model B.

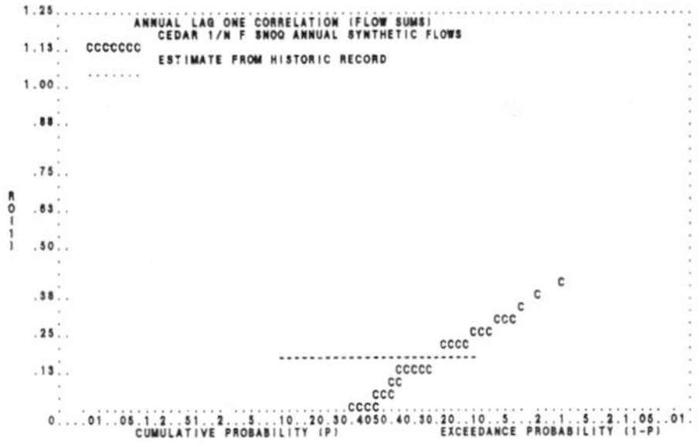

5f. Annual Lag One Correlation, Model C.

Figure 5. Annual lag zero cross correlation coefficients, Model A-C (Figures 5a-5c) and annual lag correlation coefficients, aggregated flows, Models A-C (Figures 5d-5f).

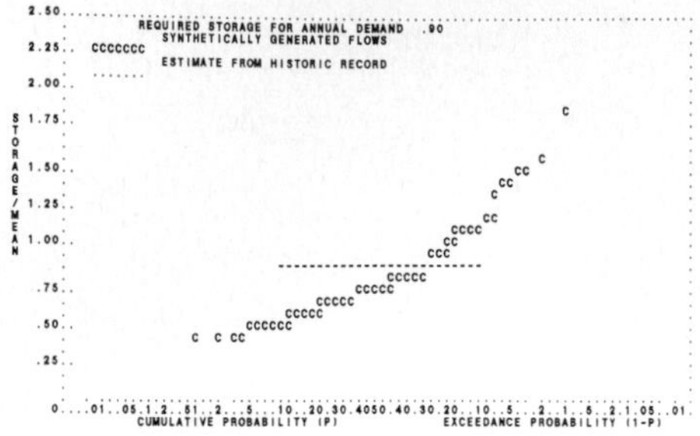

6a. Sequent Peak Storage, Model A.

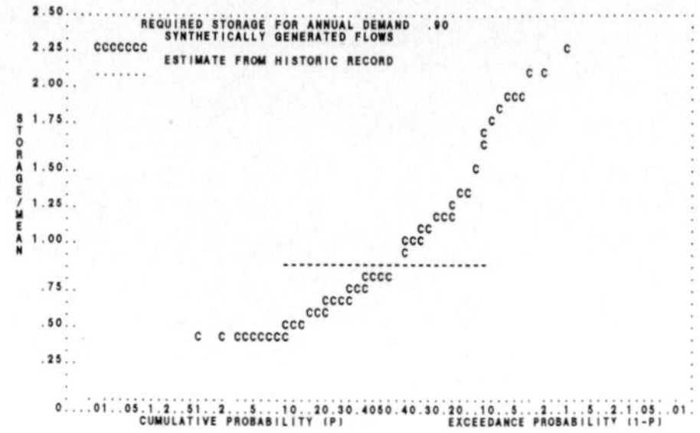

6b. Sequent Peak Storage, Model B.

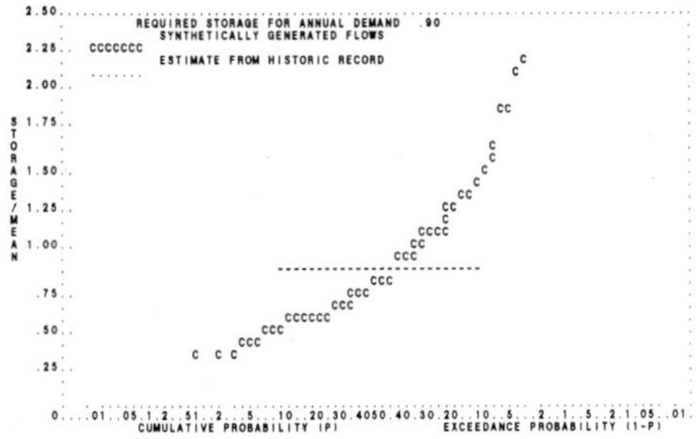

6c. Sequent Peak Storage, Model C.

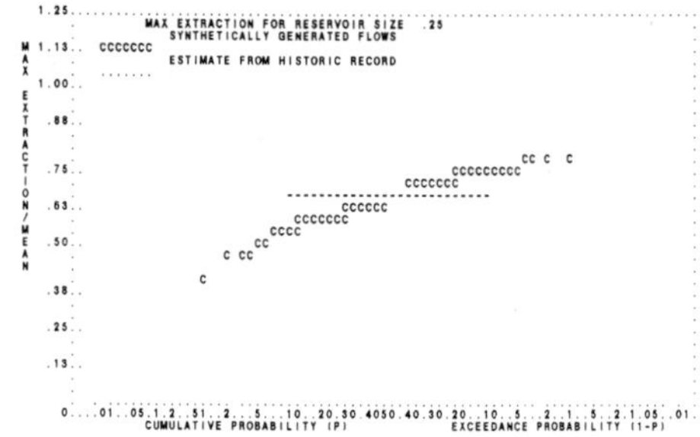

6d. Critical Extraction, Model A.

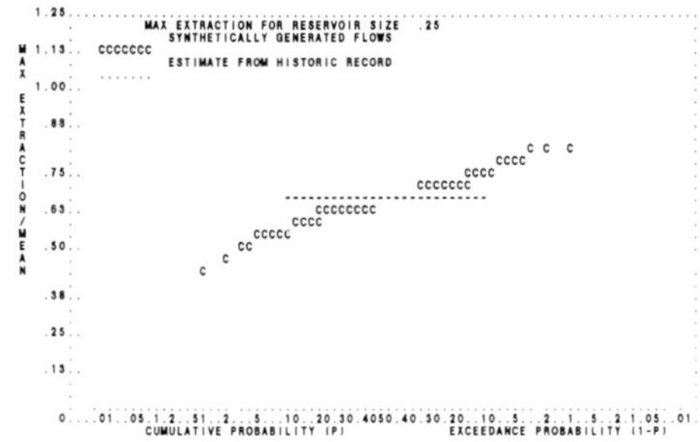

6e. Critical Extraction, Model B.

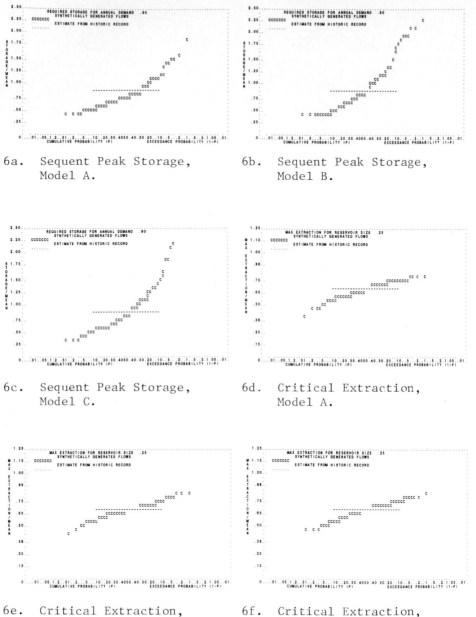

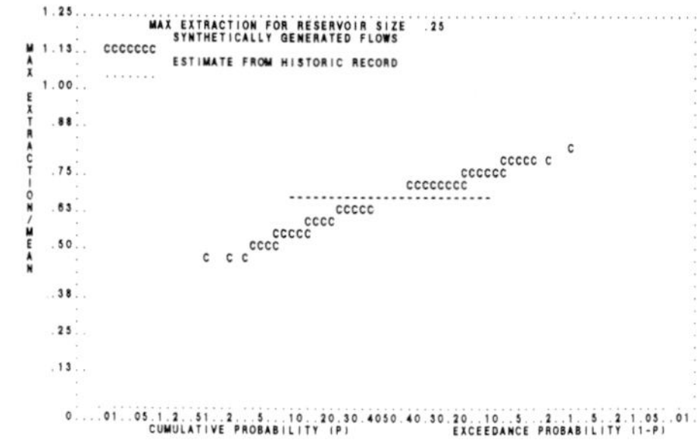

6f. Critical Extraction, Model C.

Figure 6. Empirical cumulative distribution functions of sequent peak storage, aggregated flows, Models A-C (Figures 6a-6c) and critical extraction rate (Figures 6d-6f).

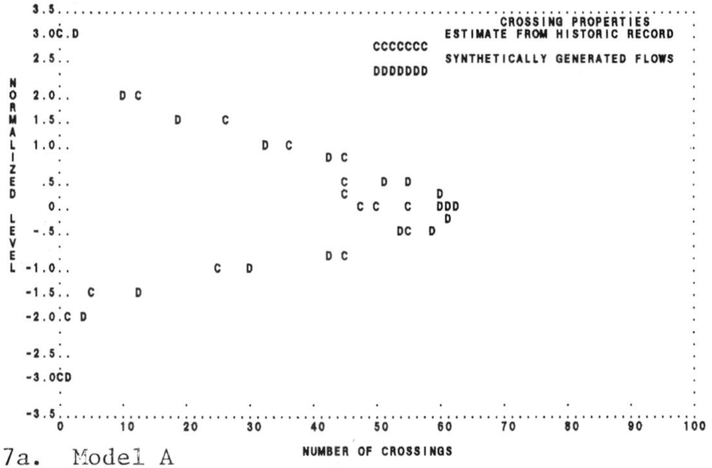

7a. Model A

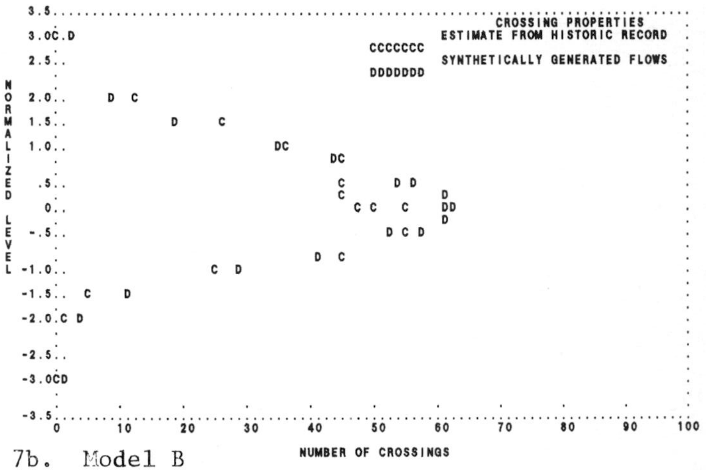

7b. Model B

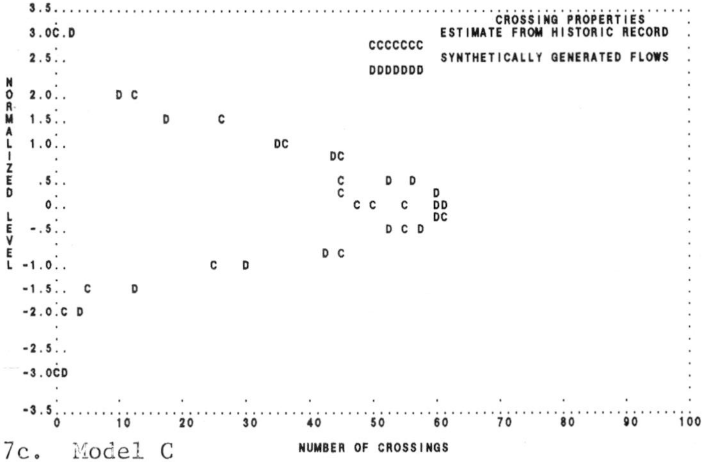

7c. Model C

Figure 7. Historic and average synthetic crossing distributions for aggregate flows, Model A–C.

significance is the steeper slope of the Model B and C distributions; this is consistent with the results of earlier work (Burges and Lettenmaier, 1977) showing that required storage at high reliability (e.g., exceedance probability 2%) can be much higher for long term persistence models. Critical extraction distributions show relatively little difference between models, most likely because the small storage size used emphasizes within-year flow properties, which are similar for all three models.

Finally, Figures 7a-c show crossing distributions for aggregate flows using all three models. Crossing distributions for each of the models are quite similar, suggesting that the form of the marginal distributions, determined by the disaggregation model, dominates. The estimated distribution for the historic data has fewer mid level crossings than the average synthetic distribution, especially in the vicinity $F^* \approx 0.5$. This may be simply a result of variability in the historic estimates. The smooth form of the average synthetic distribution, as opposed to the jagged historic estimates, supports the view that sample variability may be the most important contributor to differences in the two distributions.

SUMMARY AND CONCLUSIONS

The use of graphical techniques for validation of multivariate synthetic streamflow models is advocated. Two general types of validation measures are suggested: statistical and performance-based. Although preservation of low order moments, particularly the mean, will often be a necessary condition for model acceptance, biasing of higher order moment estimators complicates their use for validation purposes. Although moment estimators may be corrected for bias, this does not necessarily result in improvement of a stochastic model from a performance standpoint. Therefore, performance-based model validation measures, particularly sequent peak storage, may be more significant for operational validation.

Application of the techniques suggested to three two-site, three season models of the Cedar and North Fork Snoqualmie River, Washington indicated possible inadequacies in the seasonal distribution of flows, as well as differences related to long term persistence structure. The graphical results also pointed out a tradeoff in the multivariate long term persistence models between cross-site correlations and autocorrelations at the individual sites. Empirical distributions of moments and auto- and cross-correlations at the seasonal level were useful in validating the multi-site disaggregation model, while the sequent peak algorithm was most useful for overyear validation. The latter indicator is, however, sensitive to the demand pattern imposed. Critical extraction rate and crossing distributions were less useful model validation measures.

REFERENCES

Akaike, H., "A New Look at Statistical Model Identification", IEEE Transactions on Automatic Control, Vol. AC-19, No. 6, Dec. 1974, pp. 716-723.

Burges, S.J. and D.P. Lettenmaier, "A Comparison of Annual Streamflow Models", Journal of the Hydraulics Division, ASCE, Vol. 103, No. HY9, 1977, pp. 991-1006.

Burges, S.J. and D.P. Lettenmaier, "Reliability Measures for Water Supply Reservoirs and the Significance of Long-Term Persistence", Paper presented at International Symposium on Real Time Operation of Hydrosystems, University of Waterloo, June 1981.

Clarke, R.T., Mathematical Models in Hydrology, Irrigation and Drainage Paper No. 19, Food and Agriculture Organization, United Nations, Rome, 1973.

Fiering, M.B., Streamflow Synthesis, Harvard University Press, Cambridge, Massachusetts, 1967, p. 11.

Jones, D.A., P.E. O'Connell and E. Todini, "A Model Validation Framework for Synthetic Hydrology", Paper presented at Conference on Water Resources Planning in Egypt, Cairo, June 1979.

Klemes, V., R. Srikanthan and T.A. McMahon, "Long Memory Flow Models in Reservoir Analysis: What Is Their Practical Value?" Water Resources Research, Vol. 17, No. 3, pp. 737-751, June 1981.

Lane, W., "Applied Stochastic Techniques User Manual", U.S. Bureau of Reclamation, Denver 1979.

Ledolter, J., "The Analysis of Multivariate Time Series Applied to Problems in Hydrology", Journal of Hydrology, Vol. 36, pp. 327-352, 1978.

Lettenmaier, D.P., "Parameter Estimation for Multivariate Streamflow Synthesis", Proceedings, Joint Automatic Control Conference, San Francisco, August 1980.

Palmer, R.N. and D.P. Lettenmaier, "Indexing Multiple Site Synthetic Streamflow Sequences Using Reliability Measures", Paper presented at ASCE Specialty Conference, Technical State of the Art Exchange, San Francisco, August 1981.

Stedinger, J.R., "Parameter Estimation, Streamflow Model Validation, and the Effects of Parameter Error and Model Choice on Derived Distributions", Paper presented at American Geophysical Union Fall Meeting, San Francisco, December 1979.

Stedinger, J.R., "Fitting Log Normal Distributions to Hydrologic Data", Water Resources Research, Vol. 16, No. 3, pp. 481-490, June 1980.

Wallis, J.R. and N.C. Matalas, "Small Sample Properties of H and K - Estimators of the Hurst Coefficient h", Water Resources Research, Vol. 6, No. 6, December 1970, pp. 1583-1594.

Wallis, J.R., N.C. Matalas, and J.R. Slack, "Just a Moment", Water Resources Research, Vol. 10, No. 2, April 1974, pp. 211-219.

OBSERVATION AND SIMULATION OF THE SOOKE HARBOUR SYSTEM

D.P. KRAUEL, F. MILINAZZO, M. PRESS, AND W.W. WOLFE
Royal Roads Military College, Victoria, B.C. (Canada)

ABSTRACT

The findings of a continuing physical oceanographic study of the Sooke Inlet System on the West Coast of Vancouver Island are described. The system consists of a shallow harbour, freely connected to an inland basin with seasonally varying fresh water inflow at the mouth of the basin. A summary of salinity, temperature, water current, and tidal elevation data is presented. A two-dimensional, barotropic tidal model is used to predict circulation within the basin. A comparison between these calculations and the observed currents is made.

INTRODUCTION

The Sooke Harbour-Basin system is a small inlet about 30 km west of Victoria, B.C. on the Strait of Juan de Fuca (Fig 1). The Basin is about 4 km long and 3 km wide with a depth varying from a 37 m deep hole near the mouth of the Basin to tidal mud flats at the mouth. The average depth is 17 m. The Basin is connected to Juan de Fuca Strait via Sooke Harbour, a broad (≈1 km), shallow sill about 3 km long, having a mean depth of about 3.5 m. The Sooke River provides a source of fresh water at Billings Spit, the boundary between the Basin and the Harbour. There are no significant fresh water inflows directly into the Basin itself. The river flow shows a strong winter maximum, estimated at 50 m^3s^{-1} in January and becoming negligible during August (Elliott, 1969). The tides are mixed, mainly semi-diurnal with an average range of about 2 m

Reprinted from *Time Series Methods in Hydrosciences*, by A.H. El-Shaarawi and S.R. Esterby (Editors)

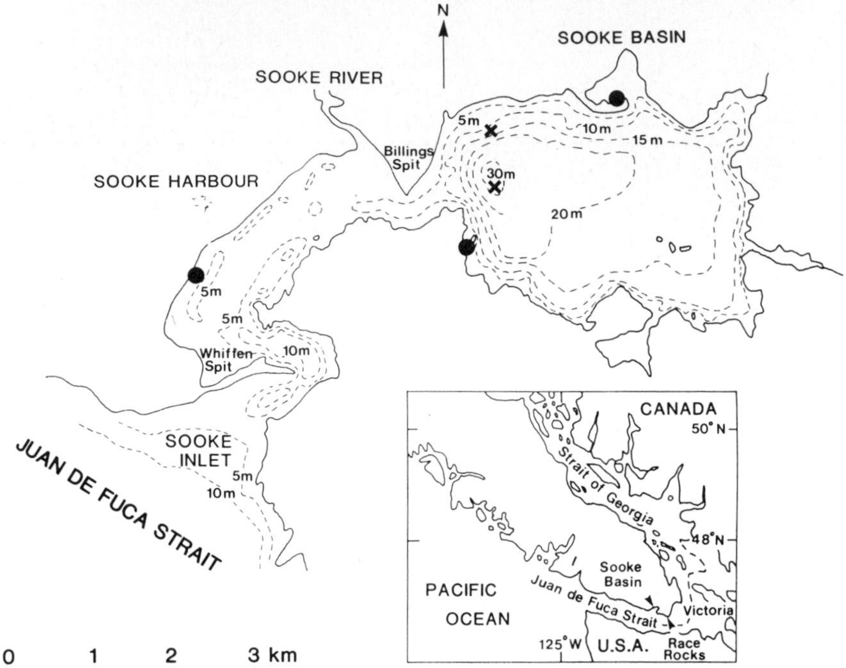

Fig. 1. Sooke Harbour and Basin; • tidal staff × current meter.

in the Strait. The hydrography permits a relatively free exchange of water between the system and the Strait and thus the tides are attenuated and delayed very little.

The system is experiencing pressures from a broad spectrum of users. There are extensive moorage and repair facilities for the West Coast fishing industry and a planned marine industrial park. Extensive log booming areas are at the north end of the Basin and at the harbour side of Wiffen Spit which contribute large quantities of wood fragments and bark to the water. The Basin provides a large harvest of shrimp, oysters, and clams. Finally, the Sooke region is an important recreation centre because of its proximity to Victoria. There are several marinas catering to the recreational fishing industry. Extensive beaches lend to beach combing and clam-digging. The effects of this rapidly developing tourist industry and the resultant disturbance to the environment are of great concern to the local governments and long term residents.

Because of these environmental and economic concerns, and because the system is of academic interest as a non-typical estuarine system, a numerical model of the system is being developed to predict the circulation, flushing, dispersion of pollutants, and the effects that dredging or construction of breakwaters and harbours might have on these parameters and on water quality. To calibrate and verify the model, we have started a series of spatial and temporal measurements of water circulation, temperature, and salinity.

Profiles of salinity and temperature as a function of depth along the axis of the inlet from the Strait to the eastern end of the Basin have been taken irregularily for a period of about a year. During January, when the fresh water run-off from the Sooke River is significant, the temperature is quite uniform at 7.5^{0}, with slightly cooler water at the surface due to the cold fresh water. The exchange of heat energy between the water and the atmosphere has been discussed by Elliott (1969). The corresponding salinity distribution has a marked stratification due to the large fresh water inflow, with salinities varying from 16‰ at the surface to 29‰ at the bottom.

During the summer months these distributions are significantly altered. The fresh water inflow is now negligible and there is a large absorption of solar energy at the surface. Thus during August the east end of the Basin, which is shallow and far removed from the large currents experienced at the Basin's mouth, can be about 4^{0} warmer than the deepest part of the Basin (17^{0} versus 13^{0}) and more than 7^{0} warmer than the Strait. The salinity in the Basin during this period is very uniform at approximately 31.5‰ and only marginally below that of the Strait.

The circulation structure in the Basin is quite complex, consisting of several counter-rotating gyres which migrate in the Basin. A series of subsurface drogues were employed to observe the circulation. A clockwise gyre was noticed in the central part of the Basin and a counter-clockwise gyre in the north-west region. Along the south-west shore the water continues flowing north and west even while the flooding tide is moving east. This circulation pattern is much more complex than that observed by Elliott (1969). Perhaps the single clockwise gyre he observed corresponds to that noted in the central part of the Basin.

The tides are being monitored at several positions in the Inlet (Fig 1). The government wharf, located on the north-west shore of the harbour is a reference tide-reporting site; thus long term records are available. In addition, tidal data are being recorded at two locations inside the Basin: one just south of the mouth and a second at the north shore. These records indicate that the tidal extrema inside the Basin are delayed by the order of an hour from those at Sooke Harbour.

Several Aanderaa recording current meters have been placed in the Basin to obtain long term records of the current, salinity and temperature. Two of these meters have been moored in the deepest part of the Basin near the mouth where the currents should be strongest (one at 10m depth, the other at 20m), and a third meter was placed east of Billings Spit, in the north-west corner of the Basin (Fig 1). These meters digitally record current speed and direction, temperature, conductivity, and pressure and can be left unattended for a 2 month period, at the end of which they are recovered to have their magnetic tapes and battery packs replaced.

The model being developed is based on the Leendertse (1967) model as modified by Willis (1977). It is 2-dimensional, based on the vertically integrated equations of motion and continuity. Thus it cannot model two-layer flow. However, during the summer, when fresh water inflow is negligible and the Basin is well-mixed, a 2-dimensional model should be adequate. Provisions are made for the effects of the earth's rotation, bottom friction, tidal forcing, and wind stress on the surface. The shoreline is approximated by a square grid with a spacing of about 210m and the dynamics are calculated with a 60 second time step. Calibration of the model is achieved by adjusting the bottom friction via a Chezy coefficient.

The response of the model to simulated tidal data and actual tidal data has been determined. The calculated flow patterns show some of the complex features expected. A comparison of the harmonic composition of the water speeds with that of the driving force has been made to determine the behaviour of the model.

THE MODEL

The model is based on the long wave approximation of the vertically integrated, shallow water equations of motion and continuity:

$$\frac{d\bar{u}}{dt} = -\frac{1}{\rho} \nabla p - g\bar{k} + \bar{F}$$

$$\text{div}\,(\rho\bar{u}) = 0$$

where $\bar{u}$ is the velocity vector, ρ is the density, g is the acceleration due to gravity, p is pressure, and $\bar{F}$ is the resultant external force composed of tide, wind, boundary friction, and Coriolis.

The bounding surface of the fluid is given by $\zeta(x,y,z,t) = 0$. The depth below the reference plain is h(x,y) while $\eta(x,y,t)$ gives the elevation of water above the plain. Note that only the free surface is time dependent. The natural boundary condition is

$$\frac{d\zeta}{dt} = 0$$

so that, at the free surface, the condition becomes the kinematic condition $\eta_t + u\eta_x + v\eta_y = w$

where u, v, and w are the velocity components in the x, y, and z directions, respectively. Since pressure is assumed to be hydrostatic and a linear function of depth, the model will be valid only for unstratified fluids.

The model also allows for viscosity terms from wind shear-stress and the effects of bottom roughness. The latter force is approximated in the equations by the Chezy coefficient, C. The derived equations can be presented as

$$\frac{\partial u}{\partial t} + u\frac{\partial u}{\partial x} + v\frac{\partial u}{\partial y} - fv + g\frac{\partial \eta}{\partial x} + g\frac{u(u^2+v^2)^{\frac{1}{2}}}{C^2\,(h+\eta)} = F^{(x)}$$

$$\frac{\partial v}{\partial t} + u\frac{\partial v}{\partial x} + v\frac{\partial v}{\partial y} + fu + g\frac{\partial \eta}{\partial y} + g\frac{v(u^2+v^2)^{\frac{1}{2}}}{C^2\ (h+\eta)} = F^{(y)}$$

$$\frac{\partial \eta}{\partial t} + \frac{\partial[(h+\eta)u]}{\partial x} + \frac{\partial[(h+\eta)v]}{\partial y} = 0$$

where f is the Coriolis parameter and $F^{(x)}$ and $F^{(y)}$ are the horizontal components of wind stress and barometric pressure.

The partial differential equations are approximated by a finite difference scheme over a space-staggered grid. The water elevations are calculated at the full grid steps (i,j); the u is calculated at the half horizontal and full vertical step $(i+\frac{1}{2},j)$;and the v is calculated at the full horizontal and half vertical step $(i,j+\frac{1}{2})$.

The scheme is semi-implicit, multi-operational, using a double time-step. On the first half time-step, η and u are solved explicitly and v implicitly. Then, on the full time-step, the calculation of the velocity components is done in reverse order. The resultant linear system is solved using a decomposition of the sparse, tri-diagonal finite-difference matrix into upper and lower triangular factors, and then performing a forward-backward substitution to solve the equations.

The closed boundary at the shorelines is assumed to be a vertical wall where the normal velocity component is zero and the depth is finite. At the forcing boundary the water levels are given as time varying water elevations. Near these boundaries, but within the computation field, the normal boundary condition cannot be applied and some terms are undefined in the differential equation. This problem is overcome by using a linear approximation.

Although the stability analysis is made difficult by the Coriolis and bottom-stress terms, the model has been shown to be stable under a number of reasonable conditions (Leendertse, 1967, Willis, 1977, Krauel and Birch, 1979).

The NRC Fortran programme (Willis, 1977) was modified to simplify I/O and to illustrate the logic flow, but the computations were not altered. The model was run on an IBM 3780 computer. The grid for Sooke Basin was developed using digitized bathymetic data and an automated interpolation for depths at the desired points. The grid boundaries were selected to reflect the region's geometry. The shoreline at the mouth was matched to a high water level to reflect the flooding during high tide when it was anticipated that most velocity features would be created. This semi-automated process enables the easy conversion to different grids. At present a 23 x 18 master grid with a grid size of 200m is being used. This gives 202 elevation points within the computation field, but is too coarse to recognize the islands within the Basin. A time-step of 1 minute was chosen as a reasonable compromise between numerical stability and computational speed.

There is some freedom in the selection of appropriate Chezy coefficients. Although the bottom of the Basin varies from silt-laden plain to a very steep, rocky hole, the long period waves may not see any small scale variations and a constant coefficient throughout the Basin may be valid. With little justification other than success by other users, a Chezy coefficient of 50 $m^{\frac{1}{2}}s^{-1}$ was chosen. The sensitivity of the model to the coefficients was measured by varying the value used by 20%. A simple sinusoidal impulse was used as the forcing tide at the mouth to the Basin. Although not completely analysed, the data show little effect due to such variation.

At selected time intervals the model outputs a record of water elevations and velocities averaged onto the elevation (η) grid. Velocities are plotted as a vector field.

The vector magnitude data at selected points has been transformed to a frequency spectrum plot using a waveform analysis package developed for the Hewlett-Packard 9825 calculator (Krauel et al, 1982).

SINUSOIDAL DRIVING FORCE

As a first approximation of the true forcing function at Sooke, the model was driven with the function

$$f(t) = 1.2 + 0.5584 \sin(0.0000729t) + 0.4415 \sin(0.0001405t)$$

This function corresponds to the two main components of the tide as identified by studies at the Institute of Ocean Sciences, Pat Bay, B.C.

The system was allowed to run for the equivalent of one tidal cycle to initiate the forcing into the field. This appears to be adequate as velocities appear to dissipate regularly with slack tides. The calculated currents display many of the large scale counter-rotating gyres observed in the Basin.

TIDAL DRIVING FORCE

Tidal records from the gauge at Sooke Harbour for the period from June 23 to July 5, 1981 were digitized at 5 minute intervals and used as a forcing function for the model. The velocities predicted by the model were plotted (Fig 2). These graphically illustrate the persistance of the counter-rotating gyres well into the ebb tide even during great tidal ranges. Also there is evidence that the outflow is mainly from the southern portion of the Basin.

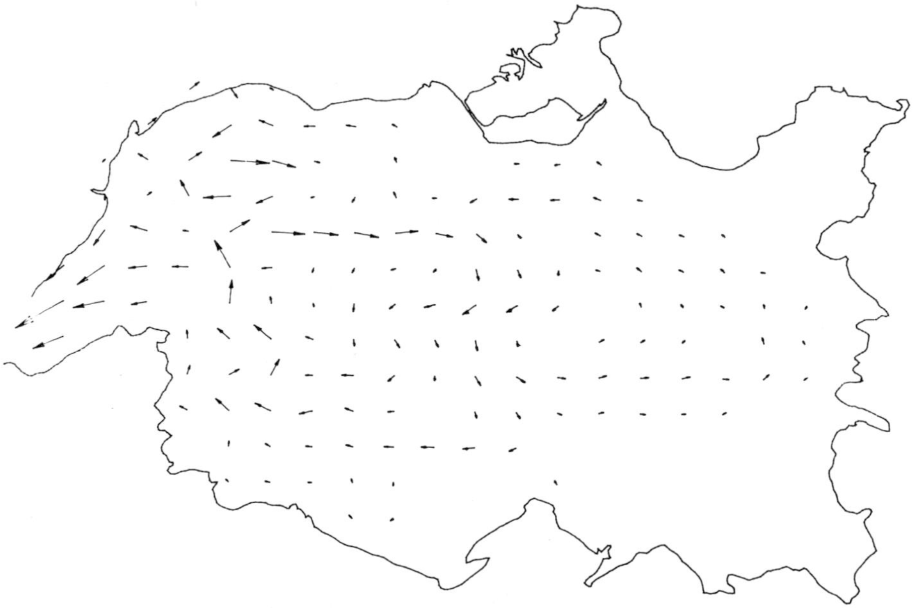

Fig. 2. Current velocities predicted by model for mid-ebb tide. The maximum speeds are in the order of 2 ms^{-1}.

A FFT of the current speeds at grid points near the location of the two current meters in the deep hole show the two main frequencies which appear in the tidal driving force and the spectra of the recorded current meter speeds (Fig 3). There is such similarity in the spectra at points across the mouth that it appears the tidal forcing occurs in a wide band within the Basin as portrayed in the vector plots.

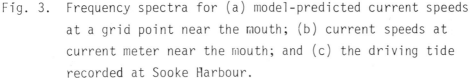

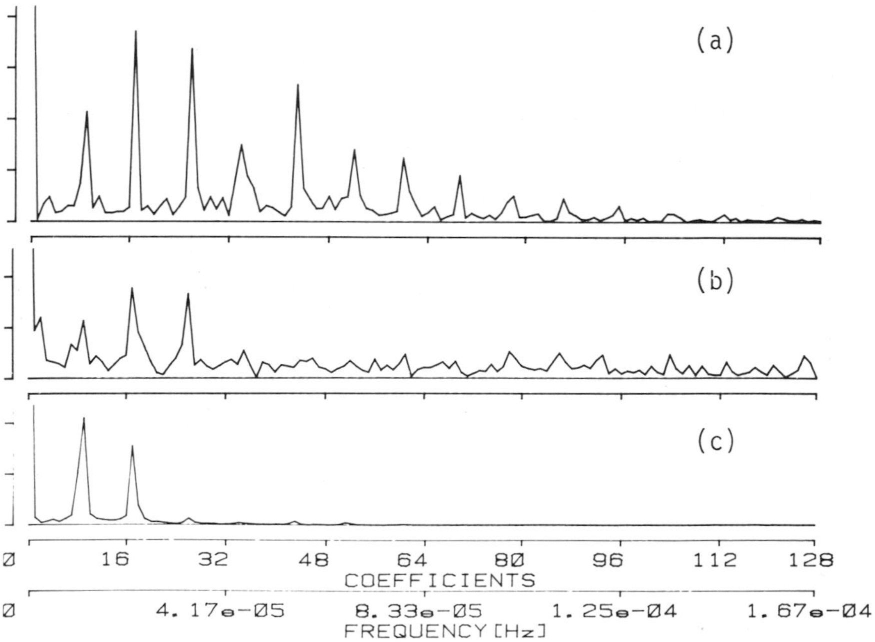

Fig. 3. Frequency spectra for (a) model-predicted current speeds at a grid point near the mouth; (b) current speeds at current meter near the mouth; and (c) the driving tide recorded at Sooke Harbour.

CONCLUSION

The model appears to be a valid predictor of currents within Sooke Basin during periods of low fresh water input and non-stratification. The system is characterized by non-stationary counter-rotating gyres which are born on the flood tide and persist into the ebb tide. Spectral analysis appears to be an effective method of verifying the model and for predicting harmonic components in the circulation. The model also predicts areas where critical current behaviour should be monitored. To further calibrate the model, data sampling will have to continue, and wind forcing and fresh water inflow data will be included.

REFERENCES

Krauel, D.P., and Birch, J.R., 1979. Wind and Fresh Water Inflow Effects on the Circulation of the Miramichi Estuary, N.B.-A Numerical Model. Coastal Marine Science Laboratory, Manuscript Report 79-2, Royal Roads Military College, FMO Victoria, B.C.

Krauel, D.P., Milinzaao, F., Press, M., and Wolfe, W.W., 1982. A Model of Sooke Harbour and Basin. Coastal Marine Science Laboratory Note, in press, Royal Roads Military College, FMO Victoria, B.C.

Leendertse, J.J., 1967. Aspects of a computational model for long-period water-wave propagation. RM-5294-PR, the Rand Corporation, Santa Monica, Ca., 165 pp.

Willis, D.H., 1977. Miramichi Channel study hydraulic investigation. Hydraulics Laboratory Technical Report LTR-HY-56, Vol. I and II, Division of Mechanical Engineering, National Research Council of Canada, Ottawa.

RAINFALL-FLOW RELATIONSHIP IN SOME ITALIAN RIVERS BY MULTIPLE STOCHASTIC MODELS *

ELPIDIO CARONI
Researcher, C.N.R./I.R.P.I., Torino

FRANCESCO MANNOCCHI
Researcher, University of Perugia, Inst. Agricultural Hydraulics

LUCIO UBERTINI
Director of C.N.R./I.R.P.I., Perugia
Professor, University of Perugia, Inst. Agricultural Hydraulics

ABSTRACT

The hourly rainfall-flow relationship was studied by multiple transfer plus noise methodologies. The formula (1) shows the general form of the models. This formula, where one or two inputs can be eliminated, was utilized to build operative models for flow simulation and flow real time forecast. In this work we present the models and the results for some events of two Italian basins (Sieve, 831 km^2; Toce, 1535 km^2) and of two experimental basins (Marchiazza, 5 km^2; Fosso degli Impiccati, 7.6 km^2). We present, besides, properties, limits and possible future developments of multiple transfer plus noise methodologies in this field.

1. INTRODUCTION

During the last few years, an effort was made in order to asses the possibilities of using stochastic models in flood simulation and real time forecast. These models are of particular interest for their capability to estimate the rainfall-flow relationships from an observed sample by means of statistical

* C.N.R. Special Project for Soil Conservation, Sub-Project Fluvial Dynamics, paper No. 155.

Reprinted from *Time Series Methods in Hydrosciences*, by A.H. El-Shaarawi and S.R. Esterby (Editors)

estimators. This fact permits the reduction of the influence of personal judgements about the physical phenomenon over the model characteristics.

In this context a criterion for parameter parsimony must be considered in order to obtain a practical operative tool.

The use of multiple linear transfer function models (Anselmo et alii, 1981) allows us to take into account in some way the nonlinearity of the phenomenon. Particular researches were carried on in this sense; in fact the series, here considered, are composed of hourly (or more frequent) data, which usually show a more evident nonlinear behaviour than other hydrologic samples with longer gaging intervals (daily, weekly and so on).

In our paper we discuss the aptitude of several input variables in a multiple transfer function plus noise model to explain the rainfall-runoff phenomenon, particularly within the peak hydrograph.

2. THE MODEL

Following the notation used by Box and Jenkins (1970) and widely adopted for stochastic modeling (Hipel et alii, 1977), the general form of the model is:

$$Y_t = v_1(B)\, X_{t-b_1} + v_2(B)\, Z_{t-b_2} + v_3(B)\, \xi^{(T)}_{t-b_3} + N_t , \qquad (1)$$

where $v_i(B)$ (for i=1,2,3) is

$$v_i(B) = (\omega_0 - \omega_1 B - \omega_2 B^2 - \dots)\,(1 - \delta_1 B - \delta_2 B^2 - \dots)^{-1}$$

and B is the backward shift operator such that $BX_t = X_{t-1}$. The ω's and δ's are called respectively input and output parameters; Y_t is the discharge at time t; X_{t-b_1} is the total rain inflow during the time interval between $t-b_1-1$ and $t-b_1$ (where b_1 is a delay parameter); Z_{t-b_2} is the cumulated rainfall effect; $\xi^{(T)}_{t-b_3}$ is the intervention variable; N_t is the noise term.

A recent work by Anselmo and Ubertini (1979) led to a simple linear model with only one input:

$$Y_t = v_1(B)\, X_{t-b_1} + N_t . \qquad (2)$$

With this model the nonlinear effects in the rainfall-flow relationships are not accounted for. In fact we found that:

- the highest values of the residuals clash with the peaks, where the effects of nonlinearity are more relevant;

- in real time forecast, underestimates of future discharges occur whenever the time origin for forecasts lies near the beginning of the hydrograph rising limb.

These underestimates diminish as long as the starting point for forecasts approaches the flood peak.

In order to improve the results of this model, two other input variables were explored. We sought a second input variable Z_t, a function of rainfall X_t, which was named "cumulated rainfall effect"; afterwards we introduced an intervention variable acting at time t=T when rainfall, X_T, or "cumulated rainfall", Z_T, exceeds a given safety threshold E, according to the formula:

$$\xi_t^{(T)} = \begin{cases} 1 & \text{if } X_T > E \text{ or } Z_T > E \\ 0 & \text{otherwise} \end{cases} \qquad (3)$$

2.1 Cumulated rainfall effect

Discharge from a river may be considered dependent on several factors besides rainfall. Among these, an important role is assumed by the global amount of rainfall fallen in the basin during a previous time period. This amount may be quantified by means of a weighted sum of antecedent precipitations in the following way:

$$Z_t = \sum_{k=0}^{\infty} w_k X_{t-k} . \qquad (4)$$

Since a decreasing importance of previous rainfalls on Z_t may be assumed going back in time, a suitable system of weights may be given by

$$w_k = (1-c)\, c^k, \text{ with } 0<c<1, \text{ for which results } \sum_{k=0}^{\infty} w_k = 1 .$$

Further, this relation may be expressed in terms of linear operators as:

$$Z_t = \sum_{k=0}^{\infty} (1-c)c^k X_{t-k} = (1-c) \sum_{k=0}^{\infty} c^k B^k X_t = (1-c)(1-cB)^{-1} X_t . \qquad (5)$$

The summation (4) can be stopped at the M-th step, choosing M so that $(1-c)c^k$ may be considered zero for k greater than M.

For the events of the Sieve river, reported in Mannocchi et alii (1981) and in Piccolo and Ubertini (1979), c was found to be equal to 0.8 and therefore M was set equal to 15. The value of c was chosen by means of a trial-and-error procedure in order to minimize the sum of squares of differences between the rainfall cumulated effect for different values of c (c=0.1,0.2,...,0.9) and the observed discharges.

2.2 Intervention modeling

The intervention variable $\xi_t^{(T)}$ defined by (3) allows a better fitting for some particular behaviours of the system (the drainage basin in our case) by means of the variations produced on the output level by the parameters of its transfer function.

Some of the effects which can be reproduced are: delayed shocks, new constant levels, temporary level variations, linear and combined linear effects, exponential decreases.

The nature of rainfall-runoff phenomenon suggested an intervention model which gives to the system a supplementary step effect with exponential decrease, or a delayed impulse at the beginning of the hydrograph rising limb.

2.3 Study cases

Some features of the models fitted, up to now, to some events in four Italian rivers are summarized in table 1; figure 1

TABLE 1.

Types of models fitted to some events in four Italian rivers.

RIVER (area)	TIME LAG (min)	NUMBER OF MODEL PARAMETERS								TOTAL	REFERENCES
		TRANSFER FUNCTIONS						NOISE			
		x_t		z_t		$\xi_t^{(T)}$		N_t			
		$\omega(B)$	$\delta(B)$	$\omega(B)$	$\delta(B)$	$\omega(B)$	$\delta(B)$	$\phi(B)$	$\theta(B)$		
F.IMPICCATI (KM^2 7.6)	30	3	1	-	-	-	-	2	-	6	ANSELMO-UBERTINI (1978)
T.MARCHIAZZA (KM^2 5.3)	30	3	1	-	-	-	-	2	-	6	(ID.)
F.SIEVE (KM^2 831)	60	3	2	-	-	-	-	2	2	9	(ID.)
	60	-	-	2	-	-	-	1	2	5	PICCOLO-UBERTINI
	60	2	-	-	-	2	1	1	1	7	(1979)
	60	2	1	-	-	-	-	2	1	6	MANNOCCHI ET AL.
	60	2	-	2	-	-	-	1	2	7	(1981)
	60	2	1	-	-	1	-	1	2	7	(ID.)
F.TOCE (KM^2 1535)	60	2	1	-	-	-	-	1	2	6	ANSELMO ET AL.
	60	-	-	2	-	-	-	1	2	5	(1981)
	60	2	1	-	-	-	-	1	2	6	(ID.)
	60	-	-	2	1	-	-	1	2	6	(ID.)
	60	2	2	-	-	-	-	1	2	7	(ID.)
	60	-	-	2	1	-	-	1	2	6	(ID.)

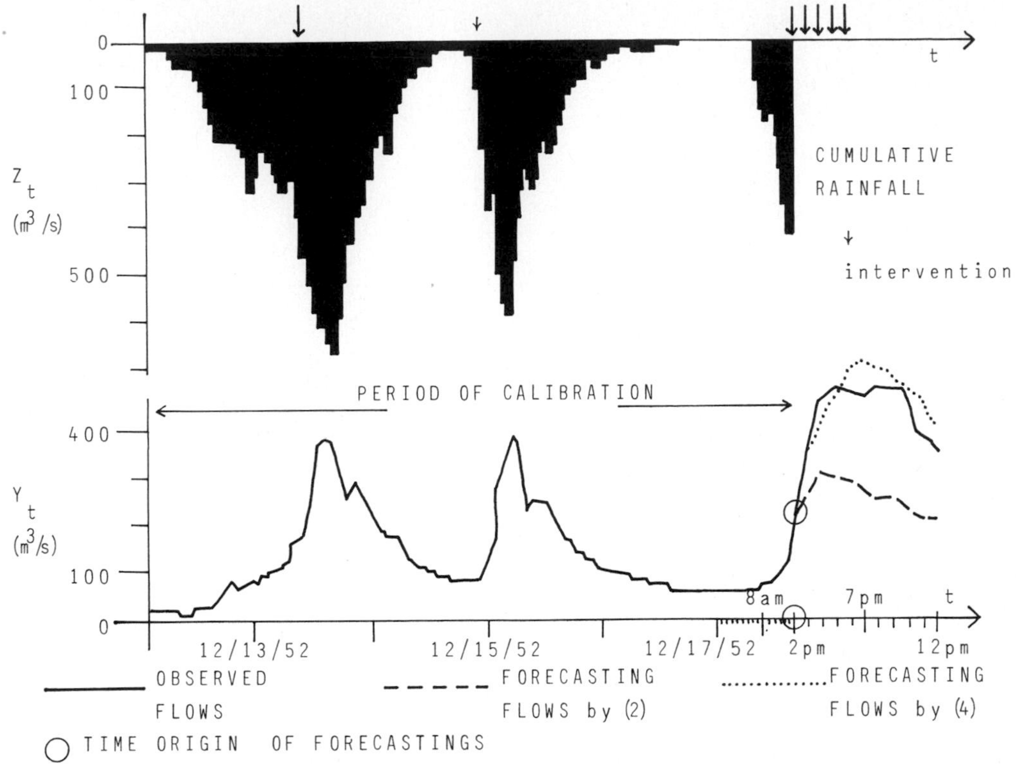

FIG. 1. Example of forecast using the intervention variable $\xi_t^{(T)}$ (from: Piccolo and Ubertini, 1979).

shows an example of real time forecast. It may be pointed out that these types of models gave good results both for medium sized basins such as Sieve and Toce rivers and for small experimental catchments such as Marchiazza and Fosso degli Impiccati.

3. DISCUSSION AND CONCLUSION

A comparison between different models for the same event may generally be made by means of some usual "goodness of fit" statistics: S, sum of residual squares, Q, Box-Pierce and Ljung-Box test of residuals, R^2, explained variance by model. Yet, in our case, owing to the observed pattern in the residuals, particular attention ought to be paid to (besides randomness), the stationarity of the residuals process, that is to controlling its evolution with time. As an example of this comparison, the event of December 12th-18th, 1952 from the Sieve river was simulated using five different models whose characteristics are summarized in tab. 2.

TABLE 2.

Different models fitted to the Sieve event of Dec. 12th-18th, 1952.

MODEL		1	2	3	4	5
X_t	ω_0	(.004) .013	--	(.009) -.051	--	(.004) .029
	ω_1	(.005) -.025	--	(.006) .017	--	(.004) -.049
	δ_1	(.015) .924	--	--	--	(.021) .853
Z_t	ω_0	--	(.020) .170	(.025) .074	(.020) .173	--
	ω_1	--	(.020) -.306	(.046) -.375	(.020) -.030	--
	δ_1	--	--	--	--	--
$\xi_t^{(T)}$	ω_0	--	--	--	(6.500) 46.543	--
	ω_1	--	--	--	(6.340) -30.317	(1.350) 5.238
	δ_1	--	--	--	(.001) .980	--
N_t	θ_1	(.005) -.495	(.060) -.670	(.047) -.606	(.060) -.631	(.049) -.651
	θ_3	--	--	(.047) .245	--	--
	θ_5	--	(.050) .224	--	--	(.049) .095
	ϕ_1	(.040) 1.004	(.030) .910	(.023) .916	(.060) .682	(.027) .902
	ϕ_3	(.039) -.145	--	--	--	--
Q		17.5	11.8	19.8	16.1	16.1
R^2		.984	.984	.985	.986	.984
σ_a^2		214.5	224.1	207.1	192.6	224.6

THE VALUES IN BRACKETS ARE THE CONFIDENCE LIMITS.

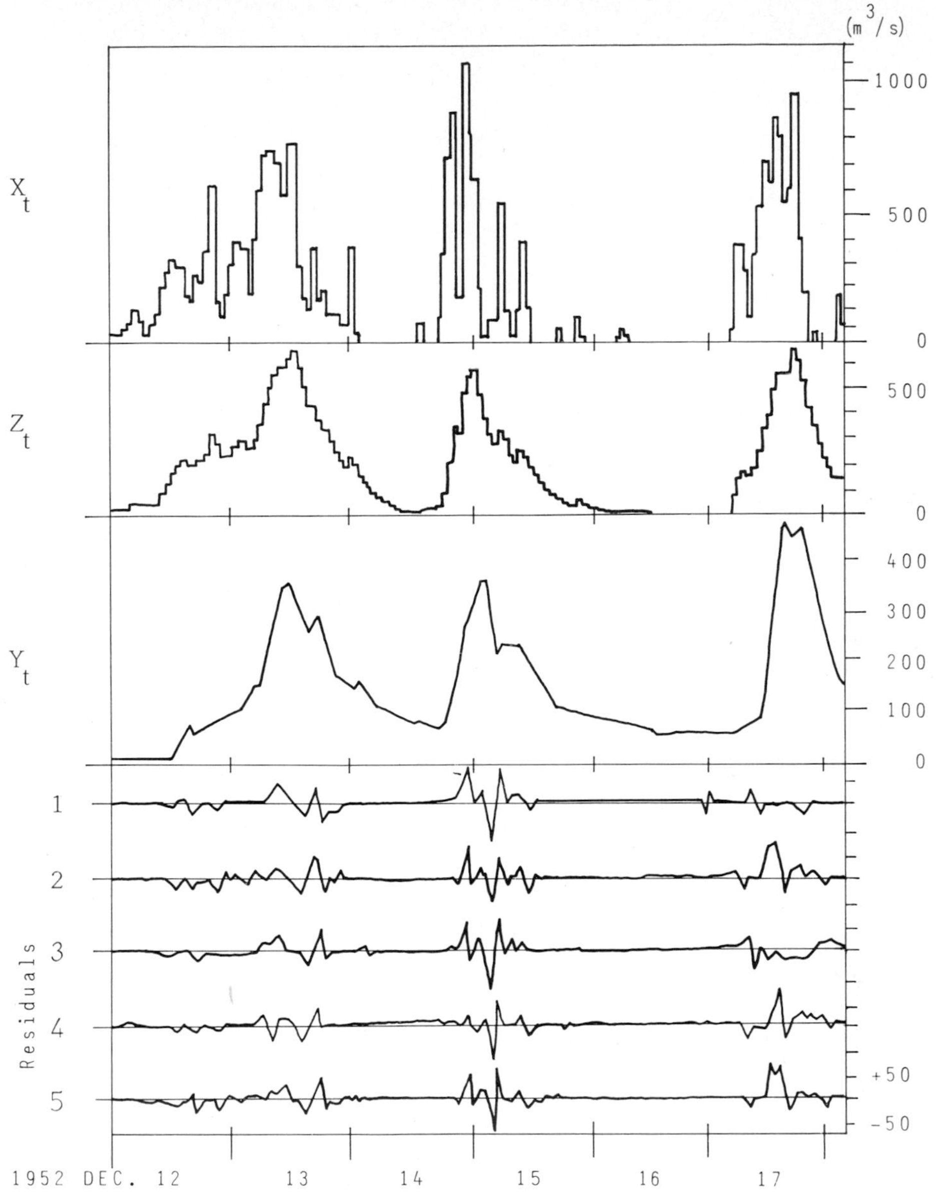

FIG. 2. Event of Sieve river, Dec. 12th-18th, 1952. Observed input (X_t and Z_t) and output (Y_t) series. Plot of the residuals from the five different models (1), (2), (3), (4), (5).

In fig. 2, both input series X_t , Z_t and observed output Y_t are plotted against time, together with the residuals from each model.

Numerical comparisons between the five models are shown in tab. 3.

For each of the three peak hydrographs constituting the event, the maximum absolute residual and the variance - for the rising and the first part of the descending limb separately - were considered, as well as the residual at peak and the total variance during the peak hydrograph.

The separation of the hydrograph into recession phase ("tail") and "peak" phase, and, within the latter, between rising and descending limb, was made to test where model improvements are more needed and what type of supplementary information would be more convenient.

TABLE 3.

Ratios of model performances referred to model (1) *.

		FIRST PEAK			SECOND PEAK			THIRD PEAK		
		R	D	P	R	D	P	R	D	P
2:1	RES.	.876	1.187	.778	.908	.628	.642	1.987	.956	2.386
	VAR.	1.023	1.327	1.179	.781	.570	.636	4.156	.586	3.178
3:1	RES.	.871	1.064	2.556	.864	.901	.498	1.177	.943	1.405
	VAR.	.902	1.025	.966	.921	.926	.925	1.431	1.623	1.483
4:1	RES.	1.124	1.003	1.389	.279	.803	.490	2.127	.933	1.171
	VAR.	1.242	1.241	1.241	.175	.644	.496	3.266	1.002	2.646
5:1	RES.	.747	1.078	.006	.733	.842	.588	2.033	.596	1.646
	VAR.	.869	1.272	1.076	.504	.780	.693	4.320	.374	3.239

*(RES.) MAXIMUM ABSOLUTE RESIDUAL AND (VAR.) VARIANCE OF RESIDUAL DURING THE RISING (R) AND THE DESCENDING (D) LIMB. RESIDUAL AT PEAK AND VARIANCE DURING THE WHOLE PEAK HYDROGRAPH (P).

The pattern of residuals shown in fig. 2 suggests that there is no real difference between the models. This is substantially confirmed by the results on table 3; although a better performance of some models in the single peaks may be seen (perhaps models (5) and (3) in the first, model (4) in the second and model (1) in the third peak) no such distinction could be made looking at the whole event. So the smoothing of errors at peak was not achieved; in table 4 the residual standard errors for "peaks" and "tails" are given: these values are almost constant from one model to another.

Formally the difference between models (1) and (2) consists in the fact that the output parameter in model (2) is masked in

TABLE 4.
Standard errors of the residuals (m^3/s).

S.E. MODEL	1	2	3	4	5
TOTAL	14.65	14.97	14.39	13.88	14.99
AT PEAKS	20.99	21.61	21.00	20.10	21.83
AT TAILS	5.92	5.59	4.64	4.98	4.94

Z_t itself: it becomes evident writing Z_t in terms of linear operators as shown by (5). So a constraint ($\delta_1 = c$) has been introduced into the model which, for this reason, diminishes its flexibility. However, during the parameter estimation stage, a less scattered series, like Z_t, provides better results.

The use of the intervention variable $\xi_t^{(T)}$ led to no appreciable improvements in the simulation step, if compared to the more simple and parsimonious model (1) (with rainfalls only in input). Nevertheless a fair improvement in forecasting could be observed (see fig. 1).

The Q test values in tab. 2 are well below the critical threshold $\chi^2 = 28.4$, with 20 degrees of freedom at 90% probability level, so that the hypothesis of randomness of the residuals would not be refused. However the residual series cannot be thought of as stationary in fact the variance of residuals proved to be appreciably different by subdividing the sample in "peak" and "tail" periods (tab. 4).

At present two main developments may be foreseen in future research. By first new auxiliary input variables might be explored in order to reduce errors during the peaks (high-water phase), such as an antecedent precipitation index or a variable which should be capable of "perceiving" the rising of discharges; in this last sense a use of "perlog" function (Appleby, 1965; O'Connell, 1980):

$$\eta_t = \frac{1}{Y_t} \frac{dY_t}{dt}$$

can be interesting.

Secondarily it would be practical to introduce some hydrological insight by dividing the model in two parts (or more as in Hino and Hasebe, 1980), base-flow and runoff for instance, as the difference in residual variances may suggest two different behaviours in the rainfall-flow relationship. These improvements are related mainly to the structure of the model. Yet, a very important problem is to "conceptualize" the model (Klemes, 1981), that is to find relationships between the parameters of the model and the physical characteristics of the drainage basin. Conceptualization would allow a wider use of these models by extrapolation and similitude. By now, this stage has been scantily tackled since it requires a good deal of insight both in the physics of the phenomenon and in the structure of the model: this would constitute a long-term field for future research.

REFERENCES

Amorocho, J., 1973. Nonlinear hydrologic analysis. Advances in Hydroscience, 9: 203-251.

Anselmo, V., Ubertini, L., 1979. Transfer function-noise model applied to flow forecasting. Hydr. Sc. Bull., 24: 353-359.

Anselmo, V., Melone, F., Ubertini, L., 1981. Space-time distribution of rainfall-runoff relations by multiple transfer plus noise model. Int. Symp. on Rainfall-runoff Modeling, Mississippi State University, USA. (In print).

Appleby, F.V., 1965. Unpublished D.I.C. Dissertation, Imperial College, London University.

Bidwell, V.J., 1971. Regression analysis of nonlinear catchment system. Water Res. Research, 7(5): 1118-1126.

Box, G.E.P., Jenkins, G.M., 1970. Time series analysis forecasting and control. Holden Day, S. Francisco, 553 pp.

Box, G.E.P., Tiao, G.C., 1975. Intervention analysis with applications to economic and environmental problems. Journ. Americ. Statist. Assoc., 70: 70-79.

Gallati, M., Greco, F., Maione, U., Martelli, S., Natale, L., Panattoni, L., Todini, E., 1977. Modello matematico delle piene dell'Arno. I.B.M. Italia, Centro Scientifico di Pisa, 1: 210pp, 2: 496pp, 3: 79pp.

Hino, M., Hasebe, M., 1980. Further test of applicability of the inverse detection method and extension to hourly hydrologic data. Proc. of the 3rd Int. Symp. on Stochastic Hydraulics of IAHR, Tokyo, 129-140.

Hipel, K.W., McLeod, A.I., Lennox, W.C., 1977. Advances in Box-Jenkins modeling: (1) model construction; (2) applications. Water Resour. Research, 13: 567-586.

Klemes, V., 1981. Stochastic models of rainfall-runoff relationship. Int. Symp. on Rainfall-runoff Modeling, Mississippi State University, USA. (In print).

Mannocchi, F., Melone, F., Ubertini, L., 1981. Rainfall-flow processes by the multiple transfer-noise models. XIX Int. IAHR Congress, New Delhi, 4: 29-36.

O'Connell, P.E. (Editor), 1980. Real time hydrological forecasting and control. Inst. of Hydrology, Wallingford, U.K., 264 pp.

Piccolo, D., Ubertini, L., 1979. Flood forecasting by intervention transfer stochastic models. Proc. XVIII IAHR Congress, Cagliari, Italy, 5: 319-326.

Sabatini, P., Ubertini, L., 1978. Ricostruzione e previsione dei deflussi in un piccolo bacino. XVI Conv. Idraulica e Costr. Idrauliche, Torino, B27: 1-12.

ANALYSIS OF WATER TEMPERATURE RECORDS USING A DETERMINISTIC - STOCHASTIC MODEL[1]

BRUCE J. NEILSON AND BERNARD B. HSIEH[2]

Virginia Institute of Marine Science and School of Marine Science, The College of William and Mary, Gloucester Point, VA 23062

INTRODUCTION

Since 1954 daily observations of the water temperature in the York River estuary have been made at Gloucester Point, Virginia by the staff of the Virginia Institute of Marine Science of the College of William and Mary. These observations were made at the end of a pier which extends 115 m. from the shoreline and is located approximately 200 m. downriver of the Gloucester Point-Yorktown constriction. (See Figure 1) Water depth at this station is 4.2 m. at mean low water and measurements were made at mid-depth.

The time series record varies over the period of observations. Initially daily temperature extremes were monitored using a mercury maximum and minimum thermometer. In 1972 a continuously recording conductivity-salinity-temperature probe was installed. In order to have the maximum record length possible and to have compatible data sets, only the maximum and minimum temperatures from this latter period were used. Thus, the primary data were daily extreme temperatures for the years 1954 to 1977 inclusive. Daily mean temperatures were calculated by averaging the daily extremes. For short data gaps (say a few days) data from a back-up instrument were used. For more lengthy periods with missing values, the data gaps were left in the record. Data gaps were greatest for the years 1964, 1968 and 1972.

---

[1]Contribution number 1038 from the Virginia Institute of Marine Science

[2]Present Address: Clemson University, Clemson, S.C. 29632

Reprinted from *Time Series Methods in Hydrosciences*, by A.H. El-Shaarawi and S.R. Esterby (Editors)

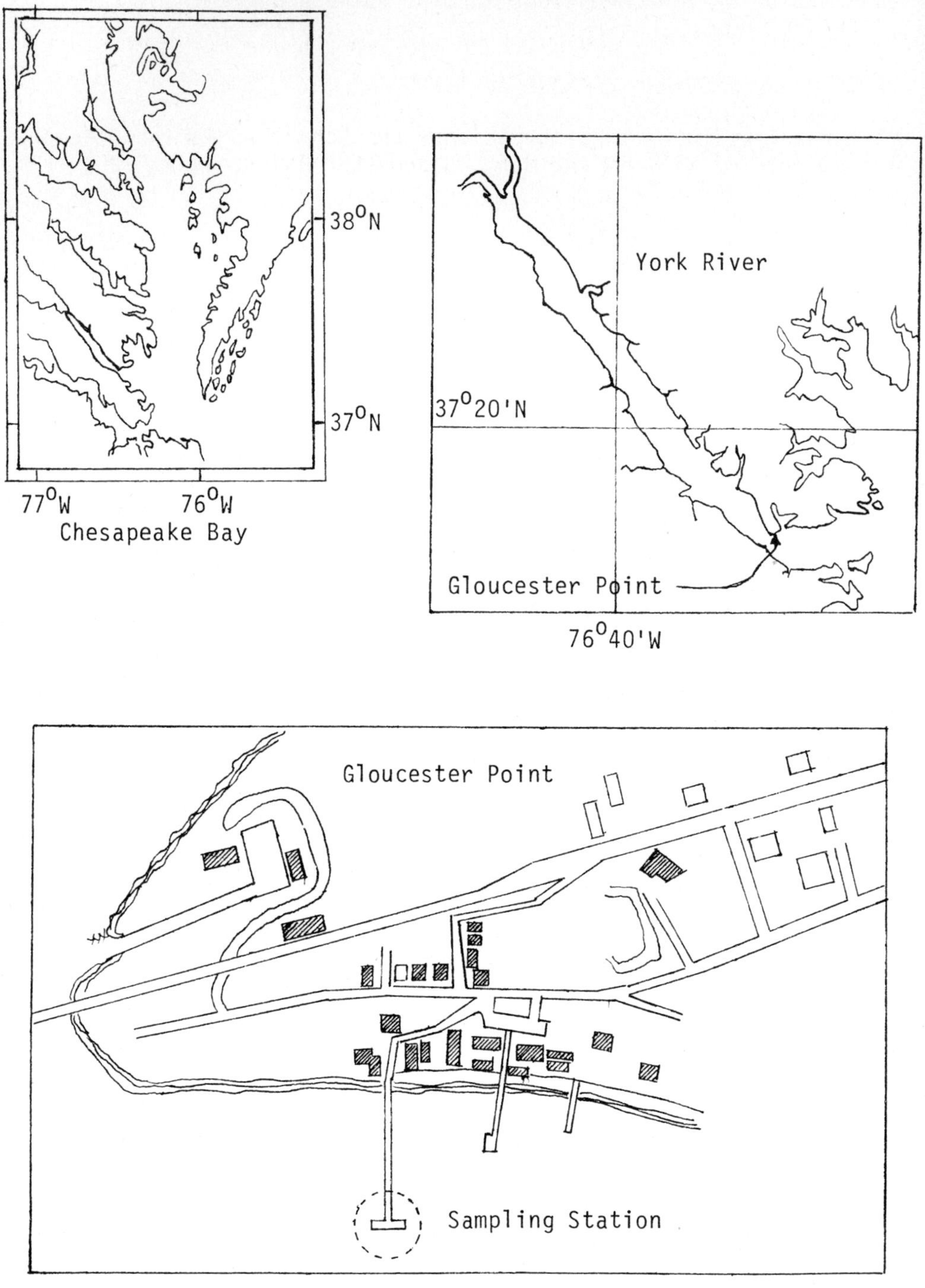

FIGURE 1. Chesapeake Bay, the York River and the location of the sampling station.

For specific days of the year, the daily temperatures for the 24 years generally were normally distributed. A few years were warmer (1959, 1975, 1977) or colder (1968, 1976) than normal and produced a large portion of the extreme values for the individual days of the year. The authors are aware of no anomalies in the data or of any systematic errors in the temperature measurements. Therefore, lacking evidence to the contrary, we assumed that the data set was suitable for analysis both in calendar year segments and as a single, continuous 24 year record.

PRELIMINARY ANALYSIS

Prior studies of a single year or a few years of water temperature measurements have shown that the annual cycle is an extremely important component of water temperature variations in rivers (Kothandaraman, 1971) and estuaries (Thomann, 1967) and that a harmonic with a period of one year accounts for roughly 95% of the total variance of the record. For that reason, the first step in our study was to examine the time series using harmonic analysis. When the data were examined on a calendar year basis, we found that the mean temperatures and the amplitudes and phase angles of the annual cycle were relatively constant from year to year, as shown in Table 1. For higher order harmonics, the amplitude was an order of magnitude smaller and both the amplitude and the phase angle varied widely.

The first harmonic or annual component accounted for between 95% and 98.6% of the variance, as can be seen in Table 2. No higher order harmonic consistently accounted for a significant portion of the residual variance. When the 24-year record was analyzed as a single time series, the annual cycle was found to account for 95.8% of the variance. The only harmonics which contributed more than 0.1% of the variance were the first (24 yr. period; 0.2%), second (12 yr. period; 0.15%), tenth (29 month period; 0.1%), twenty-first (14 month period; 0.13%), and forty-eighth (6 month period; 0.2%).

Because the periods of the harmonics are determined by the record length, the power spectrum also was examined to see if any components had been hidden in the harmonic analysis. The results of these

TABLE 1. Mean value, range of values, standard deviation and coefficient of variation for the record mean and the amplitude and phase angle of the first five harmonics determined by examination of the 24-year water temperature record on a calendar year basis.

	Water Temperature (°C)				
	Mean	Range Max	Range Min	Standard Deviation	Coefficient of Variation
Year Average	15.57	16.84	14.66	0.647	0.042
1st Harmonic					
Amplitude	11.61	12.98	10.01	0.702	0.060
Phase Angle	4.91	4.12	4.36	0.057	0.014
2nd Harmonic					
Amplitude	0.794	1.82	0.14	0.423	0.53
Phase Angle	3.90	6.25	0.04	1.37	0.35
3rd Harmonic					
Amplitude	0.53	1.27	0.09	0.315	0.59
Phase Angle	2.68	6.19	0.11	2.017	0.75
4th Harmonic					
Amplitude	0.486	0.96	0.08	0.234	0.48
Phase Angle	3.19	6.22	0.03	1.875	0.59
5th Harmonic					
Amplitude	0.44	0.99	0.06	0.221	0.50
Phase Angle	3.14	6.09	0.37	1.746	0.56

efforts, summarized in Table 3, suggest that sunspots and other solar phenomena might be affecting the water temperature record slightly, but the data are hardly conclusive (Hsieh, 1979). There is much stronger evidence to document the lack of any significant peak in the variance spectrum corresponding to the lunar cycle (29.5 day period) or the semi-lunar cycle. This is somewhat surprising since other studies have shown that stratification, and the lack of it, in the York River estuary is tied to the spring tide to neap tide lunar cycle (Haas *et al*, 1981). It had been expected that this stratification-mixing cycle would affect the water temperatures as well.

In short, the dominant periodic feature of the time series is the seasonal variation in water temperatures. The residual signal apparently is stochastic in nature since neither harmonic analysis nor spectrum analysis was able to show any consistent behavior for other

TABLE 2. Variance (in degrees centigrade squared) and percent of variance attributed to the first harmonic, second to fifth harmonics, and sixth and higher order harmonics determined by analysis of the 24-year water temperature record on a calendar year basis.

Variance ($^oC^2$)	Average	Maximum	Minimum
Total Variance	69.77	88.64	51.70
Variance contributed by:			
1st Harmonic	67.72	84.32	50.10
2nd-5th Harmonics	0.86	2.33	0.21
6th and Higher Harmonics	1.19	1.99	0.63
% of Total Variance (%)			
1st Harmonic	97.06	98.58	95.13
2nd-5th Harmonics	1.22	3.40	0.30
6th and Higher Harmonics	1.72	2.68	0.87

components or to relate these to physical processes previously believed to be important.

DETERMINISTIC - STOCHASTIC MODEL

The preliminary analysis showed that the seasonal variation of water temperatures could be approximated by a sinusoidal function, but that this simple deterministic model did not account for a portion (about 5%) of the variance in the record. On the other hand a purely stochastic approach does not seem appropriate for a time series which consistently shows a strong annual cycle. McMichael and Hunter (1972) found that a combination of deterministic and stochastic models "provided succinct and useful forecast functions for the temperature and flow changes in the Ohio River". Their approach, which was based on the work of Box and Jenkins (1976), was used to further study the York River water temperature record.

The deterministic portion of the model was taken to be the record mean (15.6^{o}C) and the annual harmonic having an amplitude of 11.6^{o}C and a phase lag of 240°. ARIMA (AutoRegressive, Integrated Moving Average) models were tested to see if they fit the residual series which was obtained by subtracting the deterministic model values from the actual time series record. The autocorrelation function (ACF) for the residual series exhibited an exponential decay for the first 80

TABLE 3. The portion of the total variance contributed by the cyclical, seasonal and irregular components of the 24-year water temperature record (after Hsieh, 1979).

Component*	Period	Intensity (% of Total Variance)
Cyclic	∿24 Year	0.44
	26 Months	0.21
	13-14 Months	0.30
	6 Months	0.24
	∿2 Months	0.06
Seasonal	12 Months	95.84
Irregular	-	2.81

*Only those harmonics which were significant at 95% probability limits have been included.

values followed by a sine wave-like oscillation at low values. The ACF for the series modified by the first difference operator had no values greater than 0.05 function units after the first four lags. The second difference series had an initial value of about 0.5 with all subsequent values less than 0.04. The characteristics of the four models identified as being appropriate for the given ACF patterns (Box and Jenkins, 1976) are summarized in Table 4.

All four models were found to simulate the data reasonably well and to incorporate essentially the same portion of the variance. All four residual series, which resulted when both the deterministic and stochastic model values were subtracted from the record, approximated white noise. About one-tenth of the ACF values were outside the 95% confidence limits. Q values for the four models also were somewhat greater than the 90% limits (Hsieh, 1979). Considering the principle of maximum simplicity, or Occam's Razor, the best choice is the (1,0,0) model. This finding is similar to that of Mehta _et al_ (1975) who found that the "combination of first-harmonic elimination and first-order autoregressive model produces a 99% reduction in the variance" of temperature data for the Passaic River. The ϕ_1 value for this study (0.919) was higher than they found appropriate for the Passaic

TABLE 4. Characteristics of the ARIMA models used to simulate the stochastic portion of the time series record. (After Hsieh, 1979)

ARIMA Type	Parameter Values	Model	Percent of Total Variance Included in Model
(1,0,0)	$\phi_1 = 0.919$	$(1 - \phi_1 B)\tilde{Z}_t = a_t$	99.42
(2,0,0)	$\phi_1 = 0.910$	$(1 - \phi_1 B - \phi_2 B^2)\tilde{Z}_t = a_t$	99.42
	$\phi_2 = 0.00919$		
(1,0,1)	$\phi_1 = 0.92$	$(1 - \phi_1 B)\tilde{Z}_t = (1 + \theta_1 B) a_t$	99.42
	$\theta_1 = -0.008$		
(0,2,1)	$\theta_1 = 0.99$	$\nabla^2 \tilde{Z}_t = (1 - \theta_1) a_t$	99.33

River (0.818 to 0.861), but was close to that for the Ohio River (0.915) (McMichael and Hunter, 1972).

The (1,0,0) model implies that on any given day the deviation from the normal annual cycle is a weighted function of all previous deviations from that cycle plus any new "shock" received that day. Furthermore, the weighting functions for this particular case decrease as the time interval increases with a factor of 0.919.

Although the combined deterministic-stochastic model can produce very realistic synthetic records (McMichael and Hunter, 1972), it is less well suited for predicting actual temperatures. Since future shocks or random factors can be simulated but not forecast, the model predicts decreasing deviations from the seasonal cycle as the length of the prediction increases. For the York River the predictions approach the harmonic curve after about 15 days lead time. Within that time frame, projections are rather good. Fifteen day forecasts made using actual temperature records and starting with arbitrarily selected starting dates in winter, spring and summer are shown in Figure 2. The actual observations usually fell within the 50% probability limits.

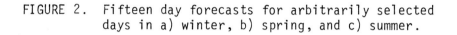

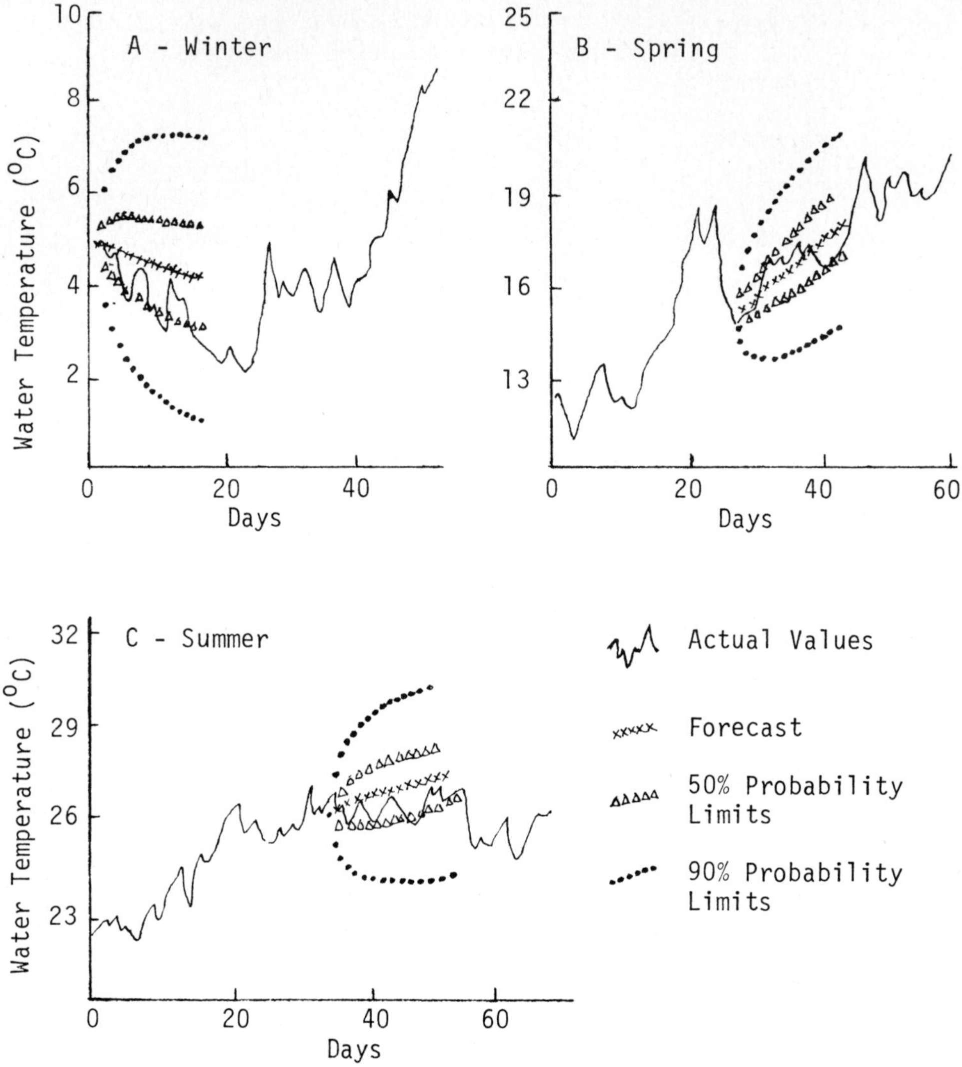

FIGURE 2. Fifteen day forecasts for arbitrarily selected days in a) winter, b) spring, and c) summer.

CONCLUSIONS

Box-Jenkins models are useful tools for the analysis and interpretation of water temperature time series for estuaries, and for creating synthetic temperature series. For the York River estuary the earth's rotation about the sun results in a strong seasonal variation which can be approximated by a single sine wave. This behavior was shown to be consistent over the 24 years of record. Therefore the best prediction of water temperatures far into the future (say months or years) is that obtained using the record mean and the annual cycle.

A number of factors introduce a stochastic aspect to the record. For the York River estuary four ARIMA models provided a good fit to the data. The simplest of these, the first order autoregressive process, was selected as the best. This model is capable of projecting future water temperatures up to about 15 days in advance. For longer periods the deviation decreases and the prediction approaches the annual harmonic.

REFERENCES

Box, George E. P. and Gwilym M. Jenkins, Time Series Analysis: forecasting and control, Revised Edition, 575 pp., Holden-Day, San Francisco, 1976.

Haas, Leonard W., Frederick J. Holden and Christopher S. Welch, "Short Term Changes in the Vertical Salinity Distribution of the York River Estuary Associated with Neap-Spring Tidal Cycle". In: Estuaries and Nutrients, B. J. Neilson and L. E. Cronin, Editors, Humana Press, Clifton, New Jersey, 1981.

Hsieh, Bernard B., Variation and Prediction of Water Temperature in the York River Estuary at Gloucester Point, Virginia, M.A. Thesis, College of William and Mary, Williamsburg, VA, 1979.

Kothandaraman, Veerasamy, Analysis of Water Temperature Variations in Large River, Jour. San. Eng'g. Division Amer. Soc. Civil Eng., 97 (SA1), 19-31, 1971.

Mehta, B. M., R. C. Ahlert and S. L. Yu, Stochastic Variation of Water Quality of the Passaic River, Water Resour. Res., 11 (2), 300-308, 1975.

McMichael, Francis Clay and J. Stuart Hunter, Stochastic Modeling of Temperature and Flow in Rivers, Water Resour. Res., 8 (1), 87-98, 1972.

Thomann, Robert V., Time-Series Analyses of Water Quality Data, Jour. San. Eng. Div. Amer. Soc. Civil Eng., 93 (SA1), 1-23, 1967.

STOCHASTIC ARIMA MODELS FOR MONTHLY STREAMFLOWS

SRINIVAS G. RAO AND EDWIN W. QUILLAN

School of Civil Engineering, Georgia Institute of Technology, Atlanta, Georgia 30332

ABSTRACT

The non-stationary and seasonal nature of the monthly flows are modeled using ARIMA class of models. A step-by-step procedure is designed to obtain valid models through proper model identification, parameter estimation, performance evaluation, model parsimony and validation of residuals. Several statistical indices of performance, and statistical tests are used to properly screen the candidate models for obtaining the best models. The importance of obtaining parsimonious models and validation of residuals for whiteness is emphasized. The procedure is demonstrated for the monthly flows of the Chattahoochee River in Georgia in obtaining a valid model capable of acceptable prediction results.

INTRODUCTION

The world today faces an expanding population and an accompanying increasing demand for water resources. Effective design and operation of water resources systems for efficient use of available water requires the ability to forecast streamflows. Of the several methods available for streamflow forecasting, stochastic models seem attractive and appropriate in dealing with the uncertainty in measurement and randomness in the process. Construction of valid stochastic models for forecasting monthly streamflow is the subject of this paper.

Monthly flows exhibiting nonstationarity and seasonality can be modeled using ARIMA models described by Box and Jenkins (1976).

Such models have been used in the past, to a limited extent to model monthly and annual streamflows (McKerchar and Delleur, 1972; Hipel et al., 1977). These models abreviated ARIMA (p, d, q)x (P, D, Q)$_S$ consists of a seasonal ARMA (P, Q) model fitted to the D-th seasonal difference of data coupled with an ARMA (p, q) model fitted to the d-th difference of the residuals of the former model. Following the three-step procedure of identification, estimation and checking as given by Box and Jenkins, we will illustrate the results of building a valid, accurate and parsimonious ARIMA model for the monthly streamflows of the Chattahoochee River near Atlanta, Georgia.

DATA USED IN THE STUDY

The data used in this study were the monthly mean flows of the Chattahoochee River at West Point near Atlanta, Georgia for water years 1898-1978. Buford Dam, 149 miles upstream from the gage at West Point, began regulation in January 1956. To determine if this dam had affected the homogeneity of the data, the data were analyzed in three periods, 1898-1978, 1898-1955, and 1957-1978. Table 1 includes several basic statistics of the observed data for the three periods. The mean value has not changed significantly while the standard deviation of the flows after the beginning of regula-

TABLE 1: STATISTICS OF OBSERVED MONTHLY FLOWS

Period of record	Number of data	Mean cfs	Standard deviation	Skewness	Kurtosis
1898-1978	972	5665	3763	1.79	5.02
1898-1955	696	5644	3986	1.85	5.18
1957-1978	264	5841	3126	1.39	2.07

tion by the Dam is decreased by 22 percent. However plots of continuous means and standard deviations show slight drift around their long-term values. A comparison of histograms for the two periods (before and after the continuation of the dam) indicate an increase in the frequency of occurrence in the mid-range and a decrease in the frequency of occurrence in the high values of flows. The correlograms and power spectra of the data for the three periods are approximately similar and do not indicate any changes in these flow characteristics due to the dam. A strong periodicity of 12 months is evident in these plots for all the three periods. These and other details can be found in Quillian (1980).

Thus some of these data analyses indicate that the Buford Dam has changed the flow characteristics, particularly the standard deviation of flows. Whether these changes are significant to cause nonhomogeneity in the data need to be pursued further and is not discussed here. Because of these observed changes due to construction of the dam, data prior to 1956 which is homogeneous is used further in this paper.

MODEL DESCRIPTION

The general form of a seasonal ARIMA model takes the form

$$\phi\ (B)\ \Phi\ (B)\ \ (\nabla^d \nabla_s^D\ Z_t\ -\ \mu) = \theta\ (B)\ \Theta\ (B^s)\ a_t$$

where Z_t is the value of time series, t = 1, 2, ... N; a_t i.i.d normal variate with zero mean and variance $\sigma_a{}^2$; B is the backward shift operator, s is the seasonality, ∇^d and ∇_s^D are respectively the regular and seasonal differencing operators, d and D are respectively the degrees of regular and seasonal differencing, ϕ (B), Φ (B), θ (B) and Θ (B) are respectively the regular AR polynomial of degree p, seasonal AR polynomial of degree P, regular MA polynomial of degree q, and seasonal MA polynomial of degree Q.

RESULTS OF MODEL BUILDING AND DIAGNOSTIC CHECKING

A step-by-step procedure for model building and diagnostic checking was applied for selecting a seasonal ARIMA model to the monthly flow of the Chattahoochee River. The period of water years 1913-1953 gives a homogeneous data for estimating and forecasting and excluded the effect of Buford Dam which was built in 1956. The step-by-step procedure was applied as follows:

Step 1 SELECTION OF s,d and D

The ACF and PACF of the monthly mean flow of the Chattahoochee River, 1913-1952 were calculated up to 48 lags (Figure 1). Since the ACF has a recurring sinusoidal pattern with a period of 12 months, it was concluded that seasonal ARIMA models with seasonality, s, equal to 12 would be adequate. Since the ACF of the observed time series did not die out quickly by exponential decay, damped since wave, or decisive cut-off it was expected that a non-stationary model was necessary. Therefore differencing was applied with d=0, 1,2 and D = 0, 1,2. The ACF's and PACF's of the differenced series showed improvement in that they came closer to the use of pure AR, MA or ARMA processes. The combination d=0, D=1 seemed best of all because its ACF and PACF, which are shown in Figure 1 die out somewhat more thoroughly than those of the other combinations.

Step 2 SELECTION OF p,q, P and Q

Although one combination of degrees of differencing was suggested as best values for p,q, P and Q were chosen for each of the four combinations by examining the plots of ACF and PACF. These combinations included d=0, D=0, d=0 D=1, d=1 D=1. In addition, several other processes such as AR(1), ARMA(2,2), ARIMA(1,0, 1) x $(1,0,1)_{12}$, and ARIMA (1,1,1) x $(1,1,1)_{12}$ were chosen.

Step 3 SELECTION OF INITIAL PARAMETER ESTIMATES

It was found that the method used for final estimation of parameters was insensitive to the value of the initial parameter estimates (Rao and Rao, 1974), therefore initial estimates were chosen arbitarily and had an absolute value less than one. For the initial estimate of the mean, which was required for models without differencing, the value of the sample mean of the monthly flow for the period 1913-1952 (5574 cfs) was used.

Step 4 FINAL ESTIMATION OF PARAMETER VALUES

For each model final estimates of parameter values and their standard errors were obtained using the Marquardt nonlinear least squares procedure. The final estimates of some of the various models are shown in Table 2.

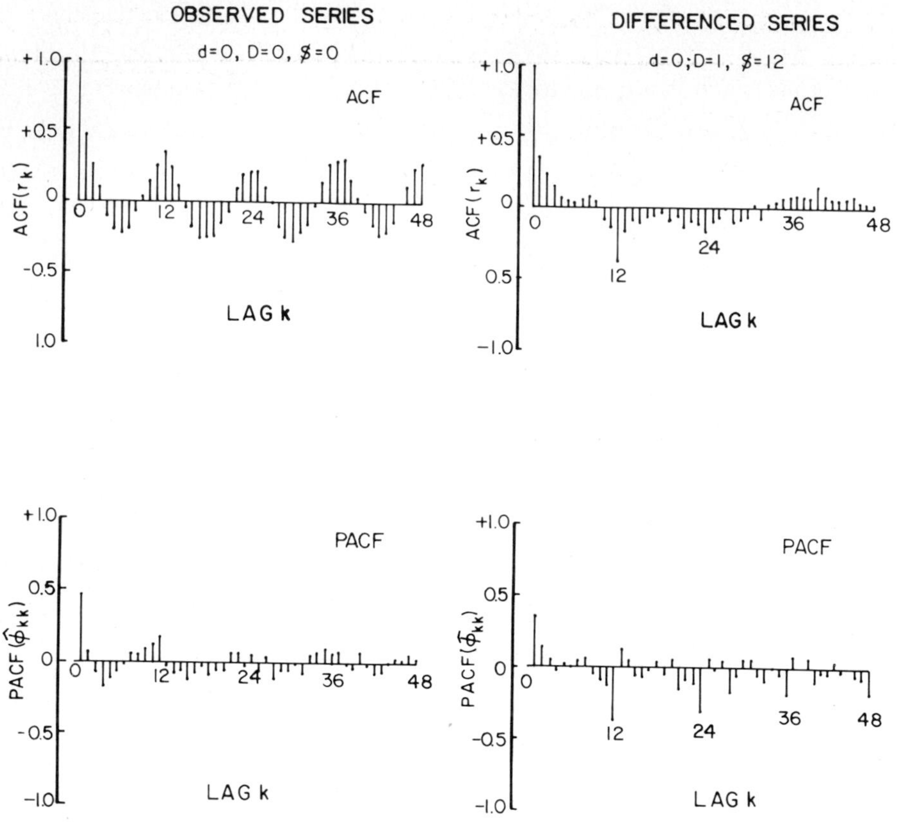

Figure 1: ACF and PACF of observed and differenced series

Step 5 STUDY OF RESIDUAL AND FORECAST ERROR PROPERTIES

The ratios of mean, mean square and variance of residuals to the corresponding quantities of Z(t) (R_1, R_2, and R_3 respectively) were calculated for residuals and for forecast errors. The parameters were estimated using data for 1913-1952 and the one-step-ahead forecasts (Box and Jenkins, 1976) were obtained for the twelve months of water year 1953. The R ratios for some of the models considered are given in Table 2.

TABLE 2. FINAL PARAMETER ESTIMATES

Model No.	Model Name	No. of Parameters*	Regular & Seasonal AR Parameters $\hat{\phi}_1$	$\hat{\phi}_2$	$\hat{\Phi}_1$	Regular & Seasonal MA Parameters $\hat{\theta}_1$	$\hat{\theta}_2$	$\hat{\Theta}_1$	Residuals R_1	R_2	R_3	Forecast Errors R_1'	R_2'	R_3'
1	AR (1)	2	.467	-	-	-	-.598	-	.000	.267	.787	.022	.152	.44
2	ARMA (2,2)	5	1.576	-.841	-	1.254	-	-	.001	.252	.744	.063	.100	.28
9	ARIMA $(1,0,0) \times (1,0,0)_{12}$	3	.416	-	.260	-	-	-	.000	.251	.740	.031	.166	.48
10	ARIMA $(1,0,1) \times (1,0,1)_{12}$	5	.633	-	.970	.363	-	.890	.008	.237	.700	.015	.108	.31
13	ARIMA $(0,1,1) \times (1,0,0)_{12}$	2	-	-	.336	.613	-	-	.001	.291	.857	-.016	.213	.62
16	ARIMA $(2,0,0) \times (0,1,1)_{12}$	3	.259	.118	-	-	-	.902	.008	.241	.712	.015	.110	.32
19	ARIMA $(0,1,1) \times (0,1,1)_{12}$	2	-	-	-	.794	-	.905	.032	.258	.759	-.031	.108	.31
20	ARIMA $(1,1,1) \times (1,1,1)_{12}$	4	.233	-	-.022	.906	-	.922	.000	.247	.728	-.027	.104	.30

*Including the mean, if d=D=0, S=12

Sample Mean = 5574, Sample Mean Square = 4.700 E7, Sample Variance = 1.593 E7

TABLE 3. STANDARD ERROR TEST

Model No.	Model Name	AR Parameters $\hat{\phi}_1$ (2SE)	$\hat{\phi}_2$ (2SE)	$\hat{\Phi}_1$ (2SE)	MA Parameters $\hat{\theta}_1$ (2SE)	$\hat{\theta}_2$ (2SE)	$\hat{\Theta}_1$ (2SE)
6	ARMA (2,2)	1.576 (.100) S	-.841 (.083) S	-	1.254 (.137) S	-.598 (.127) S	-
10	ARIMA $(1,0,1) \times (1,0,1)_{12}$	.633 (.210) S	-	.970 (.024) S	.363 (.254)	-	.890 (.092) S
16	ARIMA $(2,0,0,) \times (0,1,1)_{12}$	.259 (.095) S	.118 (.094) S	-	-	-	.902 (.040) S
20	ARIMA $(1,1,1) \times (1,1,1)_{12}$	.233 (.108) S	-	-.022 (.096) NS	.906 (.044) S	-	.922 (.038) S
19	ARIMA $(0,1,1) \times (0,1,1)_{12}$	-	-	-	0.794 (0.056) S	-	0.905 (0.028) S

Code: S = The parameter is significant; NS = The parameter is not signficant

Step 6 INITIAL SCREENING OF CANDIDATE MODELS

The candidate models selected for further study were those with the smallest R values. Of the three R values, R_2 and R_3 are better indicators of accuracy of models and accordingly the three models with the best (smallest) R_2 and R_3 values for the residuals were

No. 10 ARIMA $(1,0,0) \times (1,0,0)_{12}$

No. 16 ARIMA $(2,0,0) \times (0,1,1)_{12}$

The three models with the best R_2' and R_3' values for the forecast errors were

No. 6 ARIMA (2,2)

No. 19 ARIMA $(0,1,1) \times (0,1,1)_{12}$

No. 20 ARIMA $(1,1,1) \times (1,1,1)_{12}$

Consideration of both residual and forecast error properties led to models, 6, 10, 16, 19, and 20 for further study.

Step 7 SELECTION OF PARSIMONOUS MODELS

Parsimony is the degree to which a model uses the smallest number of parameters to give the best possible results. To determine if a parameter is extraneous (a) Standard Error Test (b) Variance Ratio Test and (c) Akaike Information criteria are used.

a. Standard Error Test

The standard error of each parameter is compared with the magnitude of that parameter. If the magnitude of a parameter is less than twice its standard error then it is declared not significant and is deleted to obtain a more parasimonous model.

Table 3 summarizes the results of Standard Error Test. For Model No. 20, ARIMA $(1,1,1) \times (1,1,1)_{12}$, the seasonal AR parameter was not significant therefore it was deleted resulting in the Model No. 20a ARIMA $(1,1,1) \times (0,1,1)_{12}$ with parameters $\hat{\phi}_1 = .241$, $\hat{\theta}_1 = .927$, and $\hat{\Theta}_1 = .902$. All other parameters tested as significant.

b. Variance Ratio Test

The Variance Ratio Test is a method of comparing two models, one having fewer parameters than the other (Bartlett, 1966). A model having n parameters is considered adequate in comparison with a model having n+m parameters ($m \geq 1$) if the statistic $S_1 \leq \chi^2 m, 1-\alpha$ where S_1

was computed as given in Table 4.

The results of the Variance Ratio Test for all of the comparisons are summarized in Table 4. In all cases the model with the greater number of parameters gave a significantly better fit. This left the following models for further consideration:

No. 6 ARMA (2,2)

No. 10 ARIMA $(1,0,1) \times (1,0,1)_{12}$

No. 16 ARIMA $(2,0,1) \times (0,1,1)_{12}$

No. 20_aARIMA $(1,1,1) \times (0,1,1)_{12}$

c. Akaike Information Criteria

Akaike information criteria (AIC) can be computed using the estimated residual variance and the model which gives minimum AIC is selected as the parsimonious model. The AIC was computed using

$$AIC = N\, Ln(\sigma_a^2) + 2\,(p+q+P+Q)$$

The results of AIC test were similar to the above tests.

TABLE 4. VARIANCE RATIO TEST

Models Being Compared		No. of Parameters*	Deg. of Freedom	Statistic S_1	X^2	Decision
No. 6b	ARMA (2,1)	4				R
vs			1	26.773	3.841	
No. 6	ARMA (2,2)	5				A
No. 10b	ARIMA $(1,0,0) \times (1,0,1)_{12}$	4				R
vs			1	5.972	3.841	
No. 10	ARIMA $(1,0,1) \times (1,0,1)_{12}$	5				A
No. 16b	ARIMA $(1,0,0) \times (0,1,1)_{12}$	2				R
vs			1	4.192	3.841	
No. 16	ARIMA $(2,0,0) \times (0,1,1)_{12}$	3				A
No. 19	ARIMA $(0,1,1) \times (0,1,1)_{12}$	2				R
vs				13.84	3.841	
No. 20a	ARIMA $(1,1,1) \times (0,1,1)_{12}$	3				A

$S_1 = N \dfrac{(Var_n - Var_{n+m})}{Var_n}$, accept if $S_1 \leq X^2_{m,\,1-\alpha}$; A = accept, R = reject, α = 5%

Step 8 CHECKING FOR WHITENESS OF RESIDUALS

For the model to be valid, the residual should be uncorrelated, i.e., white and should not contain any periodicities. If residuals are not white and/or contain periodicities this indicates that the time series has not been adequately modeled and that the residuals still contain process information. The Portmanteau Lack of Fit Test, the Autocorrelation Check and the Cumulative Periodgram Check are used to test for residual whiteness. Details of these tests can be found in Box and Jenkins (1976).

Table 5 summarizes the results of Portmanteau Test. Except for the case of Model No. 6, the residuals of all the models tested as white. For Model No. 6, ARMA (2,2), the Portmanteau Test indicates the residuals are not white and thus still contain unused information.

TABLE 5. PORTMANTEAU LACK OF FIT TEST

Model No.	Model Name	Degrees of Freedom	Statistic S_2	X^2	Decision
6	ARMA (2,2)	43	67.746	59.304	NW
10	ARIMA $(1,0,1) \times (1,0,1)_{12}$	43	32.708	59.304	W
16	ARIMA $(2,0,0) \times (0,1,1)_{12}$	45	36.622	61.656	W
20a	ARIMA $(1,1,1) \times (0,1,1)_{12}$	45	43.112	61.656	W

W = The residuals are white, NW = non white. White if $S_2 \le X^2_{f,\ 1-\alpha}$

S_2 = (No. of parameters)x $\sum_{k=1}^{L} r_k^2$; L = max No. of Lags, r_k = ACF at lag K

Figure 2 shows plots of the residuals ACF's with $\pm$2SE bounds where SE has been defined as $\sqrt{1/N}$. For Model No. 10 and Model No. 16 all the residual ACF values fell on or within the SE bounds indicating that the residuals for these two models are white. For Model No. 6 the residual ACF value at lag 12 greatly exceeded the bound and for Model No. 20a the residual ACF value at lag 24 lie outside the bound therefore the residuals of these two models were indicated to be not white.

The normalized cumulative periodograms of the residuals of Model No. 10 and 16 are shown in Fig. 2. Also shown in this figure are the Kolmogorov-Smirnov probability limit lines at 5% significance level. Since all the estimated cumulative periodogrom fall within these limits, it is concluded that the residuals from these models can be considered white and do not contain any significant periodicities.

Based on the results of the above tests the following models were accepted as valid models.

Model No. 10 ARIMA $(1,0,1) \times (1,0,1)_{12}$

Model No. 16 ARIMA $(2,0,0) \times (0,1,1)_{12}$

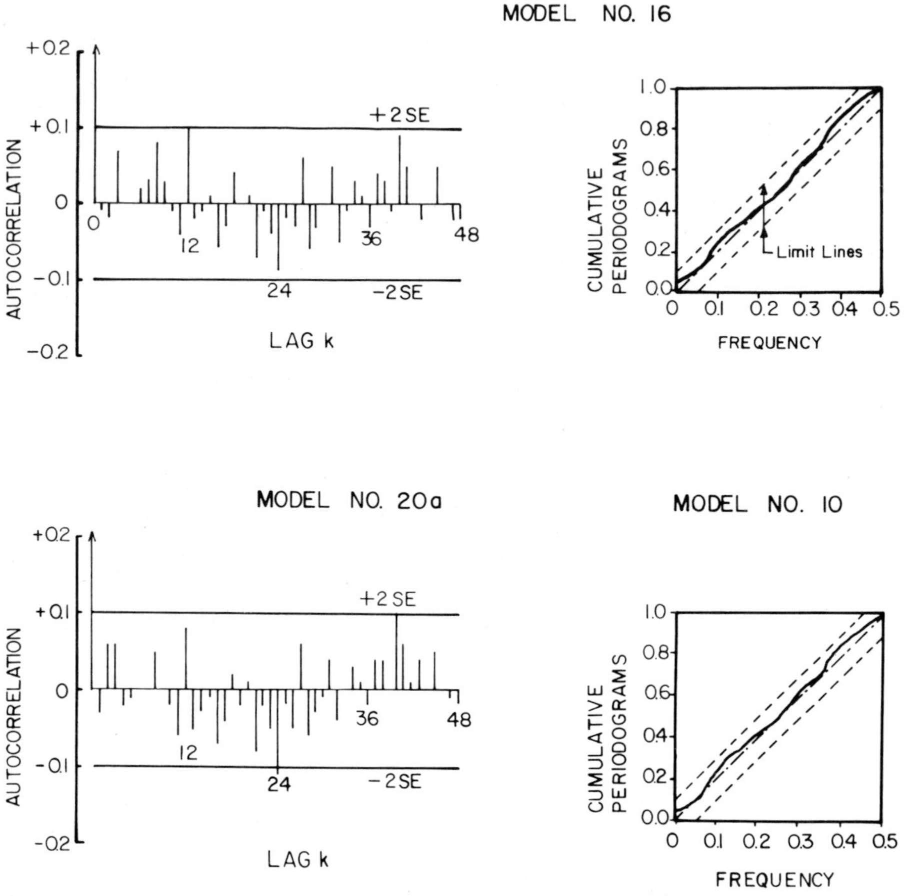

FIGURE 2. CORRELOGRAMS AND CUMULATIVE PERIODOGRAMS OF RESIDUALS

Step 9 FINAL MODEL SELECTION

A final model selection was made between the remaining two models. Model No. 10 had a smaller residual variance but more parameters than Model No. 16. The Variance Ratio Test as seen earlier indicated that Model No. 10 did indeed give a significantly better fit at the 5% level. Therefore, Model No. 10 was declared the best model and Model No. 16, as an alternative best model.

The performance of Model No. 10 and Model No. 16 in forecasting monthly flows up to 12 months for the water year 1953 was found to be very good (Quillian, 1980). The 95% confidence intervals of the forecasts included the observed values in all cases. Also the seasonal rise and fall of the monthly flow was effectively forecasted. Thus both models show promise as predictive tools for the monthly mean flow of the Chattahoochee River.

V. SUMMARY AND CONCLUSIONS

A nine-step procedure for model identification, parameter estimation, performance evaluation, model parsimony and validation of residuals which emphasis on the latter three items was implemented for obtaining valid stochaslis models for the monthly flows of the Chattahoochee River at West Point, Georgia. The ARIMA $(1,0,1) \times (1,0,1)_{12}$ and $(2,0,0) \times (0,1,1)_{12}$ model are obtained as valid models. It is interesting to note that McKerchar and Delleur (1972) fitted an ARIMA $(2,0,0) \times (0,1,1)_{12}$ model to the monthly mean flows of the Blue River in Indiana and Hipel, et al. (1977) fit an AR (3) model to the annual mean flows of the St. Lawrence River.

REFERENCES

Bartlett, M.S., Stochastic Processes. New York: Cambridge University Press, 1966.

Box, G.E.P. and Jenkins, G.M., "Time Series Analysis" Forecasting and Control San Francisco, Holden-Day, Inc., 1976.

Hipel, K.E., McLeod, A.I., and Lennox, W.C., "Advances in Box-Jenkins Modeling. L. Model Construction: and "2. Applications." Water Resources Research. 13 (june 1977), 567-586.

McKerchar, A.I., and Delleur, J.W., "Stochasitc Analysis of Monthly Flow Data Application to Lower Ohio Tributaries. Technical Report No. 26. LaFayette, Indiana, Water Resources Research Center, Purdue University, 1972.

Quillian, E.W., "Stochastic ARIMA Models for Monthly Streamflows of the Chattahoochee River," Georgia Institute of Technology, Atlanta, Georgia (unpublished Special Project for M.S. degree), 1980.

Rao, R.A. and Rao, R.G.S., Analyses of the Effect of Urbanization on Rainfall Characteristics - I. Technical Report No. 50. West LaFayette, Indiana: Water Resources Research Center, Purdue University, 1974.

THE LINEAR RESERVOIR WITH SEASONAL GAMMA-DISTRIBUTED MARKOVIAN INFLOWS

E.H. LLOYD AND D. WARREN (UNIVERSITY OF LANCASTER, U.K.)

ABSTRACT

Earlier work has established a method of tabulating the outflow distribution (by numerically inverting its Laplace Transform) from a linear reservoir fed by a discrete-time gamma-distributed inflow process. The present paper is concerned with an extension of this line of enquiry to the case where the inflow process is seasonal. Instead, however of seeking to tabulate the outflow distribution, it concentrates on the outflow skewness, and develops a technique for obtaining this by direct calculations. A detailed study is made of a special 2-season case, and the resulting skewness values are tabulated in terms of the reservoir constant, the ratio of the two seasonal mean inflows, and the inflow correlation coefficient.

1. INTRODUCTION

This is a drastically abbreviated version (necessitated by the exigencies of publication) of an examination of a simplified model of a Klemes catchment (Klemes, 1973), with particular reference to its behaviour when the inflows are skewed, autocorrelated, and seasonal. The simplifications referred to make the model operate in discrete time, with a particular continuous-valued two-season Markovian inflow process, and with outflow proportional to storage. This is thus an extension of the considerable body of existing work on the linear reservoir with independent non-seasonal inflows (Langbein, 1958; Moran, 1967; Brockwell, 1972; Klemes, 1973 and 1974; Klemes and Boruvka, 1975) and of a study of the case of skewed autocorrelated non-seasonal inflows (Anis, Lloyd and Saleem, 1979).

Reprinted from *Time Series Methods in Hydrosciences*, by A.H. El-Shaarawi and S.R. Esterby (Editors)

2. THE LINEAR RESERVOIR IN DISCRETE TIME

If we denote the quantity of water in store at time t by S(t), the inflow and outflow quantities in the interval (t,t+1) by X(t), Y(t) respectively, (t=0,1,...), with the linear response

$Y(t) = (1-b)S(t) \quad (0<b<1),$

the water balance equation $S(t+1)-S(t) = X(t)-Y(t)$ becomes

$$Y(t+1) = bY(t)+cX(t), \quad t = 0,1,..., \quad 0<b<1, \quad c = 1-b . \tag{2.1}$$

The object is to study the properties of the outflow sequence $\{Y(t)\}_{t=0}^{\infty}$ in terms of the assumed properties of the inflows $\{X(t)\}$. The structural simplicity of the present model enables us to work in terms of the explicit formal solution of the stochastic difference equation (2.1), namely

$$Y(t+1) = b^{t+1}Y(0) + c\sum_0^t b^r X(t-r). \tag{2.2}$$

For a process that has been running for a long time we may replace this by the asymptotic version

$$Y(t+1) = c\sum_{r=0}^{\infty} b^r X(t-r). \tag{2.3}$$

Our problem is then reduced to that of studying the weighted sums $\sum_r b^r X(t-r)$ of the elements of the Markov Chain $\{X(t)\}$.

One method would be to utilise the fact that, for the particular Markov Chain chosen, the Laplace Transform of the weighted sums has a comparatively simple form (Phatarfod, 1976; Lloyd and Saleem, 1979), and tabulate that transform and invert it numerically, thus obtaining a table of the outflow probability distribution for each season. This was the method used (in a non-seasonal case) by Anis, Lloyd and Saleem (1979).

Instead, however, in the present work we have concentrated on the skewness of the seasonal outflow distributions, and investigated the feasibility of obtaining these by direct calculation from (2.3).

3. THE NON-SEASONAL VERSION OF THE INFLOW MARKOV CHAIN

The non-seasonal version of the inflow process with which we have worked is a homogeneous first-order Markov Chain in which the marginal distribution of each $X(t)$ is of the 2-parameter gamma form, with scale parameter α and shape parameter p, the p.d.f. at x being

$$x^{p-1}e^{-x/\alpha}/\alpha^p\Gamma(p) \qquad \alpha>0,\ p>0, \quad (x>0) \tag{3.1}$$

with Laplace Transform (L.T.).

$$E\left(e^{-\theta X(t)}\right) = (1+\alpha\theta)^{-p}. \tag{3.2}$$

The first three moments of $X(t)$ are

$$E\{X(t)\} = p\alpha,\ E\{X(t)\}^2 = p(p+1)\alpha^2,\ E\{X(t)\}^3 = p(p+1)(p+2)\alpha^3 \tag{3.3}$$

while the variance and the coefficient of skewness are

$$\text{var}\{X(t)\} = p\alpha^2, \quad \text{skew}\{X(t)\} = 2p^{-\frac{1}{2}}. \tag{3.4}$$

The transition density is defined in terms of its L.T. as

$$E\{e^{-\theta X(t)} \mid X(t-1)=x\} = H(\theta)\exp\{-G(\theta)x\}, \tag{3.5}$$

where

$$H(\theta) = \{1+\alpha(1-\rho)\theta\}^{-b}, \quad G(\theta) = \rho\theta/\{1+\alpha(1-\rho)\theta\}, \tag{3.6}$$

the parameter ρ $(0<\rho<1)$ denoting the correlation coefficient between consecutive X-values.

The principal properties of the process that we shall need may be obtained by repeated use of the conditional expectation theorem. In particular, for $r=1,2,\ldots,$

$$\left.\begin{aligned} &E\{X(t) \mid X(t-r)=x\} = p\alpha(1-\rho^r)+\rho^r x,\\ &E\{X(t)X(t-r)\} = p\alpha^2(p+\rho^r) \quad ,\\ &\text{corr}\{X(t),X(t-r)\} = \rho^r\\ &E\{X^2(t)X(t-r)\} = E\{X(t)X^2(t-r)\} = \alpha^3 p(p+1)(p+2\rho^r),\\ &\text{and}\\ &E\{X(t-r)X(t-s)X(t-u)\} = \alpha^3\{p^3+p^2(\rho^{s-r}+\rho^{u-s})+p(p+2)\rho^{u-r}\},\\ &\qquad\qquad\qquad\qquad\qquad\qquad r<s<u \end{aligned}\right\} \tag{3.7}$$

4. OUTFLOW SKEWNESS IN THE NON-SEASONAL CASE

With the aid of the results summarised in (3.7) it is not difficult to obtain the outflow skewness for the linear reservoir of §2 subject to the non-seasonal inflow process described in §3. To avoid the frequent repetition of constant multipliers we make a change of scale, setting

$$U(j) = X(j)/\alpha, \quad j=1,2,\ldots; \quad Z(k) = Y(k)/c\alpha, \quad k=1,2,\ldots \tag{4.1}$$

It should be noted that skew $\{Z(k)\}$ = skew $\{Y(k)\}$, $k=1,2,\ldots$.
We then have, instead of (2.3),

$$Z(t+1) = \sum_{r=0}^{\infty} b^r U(t-r) \tag{4.2}$$

with $E\{U(j)\}=p$, $E\{U^2(j)\}=p(p+1)$, $E\{U^3(j)\}=p(p+1)(p+2)$,

and $\operatorname{corr}\{U(j),U(j-k)\} = \rho^k \quad (j,k = 1,2,\ldots)$.

The mean value of $Z(t+1)$ is thus

$$\lambda(1) = E\{Z(t+1)\} = p\Sigma b^r = p/(1-b) . \tag{4.3}$$

The second moment is

$$\begin{aligned}\lambda(2) = E\{Z^2(t+1)\} &= \sum_{r=0}^{\infty}\sum_{s=0}^{\infty} b^{r+s} E\{U(t-r)U(t-s)\} \\ &= \sum_r b^{2r} E\{U^2(t-r)\} + 2\sum_{\substack{r \\ (r<s)}}\sum_s b^{r+s} E\{U(t-r)U(t-s)\} \\ &= p(p+1)\sum_{r=0}^{\infty} b^{2r} + 2\sum_{r=0}^{\infty}\sum_{s=r+1}^{\infty} b^{r+s}(p^2+p\rho^{s-r})\end{aligned}$$

by (3.7). This reduces to

$$\lambda(2) = p^2/(1-b)^2 + p(1+b\rho)/(1-b^2)(1-b\rho). \tag{4.4}$$

Equivalently the variance is

$$\operatorname{var}\{Z(t+1)\} = \omega^2, \text{ say,} = p(1+b\rho)/(1-b^2)(1-b\rho). \tag{4.5}$$

The third moment is $E\{Z^3(t+1)\} = \sum_r\sum_s\sum_u b^{r+s+u} E\{U(t-r)U(t-s)U(t-u)\}$
This may be evaluated by a technique similar to that employed for $\lambda(2)$, that is, by splitting the triple sum into a single sum involving $E\{U^3(t-r)\}$, a double sum involving $E\{U^2(t-r)U(t-u)\}$ (with $r\neq u$), and a triple sum involving $E\{U(t-r)U(t-s)U(t-u)\}$ (with $r\neq s\neq u\neq r$). After not inconsiderable reduction we are left with the result

$$\text{skew } \{Z(t+1)\} = \omega^{-3/2}\{\lambda(3)-3\lambda(2)\lambda(1) + 2\lambda^3(1)\}$$

where $\lambda(1)$ and $\lambda(2)$ are as defined in (4.3) and (4.4), while

$$(1-b^3)\lambda(3) = p(p+1)(p+2) + 3p(p+1)b\{p/(1-b) + pb/(1-b^2) +2\rho/(1-b\rho) + 2b\rho/(1-b^2\rho)\} + 6pb^3\{p^2/(1-b)(1-b^2) + p\rho/(1-b)(1-b^2\rho) + p\rho/(1-b\rho)(1-b^2) + (p+2)\rho^2/(1-b\rho)(1-b^2\rho)\}.$$

5. THE SEASONAL GAMMA-DISTRIBUTED MARKOV CHAIN

The process described in §3 generalizes easily to a seasonal version with k seasons (k=2,3,...). For example, the 3-season version has the three season-to-season transition L.T.'s

$$E(e^{-\theta X(t)}|X(t-1)=u) = H(t,\theta)\exp\{-G(t,\theta)u\},$$

where, for n=0,1,2,..., we have

$$H(3n+j,\theta) = H(j,\theta) = \{1+\alpha_j(1-\rho_j)\theta\}^{-p},$$

and

$$G(3n+j,\theta) = G(j,\theta) = \alpha_j\rho_j\theta/\alpha_{j-1}\{1+\alpha_j(1-\rho_j)\theta\}, \quad j=0,1,...,$$

where $\alpha_{-1} = \alpha_2$. Then, for n=0,1,2,..., the marginal distribution of X(3n+j) (i.e. the j-season inflow) is gamma with shape parameter α_j and shape parameter p, (the shape parameter not being seasonalisable in this process), while

$$\mathrm{corr}\{X(3n+j),X(3n+j-1)\} = \rho_j$$
$$\mathrm{corr}\{X(3n+j),X(3n+j-2)\} = \rho_j\rho_{j-1}$$
$$\mathrm{corr}\{X(3n+j),X(3n+j-3)\} = \rho_j\rho_{j-1}\rho_{j-2}$$

etc. (where $\rho_{-1} = \rho_2$, $\rho_{-2} = \rho_1$, etc.).

We have to accept the restriction that, necessarily, in a k-season year, the lag-k correlation coefficient is a constant; thus if for example the "seasons" are months, the January-January correlation is the same as the May-May correlation, etc.

6. OUTFLOW SKEWNESS INDUCED BY A 2-SEASON INFLOW

In this exploratory study we take as inflow process the seasonal chain described in §5, taking however only two seasons and taking $\rho_1=\rho_2$=constant (=ρ). The obvious extension of the techniques used in §4 to study the outflow skewness induced by non-seasonal gamma-distributed Markovian inflows may now be applied. As before we replace Y(t) by the standardized version Z(t) = Y(t)/c. We regard X(t), X(t-2), X(t-4), ... as occurring in season "1", and X(t-1), X(t-3), ... as occurring in season "2". The appropriate standardized versions are

$$U(t-2r) = X(t-2r)/\alpha_1, r=0,1,...$$

and

$$V(t-2s-1) = X(t-2s-1)/\alpha_2, \; s=0,1,... \; .$$

Then the U's and V's all have gamma distributions with unit scale parameter and with shape parameter p, and the sequence

U(0),V(1),U(2),V(3),U(4)...

is a first-order homogeneous Markov Chain with transition L.T. as in (3.5). We then have, as the 1-season outflow,

$$Z(t+1) = \alpha_1 \sum_r b^{2r} U(t-2r) + \alpha_2 \sum_s b^{2s+1} V(t-2s-1) \tag{6.1}$$

from which to evaluate the skewness. Utilising results from §§3,4 for non-seasonal (i.e. homogeneous) inflows, we obtain the first moment of the 1-season outflow with no difficulty as

$$\lambda_1 = E\{Z(t+1)\} = p(\alpha_1 + b\alpha_2)/(1-b^2). \tag{6.2}$$

The second moment requires a little calculation; it turns out to be

$$\lambda_2 = E\{Z^2(t+1)\} = (\alpha_1{}^2 + b^2\alpha_2{}^2)\{p^2/(1-b^2)^2 + p(1+b^2\rho^2)/(1-b^4)(1+b^2\rho^2)\} + \\ + 2\{\alpha_1\alpha_2/(1-b^4)\}\{p(p+\rho)b + 2b^3p^2/(1-b^2) + b^3p\rho(1+\rho^2)/(1-b^2\rho^2)\}. \tag{6.3}$$

The labour required to calculate the third moment, and hence the skewness, is of a totally different order of magnitude, being in fact sufficient to discourage rather effectively the authors' original hope of studying the three-season case. In outline, on cubing (6.1) we have

$$E\{Z^3(t+1)\} = \alpha_1{}^3A + \alpha_2{}^3b^3A + 3\alpha_1{}^2\alpha_2B + 3\alpha_1\alpha_2{}^2C$$

where

$A=E\{\Sigma b^{2r}U(t-2r)\}^3$, $B=E[\{\Sigma b^{2r}U(t-2r)\}^2\{\Sigma b^{2s+1}V(t-2s-1)\}]$

$$C=E[\{\Sigma b^{2r}U(t-2r)\}\{\Sigma b^{2s+1}V_t(t-2s-1)\}^2] \tag{6.4}$$

The terms A, B and C must be separately evaluated. The results, expressed in terms of the auxiliary function

$$f(u,v) = 1/(1-b^2u^2)(1-b^4u^2)(1-b^6). \tag{6.5}$$

are

$$A=p(p+1)[(p+2)f(0,0)+3b^2\{pf(1,0)+2\rho^2f(\rho,0)\}+3b^4\{pf(0,1)+2\rho^2f(0,\rho)\}]$$
$$+6b^6p\{p^2f(1,1)+p\rho^2f(1,\rho)+p\rho^2f(\rho,1)+(p+2)\rho^4f(\rho,\rho)\} \tag{6.6}$$

$$B=p(p+1)b[b^2\{pf(1,0)+2\rho^3f(\rho,0)\}+b^4\{pf(0,1)+2\rho+f(0,\rho)\}+(p+2\rho)f(0,0)]$$
$$+2b^2p[3p^2f(1,1)+p(\rho+\rho^2+\rho^3)\{f(1,\rho)+f(\rho,1)\}+(p+2)(\rho^3+\rho^4+\rho^5)f(\rho,\rho)]$$
$$+2b^3p[p(p+\rho)\{f(1,0)+b^2f(0,1)\}+\rho\{p+(p+2)\rho\}\{f(\rho,0)+b^2\rho f(0,\rho)\}] \tag{6.7}$$

$$C=p(p+1)b^2[b^2\{pf(1,0)+2\rho f(\rho,0)\}+b^4\{pf(0,1)+2\rho^3f(0,\rho)\}+(p+2\rho)f(0,0)]$$
$$+2b^8p[3p^2f(1,1)+p(\rho+\rho^2+\rho^3)\{f(1,\rho)+f(\rho,1)\}+(p+2)(\rho^3+\rho^4+\rho^5)f(\rho,\rho)]$$
$$+2b^4p[p(p+\rho)\{f(1,0)+b^2f(0,1)\}+\rho\{p+(p+2)\rho\}\{\rho f(\rho,0)+b^2f(0,\rho)\}]. \tag{6.8}$$

7. CONCLUSIONS

An abbreviated representative selection of tabulated values of the outflow skewness is presented in the Table. Whilst in principle these values are functions of the five parameters $\alpha_1,\alpha_2,b,p,\rho$, there are two simplifying features which reduce the number of parameters to three.

One of these features refers to the dependence on α_1 and α_2. It will be seen from (6.2) that, writing λ_1 as $\lambda_1(\alpha_1,\alpha_2)$, we have

$$\lambda_1(\alpha_1,\alpha_2) = \alpha_2\lambda_1(\alpha_1/\alpha_2,1).$$

Similarly, from (6.3) and (6.4),

$$\lambda_2(\alpha_1,\alpha_2) = \alpha_2^{\,2}\lambda_2(\alpha_1/\alpha_2,1)$$

and

$\lambda_3(\alpha_1,\alpha_2) = \alpha_2{}^3\lambda_2(\alpha_1/\alpha_2,1)$.

It follows that the outflow skewness depends not on α_1 and α_2 separately but on the <u>seasonal index</u> ratio α_1/α_2, the ratio of the mean inflow in season 1 to that in season 2.

The second simplifying feature relates to the dependence of the outflow skewness on the parameter p. In the special case where the inflows are mutually independent, simple calculations show that

$$\text{skew}\{Y(t+1)\} = \frac{2}{\sqrt{p}}\, h(\alpha_1/\alpha_2, b), \tag{7.1}$$

where

$$\{h(a,b)\}^2 = \left(\frac{a^3+b^3}{1-b^6}\right)^2 \Bigg/ \left(\frac{a^2+b^2}{1-b^4}\right)^3$$

Since $2/\sqrt{p} = \text{skew}(X_{t+1})$, (7.1) shows that the ratio of outflow skewness to inflow skewness is independent of p when the inflows are independent. It turns out that this independence of p still holds when the inflows are correlated. This result only emerged when a detailed tabulation of the outflow skewness values was studied : for all values of α_1/α_2, b and ρ, the ratio

$$\frac{\text{outflow skewness}}{\text{inflow skewness}} \tag{7.2}$$

was independent of p. Thus by presenting our results in terms of this ratio it is possible to eliminate the parameter p.

The Table gives values of the ratio (7.2) for various values of the reservoir constant b (introduced in (2.1)), the lag-1 season-to-season correlation coefficient ρ, and the seasonal index ratio α_1/α_2.

In reading the Table care is needed in entering the appropriate value of α_1/α_2. The convention adopted in Chapter 6 made α_1 the α-parameter of X_t, X_{t-2}, X_{t-4}, ..., and α_2 that of X_{t+1}, X_{t-1}, X_{t-3}, ... Now the Table gives the skewness ratio of Y_{t+1}, i.e. of the outflow during a season during which the mean inflow is proportional (in our convention) to α_2. If this is the drier of the two seasons we must have $\alpha_2 < \alpha_1$, or $\alpha_1/\alpha_2 > 1$. In other words the tabulated outflow skewness ratio is to be read off against values of α_1/α_2 such that $\alpha_1/\alpha_2 > 1$ if one is concerned with the "drier" season i.e. the one which has smaller mean inflow; while if one is concerned with the outflow during the "wetter" season the appropriate value of α_1/α_2 satisfies $\alpha_1/\alpha_2 < 1$. (The Table also gives values for $\alpha_1/\alpha_2 = 1$, which is of course the non-seasonal case.)

It will be seen from these that the skewness of the outflow is always less than that of the inflow, and that the ratio of outflow to inflow skewness increases with increasing correlation coefficient ρ and decreases with decreasing ρ. At the extremes, the case $\rho=0$ corresponds to independent inflows, for which skew $(Y_{t+1})/\text{skew}(X_{t+1})$ is given by the expression $h(\alpha_1/\alpha_2, b)$ of (7.1). In the case $\rho=1$, the 2-season version of (5.1) shows that the transition Laplace Transform then reduces to

$$E(e^{-\theta X_t}|X_{t-1} = u)\Big|_{\rho=1} = e^{-\theta\alpha_1 u/\alpha_2}$$

(using the convention that the "t" season is an α_1 season) whence

$$X_t = \alpha_1 X_{t-1}/\alpha_2, \; X_{t-2} = \alpha_1 X_{t-3}/\alpha_2,$$

and so on. The inflow sequence thus degenerates when $\rho=1$ to an alternating sequence of constants. The outflow sequence behaves similarly. Then, for each t, Y_{t+1} and X_{t+1} both have zero skewness, the ratio tending to unity when $b\neq 1$.

Other limiting values of interest include the cases $b=0$, $b=1$, $\alpha_1/\alpha_2=0$, $\alpha_2/\alpha_1=0$. In the case $b=0$, (2.1) reduces to

$$Y_{t+1} = X_t :$$

the outflow is identically the same as the preceding inflow. The skewness ratio is therefore equal to unity for all values of α_1/α_2. When $b=1$, the outflow is identically zero and the outflow skewness ratio tends to zero. When $\alpha_1/\alpha_2=0$ or when $\alpha_2/\alpha_1=0$ we have inflows in only one of the two seasons in each year. Then (taking $X_t, X_{t-2}, X_{t-4}, \ldots$ to be the non-zero inflows) the outflow becomes

$$Y_{t+1} = (1-b)\Sigma b^{2r} X_{t-2r}$$

when X_t, X_{t-2}, X_{t-4}, ... are identically distributed with $corr(X_t, X_{t-2}) = \rho^2$. Thus the outflow process in this case corresponds to that for non-seasonal inflows ($\alpha_1/\alpha_2=1$) with ρ replaced by ρ^2 and b by b^2.

TABLE

showing the ratio of outflow skewness to inflow skewness as a function of the reservoir constant b, the lag-1 season-to season inflow correlation coefficient ρ, and the inflow seasonality index α_1/α_2.

(α_1/α_2 denotes the ratio of mean inflow in season 1 to mean inflow in season 2.

$\alpha_1/\alpha_2 = 1$ corresponds to non-seasonality

$\alpha_1/\alpha_2 < 1$ corresponds to outflows occurring in the wetter season

$\alpha_1/\alpha_2 > 1$ corresponds to outflows occurring in the drier season.

α_1/α_2	ρ \ b	0.1	0.2	0.3	0.4	0.5	0.6	0.7	0.8	0.9
0.1	0	.7070	.8031	.8753	.8956	.8764	.8213	.7305	.6019	.4231
	0.5	.9622	.9714	.9771	.9722	.9546	.9214	.8675	.7806	.6224
	0.9	.9990	.9991	.9992	.9988	.9977	.9955	.9912	.9807	.9417
0.25	0	.8515	.7182	.7076	.7327	.7423	.7184	.6556	.5509	.3933
	0.5	.9768	.9622	.9592	.9552	.9423	.9148	.8663	.7838	.6276
	0.9	.9993	.9989	.9988	.9984	.9975	.9955	.9914	.9811	.9421
0.5	0	.9503	.8496	.7580	.6953	.6521	.6097	.5519	.4664	.3368
	0.5	.9903	.9749	.9619	.9497	.9334	.5069	.8620	.7838	.6308
	0.9	.9997	.9992	.9988	.9982	.9973	.9954	.9914	.9812	.9423
1	0	.9860	.9482	.8922	.8225	.7423	.6531	.5544	.4426	.3056
	0.5	.9967	.9882	.9755	.9587	.9364	.9054	.8589	.7816	.6310
	0.9	.9999	.9996	.9991	.9984	.9972	.9953	.9913	.9812	.9424
2	0	.9962	.9839	.9594	.9179	.8551	.7687	.6583	.5236	.3561
	0.5	.9989	.9948	.9858	.9699	.9456	.9105	.8597	.7795	.6281
	0.9	1.0000	.9998	.9993	.9986	.9973	.9952	.9911	.9809	.9422
4	0	.9989	.9941	.9807	.9524	.9027	.8266	.7214	.5843	.4042
	0.5	.9996	.9973	.9905	.9764	.9523	.9156	.8616	.7779	.6242
	0.9	1.0000	.9999	.9995	.9987	.9974	.9952	.9910	.9807	.9419

REFERENCES

Anis, A.A., Lloyd, E.H. and Saleem, S.D., 1979. The linear reservoir with Markovian inflows. Water Res. Research, 15, 1623-1627.

Brockwell, P.J., 1972. A storage model in which the net growth-rate is a Markov Chain. J. Appl. Prob., 9, 129-139.

Klemes, V., 1973. Watershed as semi-infinite storage reservoir. J. Irrig. Drain. Div. Amer. Soc. Civil Eng., 99, 477-491.

Klemes, V. and Boruvka, L., Output from a cascade of discrete linear reservoirs with stochastic input. J. Hydrol., 27, 1-13.

Lampard, D.G., 1968. A stochastic process whose successive intervals between events form a first order Markov Chain. J. Appl. Prob., 5, 648-668.

Langbein, W.B., 1958. Queuing theory and water storage. J. Hydraul. Div. Amer. Soc. Civil Eng. HY5, 1811/1 - 1811/24.

Lloyd, E.H., 1963. Reservoirs with serially correlated inflows. Technometrics, 5, 83-93.

Lloyd, E.H., and Saleem, S.D., 1979. A note on seasonal Markov Chains with gamma or gamma-like distributions. J. Appl. Prob., 16, 117-128.

Moran, P.A.P., 1967. Dams in series with continuous release. J. Appl. Prob., 4, 330-388.

Phatarfod, R.M., 1976. Some aspects of stochastic reservoir theory. J. Hydrol., 30, 199-217.

ON THE STORAGE SIZE-DEMAND-RELIABILITY RELATIONSHIP

RAVINDRA M. PHATARFOD

1. INTRODUCTION

Investigations by hydrologists and engineers show that in a large number of cases monthly streamflows fit a model of Markov dependence (of various orders) with monthly varying transition probabilities (see e.g. Kottegoda, 1970). The order of Markov dependence in most cases is one or two, but in some cases, three. It would appear then that if one takes time-periods or seasons of about two months, a model of Markov dependence (of order one) with seasonally varying transition probabilities would fit the flows in most cases. In this paper we consider the storage process with such an inflow model, and compare the procedures that can be used to determine the storage size-demand-reliability relationship.

Let us therefore first consider, briefly, the procedures that are being used (in practice) and can be used, to determine the relationship, and then consider in details, only those for which a meaningful comparison can be made, when the input process is of the kind described above.

It is generally accepted (see e.g. McMahon and Mein, 1978) that these procedures can be put into three distinct groups. The first group (called Critical Period Techniques) includes those procedures which rely entirely on the historical data. These procedures include Rippl's Mass Curve Method (the earliest method known), Sequent Peak Algorithm, Minimum flow method, and others (see McMahon and Mein, 1978, for details). The second group (Probability Methods) includes those procedures which use the calculus of probability to the structure of the problem, resulting in a probability distribution of the storage content. The third group (Synthetic Hydrology) consists of those procedures which are based on generated data, the relationship being obtained by simple simulation. The latter two groups of procedures depend on the formulation and fitting

Reprinted from *Time Series Methods in Hydrosciences*, by A.H. El-Shaarawi and S.R. Esterby (Editors)

of a statistical model (a stochastic process) to the given historical data, and are often regarded as being statistically more rigorous than the procedures in the first group. The synthetic hydrology procedure is just one procedure-simulation; the bulk of the research associated with this procedure being done in the area of formulating and fitting of correct models for inflows. The models in common use are the Thomas-Fiering model, Matalas log Normal model, Broken line model, etc. On the other hand, the group of probability methods includes a variety of methods with varying mathematical sophistication — from numerical to analytical, the input models possible being somewhat more restrictive than those for the synthetic hydrology group (Savarenskiy, 1940; Kritskiy and Menkel, 1940; Moran, 1954; Prabhu, 1958; Lloyd, 1963; Klemes, 1970; Phatarfod, 1981a).

We shall not be considering here any procedures belonging to the first group. A comparison of these procedures and some from the second group with worked out examples has been given in McMahon and Mein (1978). There, the comparison between the procedures is made not only in terms of their limitations, and underlying assumptions etc., but also in terms of the final answer obtained — the size of the reservoir required with a specific draft and reliability of supply. In this paper, we are interested in comparing those methods which are applicable when the input process is of the kind described earlier.

The methods compared are:

A. Simulation.
B. Probability Matrix Method (Kritskiy and Menkel, 1940; Dearlove and Harris, 1965; Klemes, 1970). This is a seasonal extension of Lloyd's (1963) procedure (for Markov but non-seasonal inputs), and involves construction of matrices. The procedure is entirely numerical.
C. Bottomless Dam analytical method Mark 1. (Phatarfod, 1981a). We assume the dam to be bottomless and derive an analytical

solution for the probability distribution of the depletion of the dam.

D. Bottomless Dam analytical method Mark 2. (Phatarfod 1981b). Here we make the further assumption that the inputs are uncorrelated from year to year, although they are correlated and (seasonally Markov) dependent within a year. This assumption makes a drastic reduction in the computational complexities involved in Method C.

A common feature of the methods B, C and D is that for these methods we consider the inputs, storage, release etc., as discrete quantities, whereas for A, these are continuous.

It is customary, in hydrology and engineering disciplines, to take a month as the time unit of operation – most records of streamflows are, in fact, published as monthly values. For reasons given before, we shall consider a period of 2 months as our unit of time. For simplicity of presentation only, we shall assume here a time unit of six months, so that we have only two seasons per year. It is fairly easy to see how the procedures can be extended to the case of six seasons.

Of course, none of the above methods will provide a correct answer – the answers they all provide are approximations. One reason for comparing them is to find out to what extent the answers differ because of the approximating devices employed in the probability methods. A comparison between the answers obtained by A and B and those by C and D would indicate how the assumption of the bottomlessness of the dam has affected the answer, and under what conditions on the draft ratio, say, the answers are comparable. A comparison between A on one hand and B, C and D on the other would indicate the effect of discretization; this should tell us how small or large the unit of measurement we should choose to give comparable results. This, in turn, would tell us something about the amount of effort required by using B, C, D. Comparing answers obtained by C and D would indicate to what extent the answers differ if we neglect inter-year dependence of flows, and so on.

For the inflow model assumed, Method B — Probability Matrix method — seems to be the mathematically correct procedure, because the answer it provides is free from the sampling errors associated with Method A. However, it seems that it is not in popular use, the reason being that it is an abstract method and the general (erroneous) impression that it involves very unwieldy matrices.

The question arises, therefore, what is the use of Methods C and D, particularly since they are mathematically more abstract than Method B. A partial answer to this question is provided by the following considerations. Whilst it is certainly true that using Method A (and possibly Method B) would give an engineer a better insight into the reservoir performance in a given situation, i.e. for particular values of the parameters of the inflow model, only an analytical method would give us an insight into the general problem considered here. We shall see, for example, that for Method D, the equilibrium distribution of the depletion of the reservoir depends on a few critical values. Specifically, it is of the form: Probability $\{\text{depletion} = j\} = C_1\theta_1^{\,j} + C_2\theta_2^{\,j} + C_3\theta_3^{\,j} + \dots$. The θ's depend upon the parameter values of the annual flows and the C's depend upon the transition probabilities of the seasonal flows and the seasonal releases. In practice three or four values each of θ and C are sufficient for reasonable accuracy.

The importance of this result is as follows. The inflow model considered here has 24 parameters (6 values of means, standard deviations, skewness, and serial correlations). It is fairly obvious that construction of charts and tables (showing the reservoir size-draft-reliability relationship for various values of the parameters), by Methods A or B, is virtually impossible. On the other hand Method D shows that the relationship is governed by a few values of annual flow parameters and some transition probabilities, and not on all the 24 seasonal parameters. It would seem therefore that this approach might be more fruitful, if not for constructing charts, etc., then at least for providing an algorithm for working out dam sizes.

2. DESCRIPTION OF THE METHODS

As mentioned in the Introduction, we shall, for simplicity of presentation, consider first the case of only two seasons.

Suppose the year is divided into two seasons, Summer (dry) and Winter (wet), say. Let the inflows (continuous random variables) during the summer and winter of the nth year be denoted by S_n and W_n respectively. Let the transition probability density of S_n given $W_{n-1} = x$ be denoted by $a(y|x)$, and the stationary marginal density of W be denoted by $k(x)$. Similarly, let $\ell(y)$ and $e(x|y)$ denote the stationary marginal of S and the conditional density of W_n given $S_n = y$. In any practical situation the functions $k(x)$, $a(y|x)$ etc., are obtained by fitting them to the historical data. For example, we may take these functions to be as given by the Thomas-Fiering (two-season) model.

Let the releases during Summer and Winter be M_1 and M_2 respectively, and let the contents of the reservoir at the beginning of Summer and Winter of the nth year be denoted by $C_{S(n)}$ and $C_{W(n)}$ respectively, so that we have the water balance equation, (with K as the reservoir size),

$$\begin{aligned} C_{W(n)} &= C_{S(n)} + S_n - M_1 && \text{if } 0 \leqslant C_{S(n)} + S_n - M_1 \leqslant K \\ &= K && \text{if } C_{S(n)} + S_n - M_1 \geqslant K \\ &= 0 && \text{if } C_{S(n)} + S_n - M_1 \leqslant 0 \end{aligned} \qquad (1)$$

and a similar relation connecting $C_{W(n-1)}$ with $C_{S(n)}$ with M_2 replacing M_1.

Method A : Simulation

In the simulation procedure, we generate a sequence of values of S_n and W_n according to the assumed inflow model, ignoring the initial values, and then using the water balance equation above, simulate the reservoir behaviour, with varying values for the initial content $C_{S(0)}$, say. The general practice is to generate about 1000 or so sequences each sequence as long as the historical sequence.

Let us now consider the three probability procedures. For all of them we approximate to the true situation by working in discrete quantities rather than in continuous ones. We take a suitable unit of water δ, and express all the quantities such as the inflows, drafts and the reservoir size, etc. in terms of this unit. Let the ranges of W_n and S_n be $(0,s+\frac{1}{2})$ and $(0,r+\frac{1}{2})$ respectively, i.e. the probabilities of W_n and S_n exceeding $s+\frac{1}{2}$ and $r+\frac{1}{2}$ respectively are negligible if not zero. Let us denote the intervals $(0, 1/2), (1/2, 3/2), \ldots, (s-1/2, s+1/2)$ of the variable W_n by $0, 1, 2, \ldots, s$, and similarly for the variable S_n. Then the expressions $k(x)$, $a(y|x)$, $\ell(y)$ and $e(x|y)$ are replaced by k_i, a_{ij}, ℓ_j and e_{ji} respectively, so that we have

$$k_i = \Pr\{W=i\} = \int_{i-\frac{1}{2}}^{i+\frac{1}{2}} k(x)\,dx, \quad \ell_j = \Pr\{S=j\} = \int_{j-\frac{1}{2}}^{j+\frac{1}{2}} \ell(y)\,dy$$

$$a_{ij} = \Pr\{S_n=j \mid W_{n-1}=i\} = \int_{j-\frac{1}{2}}^{j+\frac{1}{2}}\int_{i-\frac{1}{2}}^{i+\frac{1}{2}} k(x)\,a(y|x)\,dx\,dy/k_i$$

$$e_{ij} = \Pr\{W_n=i \mid S_n=j\} = \int_{j-\frac{1}{2}}^{j+\frac{1}{2}}\int_{i-\frac{1}{2}}^{i+\frac{1}{2}} \ell(y)\,e(x|y)\,dx\,dy/\ell_j.$$

With this discretization of the input distributions, the Markov dependence of the sequence $S_1, W_1, S_2, W_2, \ldots$ is now specified by the transition matrices $\underset{\sim}{A} = (a_{ij})$, and $\underset{\sim}{E} = (e_{ij})$ where $a_{ij} = \Pr\{W_{n-1} = i \rightarrow S_n = j\}$ and $e_{ij} = \Pr\{S_n = i \rightarrow W_n = j\}$.

Method B : Probability Matrix Method

Let the content space $[0,K]$ be discretized into $K+2$ states, $0 \equiv \{0\} \equiv$ Emptiness, $1 \equiv (0,1), \ldots\ i \equiv (i-1,i), \ldots\ K \equiv (K-1,K)$, $K+1 \equiv \{K\} \equiv$ Fullness. Let the division of the space $(0,s+\frac{1}{2})$ of W into the intervals $0 \equiv (0,\frac{1}{2})$, $1 \equiv (\frac{1}{2},3/2), \ldots, s \equiv (s-\frac{1}{2}, s+\frac{1}{2})$

be as before. In addition let us denote the intervals (0,1), (1,2),... $(s,s+\frac{1}{2})$ of W by 0', 1',...,s'. Consider now the transition $\{C_{W(n-1)},W_{n-1}\} \rightarrow \{C_{S(n)},S_n\}$ i.e. {Contents at the beginning of Winter, Winter input} → {Contents at the beginning of Summer, Summer input},with the initial states {0,0'}{0,1'} ... {0,s'}, {1,0},... {1,s},... {K,0},... {K,s}, {K+1,0'},... {K+1,s'} and with final states as above with r replaving s. We construct the transition probability matrix (t.p.m.) of the transition $\{C_{W(n-1)},W_{n-1}\}\rightarrow\{C_{S(n)},S_n\}$, a $(K+2)(s+1)X(K+2)(r+1)$ matrix $\underset{\sim}{L}_1$. We also construct the t.p.m. $\underset{\sim}{L}_2$ of the transition $\{C_{S(n)},S_n\}\rightarrow\{C_{W(n)},W_n\}$. This has the dimensions $(K+2)(r+1)X(K+2)(s+1)$. For details of construction see Phatarfod and Srikantan (1981). The product $\underset{\sim}{L}_2\underset{\sim}{L}_1$ gives the annual t.p.m. of the homogeneous Markov chain $\{C_{S(n)},S_n\}$.

Let us denote the equilibrium distribution of the pair (C_S,S) by π_{ij} i.e. π_{ij} = Pr.{Contents at the beginning of Summer = i, Summer input = j}. The equilibrium distribution vector,

$$\underset{\sim}{\pi} = (\pi_{00},\pi_{01},\ldots \pi_{0r},\pi_{10} \ldots \pi_{1r} \ldots \pi_{K+1,0},\ldots \pi_{K+1,r}),$$

is obtained by powering the matrix $\underset{\sim}{L}_2\underset{\sim}{L}_1$ till its rows have identical values. Summing over groups of r+1 values, $V_{Si} = \sum_{j=0}^{r} \pi_{ij}$ gives us $Pr.\{C_S = i\}$. To obtain the equilibrium distribution of the content at the beginning of winter we evaluate $\underset{\sim}{\mu} = \underset{\sim}{\pi}\underset{\sim}{L}_2$, a vector of $(K+2)(s+1)$ values. The sum $V_{Wi} = \sum_{j=0}^{s} \mu_{ij}$ gives us $Pr.\{C_W=i\}$.

Method C : Bottomless Dam Model (Mark 1)

Here we consider the depletions of the dam, assuming it to be bottomless. Defining $D_{S(n)},D_{W(n)}$ as the depletions corresponding to the contents $C_{S(n)},C_{W(n)}$, we have now

$$D_{W(n)} = D_{S(n)} + M_1 - S_n \quad \text{if} \quad D_{S(n)} + M_1 - S_n > 0$$

$$= 0 \qquad\qquad D_{S(n)} + M_1 - S_n \leqslant 0$$

and a similar relation connecting $D_{W(n-1)}$ with $D_{S(n)}$ with M_2 replacing M_1. The mathematical theory behind this procedure is given in Phatarfod (1981a). We give below the steps required to obtain a solution.

1. From the matrices $\underset{\sim}{A}$ and $\underset{\sim}{E}$, form the matrices $\underset{\sim}{A}(\theta) = (a_{ij}\theta^i)$, $\underset{\sim}{E}(\theta) = (e_{ij}\theta^i)$, i.e. $\underset{\sim}{A}(\theta)$ is a matrix formed from $\underset{\sim}{A}$ by multiplying each element of the ith row by θ^i, $(0 \leqslant i \leqslant s)$, and similarly for $\underset{\sim}{E}(\theta)$. Derive the non-zero solutions θ_k of the equation $\det.[\underset{\sim}{E}(\theta)\underset{\sim}{A}(\theta) - \theta^M \underset{\sim}{I}] = 0$, such that $|\theta_k| < 1$. In the above $M = M_1 + M_2$. It is known that in general there are $N = M_1(M_1+1)/2 + M_2(M_2+1)/2$ such solutions.

2. For each θ_k obtained in 1, form the matrix $\underset{\sim}{E}(\theta)\underset{\sim}{A}(\theta)$ and find its eigen-values. Take that eigenvalue which is equal to θ_k^M, and find its corresponding (normalized) eigenvector.

3. The equilibrium distribution of the depletion D_S is given by

$$\alpha_j = \Pr\{D_S=j\} = \sum_1^N Z_k \theta_k^j, \; j \geqslant 1, \; \alpha_0 = \Pr.[D_S=0] = 1 - \sum_{j=1}^{\infty} \alpha_j \qquad (2)$$

where the Z's are constants satisfying N linear equations. The distribution of the depletion D_W has a form similar to (2); see Phatarfod (1981a) for details.

Method D : Bottomless Dam Model (Mark 2)

We assume here that there is a dependence between the summer and winter flows of the same year, but that the summer flows are independent of the flow of the winter of the previous year. This may not be a terribly realistic model to assume; however, if we have six seasons, say, then it is not unrealistic to take the smallest correlation coefficient to be equal to zero; we are thus assuming that the input process starts afresh each year; the input in the first season has a certain probability distribution and the inputs in the remaining five seasons are governed by the five transition probabilities between the first and the last (sixth) season.

The theory behind this procedure is given in Phatarfod (1981b). The steps for the case of two seasons are:

1. Discretize the probability distribution of the annual inputs. Let it be denoted by $p_0, p_1, p_2, \ldots$ i.e. $p_i = \Pr\{i-\frac{1}{2} \leqslant S + W \leqslant i+\frac{1}{2}\}$.

2. Derive the non-zero solutions θ_k of the equation $p_0+p_1\theta+p_2\theta^2+\ldots = \theta^M$, such that $|\theta_k| < 1$. In general we have M such solutions.

3. The equilibrium solution is given by

$$\alpha_j = \Pr\{D_S=j\} = \sum_1^M Y_k\theta_k^j \; (j \geqslant 1), \; \alpha_0 = \Pr\{D_S=0\} = 1 - \sum_1^\infty \alpha_j \tag{3}$$

where the Y's are constants satisfying M linear equations. The distribution of D_W has a similar form; see Phatarfod (1981b) for details.

3. NUMERICAL EXAMPLE

We now use the above three methods for a specific input model. We assume that the inflows follow a Two-season Thomas Fiering model. Denoting the mean, the variance and the coefficient of skewness of the summer flows by μ_1, σ_1^2 and γ_1 respectively, those of the winter flows by μ_2, σ_2^2 and γ_2 respectively, the correlation coefficient between S_n and W_n by ρ_1 and that between W_n and S_{n+1} by ρ_2, the model is given by the equations

$$\begin{aligned} S_n &= \mu_1 + \frac{\rho_2\sigma_1}{\sigma_2}(W_{n-1}-\mu_2) + \sigma_1(1-\rho_2^2)^{\frac{1}{2}} X_n \\ W_n &= \mu_2 = \frac{\rho_1\sigma_2}{\sigma_1}(S_n-\mu_1) + \sigma_2(1-\rho_1^2)^{\frac{1}{2}} Y_n, \; n = 0,1,2,\ldots \; . \end{aligned} \tag{4}$$

For the model given in (4) the conditional distributions of S_n given W_{n-1} and of W_n given S_n are gamma. Using the tables of incomplete gamma function (or alternatively the IMSL/MDGAM (1979) Subroutine) the probabilities a_{ij}, etc. can be calculated. This exercise is not carried out here; instead, for the sake of consistency the transition probabilities a_{ij} etc. were obtained by using the same sequence of generated values as used for the simulation method

(Method A). A sequence of 2030 values (each) of S_n and W_n were obtained using the model (4). The first 30 values (each) of S_n and W_n were ignored to eliminate any effect of the initial conditions. The remaining 2000 values (each) of S_n and W_n were used to obtain the equilibrium distribution of the storage content by the Method A, with initial content 2 and reservoir sizes $K = 3,4$, as well as to estimate the transition probabilities a_{ij} etc.

For our example we have taken $\mu_1 = .6$, $\sigma_1 = .3$, $\gamma_1 = 1$, $\rho_1 = 0.5$, $\mu_2 = 2.0$, $\sigma_2 = 1.333$, $\gamma_2 = 1.0$, $\rho_2 = 0.1$. We also take $M_1 = M_2 = 1$, so that the draft-ratio is $2/2.6 = 0.77$.

The Tables below compare the equilibium (cumulative) distribution of the depletions of the dam obtained by all the methods. Table 2 shows, for example, that for $K = 4$, the probabilities of emptiness of the dam at the beginning of winter are .029 and .046, by methods A and B respectively. On the other hand, if $K = 4.5$, the probabilities obtained by Methods C and D are .048 and .052 respectively. This shows that Methods C and D do overestimate the size of a dam for a required probability of emptiness; however, the difference is not much. It is fairly obvious that if our unit of water is so chosen that we would require a dam size of about 10 units, then the four methods would give roughly equivalent answers.

TABLE 1

Equilibrium distribution of the depletion at the beginning of summer

Pr(Dep.⩽x) / x	A		B		C	D
	K = 3	K = 4	K = 3	K = 4		
0	.440	.426	.441	.423	-	-
0.5	.569	.556	-	-	.521	.505
1.0	.692	.676	.710	.681	-	-
1.5	.799	.778	-	-	.752	.727
2.0	.871	.846	.867	.831	-	-
2.5	.925	.902	-	-	.883	.865
3.0	.969	.938	.959	.923	-	-
3.5		.971		-	.954	.948
4.0		.984		.976		
4.5					.968	.959
	Prob. Empty = .031	Prob. Empty = .016	Prob. Empty =.0406	Prob. Empty = .024	Prob. Empty =.032	Prob. Empty =.041

TABLE 2

Equilibrium distribution of the depletion at the beginning of winter

Pr(Dep.⩽x) / x	A		B		C	D
	K = 3	K = 4	K = 3	K = 4		
0	.059	.056	.058	.056	-	-
0.5	.3.5	.305	-	-	.296	.284
1.0	.591	.574	.594	.570	-	-
1.5	.712	.697	-	-	.706	.697
2.0	.809	.791	.797	.765	-	-
2.5	.881	.856	-	-	.852	.838
3.0	.936	.909	.920	.882	-	-
3.5		.945		-	.923	.918
4.0		.971		.954	-	-
4.5					.952	.948
	Prob. Empty = .064	Prob. Empty = .029	Prob. Empty = .080	Prob. Empty =.046	Prob. Empty =.048	Prob. Empty =.052

REFERENCES

Dearlove, R.E. and Harris, R.A., 1965. Probability of Emptiness, III. Proc. Water Res. Assoc. Symp. on Reservoir Yield, Oxford, Pap. 7.

IMSL/GGAMS, 1976. International Mathematical and Statistical Libraries, Houston, Texas, Vol. 2, 7th ed.

Klemes, V., 1970. A two-step probabilistic model of storage reservoir with correlated inputs. Water Resour. Res., 6 (3): 756-767.

Kottegoda, N.T., 1970. Statistical methods of River flow synthesis for Water Resources Assessment. I.C.E. Supplement 18, (paper 73395).

Kritskiy, S.N. and Menkel, M.F., 1940. Obobshchennye priemy rascheta regulirovaniya stoka na usnove matematischeskoy statistiki. Gidrotekhnicheskoe stroitelstvo. 2: 19-24.

Lloyd, E.H., 1963. A probability theory of reservoirs with serially correlated inputs. J. Hydrol., 1: 99-128.

McMahon, T.A. and Mein, R.G., 1978. Reservoir capacity and yield. Elsevier, Amsterdam, 213 pp.

Moran, P.A.P., 1954. A probability theory of dams and storage systems. Aust. J. Appl. Sci. 5: 116-124.

Phatarfod, R.M., 1981a. The infinitely deep dam with Seasonal Markovian inflows. SIAM. J. Appl. Maths., 40: 400-408.

Phatarfod, R.M., 1981b. The infinitely deep dam with Seasonal Markovian inflows II. (under publication).

Phatarfod, R.M. and Srikanthan, R., 1981. Discretization in Stochastic reservoir theory with Markovian inflows. J. Hydrol., 52: 199-218.

Prabhu, N.U., 1958a. On the integral equation for the finite dam. Q.J. Math., Oxford, 9(2): 183-188.

Savarenskiy, A.D., 1940. Metod rascheta regulirovaniya stoka. Gidrotekhnicheskoe Stroitelstvo. 2: 24-28.

OPTIMAL ARMA MODELS FOR THE STATISTICAL ANALYSIS OF RESERVOIR OPERATING RULES

J.W. DELLEUR, M. GINI AND M. KARAMOUZ
Civil Engineering, Purdue University, West Lafayette, Indiana, USA

ABSTRACT

A time series of mean annual flows with a high Hurst coefficient is used. The Kitagawa search procedure for the optimal order p,q of ARMA models is seen to work well. Simulation of the annual flow series using a slightly modified McLeod and Hipel's Waterloo Simulation Procedure 1 preserved the rescaled range and the Hurst coefficient. Reservoir releases and storages were obtained for an operation rule that minimizes losses for a specified penalty function. Probability distributions of individual events and sequences 2,3,...,20 events were obtained for flows, releases and storages. Temporal disaggregation is used for the generation of monthly series. These in turn are used for the development of optimal seasonal release rules for the operation of the reservoirs. The reservoir reliability is then estimated in two different ways: in terms of the frequencies of failure years and of failure months.

DATA USED

The annual series examined is for the Blacksmith Fork near Hyrum, Utah (1913-1957 from Yevjevich (1963), 1957-1970 from USGS Water Supply Papers, 1970-1979 from Utah Power and Light Company).

The probability distribution of annual flows was found to be approximately normal. For n = 66 years of record the mean was $\bar{Q}_n$ = 127.62 cfs, the standard deviation S = 42.72 cfs, the adjusted range aR_n = 604.54 cfs, the rescaled adjusted ranges aR_n/S = 14.15 and the Hurst coefficient was estimated from $K = \log(aR_n/S) / \log(n/2) = 0.75$. The autocorrelation showed values at the 95% significance level at lags 1

Reprinted from *Time Series Methods in Hydrosciences*, by A.H. El-Shaarawi and S.R. Esterby (Editors)

and 12 and the partial autocorrelation function at lags 7 and 17.

FITTING ARMA MODELS TO THE ANNUAL SERIES

Autoregressive - moving average models were fitted to the standardized series $Z_t = (Q_t - \bar{Q}_n)/S$ where Q_t is the mean annual flow of year t. The model is

$$Z_t = \sum_{j=1}^{p} \phi_j Z_{t-j} - \sum_{j=0}^{q} \theta_j \varepsilon_{t-j} \tag{1}$$

where the ϕ_j $(j = 1,2,\ldots,p)$ are the autoregressive coefficients and the θ_j $(j = 1,2,\ldots,q)$ are the moving average coefficients and $\theta_0 = -1$.

The ARMA models were fitted by the method of maximum likelihood (mle). The search for the optimal model follows the method proposed by Kitagawa (1977) which is based on the Akaike information criterion (AIC) and the determination coefficients R^2 given by

$$AIC\ (p,q) = n \ln (\text{mle}\ \sigma_\varepsilon^2) + 2\ (p+q) \tag{2}$$

$$R^2 = 1 - (\sigma_\varepsilon^2/\sigma_z^2) \tag{3}$$

where σ_ε^2 is the variance of the residuals ε_t and σ_z^2 is the variance of the standardized annual series.

Kitagawa's (1977) procedure was used to search for the optimal order of the ARMA model. Its advantage is that the optimal values of p and q can, in general, be detected without fitting the whole set of possible models with all the combinations of the autoregressive and moving average parameters. A region defined by $0 \leq p \leq p_m$, $0 \leq q \leq q_m$ is specified where p_m and q_m are the maximum values of p and q considered. The selection of p_m and q_m is based on a study by Hashino and Delleur (1981) according to which $p_m \simeq 11$ to 15, and $q_m \simeq 2$ to 3.

In addition the models should be checked for stationarity and invertibility. Those which are not stationary or invertible or for which the model did not converge to a stable solution are eliminated. The model (11,0) was found to be the best. The better acceptable models are listed in Table 1.

TABLE 1. BETTER ARMA MODELS FOR THE BLACKSMITH FORK

ARMA	AIC	R^2	q'*
0,2	-19.74	0.302	2
4,0	-20.36	0.349	20
3,2	-21.33	0.378	52
8,0	-26.02	0.471	55
8,2	-24.57	0.491	100
11,0	-27.47	0.527	131
13,0	-24.72	0.536	143

*See following section

Before performing the simulations, diagnostic checks were performed for the whiteness and normality of the residuals. Plots of the probability distributions of the residuals on normal probability paper showed that the residuals are approximately normally distributed. The portmanteau lack of fit test passed at the 10% level in all cases indicating the adequacy of the models.

SIMULATION OF ANNUAL SERIES

The Waterloo Simulation Procedure 1 of McLeod and Hipel (1978) has been used because random realizations of the underlying stochastic process are used as initial values, thus avoiding bias in simulations. For the generation of the first r terms ($r = \max(p,q)$), the procedure requires the approximation of the ARMA (p,q) model by a MA (q') model where the order q' is selected so that the differences between the theoretical variance of the ARMA (p,q) model and the variance of the MA (q') model is less than a specified error value. The order q' is shown in Table 1 using an error level of 10^{-3}. It is seen that q' increases rapidly as p increases. In order to keep the computational burden within reason, the ARMA (4,0) model was finally selected as it requires only 20 moving average terms to be calculated. The model is

$$Z_t = 0.5125\ Z_{t-1} - 0.0316\ Z_{t-2} - 0.0388\ Z_{t-3} + 0.1239\ Z_{t-4} + \varepsilon_t \qquad (4)$$

One slight modification was introduced in the procedure: when a random variable produces a negative flow, the random variable is discarded and the next random variable is introduced in the calculations. This is equivalent to using a truncated distribution of the flows. The number of simulated series was selected as N = 3,000 and the length of each was n = 500 years.

The preservation of the Hurst coefficients is checked by using the empirical cumulated distribution function (ECDF) of K for the 3000 simulations. The mean Hurst coefficient is 0.70, its variance is 0.0017 and the probability that the simulated K be larger than the historical K is 0.112, thus the Hurst coefficient is preserved.

STATISTICAL CHARACTERISTICS OF RESERVOIR OPERATION

The generated inflow series are routed through a single reservoir of known capacity operated in accordance with a release rule designed to minimize the total losses from the operation. The loss function is defined as a piecewise exponential function. Within a specified safe release range (RLOW $\leq$ release $\leq$ RUP) there is no loss as the release is large enough to satisfy the demand and yet is small enough to prevent flooding. The loss function is thus defined as

$$\mathrm{Loss}(R_t) = A[\exp(R_t/\mathrm{RUP}) - \exp(1)] \quad \text{if } R_t \geq \mathrm{RUP} \tag{5a}$$

$$\mathrm{Loss}(R_t) = 0 \quad \text{if } \mathrm{RLOW} \leq R_t \leq \mathrm{RUP} \tag{5b}$$

$$\mathrm{Loss}(R_t) = B[\exp(-R_t/\mathrm{RLOW}) - \exp(-1)] \quad \text{if } R_t \leq \mathrm{RLOW} \tag{5c}$$

where A and B are known constants that depend on the price of the water and on how extensive the property damage is, and R_t is the release during year t. For annual flows the values of the constants are taken as follows $A = 3.88 \times 10^5$, $B = 1.58 \times 10^6$, RUP = 1.2 (mean annual flow), RLOW = 0.8 (mean annual flow). The safe range is thus within 20% of the mean annual flow, and the values of A and B result in losses of 10^6 units when the release is zero or twice the mean annual flow.

The objective function is to minimize the total losses for the T years of expected economic life of the reservoir:

$$\text{Minimize } \sum_{t=1}^{T} \mathrm{LOSS}(R_t) \tag{6}$$

subject to the following constraints:

i) the mass balance of the reservoir (continuity)

$$S_{t+1} - S_t + R_t = I_t \qquad t = 1, 2, \ldots, t \tag{7a}$$

where I_t=inflow during year t, S_t=storage at the beginning of year t,

$$\text{ii)} \;\; R_t \geq R_t^{min}, \qquad \text{iii)} \;\; R_t \leq R_t^{max}$$
$$\text{iv)} \;\; S_t \geq S_t^{min}, \qquad \text{v)} \;\; S_t \leq S_t^{max} \tag{7b}$$

where the superscripts min and max indicate the minimum or maximum.

Karamouz and Houck (1981 a) solved this problem as an iterative discrete dynamic problem and regression analysis, using 20 discrete storage volumes uniformly distributed between zero and full reservoir capacity. They regressed the optimal storage, optimal release and concurrent inflow by means of the equation

$$R_t = a\, I_t + b\, S_t + c \tag{8}$$

for different bounds on R_t^{min} and R_t^{max} as follows

$$R_t^{max} = (1 + \text{BOUND})\,(a\, I_t + b\, S_t + c) \tag{9a}$$

$$R_t^{min} = \text{maximum}\,[0;\ (1 - \text{BOUND})\,(a\, I_t + b\, S_t + c)] \tag{9b}$$

In (9 a,b) the quantity $(a\, I_t + b\, S_t + c)$ represents the release rule obtained in the previous iteration.

Combining equations (7a) and (8) one obtains

$$S_{t+1} = (1 - a)\, I_t + (1 - b)\, S_t - c \tag{10}$$

If $S_{t+1} > 0$, then R_t is given by the release rule (8), if $S_{t+1}<0$, then S_{t+1} is set equal to zero and $R_t = I_t + S_t$. The storage S_{t+1} cannot exceed the reservoir capacity, CAP, and the excess is released. Therefore if $R_t \leq I_t + S_t$ - CAP then $R_t = I_t + S_t$ - CAP and S_{t+1} = CAP.

Five hundred years of simulated annual flows were routed through reservoirs of several capacities and the annual releases and storages were obtained as explained. Figure 1 shows the empirical cumulative probability distributions of the flows and releases for 1 and 3 sequential years and of the storages for 1, 2, 3, 4, 5, 6, 10 and 20 sequential years. These are shown for a storage coefficient (storage capacity/mean total annual runoff volume) of 1.4 and two different BOUND values. Similar distributions were obtained for storage coefficients of 1.0; 0.5 and 0.2, and for three BOUND values each time.

The probability distributions of k sequential releases are seen to lie one above the other as k decreases and do not intersect each other. Thus, Prob (k + 1 sequential releases $\leq R_i$) $\leq$ Prob (k sequential releases $\leq R_i$) where R_i is a specified release value. Similar statements can be written for the flows and the storages.

Comparing the probability distributions for one year low flows, I_i, Prob (1 year release $\leq I_i$) < Prob (1 year flow $\leq I_i$), so the probability of droughts is decreased by the reservoir. Likewise, comparing the probabilities of exceedance for high flows, the probability of floods is decreased by the reservoir, as expected. Comparing the deviations between the probabilities of inflows and of releases for the same flow values for several storage coefficients, the deviations are seen to increase as the storage coefficient increases. Thus larger reservoirs provide more control of the flows.

Comparing the probability distribution of storages and the values of the BOUND it appears that for higher values of BOUND the cumulative distribution increases more rapidly than for the lower values of BOUND. For the infinite BOUND values it exhibits a larger percentage of failures (reservoir empty or full), for example, for the storage coefficient = 1.4, the reservoir is empty 14.4% and full 16.2% of the time whereas there is virtually no failure with the lower BOUND values.

ANNUAL RELIABILITY

The occurrence based annual reliability R_a, is defined as the number of non-failure years expressed as a percentage of the total number of years in the given period, it is thus equivalent to the probability that the reservoir will deliver the expected demand throughout its lifetime without incurring a deficiency. The reliability characteristics are computed for stationary conditions, that is for a long operation

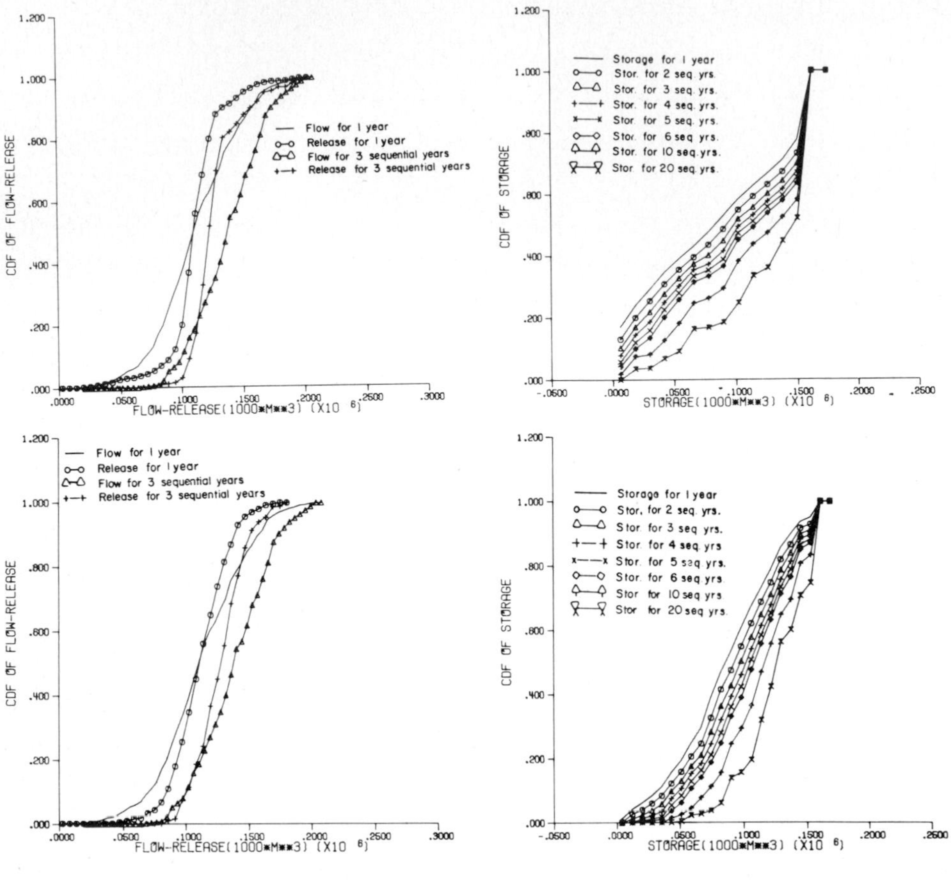

FIGURE 1. Probability distributions of flows, releases and storages with storage coefficient of 1.4 and BOUND of infinity (top) and 0.03 (bottom).

period not influenced by initial conditions of storage. This is done through the generation of 1000 replicate series of 500 years by means of the previously described ARMA model. These inflow series are routed through reservoirs of various combinations of sizes and drafts making use of the previously developed release rules and the reliability characteristics are then calculated. The averages of these annual reliabilities are shown in Fig. 2 as a function of the storage ratio (storage/mean annual runoff volume) and draft ratio (draft rate/ mean annual flow) for the several BOUND values used in the release rules. When BOUND is large, namely the maximum release is not constrained, there are many reservoir failures (reservoir empty or filled), but this rule controls the flows better by eliminating more effectively the extremes (floods and droughts). However, the amplitudes of the reservoir fluctuations are relatively large. When the BOUND is small, namely the range of permissible releases is small in the dynamic program, the number of failures is very small, but the control of the extreme flows is less effective. Usually these differences are most visible when the storage coefficient is 1.0 or larger. The annual reliability of the releases is very sensitive to the value of the draft ratio and appears to be essentially insensitive to the BOUND values.

MONTHLY RELIABILITY

The disaggregation model of Mejia and Rousselle (1976) was used to simulate the monthly flows. The procedures for the estimation of the parameters and for generation are given in Salas et al.(1980) Chapter 8.

The loss function for monthly flow is of the same form as shown in equ. (5) but with RUP = 1.2 (mean monthly flow), RLOW = 0.8 (mean monthly flow). The safe range is thus within 20% of the mean monthly flow (averaged over the 12 months). The values of A and B are the same as before and result in a loss of 10^6 units when the release is zero or twice the mean monthly flow.

The release rules for monthly flows are of the same form as in equ. (8) where I_t and R_t represent the inflow and the release during month t and S_t is the storage at the beginning of month t. As before, the release rules were obtained by regression of the optimal release vs. the optimal storage (resulting from the discrete dynamic program) and the current inflow (Karamouz and Houck, 1981 b).

Four hundred years of monthly reservoir operations have been computed by routing the monthly flows through the reservoir in accordance with the optimal release rules. The averages of the 4800 months reliabilities are plotted in Fig. 3 as a function of the storage ratio and of the draft ratio for the several BOUND values used in the definition of the release rule. The BOUND is seen to affect the reliability. In

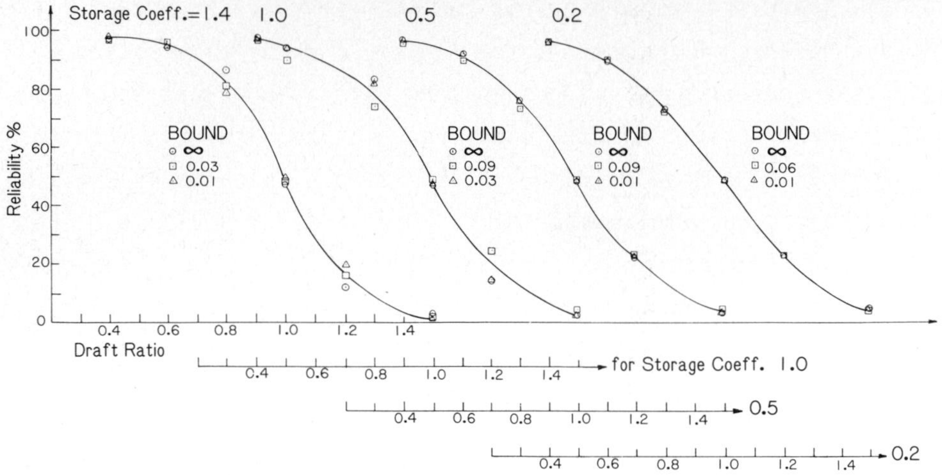

FIGURE 2. Annual reliability.

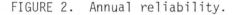

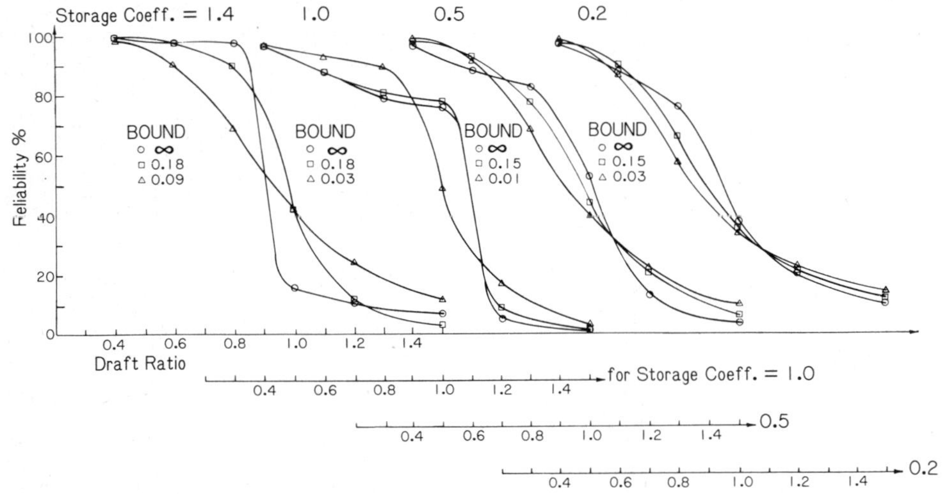

FIGURE 3. Monthly reliability.

general, for D $\geq$ 1.2 the lower BOUND results in a slightly higher reliability, whereas the higher BOUND results in a higher reliability in the vicinity of a draft ratio of 0.8 or 1.0. As the draft ratio becomes small, of the order of 0.4, the reliability tends to 100% regardless of the operating rule.

For monthly flows the number of reservoir failures is very small with small BOUND on the releases and increases as BOUND increases. The monthly reliability of the releases is very sensitive to the draft ratio but is influenced in varying ways by the BOUND values of the release rule.

ACKNOWLEDGEMENTS

This material is based upon work supported by the National Science Foundation under Grant No. CME 7916819. The writers wish to thank Professor M.H. Houck for his assistance throughout the research.

REFERENCES

Box, G.E.P. and Cox, D.R. Analysis of Transformations, J. Roy Statist. Soc. Ser. B. 26, 211-252, 1964.

Hashino, M. and Delleur, J.W., Investigation of the Hurst Coefficient and Optimization of ARMA Models for Annual River Flows, Tech. Rept. CE-HSE-81-1, School of Civil Engineering, Purdue University, 1981.

Karamouz, M. and Houck, M.H., Annual Operating Rules Generated by Deterministic Optimization for a Single Multipurpose Reservoir, School of Civil Engineering, Purdue University, Tech. Rept. CE-HSE-81-11, 1981, a.

Karamouz, M. and Houck, M.H., "Generation of Monthly and Annual Reservoir Operating Rules", Tech. Rept. CE-HSE-81-16, School of Civil Engineering, Purdue University, Dec., 1981, b.

Kitagawa, G., On a Search Procedure for the Optimal AR-MA Order, Ann. Inst., Statist. Math., Vol. 29, Part B, pp. 319-332, 1977.

McLeod, A.I. and Hipel, K.W., Simulation Procedures for Box-Jenkins Models, Water Resources Research, Vol. 14, No. 5, 1978.

Mejia, J.M. and Rousselle, J., Disaggregation Models in Hydrology Revisited, Jour. Water Res. Res. 12(2), pp. 185-186, 1976.

Salas, J.D., Delleur, J.W., Yevjevich, V., and Lane, W.L., Applied Modeling of Hydrologic Time Series, Water Resources Publications, Littleton, Colorado, 1980.

Yevjevich, V., Fluctuations of Wet and Dry Years, Part I, Research Data Assembly and Mathematical Models, Hydrology Paper No. 1, Colorado State University, 1963.

AN ANNUAL-MONTHLY STREAMFLOW MODEL FOR INCORPORATING PARAMETER UNCERTAINTY INTO RESERVOIR SIMULATION

JERY STEDINGER AND DANIEL PEI

Department of Environmental Engineering, Cornell University, Ithaca, N.Y. 14853

ABSTRACT

A monthly streamflow model is developed which can reproduce the variance and year-to-year correlation of an annual streamflow surrogate with a modest number of parameters. The model's simple structure facilitates the incorporation of streamflow-model-parameter uncertainty into synthetic streamflow sequences. The large impact of parameter uncertainty on derived reservoir capacity-reliability-demand relationships is illustrated by comparing the relationships obtained ignoring parameter uncertainty with those obtained when streamflow-model parameter uncertainty is incorporated into synthetic flow sequences.

INTRODUCTION

Synthetic streamflow sequences have long been viewed as a means of improving our ability to estimate the likely performance of reservoir systems and to free system planning studies from a total reliance on the particular flows which occurred during the period of record (Maass et al., 1962). However, the parameters of synthetic streamflow models are subject to sampling or unavoidable errors of estimation (Stedinger, 1980a, 1981; Loucks et al., 1981, Appendix 3C) and these errors can have a major impact on estimates of reservoir performance (Klemeš et al., 1981; Burges and Lettenmaier, 1981; Klemeš, 1979; Stedinger and Taylor, 1982) and the moments of generated monthly flows (Klemeš and Bulu, 1979; Stedinger, 1980b). This paper develops a new streamflow model which can be used to generate synthetic monthly streamflow sequences which incorporate the unavoidable uncertainty in streamflow model parameters.

Reprinted from *Time Series Methods in Hydrosciences*, by A.H. El-Shaarawi and S.R. Esterby (Editors)

MODEL STRUCTURE

A reasonable model of monthly and annual flows might reproduce (1) the mean, variance and other parameters of the marginal distribution of flows in each month, (2) the month-to-month correlation of flows in consecutive months, (3) the variance of the total flow within water years, and (4) the year-to-year correlation of annual flows. Such a model should be reasonable for many applications. Like the Thomas-Fiering model (Thomas and Fiering, 1962), it reproduces the correlation of flows in consecutive months. By reproducing the year-to-year correlation of annual flows it should provide a reasonable description of the persistence of those flows; high order ARMA(p,q) models are seldom necessary (Hipel and McLeod, 1978; Wallis and O'Connell, 1973; Klemeš _et al._, 1981; Burges and Lettenmaier, 1981).

It is difficult to achieve these objectives in the form articulated when monthly and annual flows have other than a normal distribution. If monthly flows q_{yt} have a three-parameter log normal distribution so that $x_{yt} = \ell n(q_{yt} - \tau_t)$ has a normal distribution, streamflow models are most conveniently formulated in terms of the x_{yt}. It is then difficult to insure that the generated annual flows have the desired properties (Loucks _et al._, 1981, p. 303). This difficulty can be circumvented if instead of modelling the annual flows, attention is focused on an annual streamflow surrogate. In particular, consider the first-order approximation of the annual flows

$$Z_y = \sum_{t=1}^{12} w_t (x_{yt} - \mu_t) + \sum_{t=1}^{12} [\tau_t + \exp(\mu_t)] \qquad (1)$$

where the weights w_t are the expected value of the derivatives dq_{yt}/dx_{yt}. For $x_{yt} = \ell n[q_{yt} - \tau_t]$, $w_t = \exp(\mu_t + \sigma_t^2/2)$.

To develop a simple model to generate x_{yt}'s which yields values of Z_y with a particular variance, note that

$$(Z_y - \mu_z)^2 = [\sum_{t=1}^{12} w_t(x_{yt} - \mu_t)]^2 = w_1^2(x_{y1} - \mu_1)^2$$

$$+ \sum_{t=2}^{12} [w_t^2(x_{yt} - \mu_t)^2 + 2w_t(x_{yt} - \mu_t) \sum_{s=1}^{t-1} w_s(x_{ys} - \mu_s)] \quad (2)$$

Hence, for a model which reproduces the variance of $E[(x_{yt} - \mu_t)^2]$ for each x_{yt}, it is only necessary to reproduce the covariance between each x_{yt} and $\sum_{s=1}^{t-1} w_s(x_{ys} - \mu_s)$ to reproduce the variance of Z_y as well.

A reasonable model of monthly flows that can reproduce the mean, variance and month-to-month correlation of monthly flows as well as the variance and year-to-year correlation of the annual streamflow surrogate is

$$x_{y,1} = \alpha_1 + \beta_1 x_{y-1,12} + \gamma_1 Z_{y-1} + v_{y,1} \quad (3)$$

$$x_{y,2} = \alpha_2 + \gamma_2 Z_{y-1} + \delta_2 x_{y,1} + v_{y,2} \quad (4)$$

and for $t \geq 3$

$$x_{y,t} = \alpha_t + \beta_t x_{y,t-1} + \gamma_t Z_{y-1} + \delta_t \sum_{s=1}^{t-1} w_s x_{ys} + v_{y,t} \quad (5)$$

where $v_{y,t}$ are independent zero-mean normal random variables. The coefficients of Z_{y-1} can be selected to reproduce the covariance between x_{yt} and Z_{y-1}, thus reproducing the covariance of annual flow surrogates:

$$E[(Z_y - \mu_z)(Z_{y-1} - \mu_z)] = \sum_{t=1}^{12} w_t E[(x_{yt} - \mu_t)(Z_{y-1} - \mu_z)] \quad (6)$$

MODEL-PARAMETER UNCERTAINTY

Given the finite and often short length of historical streamflow sequences, parameters of annual and monthly streamflow models can be estimated with only limited precision. The annual-monthly streamflow model provides a convenient structure for generating monthly streamflow sequences which incorporate the uncertainty in the parameters describing the joint distribution of monthly streamflows. Except for Beard's proposal

(Beard, 1965), earlier studies have considered only the uncertainty in statistics describing the distribution of annual flows (Vicens et al., 1975; Valdes et al., 1977; Wood, 1978; McLeod and Hipel, 1978; Stedinger and Taylor, 1982).

Methods for incorporation of model parameter uncertainty into the streamflow generation process may be developed using Bayesian inference theory for the normal regression model (Zellner, 1971, Chapter 3). Because the innovation terms $v_{y,t}$ in Equations 3 through 5 are distributed independently of one another, the parameter vectors $\underline{\beta} = (\alpha_t, \beta_t, \gamma_t, \delta_t)^T$ for each month will also be distributed independently provided their prior distributions are independent. Hence, analysis of the model required to generate x_{yt}'s for each t is essentially equivalent to analysis of individual normal regression models

$$\underline{y} = \underline{X}\,\underline{\beta} + \underline{V} \tag{7}$$

where for each $t \geq 3$, the ith row of $\underline{X}$ is $(1,\ x_{i,t-1},\ Z_{i-1},\ \sum_{s=1}^{t-1} w_s\, x_{i,s})$ and $\underline{V} = [v_{1,t},\ldots,v_{n,t}]^T$.

In this initial work, a non-information or Jeffrey's prior distribution is used to indicate that little is known about $\underline{\beta}$ and σ apart from the information provided by the historical flow record (Box and Tiao, 1973). This is a reasonable choice if available prior information is dominated by that provided by the streamflow record.

Stedinger and Pei (1981) summarize the Bayesian analysis of the model with a non-informative prior. Letting

$$\hat{\beta}_t = (\underline{X}_t^T\,\underline{X}_t)^{-1}\,\underline{X}_t^T\,\underline{y}_t \tag{8}$$

and

$$s_t^2 = (\underline{y}_t - \underline{X}_t\,\hat{\underline{\beta}}_t)^T\,(\underline{y}_t - \underline{X}_t\,\hat{\underline{\beta}}_t)/(n-k) \tag{9}$$

where k is the number of columns in $\underline{X}$, $(n-k)s_t^2/\sigma_t^2$ has a Chi-squared distribution where σ_t^2 is the unknown vari ce of the $v_{i,t}$'s. For given σ_t^2,

$$\underline{\beta}_t \sim N[\hat{\underline{\beta}}_t, \sigma_t^2(\underline{X}_t^T \underline{X}_t)^{-1}] \tag{10}$$

where $\underline{\beta}_t$ are the unknown model parameters for month t.

Streamflow sequences which reflect both the natural hydrologic variability of streamflows and what is known about the model's parameters were generated in two steps. First, N complete sets of model parameters were drawn from their posterior distribution. Each complete set of model parameters was used to generate one streamflow sequence. These flow sequences reflect both what the true values of the streamflow model's parameters may be and the characteristics of flow sequences that the model would produce with those parameter values (McLeod and Hipel, 1978; Davis, 1977).

SIMULATION RESULTS

The annual-monthly model was used to describe the character of monthly flows in the Upper Delaware River Basin in New York State. The monthly flows were modelled by a 3-parameter log normal distribution using the quantile lower bound estimator developed by Stedinger (1980a).

The 50-year historical flow record provided the statistics used to generate streamflow sequences. Flows were generated assuming the historical record was of length m = 25 or 50 years; in the former case, the values of s_t^2, $\hat{\underline{\beta}}_t$ and $(\underline{X}^T\underline{X}/n)$ were those obtained with the entire 50-year flow record. Here m may be viewed as an effective record length if informative prior distributions were used to derive the posterior distributions of $\underline{\beta}_t$ and σ_t^2.

Streamflow sequences were also generated which reflect only the hydrologic variability of flows that one would expect if the model's parameters assumed the estimated values. Two cases are considered. In the first, denoted $m = \infty_{n-1}$, each $\underline{\beta}_t$ was assigned the value $\hat{\underline{\beta}}_t$ with

$$\sigma_t^2 = (\underline{y}_t - \underline{X}_t \hat{\underline{\beta}}_t)^T (\underline{y}_t - \underline{X}_t \hat{\underline{\beta}}_t)/(n-1) \quad (11)$$

This is the value of σ_t^2 needed to reproduce the observed sample variance of the y_t's. In the second case, denoted $m = \infty_{n-k}$, the $\underline{\beta}_t$ was again assigned the value $\hat{\underline{\beta}}_t$ while the residual variance σ_t^2 equalled s_t^2 (Equation 9): Beard (1965) makes use of this unbiased estimator of the residual variance.

The sequent peak algorithm was used to determine S_{req}, the reservoir capacity required to regulate each of 1000 generated synthetic flow sequences so as to provide an annual diversion D of 30%, 50%, 70% and 90% of the historical mean annual flow, assuming the reservoir started full. A 25-year planning period is assumed. To make the results dimensionless, S_{req} is reported as a fraction of the historical mean annual flow. Table 1 reports the mean and standard deviation of S_{req}'s distribution. There is little difference between $m = \infty_{n-1}$ and ∞_{n-k}. As m goes from ∞_{n-1} to 50 and 25, the average value of S_{req} increases steadily. The standard deviation increased dramatically; except for D = 0.90, the increase for m = 50 is in excess of 75% of the value for $m = \infty_{n-1}$. Stedinger and Pei (1981) show that this increase is primarily due to increases in the upper quantiles of the distribution of S_{req}.

TABLE 1. Average and Standard Deviation of S_{req} for Various Demand Levels.

Demand Level (%MAF)	Average m =				Standard Deviation m =			
	∞_{n-1}	∞_{n-k}	50	25	∞_{n-1}	∞_{n-k}	50	25
30%	0.08	0.08	0.09	0.09	0.018	0.019	0.03	0.04
50%	0.18	0.18	0.19	0.20	0.032	0.033	0.08	0.11
70%	0.34	0.34	0.37	0.39	0.088	0.091	0.20	0.28
90%	0.89	0.89	0.92	0.97	0.394	0.395	0.53	0.67

For several reservoir capacity-demand combinations, the system's performance was also summarized by the expected value and standard deviation of two statistics: (1) the occurrence-based failure frequency F_a is the frequency of failure years during the planning period; (2) the quantity-based failure statistic V_a is the total shortfall or deficit during the planning period divided by the annual demand (Klemeš et al., 1981).

The distribution of S_{req} allows determination of the probability of failure-free reservoir operation during the entire planning period and the frequency-magnitude relationship for the worst shortfall that occurs. These quantities are of major importance in the study of reservoir systems which fail infrequently, such as municipal water supply reservoirs. Other systems meeting agriculture demands may be designed to fail, on average, one in ten years (F_a = 0.10). In this instance, use of S_{req} as a hydrologic criterion for comparing stochastic streamflow models may be inappropriate; use of F_a and V_a may provide a better assessment of the frequency and character of system failures.

Stedinger and Pei (1981) report the expected values and standard deviations of F_a and V_a obtained with the 1000 25-year synthetic streamflow sequences for storage capacities S = 0.125, 0.25, 0.50, 1.00 and 2.00 times the historic mean annual flow (MAF) and with demand levels D = 50%, 70% and 90% of the MAF. Tables 2 and 3 summarize the seven cases for which F_a (with $m = \infty_{n-1}$) fell between 0.1% and 20%, the region of greatest practical interest. In the first three cases in Table 2, F_a increased by factors of 3 to 5 for m = 25 over their values for $m = \infty_{n-1}$. In subsequent cases with initially higher failure rates, parameter uncertainty had less dramatic, though still substantial, an impact.

Table 3 shows that the impact of parameter uncertainty on the moments of V_a is much larger than it was on the moments of F_a. For S = 0.50 and D = 0.90, the average value of V_a between the $m = \infty_{n-1}$ and m = 25 cases increased by 18% of the smaller mean; the corresponding increase in F_a's mean was slightly over 1%. V_a is an important index because it is a direct measure of the magnitude of water deficits and thus perhaps of the hardship that might be incurred by those who expected that water. The large increases (generally exceeding 50% between $m = \infty_{n-1}$ and m = 25 and averaging 520%) in V_a's mean are accompanied by even larger increases in V_a's standard deviation.

TABLE 2. Average and Standard deviation of Simulated Annual Failure Frequency F_a for Selected Cases.

		Average m =			Standard Deviation m =		
S	D	∞_{n-1}	50	25	∞_{n-1}	50	25
0.25	0.5	0.001	0.005	0.007	0.007	0.024	0.032
2.00	0.9	0.002	0.005	0.009	0.015	0.037	0.055
0.50	0.7	0.004	0.008	0.012	0.016	0.036	0.051
1.00	0.9	0.037	0.041	0.047	0.073	0.083	0.098
0.25	0.7	0.112	0.133	0.144	0.076	0.097	0.104
0.50	0.9	0.163	0.163	0.165	0.120	0.129	0.132
0.13	0.5	0.171	0.192	0.204	0.087	0.105	0.107

TABLE 3. Average and Standard Deviation of V_a for Selected Cases.

		Average m =			Standard Deviation m =		
S	D	∞_{n-1}	50	25	∞_{n-1}	50	25
0.25	0.5	0.001	0.015	0.026	0.012	0.16	0.24
2.00	0.9	0.005	0.029	0.056	0.050	0.28	0.45
0.50	0.7	0.007	0.036	0.057	0.041	0.24	0.36
1.00	0.9	0.13	0.17	0.22	0.29	0.47	0.65
0.25	0.7	0.19	0.28	0.32	0.19	0.38	0.50
0.50	0.9	0.53	0.58	0.63	0.50	0.67	0.82
0.13	0.5	0.25	0.33	0.38	0.17	0.35	0.44

CONCLUSIONS

A model was developed which can reproduce the mean, variance and month-to-month correlation of monthly flows and the mean, variance and year-to-year correlation of an annual streamflow surrogate. The new model has an autoregressive structure which allows the application of Bayesian inference theory. This facilitates development of streamflow generation algorithms which incorporate the uncertainty in estimated streamflow-model-parameters. The new model was used to illustrate the impact of parameter uncertainty on derived reservoir capacity-demand-reliability relationships. With a 25 or 50-year historical flow record, streamflow-model-parameter uncertainty can have an appreciable impact on our best estimate of system reliability and especially on our assessment of possible failure magnitudes.

ACKNOWLEDGMENTS

This work was supported by NSF Grant CME-8010889.

REFERENCES

Beard, L.R., 1965, Use of interrelated records to simulate streamflow, Jour. Hydr. Div. (ASCE), 91(HY5), 13-22.

Box, G.E.P., and Tiao, G.C., 1973. Bayesian Inference in Statistical Analysis, Addison-Wesley, Reading, Mass.

Burges, S.J., and Lettenmaier, D.P., 1981. Reliability measures for water supply reservoirs and the significance of long-term persistence, International Symposium on Real-Time Operation of Hydrosystems, Waterloo, Ontario, June.

Davis, D.R., 1977. Comment on 'Bayesian generation of synthetic streamflows', by G.J. Vicens, I. Rodriguez-Iturbe, and J.C. Schaake, Jr., Water Resour. Res. 13(5), 853-854.

Hipel, K.W., and McLeod, A.I., 1978. Preservation of the rescaled adjusted range 2. Simulation studies using Box-Jenkins models, Water Resour. Res. 14(3), 509-516.

Klemeš, V., 1979. The unreliability of reliability estimates of storage reservoir performance based on short streamflow records, in Reliability in Resources Management, edited by E.A. McBean, K.W. Hipel, and T.E. Unny, Water Resources Publications, Littleton, Colorado.

Klemeš, V., and Bulu, A., 1979. Limited confidence in confidence limits derived by operational stochastic hydrologic models, Jour. of Hydrology 42, 9-22.

Klemeš, V., Srikanthan, R., and McMahon, T.A., 1981. Long-memory flow models in reservoir analysis: What is their practical value?, Water Resour. Res. 17(3), 737-751.

Loucks, D.P., Stedinger, J.R., and Haith, D.A., 1981. Water Resource Systems Planning and Analysis, Prentice-Hall,

Englewood Cliffs, New Jersey.
Maass, A., et al., 1962. Design of Water Resource Systems, Harvard University Press, Cambridge, MA.
McLeod, A.I., and Hipel, K.W., 1978. Simulation procedures for Box-Jenkins models, Water Resour. Res. 14(5), 969-975.
Stedinger, J.R., 1980a. Fitting log normal distributions to hydrologic data, Water Resour. Res. 16(3), 481-490.
Stedinger, J.R., 1980b. Comment on 'Limited confidence in confidence limits derived by operational stochastic hydrologic models', Jour. of Hydrology 43, 377-380.
Stedinger, J.R., 1981. Estimating correlations in multivariate streamflow models, Water Resour. Res. 17(1), 200-208.
Stedinger, J.R., Taylor, M.R., 1982. Synthetic streamflow generation, Part II: Parameter uncertainty, to appear in Water Resour. Res. 17( ), 000-000.
Stedinger, J.R., and Pei, D., 1981. An annual-monthly streamflow model for incorporating parameter uncertainty into reservoir simulation, Technical Report, December, Department of Environmental Engineering, Cornell University.
Thomas, H.A., and Fiering, M.B , 1962. Mathematical synthesis of streamflow sequences for the analysis of river basins by simulation, in Design of Water Resource Systems by A. Maass et al., Harvard University Press, Cambridge, Mass.
Valdes, J.R., Rodriguez-Iturbe, I., and Vicens, G.J., 1977. Bayesian generation of synthetic streamflows 2. The multivariate case, Water Resour. Res. 13(2), 291-295.
Valencia, D., and Schaake, J.C., Jr., 1973. Disaggregation process in stochastic hydrology, Water Resour. Res. 9(3), 580-585.
Vicens, G.J., Rodriguez-Iturbe, I., and Schaake, J.C., Jr., 1975. Bayesian generation of synthetic streamflows, Water Resour. Res. 11(6), 827-838.
Wallis, J.R., and O'Connell, P.E., 1973. Firm reservoir yield-how reliable are historic hydrologic records, Hydrol. Sci. Bull. 18(3), 347-365.
Wood, E.F., 1978. Analyzing hydrologic uncertainty and its impact upon decision making in water resources, Advances in Water Resources 1(5), 299-305.
Zellner, A., 1971. An Introduction to Bayesian Inference in Econometrics, J. Wiley & Sons, Inc., New York.

STOCHASTIC FLOOD PREDICTORS: EXPERIENCE IN A SMALL BASIN

P. BOLZERN, G. FRONZA AND G. GUARISO

Istituto di Elettrotecnica ed Elettronica
Centro Teoria dei Sistemi - Politecnico di Milano,
P.za Leonardo da Vinci n. 32 - 20133 Milan - Italy

ABSTRACT - The paper describes the application of three stochastic predictors of river flow-rate to a small basin (upper basin of Temo, in Sardinia $\simeq 180\ km^2$). The first predictor (ARX, AutoRegressive with eXogenous input) requires quantitative rainfall measurement, the second one (ARQX, AutoRegressive with Qualitative eXogenous input) needs only a gross information about rainfall, the third one (AR, AutoRegressive) has no rainfall input. The performance of the three predictors turns out satisfactory up to 3-4 hours, particularly by ARX. However, the gap of quality between the ARX and the other two predictors is counterbalanced by the cost of the rainfall telemetering system required by ARX.

1. INTRODUCTION

A certain number of mathematical models have been proposed in the technical literature, in order to supply real-time forecast of river flow-rates, particularly in flood situations. Most of these models, such as lumped parameter deterministic (see for instance Dooge, 1977) or stochastic, are simplified repre-

Reprinted from *Time Series Methods in Hydrosciences,* by A.H. El-Shaarawi and S.R. Esterby (Editors)

sentations of the rainfall-runoff phenomenon. As a matter of fact applying the classical partial differential equations usually raises a number of conspicuous operational difficulties, both from data and computational viewpoint.

The most common type of stochastic model (see for instance Chiu (1978), Bolzern et al. (1980)), proposed for describing the rainfall-flow-rate mechanism, is ARMAX, namely, AutoRegressive Moving Average with eXogenous input (rainfall). In an ARMAX, the flow-rate at each time step is expressed as a linear combination of previous flow-rates (AR-part), plus a linear combination of previous rainfall volumes (X-part) plus moving average noise.

Naturally, the actual implementation of an ARMAX flow-rate predictor in a basin requires to set up a rainfall telemetering system. Thus, when considering the use of ARMAX prediction, one must compare telemetering costs with forecast "benefits", at least in terms of forecast improvement with respect to more trivial predictors, which do not use quantitative information on rainfall.

Such "benefits" are pointed out in the present paper, which deals with the upper basin of river Temo ($\simeq$ 180 km^2) in the northwest of Sardinia. Specifically, in the next section an ARX predictor is illustrated, together with a pure AR and an ARQX (AutoRegressive with Qualitative eXogenous input, namely qualitative information about rainfall). The performance of these three forecast algorithms, as well as the one by the persistent predictor (PP) is described in the last section, for different forecast horizons. For instance, it turns out that .80 of correlation between forecast and reality in flood situations is obtained with a prediction horizon of 4h, 3h, 2.5h, 2h respectively by ARX, ARQX, AR and PP.

2. THE PREDICTORS

The basin under consideration is shown in Fig.1. Hourly average flow rates were available for the period 1950-1970.

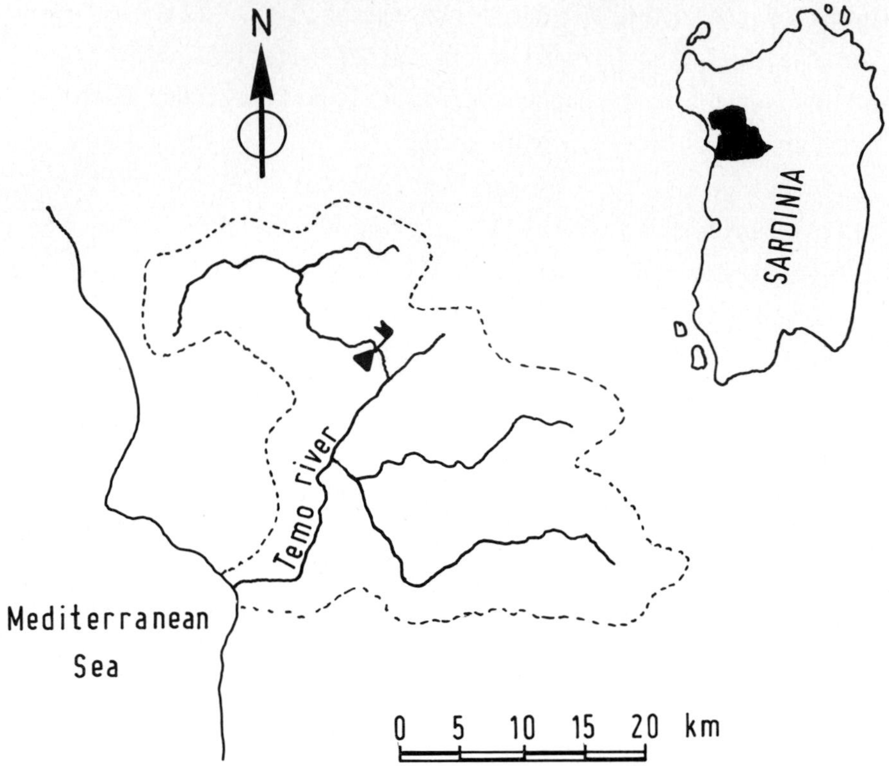

Fig.1-River Temo Basin (section considered in the present work)

Such historical record is characterized by rather prolonged "floods", with peaks often reaching more than ten times the average flow-rate.

In order of increasing complexity, the following real-time predictors of future flow-rate have been considered (Δt=1h).

2.1 The AR predictor

The pure autoregressive predictor of order p' is simply

$$\hat{q}(k+f|k) = \sum_{i=1}^{f-1} \alpha_i q(k+f-i|k) + \sum_{i=f}^{p'} \alpha_i \, q(k-i+f) \qquad (1)$$

q(k) = average flow-rate in the section pointed out in Fig.1, in the interval $[(k-1)\,\Delta t,\ k\Delta t)$;

$\hat{q}(k+f|k)$ = forecast of $q(k+f)$ made at time $k\Delta t$;

$\{\alpha_i\}_{i=1}^{p'}$ = predictor parameters.

By predictor (1) forecast is supplied only on the basis of recent flow-rate measurements, no information about rainfall on the basin is required.

2.2 The ARQX predictor

The ARQX predictor makes use of both recent flow-rate measurements and gross information about rainfall, such as distinction between "rain" and "no rain" situations. The ARQX predictor of order p" is given by

$$\hat{q}(k+1|k) = \sum_{i=1}^{p''} \beta_i \left[c(k-i+1)\right] q(k-i+1) \tag{2a}$$

$$\hat{q}(k+2|k) = \beta_1\left[c(k)\right] \hat{q}(k+1|k) + \sum_{i=2}^{p''} \beta_i\left[c(k-i+2)\right] q(k-i+2) \tag{2b}$$

$$\hat{q}(k+3|k) = \beta_1\left[c(k)\right] \hat{q}(k+2|k) + \beta_2 \left[c(k)\right] \hat{q}(k+1|k) + \sum_{i=3}^{p''} \beta_i \left[c(k-i+3)\right] q(k-i+3) \tag{2c}$$

....

where

$c(k)$ = rainfall "class" in the interval $[(k-1)\Delta t, k\Delta t)$, precisely a binary variable: $c(k) = 0$ means no rain in the upper Temo basin $c(k) = 1$ means rain (whatever its value). Naturally, the definition can be generalized to "$c(k)=0$ means rain below a certain threshold" while "$c(k)=1$ means rain above that threshold" (but in the case of river Temo, the former definition has proved more appropriate);

$\{\beta_i\}_{i=1}^{p''}$=predictor parameters.

Eqs. (2) represent an autoregressive predictor, where past flow-rates have different weights, according to the circumstance that they correspond to a rainy or a non-rainy hour. In particular, no information about c(k+1), c(k+2)... is assumed to be available at time $k\Delta t$, therefore in eqs. (2b), (2c)... all future rainfall classes are set equal to c(k).

2.3 The ARX predictor

The ARX predictor supplies forecast by using both recent flow-rate data and recent quantitative rainfall evaluation on the upper Temo basin. It is a slight variation of the forecast algorithm illustrated by Bolzern et al. (1980). This predictor is given by

$$\hat{q}(k+1|k)=\sum_{i=1}^{p'''} \phi_i q(k-i+1)+\sum_{j=1}^{r'} \psi_j u_1(k-j+1)+\sum_{j=1}^{r''} \eta_j u_2(k-j+1) \quad (3a)$$

$$\hat{q}(k+2|k)=\phi_1\hat{q}(k+1|k)+\sum_{i=2}^{p'''} \phi_i q(k-i+2)+\sum_{j=2}^{r'} \psi_j u_1(k-j+2)+ + \sum_{j=2}^{r''} \eta_j u_2(k-j+2) \quad (3b)$$

$$\hat{q}(k+3|k)=\phi_1\hat{q}(k+2|k)+\phi_2\hat{q}(k+1|k)+\sum_{i=3}^{p'''} \phi_i q(k-i+3)+ + \sum_{j=3}^{r'} \psi_j u_1(k-j+3)+\sum_{j=3}^{r''} \eta_j u_2(k-j+3) \quad (3c)$$

......

where

$\{\phi_i\}_{i=1}^{p'''}$, $\{\psi_j\}_{j=1}^{r'}$, $\{\eta_j\}_{j=1}^{r''}$ = predictor parameters;

p''', r', r'' = predictor orders;

$$u_1(k) = \begin{cases} a(k) & , \sum_{t=k-1}^{k-M} a(t) \leqslant a_s \\ 0 & , \sum_{t=k-1}^{k-M} a(t) > a_s \end{cases}$$

$$u_2(k) = \begin{cases} 0 & , \quad \sum_{t=k-1}^{k-M} a(t) \leqslant a_s \\ a(k) & , \quad \sum_{t=k-1}^{k-M} a(t) > a_s \end{cases}$$

$a(t)$ = rainfall volume (m^3s^{-1}) on the basin during the interval $[(t-1)\Delta t, t\Delta t)$;

a_s = threshold volume;

M = integer representing the "memory of soil status". If in the M steps before $(k-1)\Delta t$ the overall rainfall volume has not exceeded the threshold a_s("situation of dry soil"), the rainfall a(k) in the k-th time step is introduced into predictor (3) as u_1. In the opposite situation ("wet soil"), the rainfall a(k) is introduced as u_2.

3. RESULTS

The predictors mentioned in the previous section have been tested on river Temo for the six major floods of the period 1950-1970. Orders p', r',r", p", p"', threshold a_s and index M have been estimated by judgement or trial (see Table 1).

TABLE 1
Parameters estimated by judgement or trial

p'	p"	p"'	r'	r"	M	a_s $[m^3s^{-1}]$
2	2	2	3	3	2	300

Parameters $\emptyset_i$, ψ_j, η_j, α_i, β_i have been reestimated at each time step by the recursive extended least squares procedure (Panuska, 1969; Young, 1968). Note that flow-rate prediction has been supplied simultaneously with the updating of parameter estimates and not after the convergence of estimates, (as done by Bolzern et al. (1980)). In this sense, predictors (1) - (3) are "adaptive".

The forecast performance, particularly by ARX, has turned out satisfactory. In detail, Fig. 2 describes the overall performances (=correlation ρ between predictions and observations) by ARX, ARQX, AR for various forecast horizons f and compares them with persistence prediction PP. The figure accounts for $f \leqslant 6$: for $f = 7$ all qualities fall down to totally unacceptable values. Note that for $f \leqslant 2$ there is no significant advantage by using rainfall measurements, while for $f > 2$ such advantage can be ap-

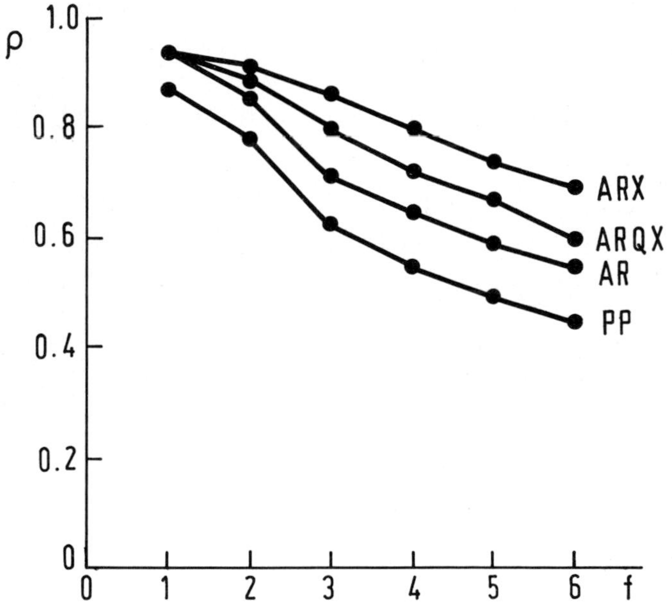

Fig. 2 - Overall forecast performance of the various predictors versus prediction horizon f.

preciated. For instance, if $\rho = 0.80$ is considered as the minimum acceptable performance, at that level of quality ARX horizon is 30% longer than ARQX, 50% longer than AR and 100% longer than PP. As stated in the introduction, the value of such extra hour (or more) of forecast given by ARX must be compared with the cost of a rainfall telemetering system.

REFERENCES

Bolzern, P., Ferrario, M., Fronza, G., 1980. Adaptive real-time forecast of river flow-rates from rainfall data. J. Hydrol., 47: 251-267.

Chiu, C.L. (Editor), 1978. Applications of Kalman filter to hydrology, hydraulics and water resources, University of Pittsburgh Press, Pittsburgh.

Dooge, J.C.I., 1977. Problems and methods of rainfall - runoff modelling. In T.A. Ciriani, U. Maione, J.R. Wallis (Editors), Mathematical Models for Surface Hydrology, John Wiley, New York, 423 pp.

Panuska, V., 1969. An adaptive recursive least squares identification algorithm, Proc. 8th IEEE Symp. on Adapt. Processes, Penns. State Univ.

Young, P.C., 1968. The use of linear regression and related procedures for the identification of dynamic processes, Proc. 7th IEEE Symp. on Adapt. Processes, Los Angeles.

TIME SERIES MULTIPLE LINEAR REGRESSION MODELS FOR EVAPORATION FROM A FREE WATER SURFACE

JAMES G. SECKLER

Associate Professor, Civil Engineering
Bradley University, Peoria, Illinois (USA)

INTRODUCTION AND SUMMARY

The objective of this study was to develop, using time series multiple linear regression techniques, mass transfer models for evaporation from a free water surface. The basic form of the mass transfer evaporation model may be expressed as

$$E_t = CN_t u^m_{z,t} \Delta e^p_{z,t}$$

where E_t = evaporation at time t, N_t = a mass transfer coefficient, $u_{z,t}$ = wind speed at elevation z above water surface at time t, $\Delta e_{z,t}$ = vapor pressure difference at elevation z at time t, and C = constant. The subscript t indicates a time dependent process whose observations are taken at equal time intervals. The terms E_t, $u_{z,t}$, $\Delta e_{z,t}$ and N_t may be considered stochastic processes. After linearization the above may be analyzed using time series multiple linear regression to obtain estimates for C, N_t, m and p based on experimental data taken for E_t, $u_{z,t}$ and $\Delta e_{z,t}$.

Using data collected at Lake Hefner, Oklahoma (Harbeck, et al, 1954) and Fort Collins, Colorado (Rohwer, 1931) the following models were obtained:

Lake Hefner: $E_t = 0.424\text{x}10^{-2}(k'/\ln(z/z_o))^{0.80} u_{z,t}^{0.80} \Delta e_{z,t}$

Fort Collins: $E_t = 0.00445 u_{z,t}^{0.50} \Delta e_{z,t}$

where k' = von Karmen's constant, and z_o = equivalent water roughness height in cm. E_t is in cm/day, $u_{z,t}$ in cm/sec, and $\Delta e_{z,t}$ is in gm/cm^2.

Using the above models the average correlation coefficients

Reprinted from *Time Series Methods in Hydrosciences*, by A.H. El-Shaarawi and S.R. Esterby (Editors)

between observed and predicted evaporation for the Lake Hefner and Fort Collins data were calculated to be 0.77 and 0.95 respectively. The Lake Hefner results compare favorably with results obtained in a previous study (Harbeck, et al, 1954). The Lake Hefner model was also applied to data from the Elephant Butte Reservoir in New Mexico (Tschantz, 1968). The correlation coefficient between observed and predicted evaporation was calculated to be 0.95.

DEVELOPMENT OF MATHEMATICAL MODEL

Dimensional Anlaysis

The primary factors affecting evaporation from a free water surface are given in Table 1. Using dimensional analysis the following equation was developed

$$E = CN\, u_z^m \Delta e_z^p \tag{1}$$

where C is a constant and N is a mass transfer coefficient given by

$$N = \frac{K_m^{1-a}}{\rho\, d_A^{1-a/2}} \left[\frac{\nu A}{K_m z_o}\right]^c \left[\frac{k'}{\ln z/z_o}\right]^m , \tag{2}$$

u_z is a wind speed at elevation z and Δe is a vapor pressure difference at elevation z. The shape factor, Sh, and wind direction, Di, were eliminated based upon the following considerations.

The shape factor is introduced primarily to account for changes in lake or reservoir depths of great magnitude. For short periods of time (e.g., a day) the amount of evaporation is of the order of one inch resulting in an insignificant change in the shape of a water body.

The wind direction, Di, is dropped based on the premise it has no affect on the vapor flux from the surface of the water body but only effects the distribution of vapor in the atmosphere above the water surface.

The wind velocity u_z is introduced into equation (2) using the logarithmic wind law given by

$$\frac{u_z}{u_*} = \frac{1}{k'} \ln z/z_o \tag{3}$$

TABLE 1

Summary of Variables Affecting Evaporation From a Water Surface

Variable	Symbol	Dimension Force-Length-Time
Evaporation per unit area per time	E	LT^{-1}
Area of water surface	A	L^2
Difference in vapor pressure	$\Delta e=e_o-e_a$	FL^{-2}
Equivalent water roughness	z_o	L
Kinematic Viscosity of air	ν	L^2T^{-1}
Coefficient of molecular diffusion	K_m	L^2T^{-1}
Density of air	ρ	FT^2L^{-4}
Shear at water surface	τ_o	FL^{-2}
Shape factor of water body	Sh	Dimensionless
Wind direction	Di	Dimensionless

where u_* = friction velocity = τ_o/ρ.

Recognition of evaporation, wind speed, vapor pressure difference and the mass transfer coefficient as stochastic processes the evaporation model may be written as

$$E_t = CN_t u^m_{z,t} \Delta e^p_{z,t} \tag{4}$$

where the subscript t indicates a time dependent process whose observations are taken at equal time intervals and the subscript z indicates the height of measurement of the mass transfer parameters.

Time Series Multiple Linear Regression Model

Equation (4) was logarithmically transformed to obtain

$$y_t = K + b_1X_{1,t} + b_2X_{2,t} + \varepsilon_t \tag{5}$$

where Y_t = log E_t, K = log C, b_1 = m, $X_{1,t}$ = log $u_{z,t}$,

$b_2 = p$, $X_{2,t} = \log(\Delta e_{z,t})$ and $\varepsilon_t = \log N_t$.

Equation (5) may be analyzed using time series multiple linear regression techniques to obtain estimates for K, b_1, b_2 and ε_t. In the analysis Y_t, $X_{1,t}$ and $X_{2,t}$ are assumed to be ergodic random processes. ε_t is assumed to be a stationary random series independent of $X_{1,t}$ and $X_{2,t}$. This last assumption may be false because (a) ε_t is the log transformation of N_t which is related to $u_{z,t}$ and $\Delta e_{z,t}$; (b) the presence of errors in observation of $u_{z,t}$ and $\Delta e_{z,t}$; and (c) non-additivity of the transformed variables $X_{1,t}$ and $X_{2,t}$. The regression constant K is obtained by

$$K = \bar{Y}_t - b_1\bar{X}_{1,t} - b_2\bar{X}_{2,t} \tag{6}$$

where the bars indicate the means of the respective series. The residual series ε_t, is determined by

$$\varepsilon_t = Y_t - K - b_1X_{1,t} - b_2X_{2,t} \tag{7}$$

The use of time series multiple linear regression techniques to determine the best estimates of the regression coefficients has one advantage over traditional regression methods. The spectral methods used to obtain estimates for b_1 and b_2 take into account the variation of the signal to noise ratio with frequency. For the evaporation model there are two signals; i.e., the covariance spectrums between wind speed and evaporation and vapor pressure difference and evaporation. The noise is represented by the variance spectrum of the residual series, ε_t. The estimates for b_1 and b_2 are weighted for frequency bands which possess high signal to noise ratios. In other words, the estimates for b_1 and b_2 are obtained by giving more weight to the frequency bands where a large degree of information is provided by wind speed or vapor pressure difference for the prediction of evaporation.

DATA ASSEMBLY

Mass transfer evaporation data from two investigations were selected for this study. The first study was performed by the United States Geological Study (USGS) at Lake Hefner, Oklahoma (Harbeck, et al, 1954). The data measured included air temperature, humidity and wind speed at two, four, eight and sixteen meters above the lake surface, and lake surface temperature. Evaporation was determined using the water budget method.

The second study was conducted by the Colorado Agricultural Experiment Station for the United States Department of Agriculture (USDA) (Rohwer, 1931). Data measured included (reported as daily averages) evaporation, water surface temperature, air temperature and wind speed at two meters elevation.

The data from both investigations was processed to give daily evaporation in cm/day, wind speed in cm/sec and vapor pressure difference in gm/cm^2. Using this data best estimates for the regression coefficients b_1 and b_2 and the regression constant K in equation (5) were obtained using time series multi-linear regression techniques.

PRESENTATION AND DISCUSSION OF RESULTS

Presented in Table 2 are the results of the regression analyses and tests of validity for the USGS and USDA data.

TABLE 2

Summary of Time Series Multilpe Linear Regression Results

USGS, Lake Hefner

Elevation meters	Model	Validity 10%	Results 25%
2	$E_t = 0.0038\ N_t u_t^{0.554} \Delta e_t^{1.122}$	Valid	Valid
4	$E_t = 0.0009\ N_t u_t^{1.293} \Delta e_t^{0.912}$	Invalid	Invalid
6	$E_t = 0.0017\ N_t u_t^{0.545} \Delta e_t^{1.431}$	Invalid	Invalid
8	$E_t = 0.0016\ N_t u_t^{0.620} \Delta e_t^{1.334}$	Invalid	Invalid

USDA Reservoir

Year	Model	Validity 10%	Results 25%
1927	$E_t = 0.0050\ N_t u_t^{0.557} \Delta e_t^{1.310}$	Valid	Valid
1928	$E_t = 0.0091\ N_t u_t^{0.464} \Delta e_t^{1.138}$	Invalid	Valid

The tests of validity indicate that for the most part the models are invalid statistically. Mathematically the tests of validity are based on the statistical independence of the residual series, ε_t, from $X_{1,t}$ and $X_{2,t}$, the independent transformed mass transfer variables. The invalid results were not entirely unexpected since ε_t is a function of atmospheric stability, kinematic viscosity of the air, area of the water body, wind speed variation with height above the water surface, water surface roughness and molecular diffusion. Therefore ε_t is not physically independent of $X_{1,t}$ and $X_{2,t}$ and should not be expected to be statistically independent. Moreover, Hamon and Hannon (1963) point out that very rarely will the tests of validity indicate independence. They state that the tests of validity for the model will detect very small discrepancies between data and hypothesis. This implies that a significant statistical result need not correspond to an operationally large discrepancy. The models for evaporation which are valid statistically are, for the most part, models where the mass transfer variables are measured at the two meter level or less. This is in the region where the effects of stability are suppressed.

It appears from Table 2 the following relationships between the exponents for wind speed (b_1) and vapor pressure difference (b_2) exist: $b_1 \neq 1$, $b_2 \neq 1$ and $b_1 \neq b_2$. Student's t tests were used to test the hypotheses (at the ten per cent significance level) $b_1 = 1$, $b_2 = 1$ and $b_1 = b_2$ with the results being summarized in Table 3 below.

For the Lake Hefner data it appears (a) b_1 is significantly different from unity, and (b) b_2 is not significantly different from unity at the ten per cent significance level. For both the Lake Hefner and the USDA reservoir data the Student t tests indicate that, in general, b_2 is significantly different from b_1. For the two periods of analysis used for the USDA reservoir data, the Student t tests indicate that both b_1 and b_2 are significantly different from unity.

TABLE 3
Summary of Student t Tests
Null Hypotheses: $b_1 = 1$, $b_2 = 1$, $b_1 = b_2$

			Null Hypothesis Acceptance?		
USGS, Lake Hefner					
Elevation, meters	b_1	b_2	$b_1 = 1$	$b_2 = 1$	$b_1 = b_2$
2	0.554	1.122	No	Yes	No
4	1.293	0.912	Yes	Yes	No
8	0.545	1.431	No	No	No
16	0.621	1.334	No	Yes	No
USDA, Reservoir					
Year	b_1	b_2	$b_1 = 1$	$b_2 = 1$	$b_1 = b_2$
1927	0.557	1.310	No	No	No
1928	0.464	1.138	No	No	No

Development of Evaporation Models

A conclusion of the USGS studies (Harbeck, et al, 1954) was the eight meter level adequately represents the upper limit of the vapor boundary layer. For this reason, the values of b_1 and b_2 for the two, four and eight meter results were averaged to give $b_1 = 0.797$ and $b_2 = 1.188$. For the USDA data average values of $b_1 = 0.510$ and $b_2 = 1.224$ for the two years of study were obtained. The value of $b_1 = 0.797$ for the exponent of wind speed for the Lake Hefner data agrees very closely with a value of $b_1 = 0.78$ as reported by Sutton (1949, 1953).

The average value for b_2 for both Lake Hefner and the USDA reservoir is close to unity, the theoretical value. However, the average values for b_2 appear to be approximately 1.20. These results indicated Student t tests should be performed to test the following null hypotheses: (1) $b_2 = 1.20$ for both Lake Hefner and USDA reservoir data; (2) $b_1 = 0.80$ for Lake Hefner data; and (c) $b_1 = 0.50$ for USDA Reservoir data. The above hypotheses were tested at the ten per cent significance level. The results are summarized in Table 4.

TABLE 4
Student's t Test Results
USGS, Lake Hefner

Elevation, meters	b_1	b_2	Null Hypothesis Accepted? $b_1 = 0.80$	$b_2 = 1.20$
2	0.554	1.122	Yes	Yes
4	1.293	0.912	No	No
8	0.545	1.431	Yes	No
	USDA Reservoir			
Year	b_1	b_2	$b_1 = 0.50$	$b_2 = 1.20$
1927	0.557	1.310	Yes	Yes
1928	0.464	1.138	Yes	Yes

The results presented in Tables 3 and 4 suggest the following evaporation model for Lake Hefner

$$E_t = CN_t\, u_{z,t}^{0.80} \Delta e_{z,t}. \tag{8}$$

The mass transfer coefficient, N_t, varies with time and with height above water surface. N_t is proportional to $(k'/\ln(z/z_o))^m$ where k' is V. Karman's constant, z is height above water surface, z_o is a water roughness parameter, and m = 0.80 for the Lake Hefner data. Using average values for N_t ($\bar{N}_t$) and C, letting

$$\bar{N}_t = C'(k'/\ln(z/z_o))^{0.80}, \tag{9}$$

and setting z_o = 0.6 cm (Harbeck et al, 1954) the following model for the Lake Hefner study was obtained

$$E_t = 0.424 \times 10^{-2}(k'/\ln(z/z_o))^{0.80} u_{z,t}^{0.80} \Delta e_{z,t}. \tag{10}$$

For the USDA reservoir Table 4 suggests b_2 should be 1.20, which is contrary to Dalton's law. To resolve this discrepancy a preliminary correlation analysis between observed and predicted evaporation was performed using the following evaporation relationships:

$$E_t = C\bar{N}_t u_{z,t}^{0.50} \Delta e_{z,t}^{1.20}; \text{ and} \tag{11}$$

$$E_t = C\bar{N}_t u_{z,t}^{0.50} \Delta e_{z,t} \tag{12}$$

where C was obtained as the antilog of K. For b_2 = 1.20 the correlation coefficients between observed and predicted evaporation were calculated to be 0.959 for the 1927 data and 0.934 for the 1928 data. For $b_2 = 1.00$ the calculated correlation coefficients were found to be 0.965 for the 1927 data, and 0.934 for the 1928 data. Since there was a small gain in the correlation coefficient between observed and predicted evaporation for the 1927 data where $b_2 = 1.00$, equation (12) was chosen as the evaporation model.

For the USDA reservoir the average value of $\bar{CN}_t$ for the two years of analysis is 0.00445. Using this value, the following model for evaporation is obtained

$$E_t = 0.0045\, u_{z,t}^{0.50} \Delta e_{z,t}. \tag{13}$$

Using the models for evaporation by equations (10) and (13) values of evaporation were predicted for Lake Hefner and the USDA reservoir. Correlation coefficients between observed and predicted evaporation are summarized in Table 5. Shown in Figures 1 and 2 are example plots of observed versus computed evaporation.

TABLE 5
Summary of Correlation Coefficients

Lake Hefner July 1, 1950 to August 31, 1951			USDA Reservoir	
Two Meter	Four Meter	Eight Meter	1927	1928
0.770	0.769	0.762	0.965	0.934

Application of Lake Hefner Stochastic Evaporation Prediction Model to Elephant Butte Reservoir

The Lake Hefner model for evaporation, equation (10), was applied to the mass transfer data reported by Tschantz (1968) for the Elephant Butte Reservoir evaporation study. The meteorological data was measured at the two meter level in the Elephant Butte study. Assuming $z_o = 0.6$ cm and $z = 200$ cm equation (10) becomes

$$E_t = 0.500 \times 10^{-3} u_{z,t}^{0.80} \Delta e_{z,t}. \tag{14}$$

Equation (14) was used to predict evaporation from Elephant Butte Reservoir. The results are summarized in Table 6 and Figure 3. Table 6 is a summary of observed and predicted evaporation for the twenty-eight thermal survey periods of the Elephant

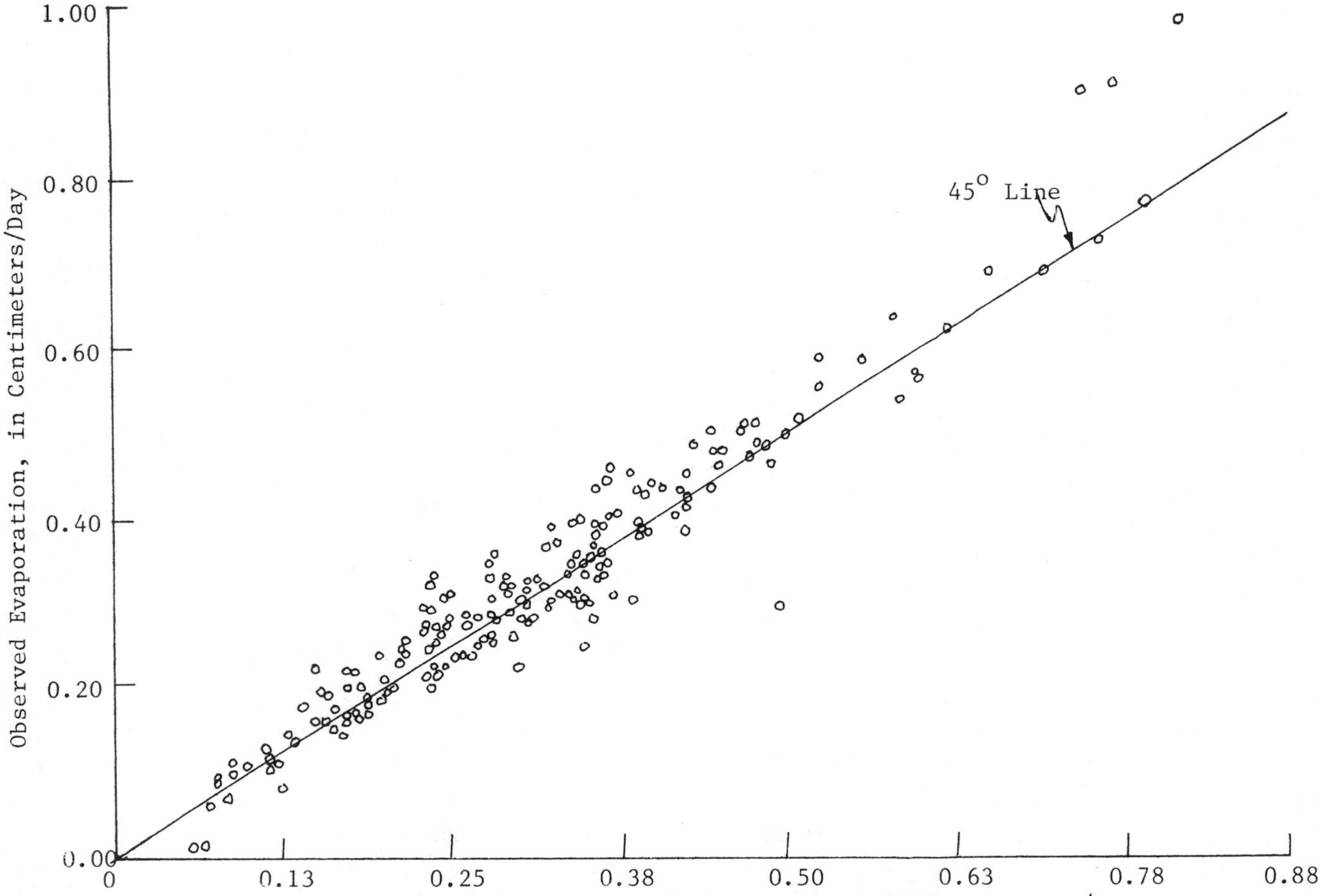

Figure 1. Computed versus Observed Evaporation for 1927 U.S.D.A. Reservoir Data

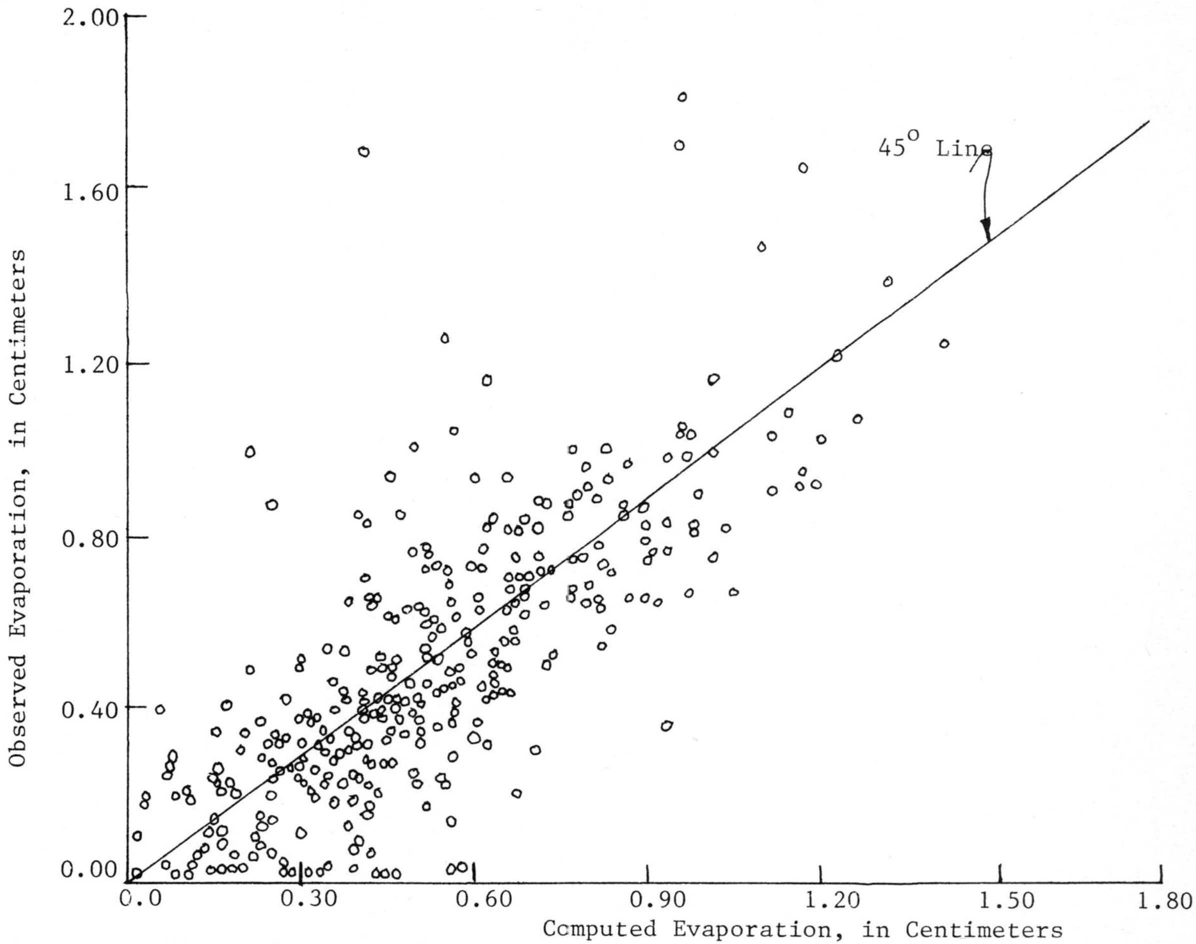

Figure 2. Computed versus Observed Evaporation for Lake Hefner Eight Meter Meteorological Level for July 1, 1950 to August 31, 1951

Butte study. Figure 3 is a plot of observed versus predicted evaporation. The correlation between observed and predicted evaporation was determined to be 0.955.

TABLE 6
Summary of Predicted vs. Observed Evaporation for Elephant Butte Study (T.S.P. = Thermal Survey Period Number)

T.S.P.	Observed Evaporation, cm/day	Predicted Evaporation, cm/day	T.S.P.	Observed Evaporation, cm/day	Predicted Evaporation, cm/day
1	0.639	0.759	15	0.202	0.289
2	0.229	0.486	16	0.268	0.414
3	0.631	0.587	17	0.365	0.473
4	0.566	0.551	18	0.378	0.414
5	0.411	0.462	19	0.671	0.666
6	0.326	0.425	20	0.757	0.844
7	0.236	0.272	21	0.637	0.599
8	0.291	0.516	22	0.804	0.716
9	0.185	0.278	23	0.943	0.996
10	0.157	0.278	24	0.886	0.766
11	0.140	0.231	25	0.695	0.768
12	0.118	0.231	26	0.748	0.723
13	0.088	0.211	27	0.686	0.868
14	0.165	0.193	28	0.780	0.883

CONCLUSIONS

Using dimensional analysis and recognizing the stochastic properties of evaporation, wind speed, and vapor pressure difference a mass transfer model for evaporation was developed and is given by

$$E_t = CN_t u_{z,t}^m \Delta e_{z,t}^p . \qquad (9)$$

This model was analyzed using time series multiple linear regression techniques to obtain estimates of C, N_t, m and p for the meteorological data acquired during the Lake Hefner Studies (Harbeck, 1954) and studies conducted by the USDA (Rohwer, 1931).

Using the eight meter level as the upper limit of the vapor boundary layer and the logarithmic wind distribution for atmospheric flow

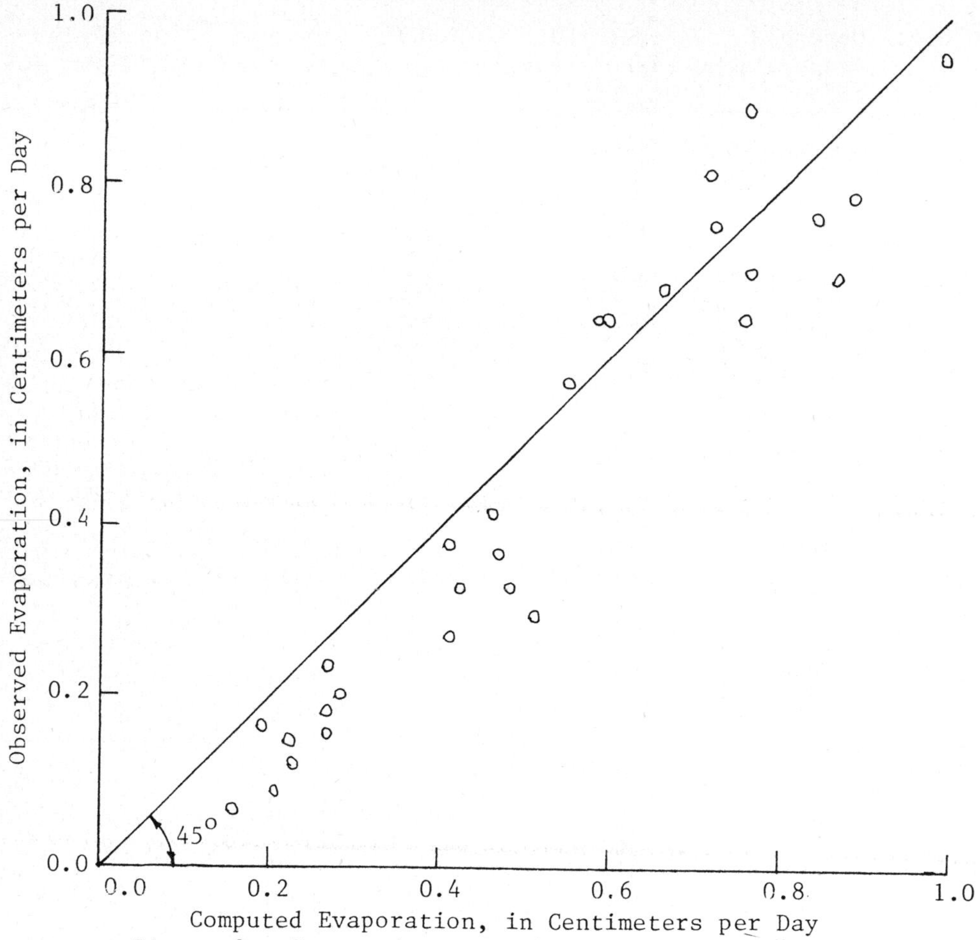

Figure 3. Computed versus Observed Evaporation for Elephant Butte Reservoir

over an aerodynamically rough surface the following model for evaporation was developed for the Lake Hefner data.

$$E_t = 0.424 \times 10^{-2}(k'/\ln(z/z_o))^{0.80} u_{z,t}^{0.80} \Delta e_{z,t} \tag{10}$$

Analysis of the USDA reservoir data produced the following model for evaporation.

$$E_t = 0.00445 u_{z,t}^{0.50} \Delta e_{z,t} \tag{13}$$

In general the exponent m for both the Lake Hefner and the USDA data was found to be significantly different from unity at the ten per cent level. The exponent for vapor pressure difference, p, was found to be, in general, non-significantly different from unity at the ten per cent significance level. Also the difference between m and p was found to be significant at the ten per cent significance level.

Equations (10) and (13) were used to predict evaporation from Lake Hefner and the USDA reservoir respectively. The average correlation coefficient between observed and predicted evaporation for the Lake Hefner data was 0.767. For the two periods of analysis for the USDA Reservoir data the average correlation coefficient between observed and predicted evaporation was 0.951.

The evaporation model developed for the Lake Hefner data was applied to the mass transfer data for the Elephant Butte Reservoir. The correlation coefficient between observed and predicted evaporation was 0.955.

REFERENCES

1. Hamon, B.V. and Hannan, E.J., "Estimating Relations Between Time Series", Journal Geophysical Research, 1963, vol. 68, no. 21, pp. 6033-6041

2. Hannan, E.J., Time Series Analysis, 1960, New York, Wiley and Sons

3. Hannan, E.J., "Regression for Time Series", Symposium of Time Series Analysis, 1963, Wiley and Sons

4. Harbeck, et al, "Water Loss Investigation: Lake Hefner Studies", U.S. Geological Survey Professional Papers 269 and 270, 1954, Washington, D.C.: U.S. Government Printing Office

5. List, R.J., Smithosonian Meteorological Tables, Smithsonian Miscellaneous Collections, Smithsonian Institution, Washington, D.C.

6. Rohwer, Carl, "Evaporation From Free Water Surfaces", Technical Bulletin No. 271, U.S. Department of Agriculture, 1931

7. Sutton, O.G., "The Application of Micrometerology to the Theory of Turbulent Flow Over Rough Surfaces", Royal Meteorological Society Quarterly Journal, 1949, Vol 74, No. 326

8. Sutton, O.G., Micrometerology, 1953, New York: McGraw-Hill Book Company, Inc.

9. Tschantz, B.A., Evaporation Investigation at Elephant Butte Res-servoir Using Energy Budget and Mass Transfer Techniques, January, 1968, Sc.D. Dissertation, New Mexico State University, Las Cruces, New Mexico

OPTIMAL MANAGEMENT OF MULTIRESERVOIR SYSTEMS USING STREAMFLOW FORECASTS

by

Rogelio C. Lazaro
Research Associate

John W. Labadie
Associate Professor

and

Jose D. Salas
Associate Professor

Department of Civil Engineering
Colorado State University, Ft. Collins, CO 80523

ABSTRACT

In developing a plan to effectively manage a complex water resource system, some form of water supply forecasting is required. In areas with a large spring snowmelt runoff, a total seasonal forecast of streamflow is often available from snow surveys as input to some type of multiple linear regression model. These seasonal estimates provide a basis for an annual operational plan. Thereafter, the operations manager monitors the daily flows and tries to meet the downstream and/or supplementary supplies. As conflicts in water use intensify, the need arises for disaggregated monthly (or even shorter term) forecasts to obtain an optimal management strategy for a system of reservoirs. This is particularly true for hydropower production planning in conjunction with other uses. A state-space forecasting model has been developed that attempts to reconstruct virgin streamflows based on historical records decomposed into low flow and high flow sequences. The Kalman filter is used to separate model and measurement errors. The resulting forecasts were found to be satisfactory primarily for the low flow months. Subsequent additional use of snow water equivalent data available prior to the start of the high flow period yielded some improvement in forecasting the important high flow period. Streamflow forecasts for up to a six month lead time were used to determine optimal water transfers and exchanges within the Cache la Poudre River, in north-central Colorado. A river basin network optimization model provided a means of processing forecasted streamflow and other system inputs to

Reprinted from *Time Series Methods in Hydrosciences*, by A.H. El-Shaarawi and S.R. Esterby (Editors)

identify the minimum cost flows to satisfy storage and diversion requirements in accordance with decreed water rights. Management strategies were generated with the objective of minimizing wasted outflows from the basin, in order to increase the likelihood of junior appropriators being served during critical low flow periods. Based on average monthly flows, improvement was significant for a typical dry year, with outflows reduced from over 42,000 acre-ft per year to around 1000 acre-ft through use of the combined forecasting/network optimization models.

INTRODUCTION

Effective operational planning of water resources systems requires knowledge of both the demand for and the supply of water resources. The demands tend to be more predictable in comparison with the supply. However, the water user must often make important planning decisions which affect his ultimate demand, in anticipation of forthcoming water supply. The normal practice is often to make an *educated guess*, based on available data, of what the total incoming supply for the season will be, and pair it with the *usual* demand. The plan implementation becomes a matter of *matching* the demand with the estimated supply. In the Colorado front range, for example, the streamflow forecast is a total seasonal aggregate based on snow pack data. A great amount of skill is needed by the water administrator in breaking this information down to something that can serve as a basis for day to day or even month to month decisions. A system with storage capacity offers some operational flexibility, depending upon the available capacity, but also greatly increases the number of decision alternatives, due to the need to define reservoir operating rules.

Computerized models are an extremely valuable way of testing out various management schemes at the river basin or subbasin level to determine their impacts on the various water users and uses; or for rapidly predicting the affects of physical and institutional changes in the system. These models can also be an important way of documenting the experience of the system administrator in anticipation of the day he retires or decides that the pressure is too great and finds another position. The new man on the job will find such a model invaluable, unless he has been extensively trained by the water manager.

Again, the computerized model can be a convenient tool for handling the large amount of data and myriad variables associated with a complex water resource system. Models can be developed which can process historical data and produce more detailed forecasts of future streamflows than are generally available. In addition, a measure of the confidence associated with these forecasts can also be estimated. This information can then be input to another model which can process these data to produce operational guidelines. As new data are received in real-time, the parameters of these models can be systematically updated to produce new streamflow forecasts and updated operational plans.

A modeling study is described herein which combines a state-space streamflow forecasting model, based on the Kalman filter, with a river basin simulation model with some optimizing capability. These models have been tested on the Cache La Poudre River basin in northcentral Colorado to determine if it is possible to improve operational efficiency. Details of the forecasting model are described elsewhere. This particular presentation focuses on how the forecasts were input into the river basin model to determine real-time operational plans.

BRIEF REVIEW OF HYDROSYSTEM MANAGEMENT TECHNIQUES

The most widely used forecasting technique in the western United States is a correlation-based streamflow forecasting method that yields only the expected total seasonal runoff volume (USBR, 1968). When hydropower is important, an attempt is made to disaggregate the total seasonal forecast into monthly estimates. This is generally accomplished by using the monthly record of flows from a previous year having a similar total runoff volume (Bellamy, 1980). After the last snowpack measurements are obtained, no further forecast updates are made. The forecasts are generally adequate for average flow years, but tend to overestimate low flows and underestimate high flows (Hawley, et. al. 1980).

In Colorado, these monthly forecasts are also used for planning annual allocation of supplemental transbasin diversion water, such as done by the Northern Colorado Water Conservancy District (NCWCD) (Barkley, 1974). These allocations are adjusted in time as new data are obtained. Within this same northern Colorado area, the Poudre River Commissioner, under the office of the State Engineer, independently prepares his own water supply forecasts using a slightly more dense network of snow courses and a simple averaging procedure (Neutze, 1980). His forecasts are utilized in determining the percentage of normal requirements that can be served in his district. Day to day decisions on how best to match demand with the observed supply are made on the basis of the established water right priorities, guided by experience and intuition.

A number of studies involving use of streamflow forecasts in river basin management and operations are reported elsewhere in the literature, (Hoshi and Burges, 1979), (Wilson and Kirdar, 1970), (Becker and Yeh, 1974), (Mejia, et. al., 1974), (Wunderluch and Giles, 1981) and (Unny, et. al., 1981).

DEVELOPMENT OF STREAMFLOW FORECASTING MODEL

Several streamflow forecasting methods have appeared in the literature in recent years, with time series analysis (e.g., McKerchar and Delleur (1974)) and state-space approaches (Movarek, et. al. (1978)) seeming to be the most popular.

Graupe (1976) suggests a possible *marriage* between time series analysis (for model form and type identification) and state-space approaches (for considering measurement errors). As is generally known, monthly streamflow records essentially consist of groups of individual data (i.e., low flow and high flow sets, or three or four seasonal groups) possessing different and distinct characteristics. Panu and Unny (1980) propose that two or more separate models, one for each group or class, be developed in lieu of a single representation for the monthly streamflow data.

It is known that monthly hydrologic time series usually have periodic components in several of their statistical characteristics such as the mean, standard deviations and correlation coefficients. This is physically justified since in late summer, fall and winter the streamflow is dominated by groundwater or baseflow contributions while in the spring and early summer, direct surface runoff from snowmelt and rainfall is the major contributor. The stochastic model considered herein includes these seasonal characteristics, in the form of a low flow sequence, where the measurement errors are relatively small and a high flow sequence where the measurement errors are usually large.

Details on the model structure identification, model parameter and noise statistics estimation as well as diagnostic checks are given in Lazaro, et. al.,(1981).

The iterative model building procedure of Box and Jenkins (1976) can still be employed in this case. The exception is that a slightly different approach in the computation of the autocorrelation and partial autocorrelation functions must be used to account for the discontinuity involved in separating the two groups of flows (Salas, et. al., 1980). The parameters for each model can be estimated using the method of maximum likelihood or a sequential least squares regression algorithm. For Gaussian time series, the latter yields estimates approaching the maximum likelihood estimates. Model adequacy diagnostic checks via the usual residual analysis for goodness of fit testing then follow.

In this modeling approach, the error term allows for the inexactness of the model to represent the true process. Any measurement errors are reflected in the estimates of the parameters. This state-space approach to modeling requires complete specification of the predictor parameters and noise terms corrupting the system dynamics and measurement models. These predictor parameters and noise statistics can be obtained from the identified time series model under an assumption of an invariant processor. The time history of measurement errors for hydrologic events are, however, difficult to obtain or rarely available. Those that are accessible are either incomplete or location and instrument specific.

RIVER BASIN NETWORK MODEL

The Texas Water Development Board (1972) has developed a model called SIMYLD that is capable of representing the physical characteristics of large-scale, complex water resource systems and selecting the optimal water allocation in a minimum cost sense. SIMYLD is basically a simulation model with optimizing capability on a month by month basis. That is, a sequential, static optimization approach is used rather than a fully dynamic global optimization.

The synthesized model used in this study, called MODSIM, essentially retains the basic structure of SIMYLD, while incorporating some additional features. Conversational, interactive coding has been incorporated to encourage use of the model by state and local water resources planners and managers in their tasks of evaluating the impacts of alternative water management policies throughout a river basin.

The model essentially considers the physical features of the system in a capacitated network form. The real system components, such as storage and non-storage points (reservoirs, demand points, canal diversions and river confluences), are represented by nodes. The river reaches, canals and closed conduits are designated as node to node linkages (links).

Mathematical Description of the Problem

The problem of network flow allocation is sequentially solved through the out-of-kilter algorithm to yield the minimum "cost" flows for each month of operation, within the confines of mass balance throughout the network (Bazarra and Jarvis, 1977). The optimization problem for any given month t is:

$$\min \sum_i \sum_j C_{ij} Q_{ij}$$

subject to node conservation or continuity of mass

$$\sum_i Q_{ij} - \sum_i Q_{ji} = 0 \quad \forall j = 1,\ldots,N$$

given bounds on flows in all links

$$L_{ij} \leq Q_{ij} \leq U_{ij} \quad \forall ij$$

where Q_{ij} = the flow in the link from node i to node j (no routing)
C_{ij} = the cost of transferring each unit of flow Q_{ij} from node i to

node j, and U_{ij} = upper bound of link from node i to node j.

Certain actual costs, such as pumping costs, can be easily assigned to appropriate links. In many situations, it may be desirable to represent the C_{ij} as weighting or priority factors rather than actual costs. Storage and diversion rights can be weighted in order of priority, with senior rights given the lower "costs". For linear objectives such as this, relative ordering of these priorities is more important than the actual magnitudes.

Though the model objective attempts to minimize costs, benefits can be represented as negative costs. In fact, desired end-of-month storage and demand links are assigned negative costs. The network is only solved period by period. Global optimal solution of the entire network for several time periods at once is extremely expensive computationally. More details on the model, the way link costs are assigned and the extended capabilities included in MODSIM can be found in Shafer, et. al., (1981).

MODSIM model applications

The reservoir operating rules as input to the program may reflect either the desire to maintain storage levels at some point below maximum capacity during certain months (flood control), maintain levels as high as possible to enhance water supply reliability (drought control), or strictly abide by the established water right priorities during the normal flow years. The priorities placed on achieving the target storage levels and/or diversion demands can be manipulated to assess different operational schemes for accomplishing these goals.

The synthesized model MODSIM has been used to evaluate the impacts of alternative management schemes for two case studies within the Cache la Poudre Poudre River basin: (1) determining if recreation opportunities could be provided in selected high mountain reservoirs by maintaining satisfactory monthly storage levels without injury to downstream water users (Shafer, et. al., (1980)), and (2) determining if sufficient reusable effluent from the City of Fort Collins, Colorado is available (given an assumed hydrological sequence) to meet monthly water demands for a proposed coal-fired power generating plant (Shafer, et. al., (1981)).

CACHE LA POUDRE CASE STUDY

The Cache la Poudre River system is a fourth order tributary of the Mississippi River, the drainage system of thelower middle portion of th North American Continent.

The basin consists of a mountainous part from which most of the water supply originates and a plains area in which the water supply is used. The river starts at Poudre Lake and several places on the Continental Divide. From the headwater, the river flows in a northeast direction to its canyon mouth for about 50 miles and then southeast for another 35 miles over an open plain towards the South Platte River..

Hydrometeorological features

The fall and winter precipitation in the basin is usually in the forr of snow, while the spring and summer precipitation generally occurs as thunderstorms. Snow accumulations in the watershed are monitored from some eight snow courses from November through April. There is one meteorological station that measures mountain precipitation year round and three stations are located on the plains at Fort Collins, Greeley and Windsor.

During the period from May to July, the spring runoff begins. About 50 to 70 percent of the surface water runoff occurs during April through July as snowmelt from the mountain tributaries. Groundwater base flow sustains the streamflow during the months from July through October. The fall and winter flows are constantly low. Two USGS streamflow gaging stations with long and continuous records are located at the mouth of the canyon at Fort Collins and at Greeley near the confluence with the South Platte River.

Water supply distribution and use system

Development of the water resources in the area parallels that in most western states of the U.S. There are several import sources diverting water to the basin to augment total supply. These are the Skyline Ditch, Grand River Ditch, Cameron Pass Ditch, Laramie-Poudre Tunnel, Michigan Ditch, Wilson Supply Ditch, and the Colorado Big Thompson project.

The snow-fed mountain streams provide excess water in the spring and inadequate amounts later in the season. A system of reservoirs was therefore developed to correct this situation. There are now some 200 reservoirs in the basin, with eleven major ones in mountain portions of the basin (i.e., Long Draw, Peterson, Joe Wright, Chambers Lake, Barnes Meadow, Big Beaver, Twin Lakes, Commanche, Worster, Haligan and Seaman). The rest are situated on the plains.

The primary use of the basin water supply is for irrigating some 240,000 acres of land on the plains. Water is distributed by about 33 irrigation companies which operate and maintain the distribution and reservoir systems. The area is within the Northern Colorado Water Conservancy District, (NCWCD) and is part of the primary service area of the Colorado-Big Thompson Project. Supplemental water from this project is delivered via Horsetooth Reservoir to the Cache la Poudre River near the point where it emerges from the mountains onto the plains.

Municipal and industrial water supply is provided to the cities of Greeley and Fort Collins with direct flow rights totaling 12.5 cfs and 19.93 cfs, respectively. Both cities also own rights which allow them to import water into the basin. They also own 13,125 and 10,250 shares, respectively, of water from the Colorado-Big Thompson (CBT) Project (Reitano and Hendricks, 1978).

Water rights: Doctrine of prior appropriation

In Colorado, the rights to the use of water are based firmly on the doctrine of prior appropriation which simply means that the *first in time* users are *first in right*. The order of priority of water right holders are specified such that at various levels of streamflow, those who are permitted to divert water can be determined (Anderson, (1975)). The rights are acquired by appropriation and are defined through the adjudication procedure conducted by the courts.

The decreed or adjudicated water rights are administered by the State Engineer and his subordinates, the division engineers and the water commissioners. A water commissioner is responsible for each major stream within a division.

Irrigation users (either individual or companies) often own a number of water rights that vary both in date of priority and in quantity of water claimed. During the season a water right holder may have some rights under which he can draw water, while other rights cannot be exercised because someone else has a higher priority. At times, water right holders will give up using water when only some of their rights are active, thereby permitting other water right holders on other ditches to divert in exchange for water at some later time.

This multiple water right makes the task of delivering water to individuals and/or companies complex for the water commissioner. He must keep all ditches informed of which of their several rights entitle them to divert water and which do not. He has to keep track of which right holders are not diverting water, even though they are entitled to. He is also responsible for allocating return flow that is available for diversions.

District operational planning practices

In preparing for the District operation plan, the river commissioner makes a forecast or estimate of expected water supply as early as the first snowfall starts. He monitors the resources by observing the snow courses established by the Snow Survey Unit of the Soil Conservation Service and maintains other courses that provide him additional confidence in estimating the snowpack accumulations in the watershed. This is done the first of every month from November through April of each year. By May 1st, the onset of the irrigation season, he has a final assessment of the snowpack accumulations. He determines the average snow water depth equivalent on some 247,000 acres of snow field, which is the approximate area of the snowline at 9000 feet elevation. With this information, and allowing for about 8% shrinkage loss, he obtains an estimate of the total volume of water that would be made available to the district for the current year operations to fill irrigation water needs and other (domestic and industrial) uses within the District. The ratio of current estimates with previous long-term records is denoted as a percentage of normal water supply available. The long-term average demands by the district is estimated at 280,000 acre-ft.

The estimated supply is then related with this demand to yield the percentage of demand that could possibly be satisfied. This *apriori* knowledge would subsequently indicate the approximate number of appropriators that could be served. This figure is of course a relative one and is highly dependent on the future weather conditions. If the temperature is cooler, the snowpack may hold and snowmelt will be slow. Conversely, if the weather is hot and dry, snowmelt occurs rapidly. The snowpack could therefore be considered as a *natural reservoir* with a certain capacity, except that the *releases* are beyond man's control.

If the expected supply is estimated to be less than the normal demand, junior appropriators who may not be served make their own assessment of the situation and may plan to buy reservoir water early in the season to augment their rights later. They may opt to run their groundwater pumps more than usual when needed. Those unsure of what their status will be may delay making decisions and wait and see what the weather conditions will be. Others may gamble and hope that localized cloudbursts will fill their needs when they require them or that surplus reservoir water will still be available when needed. On the other hand, if the expected supply is greater than the average demand, appropriators who could not be served during the average years would have a chance to exercise their rights. However, if the weather conditions are such that early and/or rapid snowmelt occurs, the low lying areas downstream are threatened by flood damages.

The forecast of the water supply is considered important in planning the operation of the reservoir in the District, but the day to day management is highly reactive to the weather conditions. If the snowpack is less than normal, reservoirs at the headwaters would be filled first to minimize transmission losses and severe shortages later in the season. The plains reservoirs generally have the higher priorities and therefore are entitled to fill first. A provision in the law, however, allows the mountain reservoirs or any other reservoirs to be filled out-of-priority to capture water that might otherwise be lost, particularly when the plains reservoir are already full (Radosevich, et. al., (1975).

Reservoir releases start at the lowest reservoirs which can be exchanged with river water upstream (since they are not direct flows) and progressively work upstream. The plains reservoirs are drawndown first to allow water exchanges while there is river water. The mountain reservoirs are drawndown last, because their water can be taken directly and is on *call* at any time.

District commissioner's management activities

The River Commissioner has a multifaceted task of: (1) satisfying as many appropriators as he can, given the current water supply; (2) minimizing the consequences of low supply from drought, as well as minimizing damages resulting from rapid release of snow accumulations, thereby causing flooding in the plains; (3) reducing outflows from the system, given available storage and canal capacities; (4) minimizing wastage and (5) maximizing utilization of river water before the supplementary supply sources are applied.

A typical day for the river commissioner involves the determination and assessment of the USGS gage flow at the mouth of the canyon in Fort Collins. Given the current flow rate and aided by long years of experience, the number of appropriators that can be served is obtained. Knowing who is currently being served, he knows who needs to continue diverting, who has to stop, who has to reduce the amount being diverted and who needs to be additionally served. Either he calls the appropriators for adjustments or they call him. When river diversion and storage rights have been accounted for, then supplemental water sources are considered next. If his transbasin supply is not enough to fill the needs, he then calls, as representative of the District water users, for quota allocation releases from the NCWCD, and/or allows pumping from groundwater resources.

Twice a week, a field inspection is done. He and his deputy divide the river approximately in half. Each one monitors and evaluates the river flows and diversions at various take-off points in some thirty three ditches. They both check if the proper amounts are in the river and if diversions are within the specified rights. Wasteways or sluiceways are likewise observed to check if water is being wasted.

Water lost due to lack of storage is also determined. This is water that could have been stored but there was no room for it since existing reservoirs were already full. The USGS gage near the City of Greeley is also read and recorded. This indicates the amount of water leaving the system.

The reservoir levels are reported to him by the owners and are generally verified once a month. Any day is open if there are complaints, or adjustments are needed, anywhere within the district, twenty-four hours a day, seven days a week. Weekly reports are prepared for the State Engineer on the flow level of the river, the right holders entitled to divert water, and the right holders actually diverting. Monthly summaries of operations are transmitted to the State Engineer through the Division engineers. At the end of the calendar year, a summary of the annual operation is prepared. The District water supply source and applications are identified along with seepage inflows and system outflows to provide indications of the end-of-the-year available storage.

It is evident that the tasks of the River Commissioner are extremely complex and challenging. It is believed that the forecasting/network optimization models presented here have the potential of being extremely useful for him, and those in similar positions of responsibility, in providing seasonal real-time management guidelines.

RESULTS OF THE MANAGEMENT STUDY

The primary objective of this management study was to determine if the combined use of forecasting/network optimization models could have reduced wasted flows (i.e., flows above required downstream releases) during a dry period. Capturing this water would imply that more appropriators could be served.

The forecast results for the low flow group were very close to the actual streamflow values, as shown in Figure 1. The forecast for the high flow group, however, tended to the mean. The model also overestimated the low flow year and underestimated the high flow year. Some other measurable input contributing to the streamflow needed to be considered to modify the model forecast. Hence, the last (April 30th) snow water equivalent measurements, considered as an accessible input, were regressed with the total seasonal streamflow. The identified

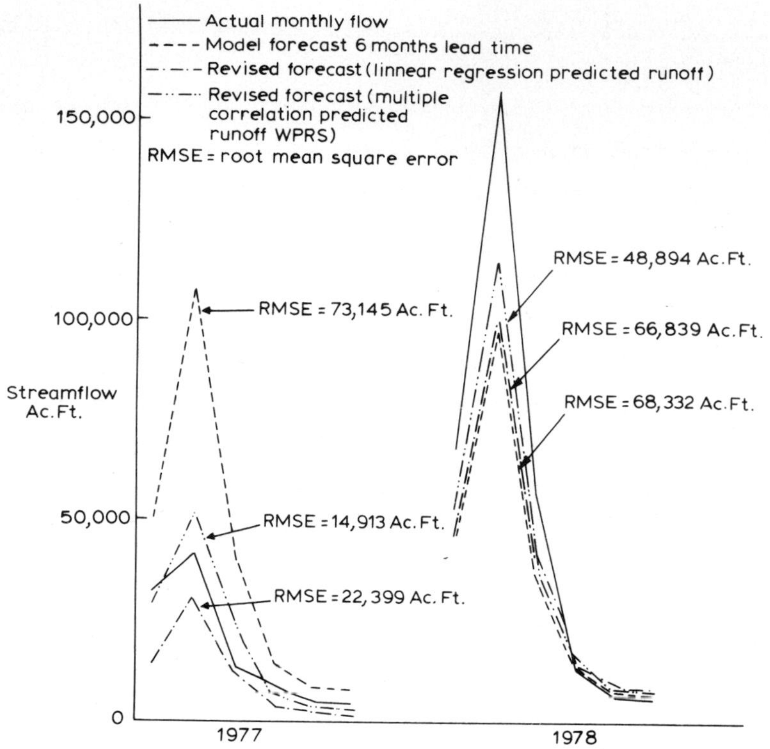

Figure 1. Revised monthly streamflow forecast using the predicted total seasonal runoff from linear regression and multiple correlation technique for the water years 1977 and 1978, high flow group sequence.

relationship was then utilized to provide indications of expected seasonal runoff volume, which in turn was disaggregated using the information available from the model representing the high flow series. Alternatively, the total seasonal streamflow forecast from the multiple correlation-based model of the Bureau of Reclamation was also used. The revised monthly forecasts for the high flow period were closer to the actual streamflow values.

System configuration and network representation

The physical features of the basin was transformed into a capacitated flow network, as shown in Figure 2. The storage (reservoirs) and non-storage (river confluence, canal diversions, and demands) points are represented as nodes while the river reaches and canals are designated as links connecting the nodes. Description of and data for the network components are given in Tables 1 and 2.

The eleven major high country reservoirs were lumped together. Some of the reservoirs on the plains were similarly aggregated by ownership and diversion canals serving them. This approach reduced the dimension or size of the network without sacrificing management flexibility, defined earlier.

All the diversion demands are individually considered. They also have higher priorities compared to ideal storage levels. Imports to the basin to augment the natural supply were also lumped together. Supplemental water supply deliveries from the Colorado-Big Thompson project via Horsetooth Reservoir was considered separately.

The two USGS gages at the mouth of the canyon near the City of Fort Collins and at the confluence of the Cache la Poudre River with the South Platte River near the City of Greeley served as an additional check in the calibration runs and an indication of the amount of water leaving the system.

Model calibration

Calibration runs of the network model were made to simulate the extremely dry '76-'77 water year. To do this, "costs" C_{ij} must be assigned. These were provided by the Poudre River Commissioner, with some adjustments. Some minor discrepancy between the historical desired

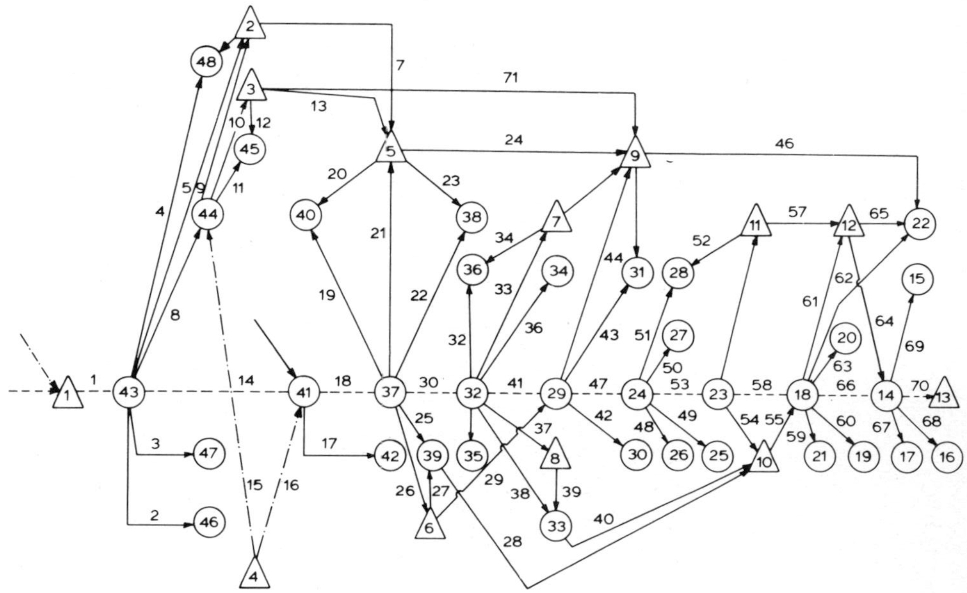

Figure 2. Network representation of the Cache la Poudre river basin's physical features.

Table 1. Network node components, Cache la Poudre River Basin.

Node #	Description	Node #	Description
1	Mountain Reservoirs (9 aggregated)	22	Greeley No. 2 Demand
		23	Demand Junction
2	North Poudre Company (20 aggregated)	24	Demand Junction
		25	Boxelder Demand
3	Windsor Reservoir and Canal (3 aggregated)	26	Chaffee Demand
		27	Coy Demand
4	Horsetooth Reservoir	28	Lake Demand
5	Water Supply and Storage (9 aggregated)	29	Demand Junction
		30	Arthur Demand
6	Claymore	31	Larimer and Weld Demand
7	Terry Lake	32	Demand Junction
8	Warren Lake	33	Larimer County No. 2
9	Windsor Reservoir	34	Taylor and Gill
10	Fossil Creek	35	New Mercer Demand
11	Timnath Reservoir	36	Little Cache La Poudre
12	Windsor and Seeley Lake (2 aggregated)	37	Demand Junction
		38	Jackson Demand
13	City of Greeley Gage (terminal node)	39	Pleasant Valley and Lake
		40	Larimer County Demand
14	Demand Junction	41	City of Fort Collins Gage
15	Ogilvy Demand	42	City of Greeley Diversion
16	Boyd and Freeman Demand	43	Demand Junction
17	Greeley No. 3 Demand	44	Demand Junction
18	Demand Junction	45	Poudre Valley Demand
19	Jones Demand	46	Fort Collins Diversion
20	Whitney Demand	47	Mountain Ditch Demand
21	Eaton Demand	48	North Poudre Canal Demand

Table 2. Network link components, Cache la Poudre system.

Link Number	From Node	To Node	Capacity Acre-Feet	Link Number	From Node	To Node	Capacity Acre-Feet
1	1	43	300,000	37	32	8	9,000
2	43	46	1,680	38	32	33	9,000
3	43	47	3,240	39	18	33	9,000
4	43	48	23,100	40	33	10	9,000
5	43	2	23,100	41	32	29	300,000
6	2	48	23,100	42	29	30	3,600
7	2	5	23,100	43	29	31	60,000
8	43	44	21,000	44	29	9	60,000
9	44	2	21,000	45	9	31	60,000
10	44	3	21,000	46	29	22	31,500
11	44	45	21,000	47	29	24	300,000
12	3	45	21,000	48	24	26	1,500
13	3	5	45,000	49	24	25	3,300
14	43	41	300,000	50	24	27	600
15	4	44	91,000	51	24	28	9,600
16	4	41	91,000	52	11	28	9,600
17	41	42	2,500	53	24	23	300,000
18	41	37	300,000	54	23	10	15,000
19	37	40	42,000	55	10	18	15,000
20	5	40	42,000	56	23	11	15,000
21	37	5	45,000	57	11	12	15,000
22	37	38	3,000	58	23	18	300,000
23	5	38	3,000	59	18	21	1,800
24	5	9	60,000	60	18	19	960
25	37	39	7,500	61	18	12	36,000
26	37	6	7,500	62	18	22	36,000
27	6	39	7,500	63	18	20	4,200
28	39	10	7,500	64	12	14	36,000
29	6	29	7,500	65	12	22	36,000
30	37	32	300,000	66	18	14	300,000
31	32	35	3,600	67	14	17	9,000
32	32	36	7,500	68	14	16	600
33	32	7	7,500	69	14	15	4,800
34	7	36	7,500	70	14	13	300,000
35	7	9	60,000	71	3	9	60,000
36	32	34	720				

storage levels and those obtained from the simulation was expected due to the adjustments made to account for reservoir evaporation and approximations of channel conveyance losses, seepage inflows and leakage outflows into and from the reservoirs. The difference between the computed and desired storage levels never exceeded 20% and were generally lower than 10%. This was deemed acceptable.

The observed '76-'77 total flow at the USGS gage near Fort Collins was miscalculated by only 3.8% on the high side. The calculated high flow months differ by 15% on the average while the low flow months are in error by 30%. The latter was not considered serious since the flow magnitudes are extremely low. Considering that some unmeasurable variables such as the losses and seepage inflows could only be roughly estimated, this calibration run was accepted without further adjustment of the priority factors. The final rankings of priorities for the various storage and diversion demands obtained from the calibration runs are shown in Tables 4 and 5.

Management objective-minimize wasted outflows

The estimate of the total seasonal streamflow based on the snowpack accumulations only allows the commissioner to determine whether that available supply is below, above or even with the normal use. This information is of benefit to appropriators who feel that they could not be served at all or that continuing service to fill up their needs could not be assured. Day to day water allocation activities of the river commissioner have been modeled by Thaemart (1976). In this model, the daily flows are simply apportioned by a direct search of users with the highest priority and deducting the amount entitled to divert in succession.

Establishing monthly targets for reservoir storage and diversions, based on forecasted monthly streamflows for the duration of the season that can be updated in real-time, is one possible additional step to take. This would maintain the efficient and effective operation of the system in anticipation of future streamflows. Through this approach, a basic framework could also be provided for real-time operations over a shorter time interval, such as daily or weekly.

Table 3. Reservoir storage target levels as percent of total (full) capacity, Cache la Poudre River Basin (water year 1977-1978).

Reservoirs	Cap.	Nov.	Dec.	Jan.	Feb.	Mar.	Apr.	May	Jun.	Jul.	Aug.	Sep.	Oct.
Mountain Reservoir (11)	42511	0.27	0.28	0.27	0.27	0.28	0.38	0.53	0.59	0.54	0.30	0.15	0.14
NPC Reservoir (20)	41121	0.62	0.61	0.63	0.63	0.61	0.61	0.58	0.56	0.54	0.30	0.28	0.29
WRC (3)	41960	0.34	0.33	0.33	0.33	0.32	0.32	0.28	0.21	0.17	0.10	0.18	0.18
WSS Reservoir (9)	25553	0.69	0.68	0.67	0.65	0.64	0.64	0.60	0.42	0.52	0.31	0.37	0.26
Claymore	954	0.64	0.62	0.63	0.67	0.76	0.85	0.71	0.34	0.24	0.43	0.02	0.00
Terry Lake	8028	0.67	0.70	0.70	0.70	0.69	0.73	0.97	0.64	0.56	0.48	0.28	0.36
Warren Lake	2089	0.66	0.61	0.59	0.56	0.54	0.50	0.58	0.42	0.68	0.83	0.57	0.51
Windsor	16786	0.41	0.45	0.49	0.54	0.58	0.62	0.75	0.59	0.36	0.11	0.11	0.14
Fossil	11100	0.64	0.48	0.58	0.66	0.79	0.86	0.82	0.74	0.74	0.37	0.25	0.36
Timnath	10070	0.00	0.00	0.00	0.00	0.00	0.00	0.00	0.00	0.04	0.15	0.28	0.31
Windsor and Seeley Reservoirs	2587	0.71	0.70	0.73	0.71	0.64	0.93	0.81	0.83	0.86	0.72	0.65	0.69

Table 4. Annual rankings for the ditch diversion demand nodes, Cache la Poudre river basin (lower numbers represent higher priority)

NODE NO. 2	YEAR 1	RANK = 66	NODE NO. 27	YEAR 1	RANK = 22
NODE NO. 3	YEAR 1	RANK = 33	NODE NO. 28	YEAR 1	RANK = 61
NODE NO. 5	YEAR 1	RANK = 64	NODE NO. 30	YEAR 1	RANK = 43
NODE NO. 8	YEAR 1	RANK = 40	NODE NO. 31	YEAR 1	RANK = 51
NODE NO. 9	YEAR 1	RANK = 51	NODE NO. 33	YEAR 1	RANK = 40
NODE NO. 12	YEAR 1	RANK = 46	NODE NO. 34	YEAR 1	RANK = 28
NODE NO. 13	YEAR 1	RANK = 75	NODE NO. 35	YEAR 1	RANK = 53
NODE NO. 15	YEAR 1	RANK = 71	NODE NO. 36	YEAR 1	RANK = 35
NODE NO. 16	YEAR 1	RANK = 15	NODE NO. 38	YEAR 1	RANK = 69
NODE NO. 17	YEAR 1	RANK = 38	NODE NO. 39	YEAR 1	RANK = 59
NODE NO. 19	YEAR 1	RANK = 30	NODE NO. 40	YEAR 1	RANK = 64
NODE NO. 20	YEAR 1	RANK = 17	NODE NO. 42	YEAR 1	RANK = 12
NODE NO. 21	YEAR 1	RANK = 20	NODE NO. 45	YEAR 1	RANK = 33
NODE NO. 22	YEAR 1	RANK = 46	NODE NO. 46	YEAR 1	RANK = 10
NODE NO. 25	YEAR 1	RANK = 25	NODE NO. 47	YEAR 1	RANK = 48
NODE NO. 26	YEAR 1	RANK = 56	NODE NO. 48	YEAR 1	RANK = 66

Table 5 Reservoir characteristics, Cache la Poudre River Basin (lower numbers represent higher priority).

Name of Reservoirs		CAPACITIES IN ACRE-FEET Starting	Minimum	Maximum	Rank Priorities
1.	Mountain Reservoir (11)	11412	0	42511	147
2.	NPC Reservoir (20)	24406	0	41121	155
3.	WRC Reservoir (3)	15029	0	41960	134
4.	WSS Reservoir (9)	17074	0	25553	166
6.	Claymore	448	0	954	100
7.	Terry Lake	4721	0	8028	141
8.	Warren Lake	1441	0	2089	113
9.	Windsor Reservoir	6019	0	16786	163
10.	Fossil Creek	5782	0	11100	152
11.	Timnath	0	0	10070	164
12.	Windsor and Seeley (2)	1779	0	2475	151
	TOTAL	88111		202647	

Table 6. Results of the management studies using the MODSIM model. Cache la Poudre River system, water year, 1976-1977.

CASES	System Outflows and Computed Reductions in Acre-Feet								
	Nov-Apr Subtotal	Irrigation Season						May-Oct Subtotal	Annual Total
		May	Jun	Jul	Aug	Sep	Oct		
Historical Gaged System Outflow Gage # 6752500	31,855 (66%)	1,701	556	6,584	1,721	1,273	4,372	16,207 (34%)	48,062 (100%)
1a/ Management Study Run #1 Computed Model Outflow	30,971 (73%)	0	0	5,938	1,721	914	2,920	11,483 (27%)	42,468 (100%)
1b/ Computed Outflow Reduction (Stored)	823 (15%)	1,701	556	646	0	359	1,452	4,714 (85%)	5,598 (100%)
2a/ Management Study Run #2 Target Minimum Outflow	360 (34%)	130	130	130	130	130	60	710 (66%)	1,070 (100%)
2b/ Computed Outflow Reduction (Stored)	31,495 (67%)	1,571	426	6,454	1,591	1,143	4,312	15,497 (33%)	46,992 (100%)

1a/ MODSIM model computed system outflow, under the policy that the outflow is allowed to fluctuate within the minimum and maximum capacity of the river reach and assigned the lowest priority demand.

1b/ Amount of water that was historically wasted but is available for storage on a month-to-month basis under the policy stated in 1a above.

2a/ Target minimum system outflow set equal to the recommended minimum streamflow for the Poudre River with assigned priority in between the lowest priority diversion and highest priority storage.

2b/ Amount of water that was historically wasted but is available for storage on a month-to-month basis under the policy stated in 2a above.

The historical summary of the District's annual operation as compiled by the Poudre River Commissioner, shows that, on an annual basis, an amount ranging from 7,000 to 151,000 acre-ft. of water (for 23 years out of 29) was lost from the District because of *lack of storage* or inability to convey to storage. Every year for the last 30 years of the District operation, an average amount of 96,000 acre-ft. of water drained to the South Platte River, as observed at the USGS gage near the City of Greeley. This amount is roughly 35% of the average annual virgin flow and represents 20% of the average total water supply from all sources. If a fraction of this amount could be saved, or stored and used, additional appropriators could be served.

To determine the worth of the monthly streamflow forecast, the model MODSIM was used to obtain monthly operation policies, that would reduce outflows from the Cache la Poudre River system, assuming historical diversion demands must be satisfied. Water year '76-'77 was again chosen, but forecasted streamflows were used rather than actual observed flows. The observed monthly streamflows for the six (6) months prior to May were used to forecast the succeeding six (6) months streamflow. The forecast was modified somewhat using the information obtained from the results of the last (April) snow course survey and input into MODSIM. The historical reservoir storage levels were specified as ideal target levels and historical diversion demands were used. However, the C_{ij} priorities from the calibration results were modified to discourage wasted outflows. This implies some foreknowledge used in the experiment. However, a specific minimum basin outflow target was set with a higher priority than the reservoirs, so use of the historical storage levels as target levels did not appreciably effect the results. As for use of the historical diversion demands, we are assuming that these can be forecasted with reasonable accuracy on a month to month basis.

Computational Results

Results are summarized in Table 6. Management run #1 allowed the basin outflow to fluctuate freely, essentially using the same C_{ij} priorities as from the calibration runs. The specified diversion demands having a higher priority were satisfied as expected. The total system computed outflows were lower by 12% compared to the observed outflows at the gage near Greeley. The total difference, 5,598 acre-ft., was stored in reservoirs with high priority and sufficient unused capacity. Notice that the observed system outflow during the irrigation season (May-Oct) is only about 34% of the annual total recorded wasted water leaving the system. This is an indication of the relatively efficient manner of water use and system management in the area, primarily based on the doctrine of prior appropriation. The opportunity to save a greater amount of the wasted water appears to be during the fall and winter months, when roughly two-thirds of the annual total has occurred.

Theoretically, the amount leaving the system can be reduced to zero (this has already occurred in some sections of the river, Coloradoan, 1974) since there is no legally binding contract or agreement for the Poudre River to pass a certain amount of water to the South Platte River. In reality, especially in the lower reaches, there would be canal conveyance losses and farm wastes from both sides of the river. For management run #2, the amount of return flow was approximated by the recommended minimum monthly streamflow, which amounts to 60 acre-ft. from October through April and 130 acre-ft. from May to September (Rhinehart, 1975). These minimum monthly flow values were therefore designated as the target minimum outflows from the Poudre River basin. The priority assigned to this flow was made lower than the lowest priority for diversion demand but higher than the highest priority for the storage demand. These results are shown in the last two rows of Table 6. The improvement is significant. The amount of water that was wasted and could have been stored appears in the reservoirs which have higher priority and sufficient capacity.

The calculated end of month storage, in some instances, is higher than the desired or target storage levels. This target storage is being utilized in the MODSIM model as a guide; hence, the water levels can fluctuate between the desired level and maximum safe level for each of the reservoirs. The amount that is stored depends on the priority assigned to the reservoir. Figure 3 shows that Management run #2 resulted in the capture and storage of considerably more water than occurred historically for Fossil Creek Reservoir. Note that the maximum reservoir storage levels were never actually reached in all months.

As an illustration of the kind of improved storage strategies the model produces, consider a reach of the Poudre River shown in Figure 4. Fossil Creek reservoir is located south of the river, but its releases to the river could serve several ditches downstream. Greeley No. 2 (or Cache la Poudre No. 2) canal is the last ditch where it is possible to divert and store river flows at either Windsor, Neff or Seeley Lakes. Any flows passing by this ditch could satisfy demands at Whitney, B. H. Eaton, Jones, Greeley No. 3, Boyd and Freeman, and Ogilvy ditches. Any excess flows beyond the last ditch, i.e., passing by Ogilvy canal, drain into the South Platte river and are lost downstream.

Out-of-priority diversion of the flows that would otherwise be lost from the system could be stored at Fossil Creek, Windsor Lake or Seeley Lake. Any shortfalls at Ogilvy could be served by Windsor or Seeley Lakes. Shortages at any of the other five (5) ditches could be easily filled by releases from Fossil Creek Reservoir. MODSIM could therefore be useful in devising voluntary exchange and trade schemes that will greatly benefit all the users in the basin.

FINAL NOTES

The results presented here need to be qualified according to the assumptions made. First, since all streamflows are in average monthly values, there is no guarantee that computed average monthly diversions could actually be realized during the month. This is because an intense thunderstorm, for example, could produce a large flow in a short period of time which could not be captured because of insufficient canal capacity; even if there is sufficient offstream storage capacity.

A large onstream reservoir could, of course, capture a significant portion of this flow. This study area does not currently have such a reservoir. This problem could be indirectly considered by reducing the effective canal capacity by an amount based on an analysis of average daily flows within the month in relation to daily available canal

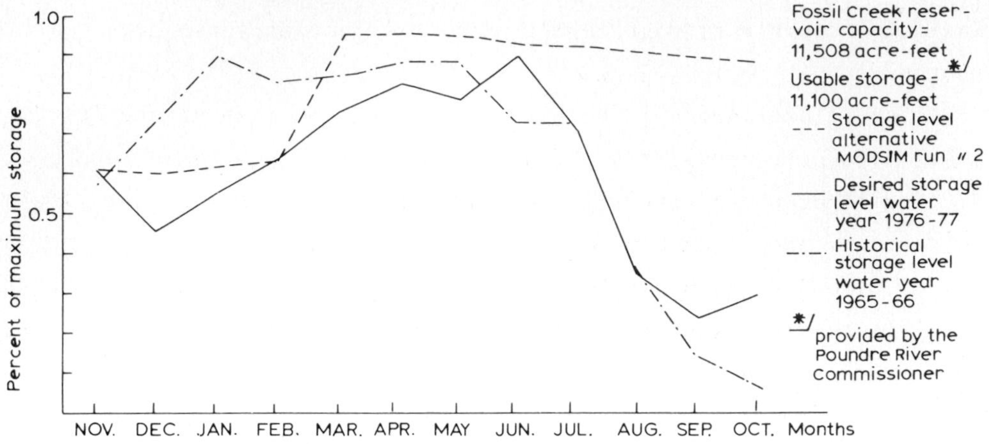

Figure 3. Reservoir storage levels under a reduced system outflow case and relationship with the historical storage levels for the lowest flow year (1976-1977) and comparable low flow year (1965-1966), Fossil Creek Reservoir, Cache la Poudre River Basin.

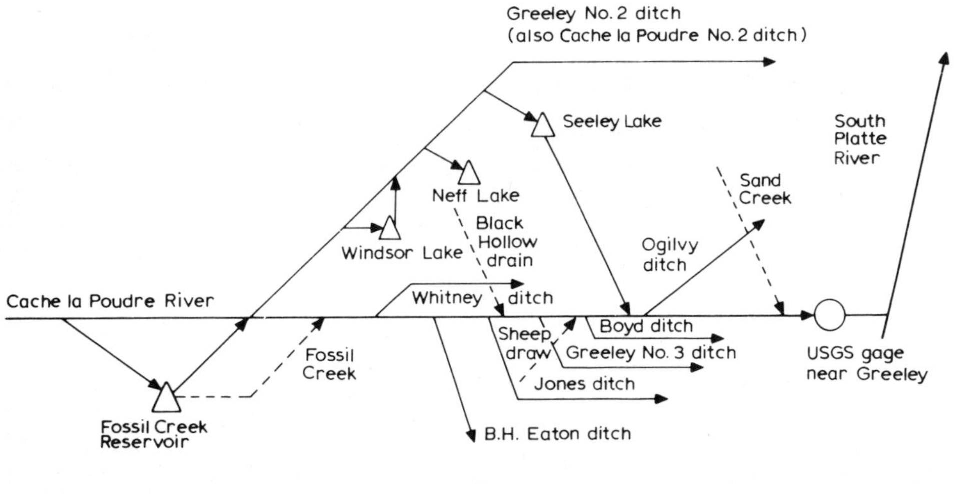

Figure 4. Schematic diagram to illustrate the out of priority diversion for storage and subsequent release strategy for ditch diversion demand satisfaction to minimize outflows from a water resource system, using the Cache la Poudre basin as an example.

capacity. An iterative procedure could be devised between the MODSIM model and a daily flow accounting model in order to determine how much the effective canal capacity should be reduced.

As mentioned earlier, there is a question about the safety of many of the offstream reservoirs. This prevents several of them from being filled. Though the target maximum storage levels used in this study are believed to be reasonable, further work is needed to refine them. They may also need to be modified in accordance with expected icing conditions during winter that would reduce effective storage capacity. Also, the model does not guarantee that storage rights will be exercised in order of priority. The weighting factors C_{ij} on carryover storage primarily control the release of water rather than priorities in initially filling the reservoirs. Further work is needed to more accurately include filling priorities in the model.

Finally, since this model uses a sequential static approach to the optimization, then it cannot fully utilize an extended, multiperiod forecast. Rather, a one-month ahead forecast is the most information it can actually use. The model then optimizes allocation of water within a given month. However, using, say, a six month forecast, the weighting factors C_{ij} can be adjusted to reflect later problems anticipated by the streamflow forecast. For example, if a drought period is forecasted, weighting factors C_{ij} on storage could be altered to take shortages earlier and retain more water in the reservoirs as a hedge on the coming dry period. Further work is needed on fully incorporating into MODSIM information obtained from the forecast model, particularly the streamflow forecast covariance estimates, in order to properly analyze the risk associated with various management scenarios.

ACKNOWLEDGEMENTS

This research was partially supported by funding provided by the Office of Water Research and Technology, U. S. Department of Interior (authorized under P. L. 95-467) as authorized by the Water Research and Development Act of 1978 and the U. S. - Spanish Project: Conjunctive Water Uses of Complex Surface and Groundwater Systems.

REFERENCES

Anderson, R. L., "The Effects of Streamflow Variation on Production and Income of Irrigated Farms Operating Under the Doctrine of Prior Appropriation," Water Resources Research, Vol. 11, No. 1, 1975.

Barkley, J. R. "The Northern Colorado Water Conservancy District," Loveland, Colorado, 1974.

Bazaraa, M. S. and J. J. Jarvis, "Linear Programming and Network Flows," John Wiley and Sons, Inc., 1977.

Becker, L and W-G Yeh, "Optimization of Real-Time Operation of a Multiple Reservoir System," Water Resources Research, Vol. 10, No. 6, December, 1974.

Bellamy, R., "Personal communication," 1980.

Box, G. E. P. and G. M. Jenkins, "Time Series Analysis, Forecasting and Control," Second Edition, Holden-Day, 1976.

Coloradoan, "140 Trout Killed in Section of Poudre River," Newspaper article on the Drying up of the Poudre River, Fort Collins, September 11, 1974.

Graupe, D., "Identification of Systems," Second Edition, Robert F. Kreiger Publishing Co., Huntington, New York, 1976.

Hawley, M. E., R. H. McCuen and A. Rango, "Comparisons of Models for Forecasting Snowmelt Runoff Volumes," Water Resources Bulletin, Vol. 16, No. 5, 1980.

Hoshi, K. and S. J. Burges, "Incorporation of Forecasted Total Seasonal Runoff Volumes into Reservoir Management Strategies," Reliability in Water Resources Management, McBean, E. A. and T. F. Unny (editors), Water Resources Publications, Fort Collins, Colorado, 1979.

Labadie, J. W., J. M. Shafer and R. Aukerman, "Recreational Enhancement of High Country Water Supply Reservoirs," Water Resources Bulletin, Vol. 16, No. 3 (June 1980).

Lazaro, R. C., Labadie, J. W. and J. D. Salas, "State-Space Streamflow Forecasting Model for Optimal River Basin Management," presented at the International Symposium on Real-Time Operation of Hydrosystems, June 24-26, University of Waterloo, Waterloo, Ontario, Canada, 1981.

McKerchar, A. I. and J. W. Delleur, "Application of Seasonal Parametric Linear Stochastic Models to Monthly Flow Data," Water Resources Research, Vol. 10, No. 3, 1974.

Mejia, J. M., P. Egli and A. Leclerc, "Evaluating Multireservoir Operating Rules," Water Resources Research, Vol. 10, No. 6, December, 1974.

Movarek, I. E., M. H. Salem and H. T. Dorrah, "Hydrological Studies on the River Nile - I. Forecasting," Research Report, Cairo University/MIT, Technological Planning Program, Cairo, 1978.

Neutze, J., "Personal communication," 1980.

Panu, U. J. and T. F. Unny, "Extension and Applications of Feature Prediction Model for Synthesis of Hydrological Records," Water Resources Research, Vol. 16, No. 4, 1980.

Radosevich, G. E., D. H. Hamburg and L. L. Swick, "Colorado Water Laws, A Compilation of Statutes, Regulations, Compacts and Selected Cases," Information Series No. 17, Center for Economic Education and Environmental Resources Center, Colorado State University, 1975.

Reitano, B. M. and D. W. Hendricks, "Input-Output Modeling of the Cache la Poudre Water System," Environmental Engineering Technical Report 78-1683-01, Dept. of Civil Engineering, Colorado State University, Ft. Collins, Colorado, Nov., 1978.

Rhinehart, C. G., "Minimum Streamflows and Lake Levels in Colorado," Environmental Resources Center, Information Series No. 18, Colorado State University, August, 1975.

Salas, J. D., Delleur, J. W. Yevjevich, V. and Lane, W., 1980, "Applied Modeling of Hydrologic Time Series," Water Resources Publications, Littleton, Colorado.

Shafer, J. M., "An Interactive River Basin Water Management Model: Synthesis and Applications," Technical Report No. 18, Colorado Water Resources Research Institute, Colorado State University, 1979.

Shafer, J. M., Labadie, J. W. and E. Bruce Jones, "Analysis of Firm Water Supply Under Complex Institutional Constraints," Water Resources Bulletin, June, 1981.

Texas Water Development Board, "Economic Optimization and Simulation Techniques for Management of Regional Water Resource Systems, River Basin Simulation Model SIMYLD-II--Program Description," Prepared by the Systems Engineering Division, Austin, Texas, July, 1972.

Thaemart, R. L., "Mathematical Model of Water Allocation Methods," Ph.D. Dissertation, Colorado State University, 1976.

United States Bureau of Reclamation, "Western Division Water Supply Forecasting: Electronic Computer Program Description," Loveland, Colorado, 1968.

Unny, T. E., Divi, R., Hinton, B. and A. Robert, Proceedings of the International Symposium on Real-Time Operation of Hydrosystems," Volumes I and II, University of Waterloo, Waterloo, Ontario, Canada, June 24-26, 1981.

Wilson, J. R. and E. Kirdar, "Use of Runoff Forecasting in Reservoir Operations," Journal of the Irrigation and Drainage Division,

Proceedings of the ASCE, Vol. 96, IR3, September, 1970.

Wunderlich, W. O. and J. E. Giles, Proceedings of the International Symposium on Real-Time Operation of Hydrosystems," Volumes I and II, University of Waterloo, Waterloo, Ontario, Canada, June 24-26, 1981.

APPROPRIATE SAMPLING PROCEDURES FOR ESTUARINE AND COASTAL ZONE WATER-QUALITY MEASUREMENTS

DONALD STEVEN GRAHAM[1] and JOHN M. HILL[2]

Introduction

Many environmental problems in estuaries involve relating changes in water quality to some hypothesized biological effect; or relating ecosystem response to forcing functions, the most important of which is usually water quality. However, the associated traditional sampling techniques often used do not include consideration of deterministic and predictable hydrodynamic and water-quality variations.

Until recently ecology was primarily a qualitative science and surveys tended to be used to assist descriptive studies. In the past few years the emphasis has been placed on statistically designed studies [see Smith (1978)] which are being used to test hypotheses regarding model validity. The degree of complexity varies greatly however, but four general categories can be defined:

i) deterministic
ii) statistical
a) biological and water-quality only
b) biological and water-quality coupled to hydrodynamics.

1/ Tudor Engineering Company, San Francisco, California, 94105
2/ Civil Engineering Department, Louisiana State University, Baton Rouge, Louisiana, 70803

Reprinted from *Time Series Methods in Hydrosciences*, by A.H. El-Shaarawi and S.R. Esterby (Editors)

The simplest type of ecological model is (iia), the most difficult is (ib). An example of the former is TECO (1980), and the latter Najarian and Harleman (1977). The advantage of the latter is that it is predictive if circumstances are altered. Use of models of type (ib) requires consideration of timescales in the sampling and analyses. Such concepts are standard in fluids engineering, but appear to be quite novel to biologists and statisticians. It is the purpose of this paper to introduce these concepts in a general way, with examples, and to illustrate the great advantages of using linked hydrodynamic and water-quality models in association with ecological studies in dynamic environments. However the relevance of discussion is not likely to be limited to biologists, for, as Najarian and Harleman (1977) have pointed out, the traditional methods of water-quality data collection in transient environments by field sampling teams may be obsolete insofar as these data are now often used as input to real-time numerical models. Because the environmental impacts of engineering works must now be studied, interaction between biologists and engineers has increased markedly. However, for many practical engineering applications where biological information is needed, such as environmental impact statements, the data provided are not compatible with the requirements of water quality models. If these data, which are essentially time series, are to be useful, then certain sampling considerations must be taken into account. The purpose of this paper is to outline in general terms, with specific examples for illustration, some considerations for acquiring useful data in estuarine and coastal environments. It is intended to be practical and illustrative, rather than rigorous.

Advection and Dispersion

Unfortunately, the traditional method of biological sampling for environmental surveys has been to measure biota and water

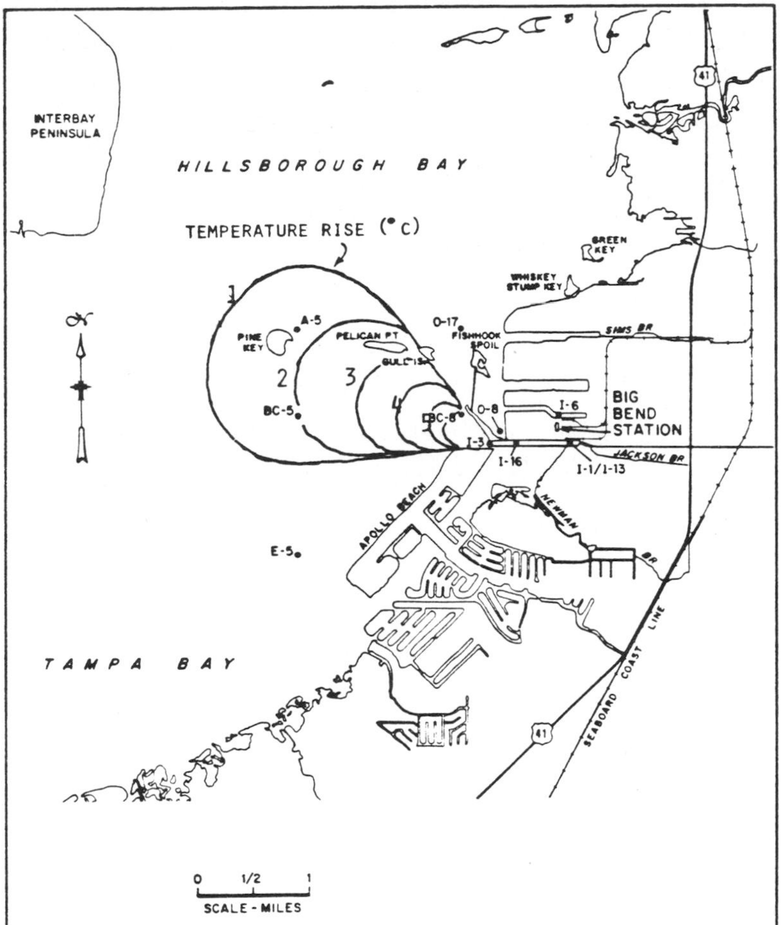

FIGURE 1
MEROPLANKTON SAMPLING STATIONS 1976-1977

Source: TECO, 1980

quality variables simultaneously at rather long time intervals. The two are sometimes then correlated as B vs. A from B(t) and A(t), although even this is often not done. An example of a typical sampling scheme is shown in Figure 1 in which ten stations were sampled at two-week intervals in order to predict the effects of an increase in the heat discharge from the Bend Bend Station in Tampa Bay, Florida. A typical time series of larvae numbers is shown in Figure 2 (TECO, 1980). This graph further demonstrates the need to change traditional sampling schemes to take into consideration the nature of an ever-changing population in a dynamic water body. Also shown in Figure 1 is the output of a simple temperature plume model from TECO (1980) showing the temperature gradient at 64% plant load under the assumption that there are no tides in Tampa Bay. Obviously, the plume is advected back and forth with the tide in reality. However, note the following facts:

- the temperature difference along the plume is of the order of 6^{0}C, which is also the order of the annual ambient temperature variation

- the temperature gradient along the plume is likely greater than the seasonal temperature gradient

- the temperature gradient across the plume is very high, relative to the along-plume or seasonal change

- because a plume is a separated-flow phenomenon, the instantaneous gradient would be much larger than a tidally-averaged one (top-hat <u>vs</u>. bell-shaped)

- the plume velocity would be of the same order as, or less than, the tidal rms velocity at a distance not far from the source

- if the plume is of width B and the rms tidal velocity is of order U_t, then

$$\frac{B}{U_t} \ll 2 \text{ weeks} \qquad (1)$$

FIGURE 2

Density of Menippe Mercenaria Larvae at Inshore Stations 1976-1977

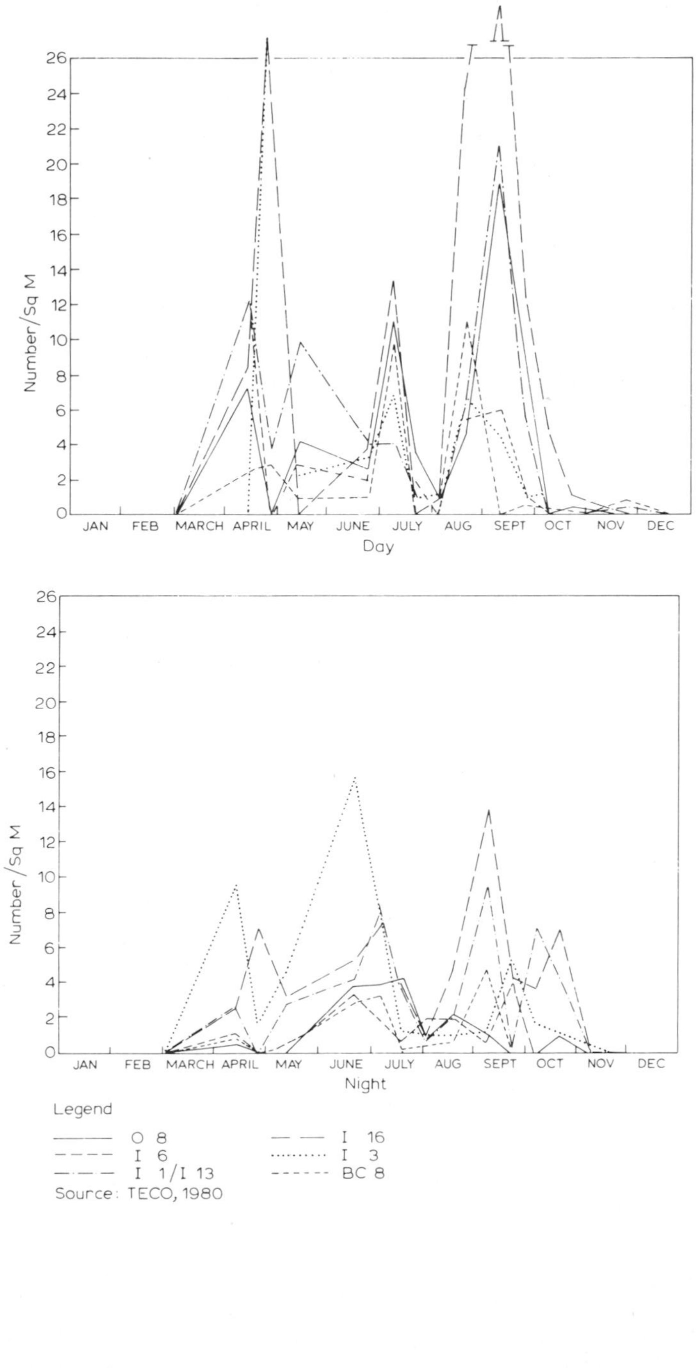

Source: TECO, 1980

It is evident then that both the temporal and spatial scales of the physical and biological sampling are not appropriate to the scales of the hydrodynamic phenomena. Furthermore, since they cannot be related, the effect of a change in the latter upon the former cannot predicted.

Using the TECO example to provide a general theme, a discussion of advective vs. diffusive effects is now in order. Assume that a water quality or biological variable is a function of another (DO as temperature for instance):

$$DO = fn(T) \tag{2}$$

Now, temperature is quasi-conservative and can be approximated in a two-dimensional Eulerian Cartesian frame in the x-direction by

$$\frac{\partial T}{\partial t} + \frac{U\partial T}{\partial x} + \frac{U\partial T}{\partial y} = \frac{\partial}{\partial x}\frac{K_x\partial T}{\partial x} + \frac{\partial}{\partial x}\frac{K_y\partial T}{\partial y} - Q(T-T_e) + T_i - T_o \tag{3}$$

where
- T - temperature
- U - velocity in x-direction
- V - velocity in y-direction
- K_i - dispersion coefficients
- Q - heat exchange coefficient
- T_e - equilibrium temperature
- T_i - sources of heat
- T_o - sinks of heat

Note that this equation is linear in T. Further, note that if T were completely conservative and steady, then

$$\frac{U\partial T}{\partial x} + \frac{U\partial T}{\partial y} = \frac{\partial}{\partial x} K_x \frac{\partial T}{\partial x} + \frac{\partial}{\partial y} K_y \frac{\partial T}{\partial y} \tag{4}$$

where the terms on the rhs represent advective transport and on the left dispersive transport. Both are different expressions of the same thing, i.e. motion. Most biological (and associated water-quality) sampling schemes assume dispersive processes to completely dominate. This is only correct for certain timescales if U, V $\neq$ 0 in an Eulerian frame however; indeed, except for molecular diffusion, the entire dispersive term is inherently fictitious and results from averaging. Using the usual conventions for disaggregating the joint time-series:

$$U = \bar{U} + U' \tag{5}$$

$$T = \bar{T} + T' \tag{6}$$

$$UT = \overline{UT} + \bar{U}T' + U'\bar{T} + U'T' \tag{7}$$

Averaging UT then results in

$$\overline{UT} = \overline{\overline{UT}} + \overline{U'\ T'} \tag{8}$$

so that the true advective term $\partial/\partial x(UT)$ in equation 3 is represented in an averaged format by

$$\frac{\partial}{\partial x}(\overline{UT}) = \frac{\partial}{\partial x}(\overline{\overline{UT}}) + \frac{\partial}{\partial x}(\overline{U'T'}) \tag{9}$$

$$= \bar{U}\frac{\partial \bar{T}}{\partial x} + \left(-\frac{\partial}{\partial x} K \frac{\partial \bar{T}}{\partial x}\right) \tag{10}$$

where the diffusion analogy is used for the second term on the rhs of equation 9 by assumption of a lengthscale. Therefore, the magnitude of the dispersive term merely represents the aliasing effects of the lengthscales and timescales by the sampling scheme used. Sampling procedures suggested by Smith (1978) do not attempt to minimize this term.

FIGURE 3

Water-Quality Data
At Apalachicola, Florida

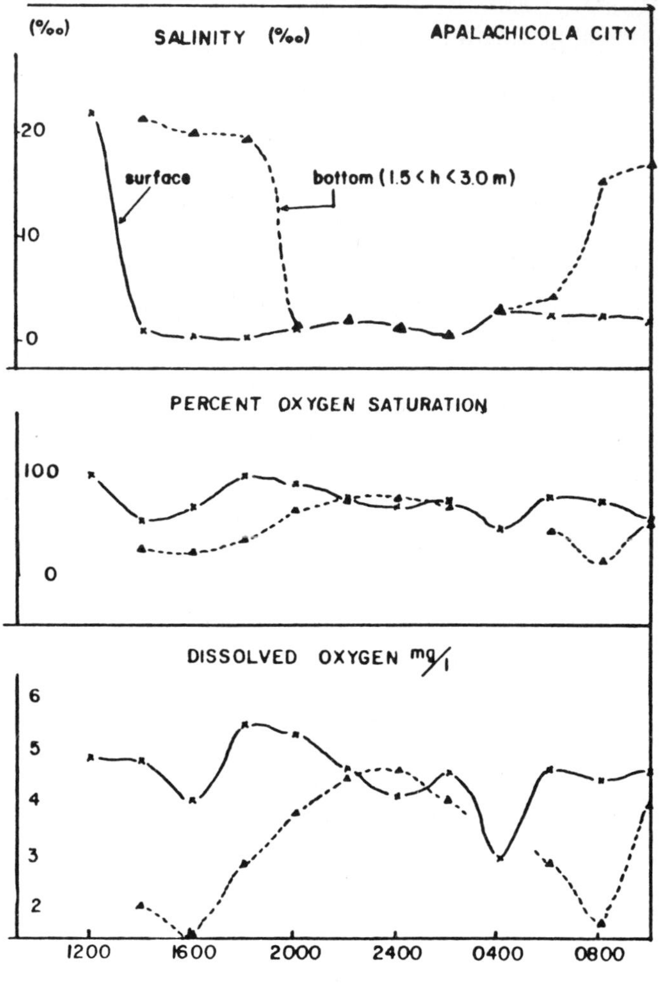

Source: Graham *et al.*, 1978

The effects due to advection often dominate at the small time and length scales of collection of the individual sample. During the sampling time at a station the tide can be either in or out, for instance. A sampling station in Figure 1 can either be inside or outside the plume while the sample is being taken. Therefore, for the data to have meaning in regions with sharp gradients at short sampling times (but not necessarily intervals), one must know the relation between the data and the velocity field.

Another example is provided from Graham et al. (1978) as Figure 3, which depicts water-quality data taken at the mouth of the Apalachicola River, Florida over a ten hour time interval. Note the rapid decrease in salinity from 20 to 0 ppt, and the associated rise in DO from 0 to 5 mg/l near the bottom as a result of tidal advection. Obviously, an animal with a short-period threshold DO tolerance of 2.5 mg/l would survive at this location on the average, but be dead in reality. A rapid sample taken here at a biweekly interval could have a very large variation, but this would be quite predictable. Finally, the data would be quite different if they were taken at the bottom, top, or depth-averaged. The point of this example is that at the timescales and lengthscales we are often most interested in, or tend to sample in, advection is important relative to diffusion-dispersion.

The effect of the advective term enters in the lhs of equation 3. In a 2-dimensional horizontal Eulerian frame the usual form of the momentum equation for tidal flow is:

$$\frac{\partial U}{\partial t} + \frac{U\partial U}{\partial x} + \frac{U\partial V}{\partial y} - fV + gH\frac{\partial \eta}{\partial x}$$

$$- \frac{\rho air}{\rho} C_D \left| U_{10} \right| U_{10x} - C_f \, (U^2 + V^2)^{1/2} \frac{U}{H^2}$$

$$- \frac{\partial}{\partial x} \left(E_{xx} \frac{\partial U}{\partial x} \right) - \frac{\partial}{\partial y} \left(E_{xy} \left(\frac{\partial U}{\partial y} + \frac{\partial V}{\partial x} \right) \right)$$

$$- M_x = 0 \qquad (11)$$

where U - vertically-averaged velocity in x-direction

V - vertically-averaged velocity in y-direction

f - Coriolis parameter, $= 2\ \omega \sin$ (latitude)

D - depth below MLW

η - height above MLW

$H = D + \eta$ (12)

ρ_{air} - density of air

ρ - density of water

U_{10} - air velocity at 3m

U_{10x} - air velocity at 3m in x-direction

C_D - air-water momentum transfer coefficient

$C_f = \frac{n^2 g}{H^{1/3}}$

E_{xy} - eddy viscosity terms

M_x - momentum addition rate per unit area

Note that equation (11) is quadradic in velocity and hence is nonlinear. The advantage of a dispersive assumption to remove the nonlinearity can certainly be appreciated, but this would lead to a nonpredictive capability in almost all circumstances.

Assuming then that it is desirable to know the velocity field in almost all practical water quality problems in coastal areas, how then can it be predicted and how well? Use of equation (11) along with an analogous equation in the y-direction and a continuity equation closes the problem. Several good computer models exist to predict the velocity field. An example showing the quality of a simulation to measured data in Apalachicola Bay from Daniels and Graham (1981) is presented as Figure 4. This illustrates that the velocity field can be represented quite nicely in real-time (1-minute increments) even in a case in which wind-imparted momentum is significant. The advective terms in equation (2) are deterministic therefore and need not be considered random in either the sampling scheme or the time-series analysis of the data. The water-quality field can then be subsequently calculated with known boundary and initial conditions, with the velocity field input. For most cases the effects of water-quality on the momentum field can be neglected.

FIGURE 4

Computer Simulation of Tidal Wave,

Apalachicola Bay, Florida

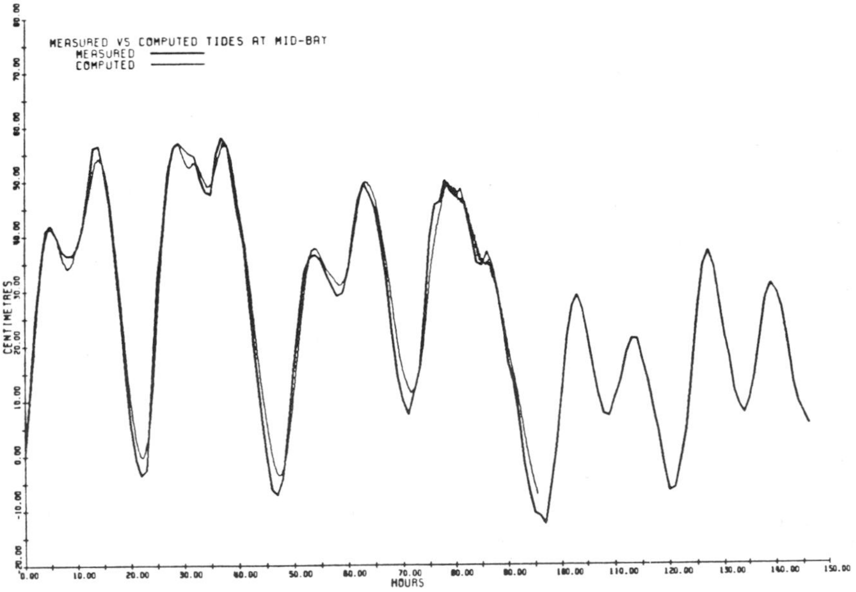

Source: Daniels and Graham (1981)

Continuous traces of water quality variables at a point are often dominated by advective effects, and this variance is predictable and, therefore, removable from the signal. In addition to Figure 3, see Figure 5 of Gunnerson (1966) and the paper by Najarian and Harleman (1977) for examples. An obvious advantage of this approach is that effects of changed circumstances upon the velocity, and hence water quality, are relatively predictable to the extent advection dominates diffusion in the transport processes. Methods of relating the value of the dispersion coefficient to the velocity field are generally empirical and nonrigorous, so the problem cannot be closed completely at this time.

The next three questions to logically arise are:

1. What is the appropriate sampling interval temporally?
2. What is the appropriate sampling interval spatially?

and

3. What is the appropriate duration of sampling?

Because the hydrodynamics of coastal areas are dominated by periodic tidal wave phenomena, the sampling interval must be able to resolve tidal effects both spatially and temporally. The usual methodology for sampling and analyzing time-series with inherent periodicities is the well-known power spectrum technique, which needs no further elaboration here. However, the best methods to obtain data which can be properly input into deterministic models needs some re-evaluation in the light of current technology. Harleman and Najarian (1977) were the first to point out this requirement:

> Large temporal variations of nutrient concentrations have been shown to occur within a tidal period at fixed

> locations. This suggests that a reevaluation of traditional estuarine field data collection techniques is necessary. More attention must be given to determining the temporal as well as the spatial variations of constituents in estuarine water quality surveys. Relatively little useful information can be expected from the random collection of samples from a moving boat. Probably the greatest limitation to the continued development and improvement of predictive water quality models is in the general lack of coordination between data needs for model variation and the data actually obtained in field surveys. (p. 537)

The typical practical sampling problem can be outlined by the following example. Apalachicola Bay, Florida, is approximately 10 km wide and 30 km long. A boat is available which can go about 10 km/h. A traverse of the estuary then takes about 8 hours, which is of the timescale of the dominant M_2 tide. Therefore, intratidal variations at any sampling point cannot be determined. Regarding Figure 3, it is known that these intratidal variations can dominate intertidal ones at a point, and therefore cannot be neglected. Further, if DO is a parameter of interest, then it is also known that DO is dependent on sunlight (through phytoplankton activity), wind and wave action, and temperature. Temperature variations at a point are influenced by the velocity field over the period of a tidal cycle in this area because the river inflow and Gulf of Mexico often have different temperatures. The M_2 tide has a dominant periodicity of about 12.42 hours, and the period of daylight at the equinox is 12 hours. To separate tidal from diurnal temperature and phytoplankton-induced changes in DO, it is necessary to devise a temporal sampling interval. In this case, the Nyquist criterion on the more frequent 12 hour signal holds, so the sampling interval must be at least 6 hours. Of course, for practical reasons sampling should be done at a rate roughly five times as dense as the Nyquist frequency. Hourly sampling is therefore appropriate. However, then the record length, ΔT, necessary to separate the two signals with frequencies $f_1 = 1/12$ h and f_2 = 1/12.42 h remains unaffected by the sampling interval, and is:

$$\Delta T \geq \frac{1}{f_2 - f_1} \tag{14}$$

$$\geq \frac{1}{.083 - .081} \tag{15}$$

$$\Delta T \geq 355 \text{ hrs} = 14.8 \text{ days} \tag{16}$$

assuming a rectangular data window. This duration is fortuitously very close to the period of the fortnightly M_1 tide (327.9 hours), so that a sampling period of a 15-day length should suffice. Therefore, as a general rule of thumb, hourly sampling for two weeks is a reasonable minimum for determination of diurnally-varying water-quality variables, and associated biological activity, in the coastal zone.

Hourly sampling for a 15-day period then entails 360 samples at each point, exclusive of the sampling taken beforehand to establish the initial conditions. Obviously, if periodicities in the data which are purely deterministic can be removed first, the length of the required sample length can be considerably reduced in most cases. If f_1 is removed in equation 14, for instance, then ΔT need only be greater than 12 hours. Further, phase information is not lost in this procedure, as is the case with variance spectrum analysis, so that the series is reconstitutable and, to some degree, predictable. This is an essential point: statistics can describe, but not be used to predict, if the forcing functions are changed.

It should be pointed out that many of these same considerations arise with respect to man-induced periodicities. The plant factor of the generating station in Figure 1 for instance will tend to have a 24-hour periodicity which will be out of phase with the solar day, and change phase with respect to the M_2 tide. The same is true of a STP, or releases through a hydroelectric project. Occasional sampling at the same time of day can result in a highly aliased record of these effects on water quality.

The number of points required to be sampled will vary with the particular location and problem, but there should be at least enough to provide the necessary spatial resolution. In coastal areas, a tidal wave is a shallow one moving at celerity $(gH)^{1/2}$. Hence if $H = 2m$, the wave moves at 4.4 m/s and travels about 100 km per one-half M_2 tidal period. This is not long relative to the 30 km length of Apalachicola Bay. The spatial resolution of the phenomenon should be the analog of the temporal Nyquist frequency, or 50 km at least and 10 km or so ideally. Obviously, the time taken to traverse 10 km on the water (1 hour) is so great that no approximation of coincident sampling can be made. Further, a 10 km grid could miss many water quality phenomena of interest whose lengthscale is much smaller than this (the plume in Figure 1, for instance, which should determine the spatial resolution of the sampling stations in this experiment, but apparently does not). In other words, numerous water mass boundaries are often missed due to lack of an appropriate sampling grid. As an attempt to address this problem we have been experimenting with use of remotely-sensed Landsat satellite images to provide synoptic resolution of the spatial field. An example of such an image for Apalachicola Bay is presented as Figure 5.[1] Field sampling is necessary to relate the

---

1 A colour version of this image can also be found on the cover of the Journal Water Pollution Control Federation, Volume 53, No. 4, April, 1981.

FIGURE 5

Enhanced Landsat Image of Apalachicola Bay, Florida, 26 February 1975

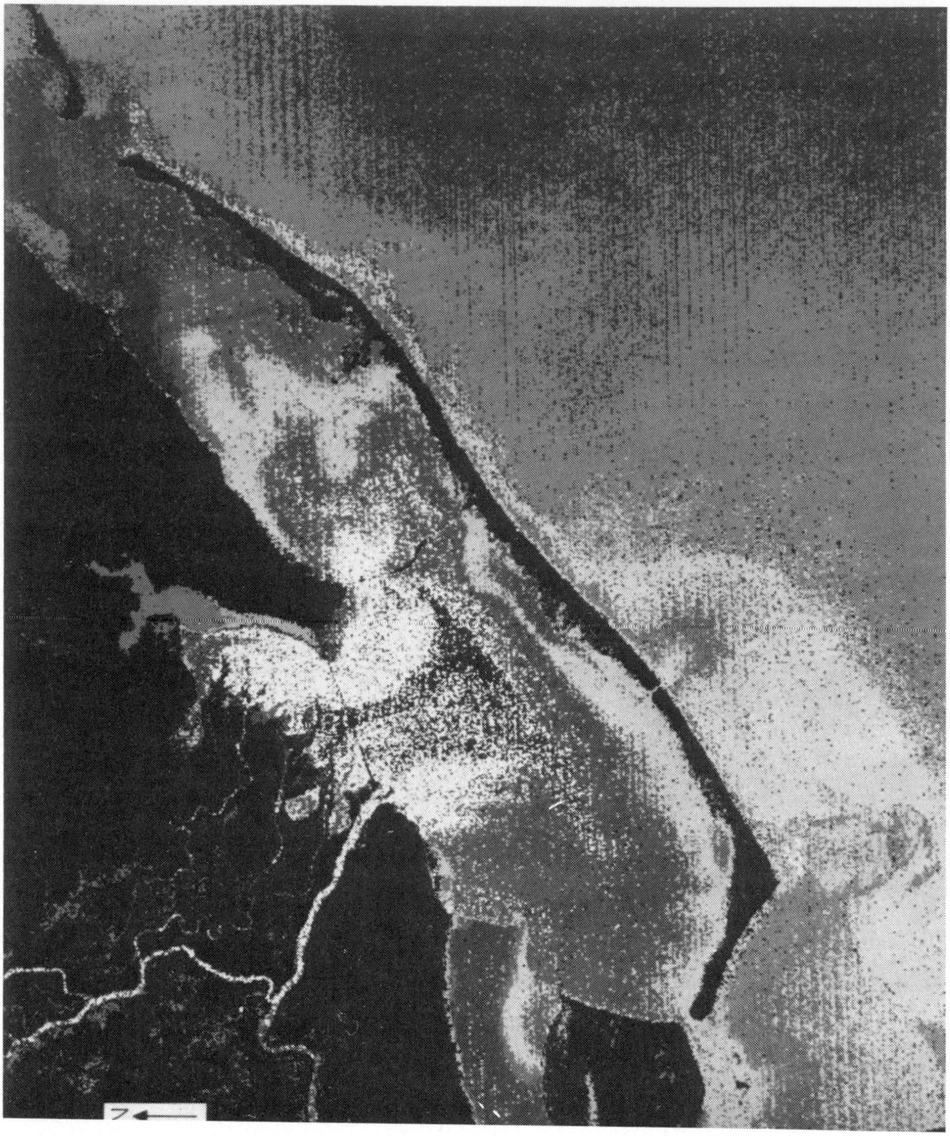

image patterns to water quality parameters (i.e., turbidity, color), but the image allows greatly expanded spatial resolution of the data-collecting capability of a ground sampler. A time and cost-efficient water sampling scheme which can rapidly sample large bodies of water, particularly in conjunction with satellite overpasses, is described in Hill and Dillon (1976). A description of an experiment to use satellite technology to provide spatial data for model verification appears in Graham and Hill (1980).

SUMMARY AND CONCLUSIONS

Biological and water quality studies in the coastal zone typically consist of gathering series of data at selected points. These sets of time-series data are then analyzed to yield conclusions which, for engineering studies, are used to solve the problem(s). The utility of these data, and hence, the ability of the analyst to solve the problem, depends greatly upon the sampling method. Many processes in the coastal environment are periodic and are relatively deterministic. Sampling must be conducted in such a manner that aliasing of periodic processes does not occur, thereby masking the underlying randomness or interrelationships between parameters. This requires the timescales and lengthscales of all the hydrodynamic water quality and biological parameters to be taken into account. Since the most useful portion of the data often lies in the residuals, these can be made most meaningful by removing significant known deterministic variations, such as tidal and diurnal effects. This can be best done using numerical models which, in turn, require data to be collected in a manner different from usual past practice. In particular high-frequency sampling at a few points and measurement of boundary conditions are required. Synoptic spatial data may have to be acquired by remotely-sensed means. Similarly biological surveys may be much more meaningful if they are taken in context of the hydrodynamics which, again, requires sampling on different time and space scales than has been the custom in the past. A discussion of characteristic biological and chemical timescales is presented by Ford and Thorton (1979).

Acknowledgements

The assistance of John Daniels with Figure 4 is gratefully acknowledged. This research was supported, in part, by funds provided by the U.S. Department of Commerce, National Oceanographic and Atmospheric Administration, Office of Sea Grant, under Grant No. 04-158-44046.

References

1. Daniels, J.P. and Graham, D.S., "Application and calibration of the CAFE-1 model to Apalachicola Bay, Florida", Proceedings of the 5th Canadian Hydrotechnical Conference, CSCE, held at Fredericton, N.B., 26-27 May 1981, Vol. 2, pp. 515-536.

2. Ford, D.E. and Thorton, K.W., "Time length scales for the one-dimensional assumption and its relation to ecological models", Water Resources Research, 15, (1), 1979, pp. 113-120.

3. Graham, D.S. and Hill, J.M., "Field Study for Landsat Water Quality Verification", in Proceedings of ASCE Symposium "Civil Engineering Applications of Remote Sensing", held at Madison, Wisconsin, August 13-14, 1980, pp. 101-117.

4. Graham, D.S., et al., Stormwater Runoff in the Apalachicola Estuary, Florida Sea Grant Report R/EM-11, March 1978, 76 pp. plus Appendices.

5. Gunnersen, C.G., "Optimizing sampling intervals in estuaries", Journal of the Sanitary Engineering Division, ASCE, 92, (SA2), Proc. Paper 4799, April 1966, pp. 103-125.

6. Hill, J.M., and Dillon, T.M., "A unique and effective oceanographic surface truth monitoring program for correlations with remotely sensed satellite and aircraft imagery," Technical Bulletin No. 76-2, Texas Engineering and Experiment Station, Texas A. and M. University, College Station, Texas, April 1976, 209 pp.

7. Najarian, T.O., and Harleman, D.R.F., "Real-time simulation of nitrogen cycle in an estuary", Journal of the Environmental Engineering Division, ASCE, 104, (EE4) Aug. 1977, pp. 523-538.

8. Smith, W., "Environmental survey design: a time series approach", Estuarine and Coastal Marine Science, 6, 1978, pp. 217-224.

9. TECO (Tampa Electric Company), 316 Demonstration, Big Bend Station - Unit 4, Aug. 1, 1980.

10. Thomann, R.V., "Time-series analysis of water-quality data", Journal of the Sanitary Engineering Division, ASCE, 93, (SA1), Proc. Paper 5108, Feb. 1967, pp. 1-23.

TIME SERIES ANALYSIS OF SOIL MOISTURE DATA

SHAW L. YU AND JAMES F. CRUISE

Department of Civil Engineering, University of Virginia,
Charlottesville, Virginia 22901

INTRODUCTION

Recent studies have shown the feasibility of statistically based investigations of infiltration and soil moisture regimes. Cordova and Bras (1981) and others have utilized probabilistic models in the analysis of soil moisture and infiltration. It has been well recognized that the soil moisture regime is a stochastic variable, consisting of both deterministic and random components. However, until now sufficient data has not been generally available to allow detailed probabilistic description of the soil moisture process. Therefore, an analysis of the variability of soil moisture based on a sufficient data sample is highly desirable.

In this study, time series analysis techniques and a linear auto-regressive prediction model are employed in an effort to examine the internal structure of the soil moisture process. The data were generated from a study of soil moisture fluctuations under various vegetative covers which was conducted during 1950-55 at the Calhoun Experimental Forest near Union, South Carolina (Kent, et al, 1981). There were a total of 5 years of data on rainfall amounts and total soil moisture to a depth of 1.68 meters. However, only for two of the years continuous daily observations were available without any missing observations. Consequently these 730 daily records were utilized in this study. The vegetation cover was a forest of Loblolly pine.

HARMONIC ANALYSIS

Initially, the monthly average and the standard deviations of the soil moisture data were computed and are listed in Table. 1. The results indicated generally high soil moisture in the winter months

Reprinted from *Time Series Methods in Hydrosciences*, by A.H. El-Shaarawi and S.R. Esterby (Editors)

TABLE 1

MONTHLY AVERAGES AND STANDARD DEVIATIONS OF SOIL MOISTURE DATA

Month	Mean cm	Standard Deviation cm
January	45.37	.356
February	47.63	.813
March	51.9	1.825
April	49.89	1.648
May	45.33	1.349
June	42.10	1.325
July	40.93	1.655
August	40.07	1.039
September	39.32	1.072
October	38.69	.335
November	40.02	.548
December	42.30	1.609

and low soil moisture in the summer months. Histograms of the summer and winter month soil moisture were plotted in Figure 1, which shows that soil moisture is higher in the winter months and also has high variability.

A harmonic analysis was then performed on the soil moisture data. The results of this analysis are shown in Figure 2. At first glance there appears to be an annual cycle present in the data. Visual inspection of the data also suggests the presence of the annual cycle.

There is, however, reason to suspect a certain degree of correlation in the data which would tend to distort the frequency spectrum. This distortion is referred to as "red noise" (Gilman, et al, 1963; Mitchell, 1964). The presence of "red noise" tends to suppress the relative variance at higher frequencies and consequently to inflate the relative variance at the lower frequencies.

The "red noise" spectrum is a function of the autoregressive coefficient of the data. Figure 3 shows a family of "red noise" spectra for various autoregressive coefficients. (Gilman, et al, 1963). It can be seen that for a substantially large autoregressive coefficient, the relative importance of the first two harmonics in the data is significantly reduced.

A periodogram analysis of the spectrum is given here rather than the Blackman-Tukey approach because in this instance the spectrum is easily smoothed and a degree of autocorrelation is present.

The advantages of the periodogram in these instances have been pointed out by Jones (1964). Among these advantages are that the periodogram spectrum is always positive and that the cross spectral density is easily calculated from it.

An analysis of the significance of the peaks involved in the first two harmonics showed, as expected--that they contributed little significance when compared with the "red noise" spectrum. Despite the apparent lack of significance in the annual cycle it was still considered advisable to remove it from the data before proceeding with the rest of the analysis. This was done because, given the presence of the "red noise" distortion, it was not possible to positively determine the real significance of this component of the data. The cycle was removed by computing the monthly averages, month by month, for the five years of available data and then subtracting the appropriate average from each daily observation. Monthly averages were used in this case instead of the recommended daily means (Jones, 1964) due to the small sample size. It was felt that daily means would be unduly biased in this case.

CROSS-CORRELATION ANALYSIS

A cross-correlation and cross-spectrum analysis was then attempted on the anomalies from the above operation and the daily rainfall observations. A cross-spectrum analysis is concerned with the contribution of each frequency to the total covariance of the two series (Jenkins, 1961). The analysis was performed by methods described by Panofsky and Brier (1958).

The results of this analysis did not reveal any significant relationship between the two series at any frequency. In order to account for any delay in the measurement of soil moisture values after a rainfall event, the soil moisture was then lagged one day, but still no significant correlation or coherence was obtained.

In order to test the hypothesis that the lack of correlation in the two series was due to very short duration perturbation, the soil moisture data was filtered by taking 7-day moving averages. This series was then correlated with the 7-day rainfall series. In this analysis a correlation coefficient of only about 0.2 was computed.

AUTOCORRELATION ANALYSIS

An autocorrelation analysis was next performed on the soil moisture data for various lags. The results are shown in Figure 4. As can be seen from the figure, an extraordinarily high degree of correlation exists in the data. A very large correlation coefficient was obtained at a 1-day lag and a large degree of persistence was observed. As can be seen, the data showed significant correlation at lags even greater than 150 days.

The results suggest a very strong "carry-over" nature of the soil moisture data.

AUTOREGRESSIVE MODEL

It seemed reasonable to assume that the strong autocorrelation evidenced by the soil moisture data was one of the reasons for the lack of correlation between it and the rainfall series. Therefore, it was decided to fit an autoregressive type model to the data. Initially, a first-order Markov type model was used. This model is of the form:

$$x(t) = \alpha x(t-1) + \eta$$

where:

$x(t)$ = soil moisture at time t

$x(t-1)$ = soil moisture at time t-1

α = autoregressive coefficient

η = random residual component

The method used to fit this model is outlined by Jones (1964) and will not be repeated here. In this analysis an autoregression or predictor constant (α) of .9856 was obtained with a standard error of prediction of .0061 and an R^2 of 97%.

This autoregression coefficient could have been obtained by direct comparison of the spectral analysis with curves of the "red noise" spectrum for different values of α given by Gilman, et al (1963) and Mitchell (1964).

CROSS-CORRELATION BETWEEN RAINFALL AND SOIL MOISTURE

The random component, or residuals from the above autoregressive model, represent the true random fluctuation in the observed soil moisture data because the deterministic components--cycles and short-term autoregressive effects--have already been eliminated. Therefore, it would seem reasonable to assume that the residuals would be better correlated to the observed rainfall series than was the original data. This supposition was born out by cross-spectral analysis. Figures 5 and 6 show the results of this analysis. The total Pearson correlation coefficient due to all frequencies was 0.46, which was significant at the 5% level. From Figure 6a and 6b it does not appear that there is any significant phase lag between the two data series.

From the results of this study it would appear that once the deterministic components are removed from the soil moisture data, the variance in that series is well explained by the occurrence or non-occurrence of precipitation over the watershed. In this analysis it is assumed that soil moisture fluctuations are due primarily to the evapotranspiration process and to precipitation. The evapotranspiration process is explained by the autoregressive effects present in the data. This seems to suggest that on the average about 1.5% of the soil moisture is lost each day due to evapotranspiration. The remaining part of the unexplained variance in the soil moisture is apparently due to measurement errors or some physical process which cannot be accounted for in this type of analysis.

CONCLUSIONS

From the results of this study it is possible to make several useful conclusions. (1) The presence of an over-powering autocorrelation in soil moisture data makes it impossible to perform a Fourier analysis on this series because of the "red noise" distortion of the frequency spectrum. For the same reason, no direct correlation can be obtained between the raw soil moisture data and the daily rainfall series. (2) This autoregressive effect or correlation in the data shows a very high persistence out to at least 150 days. Thus, care should be taken when performing statistical analysis on soil moisture data where independence assumptions are necessary. (3) A first order linear autoregressive model adequately describes the soil moisture data. This model fits the data very well and accounts for over 90% of the variance in the data. (4) Once the deterministic components of the soil moisture series have been removed, a significant correlation is obtained between the residuals and the rainfall series. (5) The autoregression analysis indicated that on the average about 1.5% of the total soil moisture is lost each day, possibly due to evapotranspiration, when the vegetation cover is Loblolly pine.

ACKNOWLEDGEMENTS

The writers would like to thank J.E. Douglass of the United States Department of Agriculture, Forest Service, Southeastern Forest Experiment Station, who provided the hydrologic data from the Calhoun Experimental Forest. The writers assume full responsibility for all conclusions drawn from these data.

REFERENCES

Cordova, Jose R., and Rafael L. Bras, "Physically Based Probabilistic Models of Infiltration, Soil Moisture, and Actual Evapotranspiration," Water Resources Research, Vol. 17, No. 1 (February 1981) pp. 93-106.

Gilman, D.L., Freglister, J., and J.M. Mitchell, Jr., "On the Power Spectrum of 'Red Noise'," Journal of the Atmospheric Sciences, Vol. 20 (March 1963) pp. 182-184.

Jenkins, G.M., "General Considerations in the Analysis of Spectra," Technometrics, Vol. 3, No. 2 (May 1961) pp. 133-143.

Jones, Richard H., "Spectral Analysis and Linear Prediction of Meteorological Time Series," Journal of Applied Meteorology, Vol. 3 (February 1964) pp. 45-52.

Kent, Edward J., Roy Burke III, and Shaw L. Yu, "Some Control of Stormwater Through Joint Use of Constructed Storage and Soil Management," Proceedings, International Symposium on Urban Hydrology, Hydraulics and Sediment Control, University of Kentucky, Lexington, KY (July 1981) pp. 453-464.

Mitchell, J. Murray, Jr., "A Critical Appraisal of Periodicities in Climate," CAED Report No. 20, Weather and Our Food Supply, Iowa State University (1964) pp. 189-227.

Panofsky, H.A., and G.W. Brier, Some Applications of Statistics to Meteorology, The Pennsylvania State University, University Park, PA (1958) pp. 224.

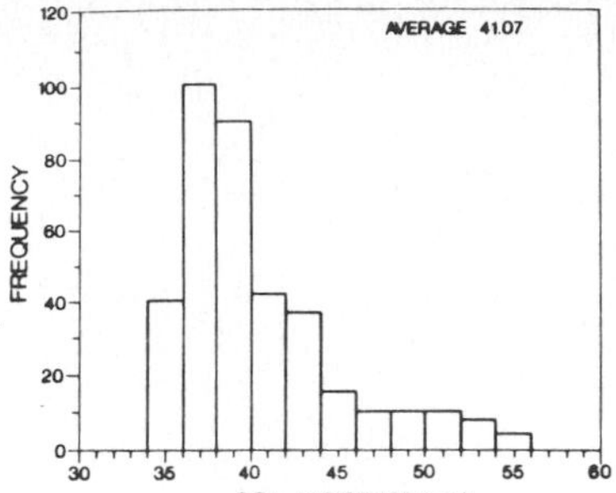

FIGURE 1. SOIL MOISTURE HISTOGRAM - SUMMER SEASON

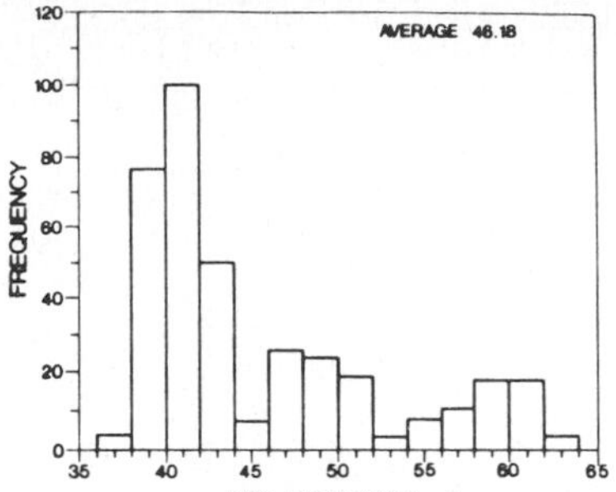

FIGURE 1a. SOIL MOISTURE HISTOGRAM - WINTER SEASON

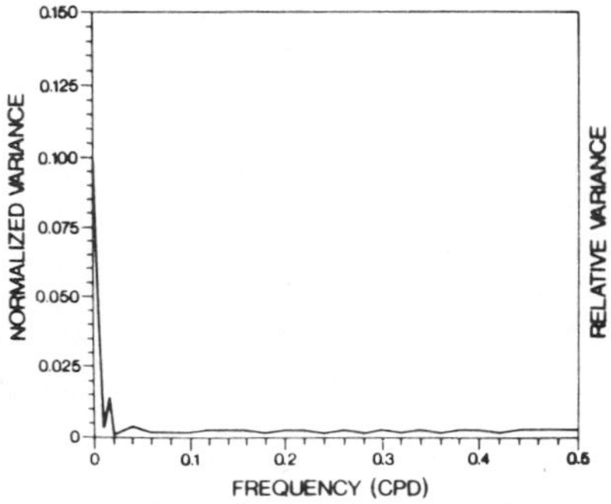

FIGURE 2. SOIL MOISTURE PERIODOGRAM

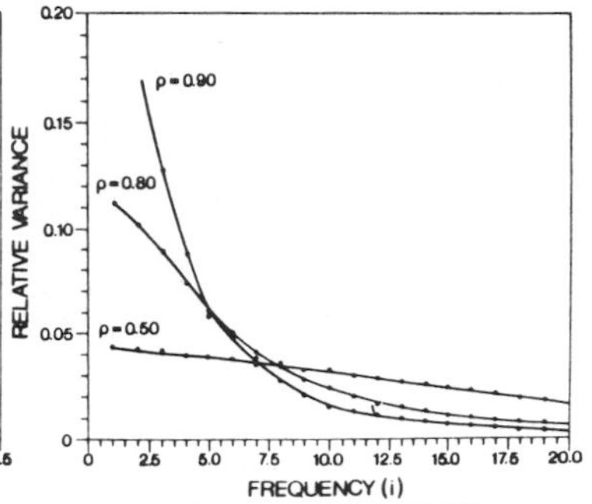

FIGURE 3. RED NOISE SPECTRUM

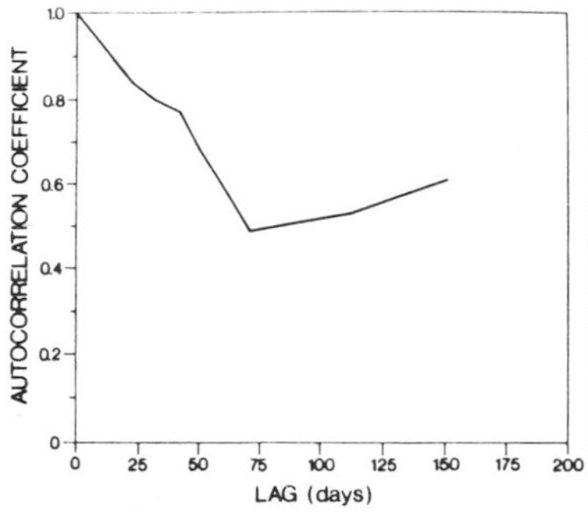

FIGURE 4. AUTOCORRELATION VS LAG

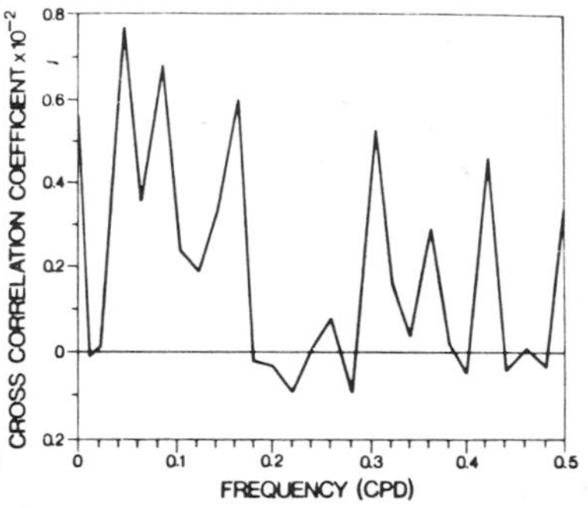

FIGURE 5. CROSS CORRELATION VS FREQUENCY

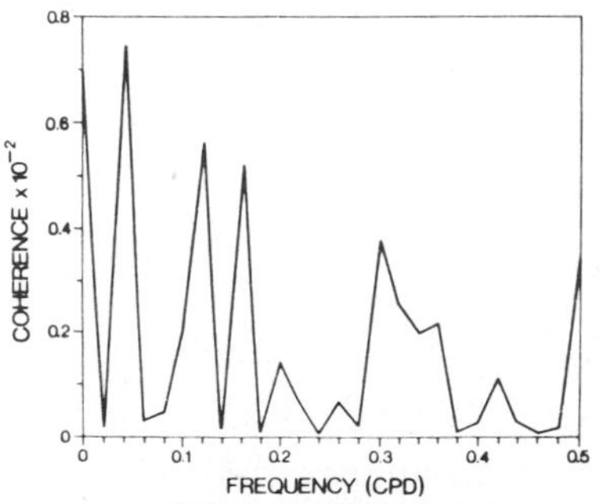

FIGURE 6a. COHERENCE (P, SM)

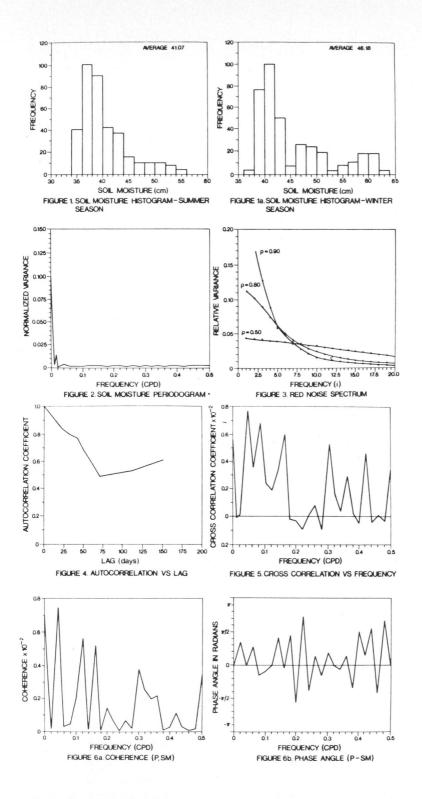

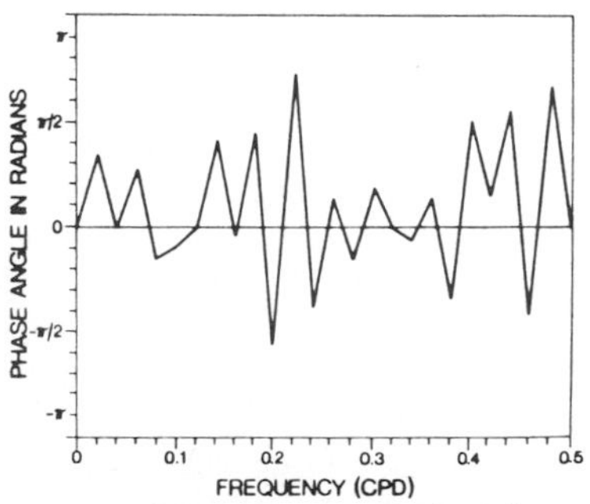

FIGURE 6b. PHASE ANGLE (P - SM)

FORECASTING UNDER LINEAR PARTIAL INFORMATION

M. BEHARA AND E. KOFLER
McMaster University and University of Zurich

ABSTRACT

In a classical forecasting procedure, using regression models, the credibility of a forecast is based on the (i) credibility of the fixed exogene and endogene variables in the observation period (ii) extrapolation of the regression lines into the forecast period (iii) credibility of the fixed exogene variables in the forecast period etc. There are also rigid conditions imposed on the residual variables. In this paper, with the help of linear-partial-information (LPI) method, we modify the classical forecasting by assuming fuzziness for the exogene and endogene variables in observation and exogene variables in the forecast-spaces which would result in greater credibility of the respective variables. Similarly, the extrapolation is assumed to be fuzzy. Further, the LPI-fuzziness of the residual variables is also considered. The better quality of the LPI-forecast is then tested in a decision-theoretic way.

1.1 INTRODUCTION

Let $(x_t, y_t)_{t=1,\ldots,n}$ R^2, where x_t and y_t denote the exogenous and the endogenous variables respectively. The method of ordinary least squares yields the regression line

$$y_t = m_y + b\,(x_t - m_x) + u_t,\quad t = 1,\ldots,n$$

where m_x and m_y are the mean values of x and y respectively; b is the regression coefficient and u_t, $t=1,\ldots,n$ are the white noise error-variables.

Based on the principle of extrapolation, the transition from the estimation-domain $t = 1,\ldots,n$ to the forecast-domain (without the consideration of u_t, at first) is given by the regression line for $t = n+1, n+2, \ldots$ and yields the corresponding point-forecast. This is the so called classical regression problem.

Reprinted from *Time Series Methods in Hydrosciences*, by A.H. El-Shaarawi and S.R. Esterby (Editors)

Seldom, in forecasting problems, an observation (x_t, y_t) is given by an exact point. We get a more credible process if the variables are allowed to assume wide range of values rather than single points. For example, it is more credible to assume the rate of inflation for a certain year to be between 10 and 13 percent rather than exactly 11 percent. We may define "credibility" of (x_t, y_t) as an open or closed neighbourhood of (x_t, y_t) in R^2 according to a prior specified metric. The credibility of an observation is, therefore, directed, to some degree, by the assumption of fuzziness for $(x_t(x_t, y_t), t = 1,\ldots,n)$ and for $x_{n+1}, x_{n+2}, \ldots$. We shall use the LPI-method of Kofler 1,2 to forecasting problems where credibility of observation (x_t, y_t), $t = 1,\ldots,n$ as well as credibility of $x_{n+1}, x_{n+2}, \ldots$ given at n have been assumed.

1.2 THE LPI-FUZZINESS

Let us consider the case of discrete endogenous variables where non-stochastic fuzziness is assumed. Let x_t be associated with the finite set $s_t = (y_t^1, \ldots, y_t^2)$, $t = 1,\ldots,n$. Let $S = \times_1^n s_t$. Then S consists of $k = k_1 k_2 \ldots k_n$ n-tuples $PW_1, PW_2, \ldots, PW_k$ where each PW_j, $j = 1,\ldots,k$ is associated with a regression line RG_j, $j = 1,\ldots,k$. Hence, we get k different forecasts for y_{n+1} by evaluating k regression lines in x_{n+1}. Clearly, for $k_1 = k_2 = \ldots = k_n = 1$, the above problem reduces to classical regression problem.

In the case of stochastic LPI for the discrete endogenous variables, on the other hand, the LPI for each s_t, $t = 1,\ldots,n$ is assumed to be known. If LPI (s_t) is given for $t = 1,\ldots,n$, then, forecast regarding Y_{n+1} is determined to be an expectation interval as follows:

$$E(y_{n+1}) \quad \underline{E}(y_{n+1}), \quad \bar{E}(y_{n+1})$$

where $\underline{E}$ and $\bar{E}$ denote the minimal and the maximal expected values respectively. This is easily seen, as the resulting LPI (S), due to the theorem of the aggregation of the LPI's, over the k states is obtained. Thus, we have

$$\text{LPI }(S) = \text{LPI }(PW_1, \ldots, PW_k).$$

Similarly, the LPI-fuzziness of other variables may be studied.

2.1 INTERVAL-UNCERTAINTY

Often the experimental results are found to lie in an interval rather than taking discrete values as discussed above. Let us consider an endogenous variable Y to assume an interval of uncertainty with respect to an exogenous variable X. Let a_t, b_t, t = 1,...,n be such an interval for y_t. At first, we assume no prior information on the distribution of the y_t's in a_t, b_t. Obviously, for a given x_{n+1}, an uncertainty-interval a_t, b_t of y_t leads to the determination of an interval-forecast for y_{n+1} as

$$y_{n+1} \in [\underline{y}_{n+1}, \bar{y}_{n+1}]$$

where $\underline{y}$ and $\bar{y}$ denote the minimum and maximum values of y.

Assuming now the density $f(y_t)$ given for the interval a_t, b_t, of y_t, t = 1,...,n, it may be easily proved that for a given x_{n+1}, there exists a forecast for $E(y_{n+1})$. This is the perfect stochastic case.

Finally, for the case of linear partial information for each $y_t \in [a_t, b_t]$, there exists an LPI($[a_t, b_t]$), t = 1,...,n. And, for a partition of the interval $[a_t, b_t]$, there exist LPI-statements on the partitions.

For LPI($[a_t, b_t]$), we have

$$E(y_{n+1}) \in [\underline{E}(y_{n+1}, \bar{E}(y_{n+1})].$$

This is obtained by the following procedure: For the interval $[a_t, b_t]$, t = 1,...,n, a finite set of states $\{z_t\}$ is assigned. To each set $\{z_t\}$ there corresponds an LPI($\{z_t\}$), t = 1,...,n. Defining the Cartesian product $\times_{t=1}^{n} z_t$, let $w_1,\ldots,w_m$ be the elements of this product. According to the theorem on the composition of the independent LPI's, the components of LPI($\{z_t\}$) determine the LPI(w_j), j = 1,...,m. Each w_j, j = 1,...,m is then associated with an interval-forecast.

3.1 EXAMPLE OF AN LPI-FUZZY ENDOGENOUS VARIABLE

In a given sample

$(x_1,y_1),\ldots,(x_j,y_j),\ldots,(x_k,y_k),\ldots,(x_n,y_n)$, y_j and y_k are considered to be LPI-fuzzy. Considering only two values for each of y_j and y_k, the LPI-assignments are as follows:

$$\begin{cases} y_j \in \{y_j^1, y_j^2\}, \text{ Prob } (y_j^1) = p_1, \text{ Po} \qquad , \text{ LPI}(y_j): = p_1 \geqslant p_2. \\ y_k \in \{y_k^1, y_k^2\}, \text{ Prob } (y_k^1) = q_1, \text{ Prob } (y_k^2) = q_2, \text{ LPI}(y_k): = q_1 \geqslant q_2. \end{cases}$$

In this case, there are 2 x 2 = 4 different sets of regression points: PW_1, PW_2, PW_3, PW_4 and hence 4 regression lines corresponding to the regression points. From the above LPI-assignment, we obtain the resulting LPI (r_1,r_2,r_3,r_4) by composition of the different LPI-components, where $r_1 = p_1q_1$, $r_2 = p_1q_2$, $r_3 = p_2q_1$, $r_4 = p_2q_2$. Clearly, we have

$$r_1 \geqslant r_2 \geqslant r_4,\ r_1 \geqslant r_3 \geqslant r_4,\ r_1 - r_2 \geqslant r_3 - r_4$$

with the matrix of the extreme points given by

$$M[\text{LPI } (r_1,r_2,r_3,r_4)] = \begin{bmatrix} 1 & 1/2 & 1/2 & 1/4 \\ 0 & 0 & 1/2 & 1/4 \\ 0 & 1/2 & 0 & 1/4 \\ 0 & 0 & 0 & 1/4 \end{bmatrix}$$

Hence we have four forecast values for y_{n+1} given by:

$RG_1(x_{n+1})$, $RG_2(x_{n+1})$, $RG_3(x_{n+1})$, $RG_4(x_{n+1})$.

In the matrix of the extreme points given above, we determine the minimal and the maximal expectation values $\underline{E}(y_{n+1})$ and $\bar{E}(y_{n+1})$. Therefore,

$$E(y_{n+1}) \in [\underline{E}(y_{n+1}),\ \bar{E}(y_{n+1})]$$

4.1 DECISION-THEORETIC EVALUATION OF FORECAST

The evaluation of forecasting problems belongs to the field of semantic information theory. The principle of evaluation given as follows:

For every multistage decision situation, each dynamic and control problem, an optimal strategy is gained by using the maxE_{min}-principle [3] with the use of LPI's. Every new forecast generally changes the information-state of the decision situation and, therefore, the $\max E_{min}$-optimal strategy. Hence, erroneous forecasts are associated with erroneous $\max E_{min}$-optimal strategy.

Let us consider the decision scheme

$$
\begin{array}{c@{\qquad}c}
p_1 & p_2 \\
z_1 & z_2
\end{array}
$$

$$
\begin{array}{c}
d_1 \\
d_2 \\
\cdot \\
\cdot \\
\cdot \\
d_m
\end{array}
\left[
\begin{array}{cc}
u_{11} & u_{12} \\
u_{21} & u_{22} \\
\cdot & \cdot \\
\cdot & \cdot \\
\cdot & \cdot \\
u_{m1} & u_{m2}
\end{array}
\right]
$$

where d_i, $i = 1,\ldots,m$ are m different money-deposits in a fund, z_1 and z_2 denote a low and a high inflation rate respectively. The distribution of z_1 and z_2 is given by p_1 and p_2 respectively.

According to the information of the past few months on p_1 (the relative frequency of z_1) we have observed, say

$$(x_1, p_1^{(1)}),\ldots,(x_n,p_1^{(n)}).$$

A certain output may be LPI-fuzzy. Let us suppose $p_1^{(s)}$ is such an output. If we do not take LPI-fuzziness of $p_1^{(s)}$ into consideration we obtain a false forecast F_2, say and determine $p_1^{(n+1)}$ from the $p_1,\ldots,p_n$ each of which is single-valued. On the other hand, if we take LPI-fuzziness into account, then the multi-valued $p_1^{(s)}$ leads to the interval $[\underline{E}(p_1^{(n+1)}), \bar{E}(p_1^{(n+1)})]$ where the true forecast F_1, say lies. Now, there are two decision situations:

(1) The correct decision, denoted by D_1, with given interval $[\underline{E}(p_1^{(n+1)}), \bar{E}(p_1^{(n+1)})]$ for $p_1^{(n+1)}$.

(2) The incorrect decision, denoted by D_2, with the false forecast $p_1^{(n+1)}$. Let d^{*1} and d^{*2} denote the optimal strategies in D_1 and D_2 respectively. The evaluation of the forecast is then given by

$$p_1^{(n+1)} \in [\underline{E}(p_1^{(n+1)}), \bar{E}(p_1^{(n+1)})] =: V_D$$

where V_D denotes the value of the strategy in the decision situation D. Thus,

$$V(F_1) = V_{D_1}(d^{*1}) - V_{D_1}(d^{*2})$$

which is nonnegative since d^{*2} is not the $\max E_{min}$-optimal strategy. $V(F_1)$ is the loss due to the exact but unfortunately false forecast F_2. It follows that by using the LPI-formulation associated with the LPI-forecast F_1 this loss may be reduced.

4.2 EXAMPLES

Example 1. Two-person zero-sum game.

	z_1	z_2	z_3	z_4
d_1	5	4	8	6
d_2	7	3	9	5
d_3	4	6	8	7

Solution:
$d^* = (0,1/3,2/3)$
value $(d^*) = 5$

LPI: $p_1 \leqslant p_2 \leqslant p_3 \leqslant p_4$

	z_1	z_2	z_3	z_4
d_1	5.75	6.00	7.00	6.00
d_2	6.00	5.66	7.00	5.00
d_3	6.25	7.00	7.50	7.00

Solution:
$d^* = d_3$
value $(d^*) = 6.25$

We observe that using the LPI, the value of the game has increased from 5 to 6.25.

Example 2.

Let the information on the states be given by an exact probability distribution $I_1 = (0.3,0.3,0.2,0.2)$. This we consider as the false information F_1. The LPI on the states is given by $I_2 = 0.2 \leqslant p_j \leqslant 0.4$, $j = 1,2,3,4$. For the utility matrix:

$[u_{ij}] =$

	z_1	z_2	z_3	z_4
d_1	0	2	1	1
d_2	2	0	2	1
d_3	1	2	0	2

The solution is given by d_3; value $(d_3) = 1.3$
The extreme-point matrix is calculated as

$$M = \begin{bmatrix} 0.4 & 0.2 & 0.2 & 0.2 \\ 0.2 & 0.2 & 0.2 & 0.2 \\ 0.2 & 0.4 & 0.2 & 0.2 \\ 0.2 & 0.2 & 0.4 & 0.2 \end{bmatrix}$$

Finally, the product of the above matrices $[u_{ij}]$ M is given by

$$A = \begin{bmatrix} 0.8 & 1.0 & 1.0 & 1.2 \\ 1.4 & 1.4 & 1.2 & 1.0 \\ 1.2 & 1.0 & 1.4 & 1.4 \end{bmatrix}$$

The solution is given by d_2 or d_3; Value (d_2) = Value $(d_3) = 1.0$
The solution with respect to I_1 for $d^* = (0,0.5,0.5)$, value $(d^*) = 1.2$
Therefore, [Value $(I_1) \rightarrow$ value (I_2)] = 1.2 - 1.0 = 0.2.

REFERENCES

Behara, M., Kofler, E. and Menges, G. (1978)
Entrophy and informativity in decision situation under partial information, Statistische Hefte, 19, p. 124-130.

Kofler, E. and Menges, G. (1976)
Entscheidungenbei unvollstandiger Information. Springer-Verlag, Berlin-Heidelberg, New York.

Kofler, E. and Menges, G. (1979)
The structuring of uncertainty and the maxE$_{min}$-principle. Operations Research Verfahren, 34.